The Biology of the Porifera

Frontispiece. A typical sponge, coral, gorgonian, algal community photographed by flashlight at ca.50 m depth on a coral fore-reef slope off Discovery Bay, Jamaica. The large orange-red sponge (top right) is *Agelas* sp., the yellow tubular sponge (centre) is *Verongia* sp., and the two polychrome purple and yellow tubes (lower right) are *Ianthella* sp. Other sponges are *Cliona* sp. and ? *Plakortis* sp. The coral is *Agaricia* sp., the gorgonian is *Pseudopterogorgia*, and a patch of *Halimeda cryptica* is visible at the top left. The horizontal distance across the centre is approximately 60 cm. Photograph T. F. Goreau.

SYMPOSIA OF THE ZOOLOGICAL SOCIETY OF LONDON
NUMBER 25

The Biology of the Porifera

(The Proceedings of a Symposium held at the Zoological Society of London on 10 and 11 September, 1968, with additional articles presented during Colloquia held at the British Museum (Natural History) on 12 and 13 September, 1968)

Edited by

W. G. FRY

Department of Science, Luton College of Technology, Luton, Bedfordshire

Published for

THE ZOOLOGICAL SOCIETY OF LONDON

BY

ACADEMIC PRESS

1970

ACADEMIC PRESS INC. (LONDON) LTD.
Berkeley Square House
Berkeley Square
London, W1X 6BA

U.S. Edition published by
ACADEMIC PRESS INC.
111 Fifth Avenue
New York, New York 10003

Library of Congress Catalog Card Number: 70-92403
SBN: 12-613325-5

PRINTED IN GREAT BRITAIN BY
J. W. ARROWSMITH LTD., BRISTOL

CONTRIBUTORS

BERGQUIST, P. R., *Zoology Department, University of Auckland, Auckland, New Zealand* (p. 247)

BOROJEVIĆ, R., *Laboratoire de Biologie des Invertébrés Marins, Muséum National d'Histoire Naturelle, Paris, France* (p. 467)

BRIEN, P., *Institut Zoologique Torley Rousseau, Université Libre de Bruxelles, Bruxelles, Belge* (p. 163)

CLIFFORD JONES, W., *Department of Zoology, University College of North Wales, Bangor, Wales* (p. 91)

CURTIS, A. S. G., *Department of Cell Biology, University of Glasgow, Glasgow, Scotland* (p. 335)

EFREMOVA, S. M., *Department of Embryology, Leningrad State University, Leningrad, U.S.S.R.* (p. 399)

FINKS, R. M., *Department of Geology, Queens College, City University of New York, U.S.A.* (p. 3)

FJERDINGSTAD, E. J., *Baylor College of Medicine, Texas Medical Center Houston, Texas, U.S.A.* (p. 125)

FRY, W. G., *Marine Science Laboratories, Menai Bridge, Anglesey, Wales* (p. 135)

GARRONE, R., *Laboratoire d'Histologie et Biologie Tissulaire, Faculté des Sciences, Université de Lyon, Lyon, France* (p. 449)

GOREAU, T. F., *Marine Laboratory, University of the West Indies, Discovery Bay, St. Anne, Jamaica, W.I.* (p. 205)

HARTMAN, W. D., *Department of Biology and Peabody Museum of Natural History, Yale University, New Haven, Connecticut, U.S.A.* (p. 205)

HOGG, J. J., *Zoology Department, University of Auckland, Auckland, New Zealand* (p. 247)

HUMPHREYS, T., *Department of Biology, Revelle College, University of California, La Jolla, California, U.S.A.* (p. 325)

KOLTUN, V. M., *Zoological Institute of the Academy of Sciences of the U.S.S.R., Leningrad, U.S.S.R.* (p. 285)

KOROTKOVA, G. P., *Department of Embryology, State University of Leningrad, Leningrad, U.S.S.R.* (p. 423)

LÉVI, C., *Laboratoire de Biologie des Invertébrés Marins, Muséum National d'Histoire Naturelle, Paris, France* (p. 353)

MACLENNAN, A. P., *Microbiological Research Establishment, Porton, Nr. Salisbury, England* (p. 299)

MERGNER, H., *Abteilung für Entwicklungslehre des Zoologischen Institutes der Justus Liebig-Universität, Giessen, West Germany* (p. 365)

PAVANS DE CECCATTY, M., *Laboratoire d'Histologie et Biologie Tissulaire, Faculté des Sciences, Université de Lyon, Lyon, France* (p. 449)

RASMONT, R., *Laboratoire de Zoologie générale, Université libre de Bruxelles, Bruxelles, Belge* (p. 415)

REID, R. E. H., *Department of Geology, The Queen's University, Belfast, Northern Ireland* (p. 63).

RIETSCHEL, S., *Frankfurt am Main, Natur-Museum und Forschungs-institut Senckenberg, Frankfurt am Main, West Germany* (p. 23)

SARÀ, M., *Istituto di Zoologia Universit` di Genova, Genova, Italy* (p. 273)

SINCLAIR, M. E., *Zoology Department, University of Auckland, Auckland, New Zealand* (p. 247)

THINEY, Y., *Laboratoire d'Histologie et Biologie Tissulaire, Faculté des Sciences, Université de Lyon, Lyon, France* (p. 449)

TUZET, O., *Laboratoire de Zoologie Générale, Université de Montpellier, Montpellier, France* (p. 437)

VACELET, J., *Station Marine d'Endoume, Faculté de Sciences de Marseille, Marseille, France* (p. 189)

ZHURAVLEVA, I. T., *Institute of Geology and Geophysics, Novosibirsk* 90, *U.S.S.R.* (p. 41)

ZIEGLER, B., *Staatliches Museum für Naturkunde, Stuttgart, West Germany* (p. 23)

PARTICIPANTS IN THE DISCUSSIONS AND COLLOQUIA

BOURY-ESNAULT, N., *Laboratoire de Biologie Marine, Banyuls-sur-mer, France*

CARMONA, M., *Zoologisches Institut der Justus Liebig-Universitat, Giessen, West Germany*

CASTLE, J. F., *Department of Zoology, British Museum (Natural History), London*

CONNES, R., *Laboratoire de Zoologie, Université de Montpellier, France*

DEBRENNE, F., *Institut de Palaontologie, Muséum National d'Histoire Naturelle, Paris*

DEVOS, L., *Faculté des Sciences, Université Libre de Bruxelles, Bruxelles, Belge*

EMSON, R. H., *Department of Zoology and Animal Biology, King's College, University of London*

GAMULIN-BRIDA, H., *Institute of Zoology, University of Zagreb, Yugoslavia*

GEORGE, D., *Department of Zoology, British Museum (Natural History), London, England*

KILIAN, E. F., *Zoologisches Institut der Justus Liebig-Universitat, Giessen, West Germany*

HAWKINS, D. C., *Microbiological Research Establishment, Porton, Nr. Salisbury, U.K.*

PARIS, J., *Station Biologique de Sète, Hérault, France*

PULITZER-FINALI, G., *Istituto di Zoologia ed Anatomia Comparata, Università di Bari, Bari, Italy*

ROZENFELD, F., *Faculté des Sciences, Université Libre de Bruxelles, Bruxelles, Belge*

SCHMIDT, I., *Faculté des Sciences, Université Libre de Bruxelles, Bruxelles, Belge*

SIMPSON, T. L., *Department of Biology, University of Hartford, U.S.A.*

SIRIBELLI, L., *Istituto i Museo di Zoologia, Università di Napoli, Italy*

STONE, A. R., *Department of Biology, The Polytechnic, London, England*

STONE, S. M. K., *Department of Zoology, British Museum (Natural History), London, England*

TENDAL, O. S., *Universitetes Zoologiskes Laboratorium, København, Denmark*

TSURNAMAL, M., *Department of Zoology, The Hebrew University of Jerusalem, Israel*

VAN DE VYVER, G., *Faculté des Sciences, Université Libre de Bruxelles, Bruxelles, Belge*

WEIDENMAYER, F., *Blumenstrasse* 48,8500 *Frauenfeld, Switzerland*

WINTERMANN-KILIAN, G., *Zoologisches Institut der Justus Liebig-Universitat, Giessen, West Germany*

ORGANIZER AND CHAIRMEN

ORGANIZER

W. G. FRY, *on behalf of the Zoological Society of London*

CHAIRMEN OF SESSIONS

P. BRIEN, *Institut Zoologique Torley Rouessau, Université Libre de Bruxelles, Bruxelles, Belge*

M. BURTON, *Weston House, Albury, Surrey, England*

W. D. HARTMAN, *Department of Biology and Peabody Museum of Natural History, Yale University, New Haven, Connecticut, U.S.A.*

R. E. H. REID, *Department of Geology, The Queen's University, Belfast, Northern Ireland*

M. F. STEMPIEN, *New York Aquarium, Coney Island, Brooklyn, New York, U.S.A.*

E. N. WILLMER, 11 *Park Terrace, Cambridge, England*

INTRODUCTION

The idea of arranging a symposium on sponges arose during an informal gathering of biologists at Roscoff in 1966. One of the purposes of this gathering was to try to standardize some of the descriptive terminology applied to the Porifera, but from the discussions came the realization that the increasing number of publications on the group made such a standardization of terminology difficult (Borojević *et al.*, 1968). The existence of conflicting contemporary usages of terminology, and the differences of interpretation which they implied, were the best possible reasons for trying to assemble as many as possible of the biologists interested in sponges, to enable them to present their views both in formal papers and in more informal discussion.

The response to the suggestion of such a meeting was overwhelming, for more than forty formal contributions were offered by biologists from sixteen nations. The embarrassment of this situation was avoided, to some extent, by arranging two extra days of Colloquia immediately following the Symposium, and by the generosity of the Zoological Society in allowing the inclusion in this Symposium volume of some of the papers presented during the Colloquia. However, this still left me in the invidious position of having to exclude from the volume many excellent articles, while the inclusion of extra articles required the omission from the text of any record of the discussion of papers presented during the Symposium.

In selecting the articles for this volume I have allowed myself to be guided by two intentions: to allow coverage of as many aspects of the biology of the Porifera as possible and to include the views of as many schools of research as possible. This method of selection has brought together conflicting points of view on several topics, notably on phylogeny, cell adhesion, and other aspects of morphogenesis and clearly denies the volume the right to be regarded as a treatise on sponges. This is, perhaps, unfortunate, for there is considerable need for new authoritative texts on the group. Since the first decade of this century only one treatise which could be regarded as comprehensive of the biology of the group has been published (Hyman, 1940), and this is almost entirely a recompilation of earlier treatise (e.g. von Lendenfeld, 1884, 1889; Vosmaer, 1882–1886; Delage and Hérouard, 1899; Minchin, 1900; Dendy and Row, 1913).

On the other hand, the profound differences of observation and opinion revealed by the controversies displayed in this Symposium volume may be taken as an indication that we are not at present in a position to expect new and widely acceptable syntheses of all the various aspects of sponge biology. Rather, our knowledge and understanding of sponges are undergoing profound changes as a result of a new interest in the group. Although it would be unwise to imagine that a larger proportion of biologists than ever before are interested now primarily in sponges, it does appear that sponges are at present attracting more attention as experimental material than they have done for some years. Dr. Burton, in his Chairman's introduction to one of the Symposium sessions, underlined this state of affairs when recalling Professor Dendy's comment, upon his beginning research into sponges, that the number of British spongologists had increased by 50%.

It seems pertinent to any synthesis of sponge biology to try to discover the causes of the recent and continuing resurgence of interest in sponges as experimental material.

A lack of obvious and consistent organization above the cellular level sets the Porifera apart from other multicellular animals. During the nineteenth century this apparent simplicity, and the primitiveness which it was taken to imply, stimulated researches into sponges for answers to questions fundamental to all biology. Data from sponges figured strongly in arguments about primordial germ-layers and the probable nature of ancestral multicellular organisms (e.g. Haeckel, 1874; Metschnikoff, 1886).

It was probably the emergence after the 1914–18 war of experimental biology as a powerful research discipline which led to a decline in interest in sponges. In the early days of its development, experimental biology relied upon more obviously differentiated tissues and organ systems as sources of its data. The apparent absence of any such structures in sponges made them unattractive experimental material. At the same time, comparatively little new and fundamental information could be obtained with the light microscope. Even the discovery of the remarkable morphogenetic powers of dissociated sponge cells, although recognized as being of profound importance (e.g., Wilson, 1907, 1910, 1925; Huxley, 1911, 1921a, b; de Beer, 1922; Galtsoff, 1925a, b, 1929; de Laubenfels, 1928; Brien, 1937), and stimulating much speculation, could not be exploited.

Undoubtedly it was the advent of the electron microscope to biology, and the data which this instrument provides for the erection and substantiation of hypotheses on the biophysics, biochemistry and other aspects of cell behaviour, which provided a fresh stimulus to the use of sponges as experimental material. The very independence of sponge cells, and their apparent inability to form spatially distinct tissues and organs, make sponges ideal material for studies of the fundamental attributes of cells.

Before data from electron microscopy were available much information about the appearance, behaviour, and differentiation potentials of different cell types in sponges had been accumulated, but this information did not fit easily into existing theories of cell biology; the uniformity of ultrastructure and the limited number of pathways of differentiation existing in biological systems awaited discovery and recognition (e.g., Willmer, 1960; Pantin, 1966). No more precise understanding of the organization and homeostasis of sponges could be found than descriptions of the microscopic appearance of cell types and the conditions under which they transformed their appearances. The integration mechanisms of sponges eluded discovery.

The contents of this volume show a very different state of our understanding of sponges and their morphogenetic and homeostatic mechanisms. Data which have been collected during the last ten years or so combine to provide a picture of sponges as populations of cells interacting both in morphogenesis and homeostasis in ways which are highly reminiscent of mixed populations of organisms. Especially since the publication of certain data by Borojević and Lévi (1962–1967), emphasis of interest has shifted away from the archaeocyte as the totipotent cell initiating and controlling morphogenesis and homeostasis, until we are left with a picture of the sponge as a population of cells differentiating and redifferentiating according to their position within the sponge and the morphogenetic and homeostatic needs of that part of the sponge in which they find themselves.

However, it did not require the publication of the proceedings of the Symposium and Colloquia for this picture to become clear. What does emerge from the contents of this volume is the fact that we are tantalizingly close to a quantitative expression of this picture. The information obtained by Professor Rasmont and his co-workers, by Professor Lévi and Dr. Borojević, by Dr. Jones, and by Dr. Efremova, indicate that soon we may be able to describe and discuss sponge morphogenesis and homeostasis in the numerical terminology which ecologists use to describe the dynamics of mixed populations of organisms (e.g., Odum *et al.* 1962; Phillipson, 1966; Wiegart, 1964).

In contrast, the contributions of Professor Mergner, Professor Tuzet, and Dr. Korotkova show that we still may obtain a useful understanding of the nature of sponges by regarding them as organisms constructed of tissue layers.

At first, these two interpretations of the nature of sponges may appear to conflict. That they need not do so is shown by Professor

Curtis', Dr. Maclennan's and Dr. Humphreys' pictures of the parts played by biophysical and biochemical phenomena in the integration of sponge cells, and by the demonstration by Professor Pavans de Ceccatty and his colleagues of the possible ultrastructural bases of the methods of intercellular communication which bring about the integration of cell behaviour and differentiation.

Thus, we are left with a picture of sponges as organisms which, for certain purposes, may best be regarded as complex cellular ecosystems while for other purposes they may be regarded in the same way as organisms with more clearly differentiated tissues, not unlike coelenterates and some higher Metazoa. Neither of these syntheses of sponge morphogenesis and homeostasis can be considered the only acceptable and correct interpretation; each synthesis has its usefulness as a basis upon which other hypotheses may be erected. In 1898 Saville–Kent (Delage *et al.*, 1899) urged 'the abandonment of the pickle bottle and the recourse to a thorough and painstaking investigation of the living aspects and phenomenon (sic) of the structures' of sponges. He considered that 'the result of such . . . investigation will proceed a long way towards releasing the sponges from that position of doubtful parentage and relationship, which it is scarcely to the credit of biologists they are permitted to occupy . . .'. In recent years there have been many painstaking and thorough investigations of living sponges, but the contents of this volume suggest that sponges must continue to occupy an anomalous position among multicellular organisms.

The contributions dealing with the taxonomy and phylogeny of sponges also indicate that sponges must continue to enjoy a position of 'doubtful parentage and relationship'. Professor Lévi argues that in sponges transformations occur continually between types of cell ultrastructure which have been held (Willmer, 1960) to be the basic types of differentiation in all animals. This suggests that in some way sponges have retained a plasticity of cell organization which is shared by few, if any, other organisms in their normal morphogenesis and homeostasis. It is not surprising therefore that Dr. Finks and Dr. Reid should be able to show so clearly that Dendy's (1921) neat picture of the evolution of the tetraxonid spicule is an over-simplification of the phylogeny of silicious sponges.

The description by Dr. Hartman and Professor Goreau, and by Dr. Vacelet, of living forms of sponge apparently scarcely differing at all from very ancient fossil forms, and Dr. Zhuravleva's tentative equation of Cambrian and Precambrian remains with organisms resembling both sponges and the hypothetical ancestral blastaea, taken together with Professor Lévi's argument, lead to the suggestion that

that part of the genetic constitution of sponges which controls cell differentiation became stable in the early Palaeozoic, or before. This would be so remarkable a phenomenon that one is tempted to speculate that sponges, rather than failing to acquire a genetic ability for persistent and far-reaching cell differentiation, actually may have acquired in their genotype a block to such a pattern of differentiation.

Whatever may be the genetic history of the differentiation potential of sponges, they remain a very important part of the marine ecosystem, in some regions even dominating the environment, both in their effects upon its physical and chemical parameters and in their biomass (Renouf, 1937; Belyaev and Ushakov, 1960; Zenkevitch *et al.* 1960; Riedl, 1966; Bullivant and Dearborn, 1967), just as they have done since Palaeozoic or earlier times. It is unfortunate, therefore, that it was not possible to include more ecological papers in the volume and that there are not more ecologists who pay attention to sponges.

Perhaps the main reasons for the shortage of works on the ecology of the Porifera are the difficulty of identifying specimens and the confused state of sponge taxonomy. Undoubtedly, the fact that until recently there has been relatively little interest in the biology of sponges has hampered the work of sponge taxonomists. As Dr. Bergquist has shown in this volume, we are still alarmingly ignorant of the reproductive mechanisms of sponges. In this state, and with a scarcity of knowledge of sponge ecology, we can have little insight into speciation mechanisms in sponges, and so taxonomists must have the greatest difficulty in producing objective classifications.

Because of the room for subjectivity that exists in sponge classification, I have tried to ensure that, in each article, every taxonomic term has been defined by the citation of its author and date of publication. Some of these definitions differ between articles, and to ensure that these differences can be appreciated, the index contains all taxonomic terms used in the volume. The inclusion of the names of the authors of taxonomic terms provides an introduction to the general literature on sponges. So that this introduction can be of use to readers who may wish to look further into particular hypotheses discussed in the volume, additional references to the Porifera have been included at the end of this introduction and an index of authors' names placed at the end of the volume.

Many people have helped in the organization of the meetings and in the editing of the volume, but I owe an especial debt of gratitude to Dr. H. G. Vevers and Mrs. Linda Réné-Martin of the Zoological Society of London for the care and attention to detail with which they ensured the efficient organization of the Symposium. I am extremely grateful also to the Director of the British Museum (Natural History) for permission to hold the Colloquia in the Museum, and for allowing Miss Shirley Stone and Mr. J. F. Castle to contribute so much of their time to ensuring the smooth running of the meetings.

Miss Stone, with her profound knowledge of the literature on sponges, has provided invaluable help during the editing of the volume, while my colleague Dr. W. Clifford Jones has given much useful advice and encouragement. I also wish to thank Mrs. Frances Simpson for her assistance with the multilingual correspondence which this editorial work has involved.

I must also thank Pinchin Johnson Associates Limited because the organization of the meeting and the editing of the volume occurred during my tenure of their Research Fellowship at Menai Bridge.

William G. Fry
Luton College of Technology,
Park Square, Luton, Bedfordshire, England.

References

(including some not cited elsewhere in the volume)

Arndt, W. (1935) Porifera. *Tierwelt N.-u. Ostsee.* Teil **3a**: 1–140.

Beer, G. R. de (1922). Notes sur les colonies artificielles de choanocytes. *Arch. Zool. exp. gén.* **61** (*Notes et Revs.*): 47–49.

Belyaev, G. M. and Ushakov, P. B. (1960). Quantitative study of the benthos of the Tasman Sea and Antarctic waters south of New Zealand. *Dokl. Akad. Nauk SSSR* **130** (4).

Borojević, R. (1966). Etude expérimentale de la différenciation des cellules de l'éponge au cours de son développment. *Devl. Biol.* **14**: 130–153.

Borojević, R. (1967). La ponte et le développment de *Polymastia robusta* (Démosponges). *Cah. Biol. Mar.* **8**: 1–6.

Borojević, R., Fry, W. G., Jones, W. C., Lévi, C., Rasmont, R., Sarà, M. and Vacelet, J. (1968). Mise au point actuelle de la terminologie des Eponges. *Bull. Mus. natn. Hist. nat., Paris* (2) **39**: 1224–1235.

Borojević, R. and Lévi, C. (1964). Ultrastructure des cellules d'éponge en suspension cellulaire. *3rd European Reg. Conf. Electr. Microscopy*: 561–562.

Borojević, R. and Lévi, C. (1964). Etude au microscope électronique des cellules de l'éponge: *Ophlitaspongia seriata* (Grant), au cours de la réorganisation après dissociation. *Z. Zellforsch. mikrosk. Anat.* **64**: 708–725.

Borojević, R. and Lévi, C. (1964). Métamorphose artificielle de larves d'Eponges, après dissociation et réagrégation des cellules larvaires. *C.r. hebd. Séanc. Acad. sci. Paris* **259**: 4364–4366.

Borojević, R. and Lévi, C. (1965). Morphogénèse expérimentale d'une Eponge à partir de cellules de la larve nageante dissociée. *Z. Zellforsch. mikrosk. Anat.* **68**: 57–69.

Brien, P. (1937). La réorganisation de l'Eponge après la dissociation par la filtration et phénomènes d'involution chez *Ephydatia fluviatilis*. *Archs Biol., Leige* **48**: 185–268.

Brien, P. (1968). The Sponges or Porifera, in *Chemical Zoology*, **2**, Section 1, Porifera: 1–30. Florkin, M. and Scheer, B. T. (Ed.). New York and London: Academic Press.

Bullivant, J. S. and Dearborn, J. H. (1967). General Accounts, Station Lists, and Benthic Ecology, Part **5**, *Fauna of the Ross Sea. Mem. N.Z. oceanogr. Inst.* **32**: 1–77.

Delage, Y. and Hérouard, E. (1899). *Mésozoaires et Spongiaires. Traité de zoologie concrète.* Paris.

Delage, Y., Minchin, E. A., Vosmaer, G. C. J. and Saville-Kent, W. (1899). *Int. Congr. Zool.* **4**: 57–68.

Dendy, A. (1921). The tetraxonid sponge spicule; a study in evolution. *Acta zool., Stockh.* **2**: 95–152.

Dendy, A. and Row, R. W. H. (1913). The classification and phylogeny of the Calcareous Sponges. *Proc. zool. Soc., Lond.*: 704–813.

Florkin, M. (1968). Skeletal structures of Porifera, in *Chemical zoology*, **2**, Section 1, Porifera: 31–35. Florkin, M. and Scheer, B. T. (Ed.). New York and London: Academic Press.

Galtsoff, P. S. (1925a). Regeneration after dissociation. I. Behaviour of dissociated cells of *Microciona prolifera* Verr. under normal and altered conditions. *J. exp. Zool.* **42**: 183–221.

Galtsoff, P. S. (1925b). Regeneration after dissociation. II. Histogenesis of *Microciona prolifera* Verr. *J. exp. Zool.* **42**: 223–251.

Galtsoff, P. S. (1929). Heteroagglutination of dissociated sponge cells. *Biol. Bull. mar. biol. Lab. Woods Hole* **57**: 250–260.

Goodwin, T. W. (1968). Pigments of Porifera, in *Chemical zoology*, **2**, Section 1, Porifera: 37–41. Florkin, M. and Scheer, B. T. (Ed.). New York and London: Academic Press.

Hammen, C. S. and Florkin, M. (1968). Chemical composition and intermediary metabolism—Porifera, in *Chemical zoology*, **2**, Section 1, Porifera: 53–63. Florkin, M. and Scheer, B. T. (Ed.). New York and London: Academic Press.

Haeckel, E. (1874). Die Gastraea-Theorie, die phylogenetische Klassification des Thierreichs und Homologie der Kemiblatter. *Jena. Z. naturw.* **8**: 1–55.

Huxley, J. S. (1911). Some phenomena of regeneration in *Sycon*. *Phil. Trans. R. Soc. B* **202**: 165–189.

Huxley, J. S. (1921a). Differences in viability in different types of regenerates from dissociated sponges, with a note on entry of some somatic cells by spermatozoa. *Biol. Bull. mar. biol. Lab. Woods Hole* **40**: 127–129.

Huxley, J. S. (1921b). Further studies on restitution-bodies and free tissue-culture in *Sycon*. *Q. Jl microsc. Soc.* **65**: 293–322.

Hyman, L. (1940). *The Invertebrates*. **1**, *Protozoa through Ctenophora*, Porifera: 284–364. New York and London: McGraw Hill.

Lendenfeld, R. von (1884). Das System der Monactinellidae. *Zool. Anz.* **7**: 201–206.

Lendenfeld, R. von. (1889). *A monograph of the Horny Sponges*. London.

Laubenfels, M. W. de (1928). Interspecific grafting using sponge cells. *J. Elisha Mitchell scient. Soc.* **44**: 293–322.

Lévi, C. (1964). Observations au microscope électronique de fragmentations nucléaires et cellulaires de cellules de larves de Spongiaires. *C.r. Séanc. Soc. Biol.* **158**: 1572–1576.

Lévi, C. (1964). Ultrastructure de la larve parenchymella de Démosponge. I. *Mycale contarenii* (Martens). *Cah. Biol. Mar.* **5**: 97–104.

Lévi, C. (1965). Morphogénèse normale et expérimentale des Eponges. *Atti Semin. Stud. Biol.* **2**: 1–11.

Lévi, C. (1965). La microscopie électronique et l'étude des Eponges. *Atti Semin. Stud. Biol.* **2**: 1–15.

Lévi, C. and Porte, A. (1962). Etude au microscope électronique de l'éponge *Oscarella lobularis* Schmidt et de sa larve Amphiblastula. *Cah. Biol. Mar.* **3**: 307–315.

Metschnikoff, E. (1886). *Embryologische Studien*. Wien: A. Hölder.

Minchin, E. A. (1900). Porifera, in *A Treatise on Zoology*, E. Ray Lankester (Ed.). Part 2, the Porifera and Coelenterata: 1–178. London.

Odum, E. P., Connell, C. E. and Davenport, L. B. (1962). Population energy flow of three primary consumer components of old-field ecosystems. *Ecology* **43**: 88–96.

Pantin, C. F. A. (1966). Homology, analogy and chemical identity in the Cnidaria, *Symp. zool. Soc. Lond.* No. **16**: 1–16.

Phillipson, J. (1966). Ecological energetics. *Inst. Biol. Stud. Biol.* **1**.

Rasmont, R. (1968). Nutrition and digestion, in *Chemical zoology* **2**, Section 1, Porifera: 43–51. Florkin, M. and Scheer, B. T. (Ed.). New York and London: Academic Press.

Rasmont, R. (1968). Chemical aspects of hibernation, in *Chemical zoology*, **2**, Section 1, Porifera: 65–77. Florkin, M. and Scheer, B. T. (Ed.). New York and London: Academic Press.

Renouf, L. P. W. (1937). Invertebrates, Porifera and Tunicata in particular, as reef-builders on the South Coast of Co. Cork, I.F.S. *Festschr. Embrik Strand Riga*: 5–54.

Reidl, R. (1966). *Biologie der Meereshöhlen*. Hamburg and Berlin: Paul Parey.

Simpson, T. L. (1968). The structure and function of sponge cells: New criteria for the taxonomy of poecilosclerid sponges (Demospongiae). *Bull. Peabody Mus. nat. Hist.* **25**: 1–141.

Vosmaer, G. C. J. (1882–1886). Porifera. *Bronn's Kl. Ordn. Tierreichs* **2**.

Wiegert, R. G. (1964). Population energetics of meadow spittle bugs (*Philaenus spumarius* L.) as affected by migration and habitat. *Ecol. Monogr.* **34**: 217–241.

Willmer, E. N. (1960). *Cytology and evolution*. New York and London: Academic Press.

Wilson, H. V. (1907). On some phenomena of coalescence and regeneration in sponges. *J. exp. Zool.* **5**: 245–258.

Wilson, H. V. (1910). Development of Sponges from dissociated tissue-cells. *Bull. Bur. U.S. Fish.* **30**: 1–30.

Wilson, H. V. (1925). Studies on dissociated sponge cells. *Yb. Carnegie Instn Wash.* **24**: 242–246.

Zenkevitch, L. N., Barsanova, N. G. and Belyaev, G. M. (1960). Quantitative distribution of benthos in the abyssal region of the world ocean. *Dokl. Akad. Nauk SSSR* **130**: 183–186.

CONTENTS

THE PALAEONTOLOGICAL EVIDENCE

The Evolution and Ecologic History of Sponges during Palaeozoic Times

R. M. FINKS

Phylogenetic Relationships of Fossil Calcisponges

B. ZIEGLER and S. RIETSCHEL

Porifera, Sphinctozoa, Archaeocyathi—their Connections

I. T. ZHURAVLEVA

SPICULES AND EVOLUTION

Tetraxons and Demosponge Phylogeny

R. E. H. REID

The Composition, Development, Form and Orientation of Calcareous Sponge Spicules

W. CLIFFORD JONES

Ultrastructure of the Spicules of *Spongilla lacustris* (L.)

E. J. FJERDINGSTAD

The Sponge as a Population: a Biometric Approach

W. G. FRY

Les Potamolepides Africaines Nouvelles du Luapula et du lac Moero

P. BRIEN

Les Eponges Pharetronides Actuelles

J. VACELET

Jamaican Coralline Sponges: their Morphology, Ecology and Fossil Relatives

W. D. HARTMAN and T. F. GOREAU

ECOLOGY AND BIOGEOGRAPHY

Adaptation to Intertidal Existence: Reproductive Cycles and Larval Behaviour in Demospongiae

P. R. BERGQUIST, M. E. SINCLAIR and J. J. HOGG

Competition and Cooperation in Sponge Populations

M. SARÀ

Sponges of the Arctic and Antarctic; A Faunistic Review

V. M. KOLTUN

BIOCHEMISTRY AND THE CELL SURFACE

Polysaccharides from Sponges and their Possible Significance in Cellular Aggregation

A. P. MacLENNAN

Biochemical Analysis of Sponge Cell Aggregation

T. HUMPHREYS

Problems and some solutions in the Study of Cellular Aggregation

A. S. G. CURTIS

CELLS, INTEGRATION AND MORPHOGENESIS

Les Cellules des Éponges

C. LÉVI

Ergebnisse der Entwicklungsphysiologie bei Spongilliden

H. MERGNER

Proliferation Activity and Synthesis of Protein in the Cells of Fresh-water Sponges during Development after Dissociation

S. M. EFREMOVA

Some New Aspects of the Physiology of Fresh-water Sponges

R. RASMONT

Regeneration and Somatic Embryogenesis in Sponges

G. P. KOROTKOVA

La Polarité de L'Oeuf et La Symétrie de La Larve des Éponges Calcaires

O. TUZET

Les Bases Ultrastructurales des Communications Intercellulaires dans Les Oscules de quelques Éponges

M. PAVANS DE CECCATTY, Y. THINEY and R. GARRONE

Différenciation Cellulaire dans L'Embryogenèse et La Morphogenèse chez Les Spongiaires

R. BOROJEVIĆ

THE PALAEONTOLOGICAL EVIDENCE

Symp. zool. Soc. Lond. (1970) No. 25, 3–22

THE EVOLUTION AND ECOLOGIC HISTORY OF SPONGES DURING PALAEOZOIC TIMES

ROBERT M. FINKS

Department of Geology, Queens College of the City University of New York, U.S.A.

SYNOPSIS

The earliest Hexactinellida (early Cambrian) preserved entire have single-layered walls of stauractins while later Cambrian forms have multiple-layered walls of hexactins. This may reflect less complex folding of choanocyte-bearing surfaces in the earliest history of the group. General Palaeozoic trends, possibly polyphyletic, include: (1) thickening of body wall, (2) enlargement of hypodermal spicules, culminating in a rigid hypodermal skeleton and (3) fusion of parenchymalia. Amphidiscs and hexasters appear in the Carboniferous, but not necessarily the living orders based on them.

Cambrian Demospongea are monaxonic. By Middle Ordovician three major groups are present: (1) monocrepid lithistids (Orchocladina, Rhizomorina), (2) monaxonic sublithistids (Dystactospongiidae), possibly ancestral to Ceractinomorpha and Clavaxinellida and (3) eutaxicladine lithistids, which grade into some Mesozoic lithistid Tetractinomorpha. If these are true clades, then the tetraxon, first appearing in the Carboniferous, is a grade attained in several lines.

Calcarea appear in the Carboniferous (Sphinctozoa and possibly other non-pharetronids) and the pharetronids in the Permian. Silurian Aphrosalpingoidea and Cambrian Archaeocyatha show some resemblance to Sphinctozoa. Stromatoporoids, differentiated since the Cambrian and probably persisting today, show affinities to Pharetronida and Demospongea. It is possible that all these groups, as well as the Cambrian to Permian Heteractinida, have a common Precambrian ancestry separate from that of the Hexactinellida.

As shown by reefs, the strongest competitors of Palaeozoic sponges were corals and bryozoans. Sponges dominated reefs chiefly before (Ordovician lithistids) and after (Permian Sphinctozoa) the main period of rugose and tabulate abundance. Stromatoporoids disputed the reef environment with corals during their Silurian and Devonian heyday, but tended to be mutually exclusive. Sponges and corals tend to mutual exclusion outside reefs as well. On the basis of feeding habits among modern forms, the competition was not for food. Calcareous sponges and sponge-like organisms are confined to very shallow water, evidenced by persistent association with algae and reefs. Siliceous sponges tend to avoid reefs (except Ordovician lithistids) but occur in off-reef shallow facies as well as in deeper water. Late Palaeozoic Calcarea have a distinctly Tethyan distribution in contrast to the more widespread siliceous sponges.

The best sample we possess of the earliest sponge faunas is that of the middle Cambrian Burgess Shale of western Canada. Although it is not the earliest record of sponges, no major types have been recovered from early Cambrian strata that are not represented in the Burgess Shale.

Three major classes of the Phylum Porifera are already present in the middle Cambrian: the Hexactinellida Schmidt, 1870, the Heteractinida de Laubenfels, 1955 and the Demospongea Sollas, 1885. It is

FIG. 1. (Lower) *Eiffelia globosa* Walcott, 1920, syntype, U.S. Nat. Mus. No. 66522; (Upper) *Protospongia hicksi* Hinde (Walcott, 1920), hypotype, U.S. Nat. Mus. No. 66502. Both are from the Burgess Shale at its type locality, near Field, British Columbia, Canada. Both ×3.

apparent that these classes differentiated very early. The first hexactinellids and heteractinids go back to the early Cambrian. The first securely determinable demosponges are from the Burgess Shale but numerous monaxons in lower Cambrian strata may belong to this class. There is no certain record of the phylum before the Cambrian.

The hexactinellids (*Protospongia* Salter, 1864) and at least some of the heteractinids (*Eiffelia* Walcott, 1920) of the Burgess Shale are extremely thin-walled, sac-like sponges in which the spicules form essentially a single layer (Fig. 1). The rays of the principal spicules are paratangential to the body wall of the sponge. Such a skeleton could not have supported a thick-walled sponge and in the case of the hexactinellids there would have been no mesogloea for support either. Thus these early hexactinellids of the *Protospongia* type may well have had an asconoid structure, a possibility strengthed by the fact that embryos of living hexactinellids are asconoid and have a monolayered spiculation

FIG. 2. *Multivasculatus ovatus* Howell and Van Houten, 1940, hypotype U.S.Nat. Mus. No. 163626. Windfall Formation, Bullwhacker Member (U. Cambrian), Windfall Canyon, Eureka, Nevada. ×30.

of stauractins like that of *Protospongia*. The earliest heteractinids may also have been asconoid.

All early Cambrian hexactinellid material in North America is of *Protospongia* type, or consists of isolated stauractins. The first hexactins, implying a thicker body-wall, appear in the middle Cambrian (*Bolaspis*-zone in North America—A. R. Palmer, personal communication) although it is not until the late Cambrian *Multivasculatus* Howell and Van Houten, 1940 that a coherent skeleton of hexactins is preserved to us (Fig. 2). However, Dr. R. E. H. Reid reports to me (personal communication, 1968) that he and Dr. Douglas Bassett have recently found isolated hexactins, as well as stauractins, in the Lower Cambrian of Wales. Nevertheless, there seems to be a genuine preponderance of thin-walled, possibly asconoid, hexactinellids in the early Cambrian. The heteractinids are all thick-walled and presumably leuconoid from Ordovician times on. Among the hexactinellids, possibly asconoid *Protospongia*-like forms persist into the Ordovician, being found in black-shale facies.

The earliest demosponges from the Burgess Shale are thin-walled tubular forms but obviously thick enough to have been of leuconoid architecture. In view of the fact that living demosponge embryos develop a leuconoid canal system directly, members of this class may never have possessed asconoid or syconoid forms.

The spiculation of the Burgess Shale demosponges consists entirely of oxeas. They are arranged either in an isodictyal net, with mesh spaces as wide as one spicule-length, or in ramifying, subparallel, anastomosing tracts in which the fascicles of oxeas may have a slightly plumose arrangement. Both types of organization are to be seen in the genus *Hazelia* Walcott, 1920. It is likely that spongin was present to hold the spicules together. Tangential or perpendicular oxeas supporting a dermal membrane may have been present in some forms but no specialized spicules are known.

The demosponges are the first group to undergo a major expansion or adaptive radiation in the Ordovician. Perhaps arising from the poorly-known late Cambrian genera *Wilbernicyathus* Wilson, 1950 and *Gallatinospongia* Okulitch and Bell, 1955, and certainly appearing in the early Ordovician *Archaeoscyphia* Hinde, 1889, the lithistid family Anthaspidellidae Miller, 1889 expands to some seventeen described genera from lower middle Ordovician rocks (Llanvirnian and Llandeilan). The skeleton is essentially an isodictyal net, of the sort seen in the Cambrian *Hazelia*, in which the constituent monaxons have developed terminal zygoses. Sponges of this family participated in the building of early (Arenigian) and middle (Llanvirnian) Ordovician reefs

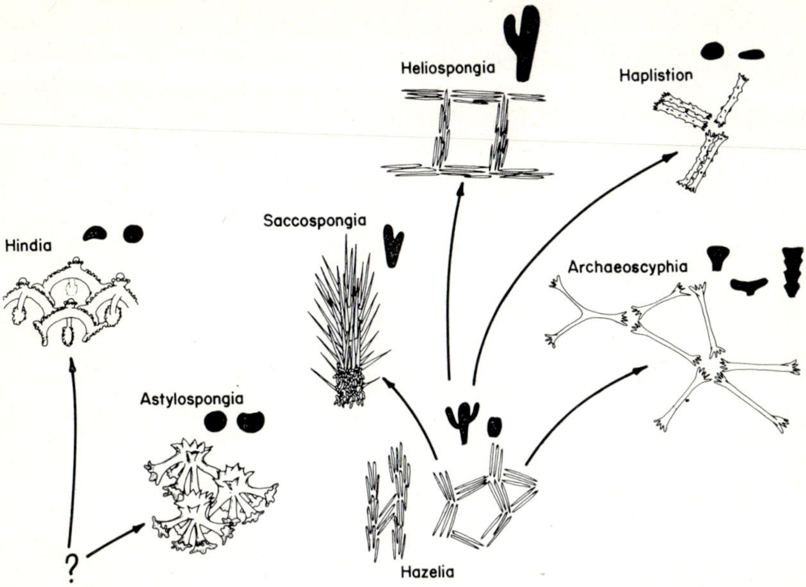

FIG. 3. Principal types of skeletal net amongst Palaeozoic demosponges, showing postulated relationships. Form of whole sponge shown in silhouette.

and elsewhere occur in shallow-water, shelly-facies deposits. The lithistid modification is undoubtedly adaptive for such an environment, conferring strength and resistance to mechanical disruption in rough water. Indeed it may be that the availability of such niches gave direction to this branch of demosponge evolution. The adaptive radiation was cut short in the later Ordovician as was the participation of lithistid sponges in reefs, although the family persists to the Permian.

The rhizomorine type of lithistid skeleton, in which zygoses develop along the sides as well as at the ends of the spicules, may also have appeared in the Ordovician (*Nipterella* Hinde, 1889, *Patellispongia* Bassler, 1927).

With the later middle Ordovician (Caradocian/Trentonian) at least three other major types of demosponges appear, constituting as many separate lineages that continue into later times. One of these lineages is represented by *Saccospongia* Ulrich, 1889 and related dystactospongiids, built of subparallel plumose bundles of styles, coated with monocrepid desmoids. Very similar structures are present in living sublithistid Sigmatosclerophora Burton, 1929 and a direct line of descent is quite possible (Finks, 1967). One could even see in a form like *Hazelia palmata* or "*Vauxia*" *dignata* (Walcott, 1920) the Cambrian roots of this lineage.

Two additional types, rather different from the foregoing, first appear at this time, namely, *Hindia* Duncan, 1879 and *Astylospongia* Roemer, 1860. The first is built of concentric shells of tripodal spicules, the second of a three-dimensional net of hexapodal ones. Both have, in addition, radial and tangential oxeas that probably supported a dermal membrane (Fig. 4). Both develop toward Mesozoic and later forms (Dicranocladina Schrammen, 1924 and Sphaerocladina Schrammen, 1910, respectively) that possess tetraxonic dermalia and astrose microscleres.

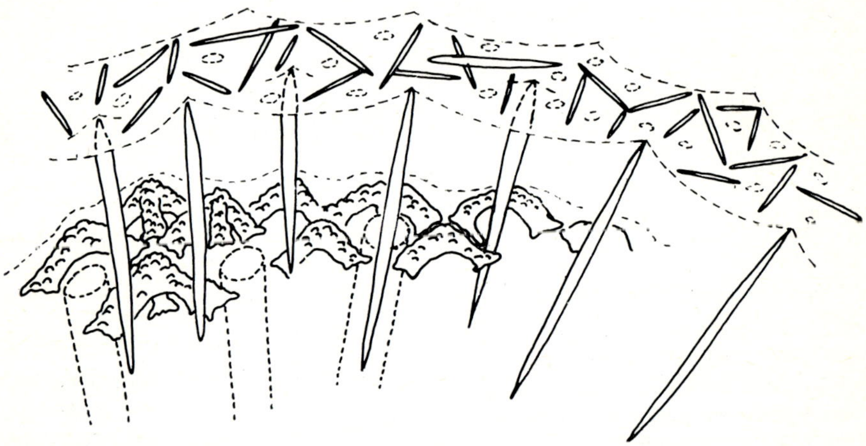

FIG. 4. Skeleton of the family Hindiidae shown in its postulated relationship to the dermal membrane, subdermal space, and choanosome with its radial canals. Outlines of flesh dotted. (*Scheiia tuberosa* Tschernyschew and Stepanov, 1916.)

At this juncture in demosponge history it does not seem unreasonable to see already present the roots of the Clavaxinellida Lévi, 1953 and Ceractinomorpha Lévi, 1953 in *Saccospongia* and its non-lithistid relatives and at least some of the roots of the Tetractinomorpha Lévi, 1953 in the hindiids and astylospongiids (Fig. 5).

Recent work by Hartman and Goreau (1966), reported upon further in this volume, has strongly suggested that at least some of the Palaeozoic Stromatoporoidea Nicholson and Murie, 1878 are ancestors of the living aberrant, lime-secreting demosponges *Ceratoporella* Hickson, 1912, *Astrosclera* Lister, 1900 and *Merlia* Kirkpatrick, 1911. If so, this line of sponge evolution diverged very early, for the first stromatoporoids appear in the Cambrian.* Furthermore, they antedate by far any of the

* Here I follow the *Treatise on Invertebrate Paleontology* (Lecompte, 1956). Some stromatoporoid specialists, for example, Flügel and Flügel-Kahler (1968) consider the earliest to be Middle Ordovician, but this is still prior to the oldest Calcarea.

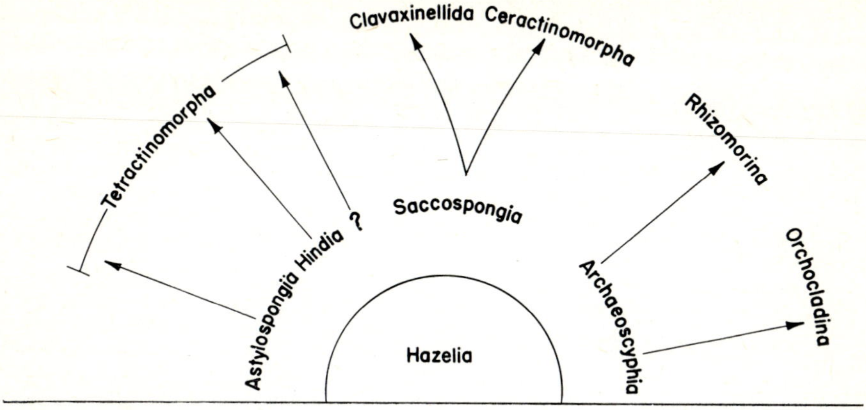

FIG. 5. Postulated lines of descent of living demosponge taxa from Cambrian and Ordovician genera and their relatives.

forms that have been referred, even remotely, to the Calcarea (except for the Archaeocyatha). The Stromatoporoidea, it may be noted, had their acme of development in the Silurian and Devonian, during which time they and the rugose and tabulate corals were the chief reef-building animals.

The mid-Ordovician also marks the first appearance of an advanced hexactinellid, *Brachiospongia* Marsh, 1867 (Fig. 6), which has large, specialized, hypodermal spicules, covered by smaller autodermalia. Its body wall is much thicker than that of any of the known Cambrian sponges and the spicules in this wall lack the regular parallel orientation seen in Cambrian forms. Silurian forms such as *Oncosella* Rauff, 1894 may continue this line and a late Paleozoic form (*Stioderma* Finks, 1960) has a remarkably similar complement of spicules, especially in the form of the hypodermalia and the interior spinose hexactins. Other Permian forms, such as *Docoderma* Finks, 1960, have a similar construction. In the late Paleozoic end-members of this lineage, if such it is, the hypodermalia and autodermalia tend to fuse in the adult sponges, forming a rigid skeleton (Fig. 7). In *Docoderma* there is a development of interlocking processes on the hypodermal spicules, a functional parallel to the skeleton of the lithistid demosponges and apparently unique among the hexactinellids (see Finks, 1960).

The Ordovician also contains the beginnings of a second hexactinellid lineage springing directly from the Cambrian *Protospongia* (Fig. 8). *Cyathophycus* Walcott, 1879 of the Ordovician is like a thicker-walled *Protospongia* and is not too different from the simplest of the Devonian dictyosponges namely, *Dictyospongia* Hall and Clarke, 1899 itself (Hall and Clarke, 1899). In this line of development the spiculation remains

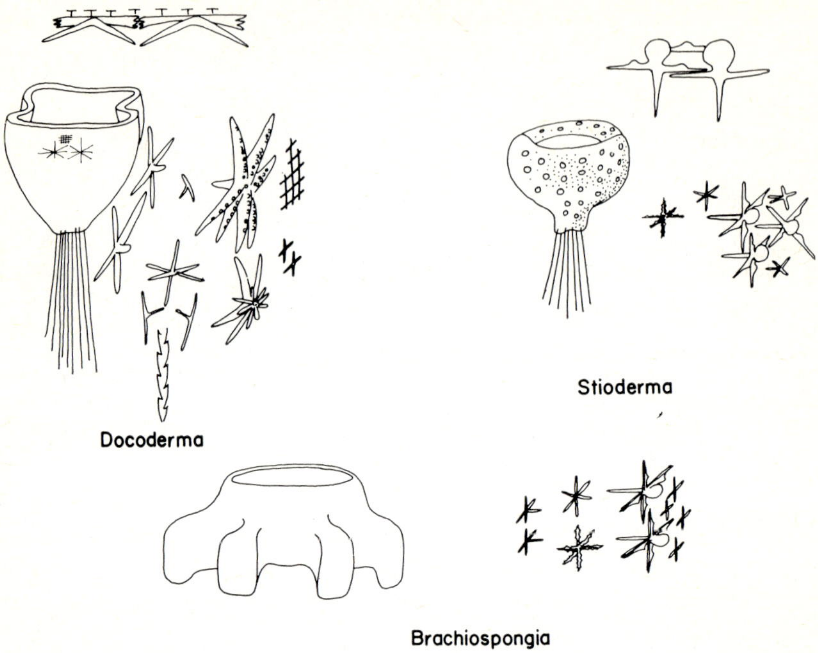

FIG. 6. Palaeozoic hexactinellids related to *Brachiospongia*. *Brachiospongia* is Ordovician, *Stioderma* Carboniferous and Permian, *Dococerma* Permian. The spicular skeleton is shown in "exploded" view, the outside of the sponge to the right. Above *Dococerma* and *Stioderma* are side views of the dermal layer showing the manner of joining of the hypodermalia and autodermalia. Scale is variable.

simple hexactins and their derivatives, in parallel orientation, without enlarged hypodermalia. The structural type, and presumed lineage, can be traced through to the Permian (*Microstaura* Finks, 1960) with a curious thick-walled offshoot (*Stereodictyum* Finks, 1960). The dictyonine type of rigid net, which first appears in the Triassic, could have been derived from this lineage and the presence of hexasters in a Carboniferous dictyosponge (*Griphodictya* Hall and Clarke, 1899) does no harm to this hypothesis.

The Palaeozoic history of hexactinellid microscleres is worth noting, inasmuch as the classification of living hexactinellids is based on the mutual exclusion of hexasters and amphidiscs. Besides hexasters, *Griphodictya* possesses paraclavules, as do several other dictyosponge genera. Paraclavules are shaped like one-ended amphidiscs. Unequal-ended amphidiscs, called hemidiscs, are known as isolated spicules from the Cretaceous (Ortmann, 1912; Schrammen, 1924). True amphidiscs are known almost as early as hexasters, having been reported from the late Carboniferous (*Uralonema* Librovitch, 1929). Recently Kling and

Reif (1969) have reported two late Carboniferous species, one bearing amphidiscs, the other bearing hemidiscs. It is worth noting that at least one of their amphidiscs is slightly unequal-ended. Figure 9 shows the ranges of diagnostic hexactinellid microscleres.

The graded series paraclavule-hemidisc-amphidisc, and the occurrence of paraclavules with hexasters, suggest that the Hexasterophora

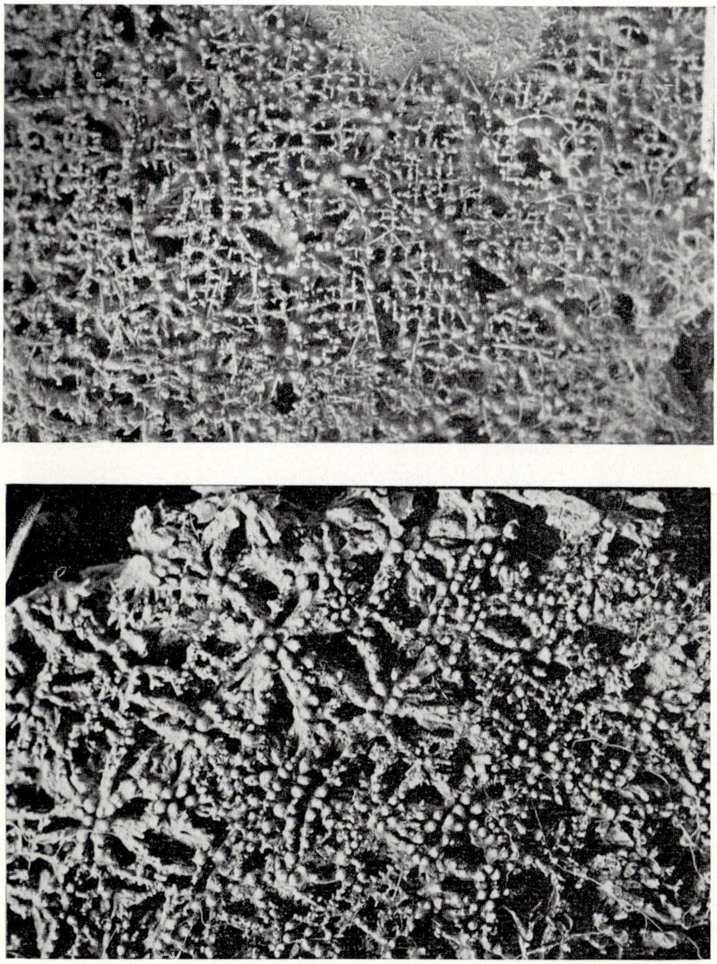

FIG. 7. *Docoderma rigidum* Finks, 1960, showing fusion of hypodermal and autodermal spicules. Photograph below shows outer surface of fused hypodermal spicules; that above shows another part of the same specimen in which fused autodermalia overlie the hypodermalia. Holotype, U.S. Nat. Mus. No. 127659, from Road Canyon Formation (Permian), Word Ranch, Glass Mountains, Texas, U.S.A. Both are ×5.

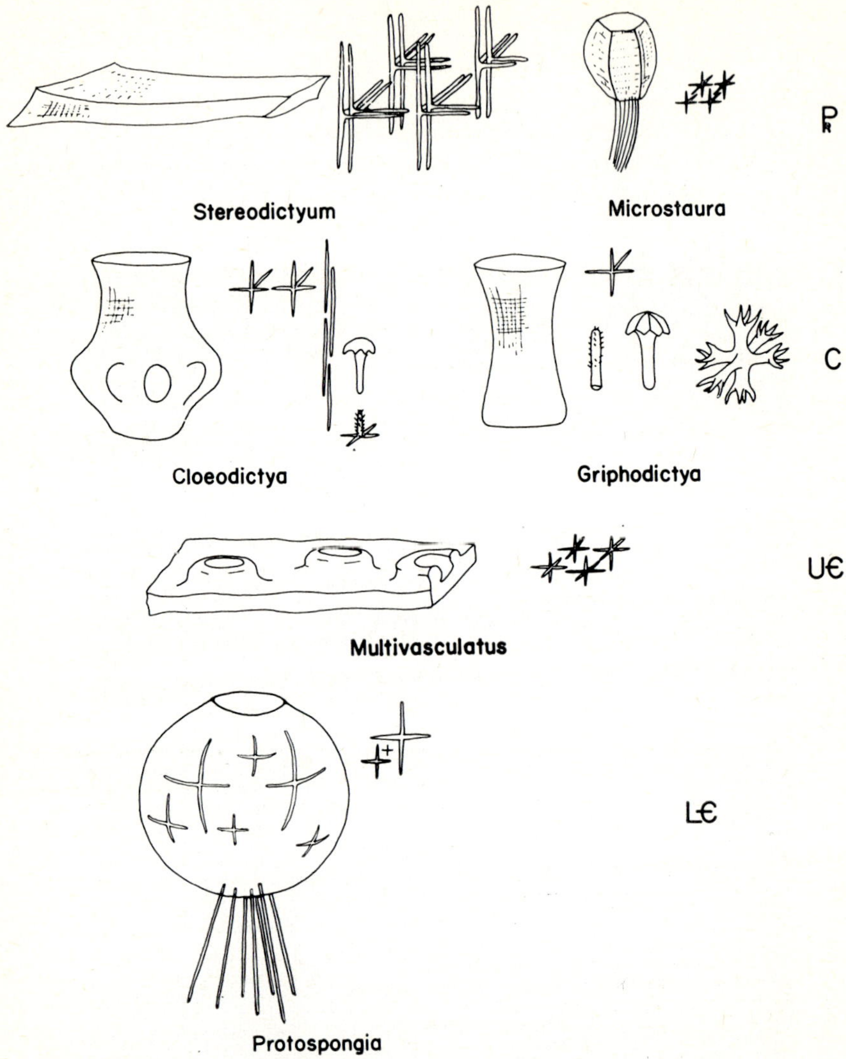

FIG. 8. Palaeozoic hexactinellids showing variants of the simple, parallel spicule arrangement characteristic of *Protospongia*, from which they may have descended. Geologic periods shown along side. Scale is variable; the two right-hand-most spicules beside each of the Carboniferous dictyosponges are microscleres (×90 or so). P, Permian; C, Carboniferous; U∈, Upper Cambrian; L∈, Lower Cambrian.

Schulze, 1887 and Amphidiscophora Schulze, 1887 could have diverged from a common stock during the Palaeozoic and that they could be the sole survivors of a number of early lineages in some of which both hexasters and amphidiscs conceivably could have occurred together.

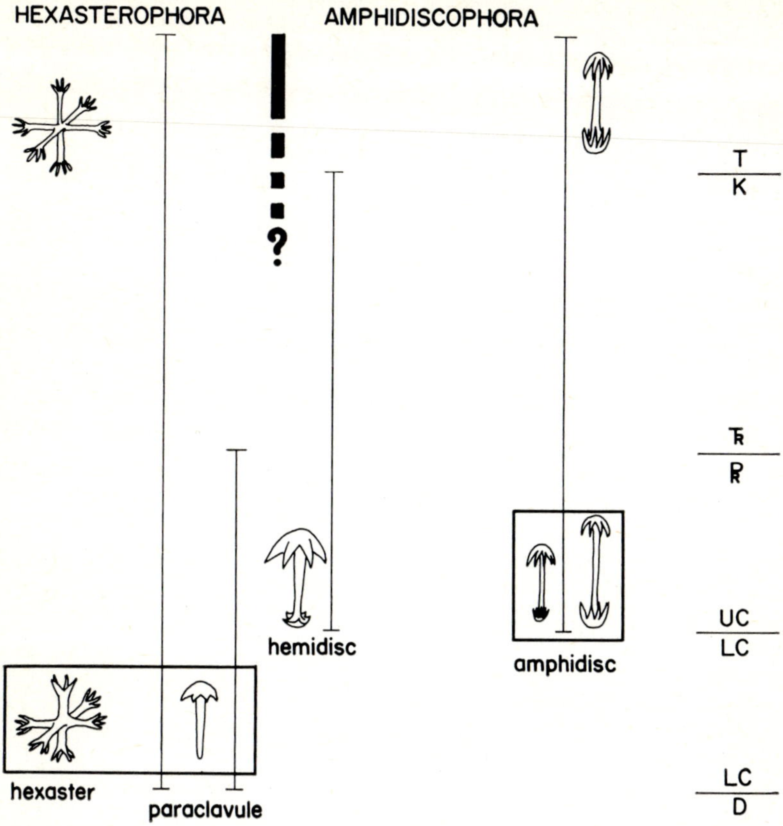

FIG. 9. Ranges of diagnostic microscleres in the Class Hexactinellida. Two spicules enclosed by a rectangle occur in the same specimen. T, Tertiary; K, Cretaceous; Tr. Triassic; UC, Upper Carboniferous; LC, Lower Carboniferous; D, Devonian.

The Heteractinida, in their relatively modest history, show a general trend toward increasing complexity and diversity of the spicules (Fig. 10). The Cambrian *Chancelloria* Walcott, 1920 with its articulated spicule rays of variable number, stands somewhat apart from the main line of heteractinid evolution, which seems to start with the Cambrian *Eiffelia*. The symmetrical, two-dimensional sexiradiates of *Eiffelia* continue in the Ordovician thicker-walled sponge *Astraeoconus* Rietschel, 1968 but in the Silurian and Devonian *Astraeospongia* F. Roemer, 1860 short proximal and distal rays may be developed. The Devonian *Ensiferites* Reimann, 1945 has a greatly enlarged proximal ray and the distal ray may dissolve into a rosette of short branches. The early Carboniferous *Asteractinella* Hinde, 1887, as described by Vandercammen

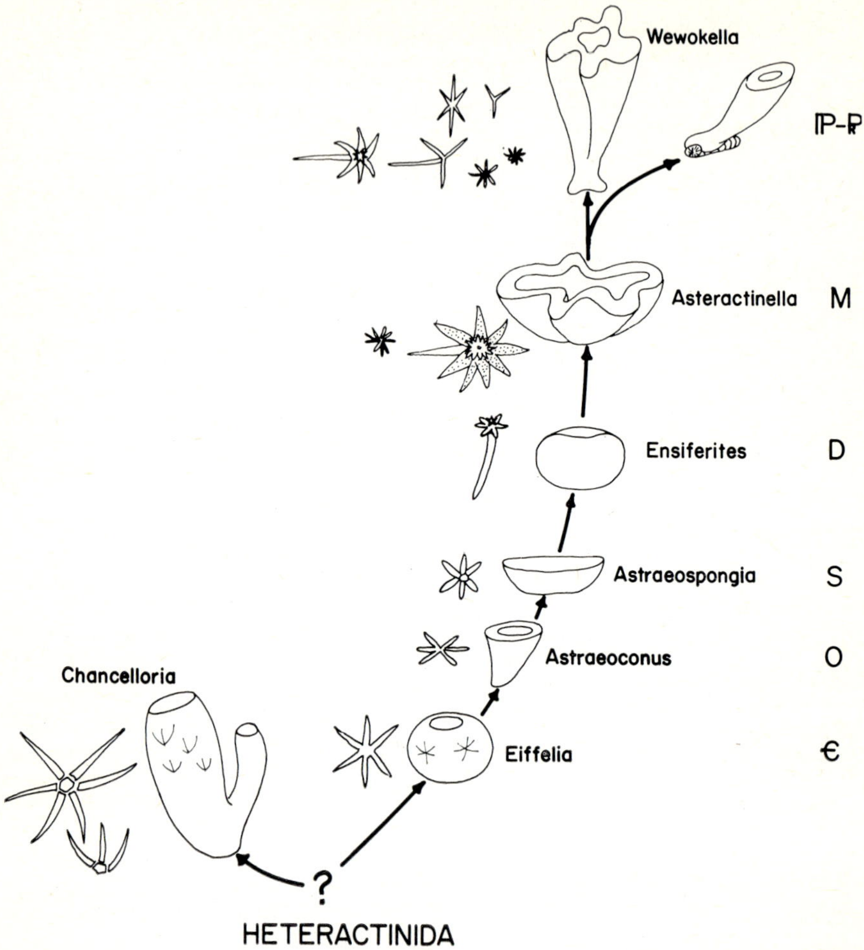

FIG. 10. Evolution of the Heteractinida. Geologic periods along right-hand side. Spicule complement shown to left of each sponge. Scale variable. IP, Pennsylvanian; M, Mississipian; S, Silurian; O, Ordovician.

(1950), continues this trend with spicules having recurved, umbrella-like, paratangential rays, often more than six in number, and bearing a distal rosette or tubercles, to which is added a new type of spicule, the globular polyactin. The later Carboniferous and Permian *Wewokella* Girty, 1911 has the same type of spicules as *Asteractinella* but the paratangential rays of its principal spicules are often reduced in number to three. Simple triradiates and sexiradiates are also present.

The Calcarea in the strict sense, that is, excluding the Archaeocyatha Vologdin, 1937, Aphrosalpingoidea Miagkova, 1955, Heteractinida de

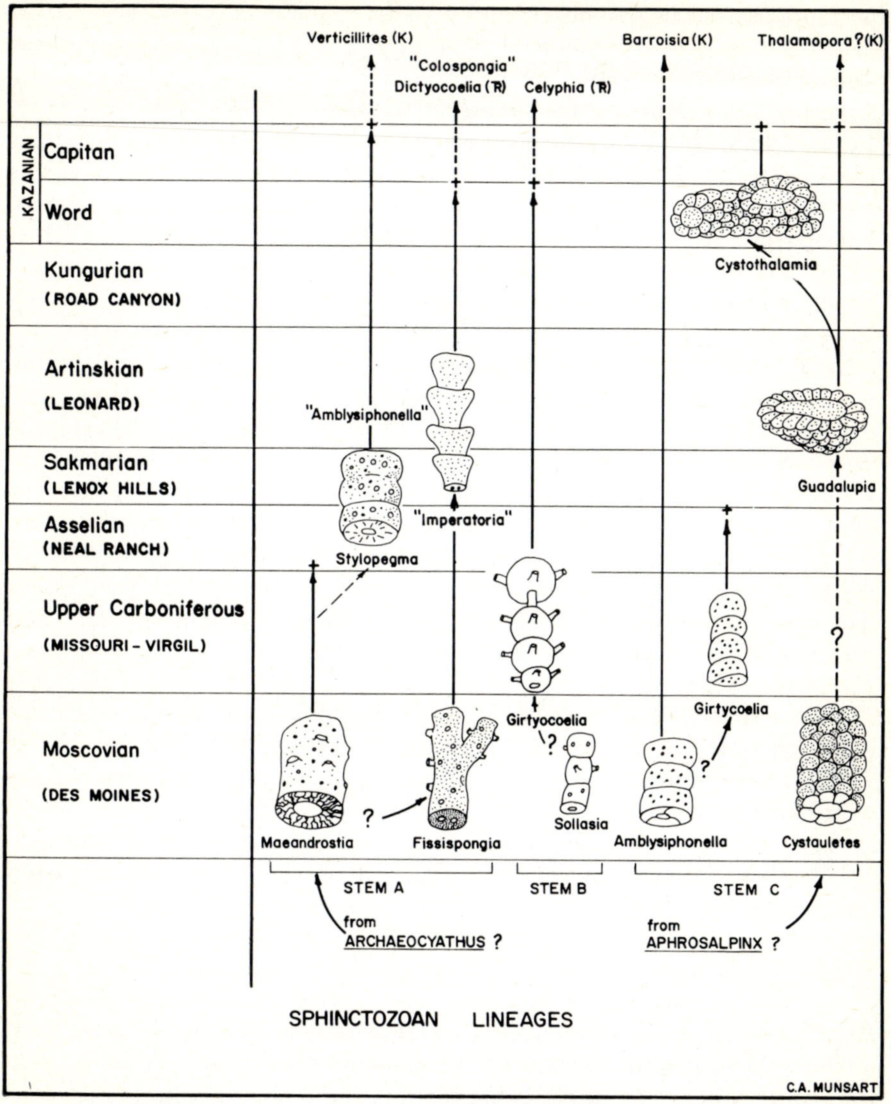

FIG. 11. Postulated lineages among late Palaeozoic Sphinctozoa, from the time of their first appearance. Relative scale of sponge drawings approximately uniform (Finks, 1967).

Laubenfels, 1955 and Stromatoporoidea, appear later in the record than any other class of sponges. The Devonian *Protoleucon* (Bolkhovitinova, 1923) is not indubitably a calcareous sponge. The isolated triradiates and tetraradiates from the lower Carboniferous described as *Peronidella sparsa* by Hinde (1888) may belong to the Calcarea (although Heteractinida cannot be excluded) and if so are the oldest described spicules of the class. The Sphinctozoa Steinmann, 1882 make an undoubted appearance in the middle Carboniferous in Spain (Bashkirian) and North America (Des Moines/Moscovian) (Steinmann, 1882; Barrois, 1882; Finks, 1960). Among these early Sphinctozoa (Fig. 11) we can recognize three major structural types: (1) those built of porous-walled chambers containing internal solid partitions (*Amblysiphonella* Steinmann, 1882, *Sebargasia* Steinmann, 1882, *Girtycoelia* King, 1933, *Cystauletes* King, 1943); (2) those built of imperforate-walled, hollow chambers with a small number of spout-like openings (*Sollasia* Steinmann, 1882, *Girtyocoelia* Cossmann, 1909) and (3) those containing a trabecular internal skeleton between well-defined, coarsely-perforate, dermal and cloacal walls (*Maeandrostia* Girty, 1908, *Fissispongia* King, 1938). Because each of these structural types can be traced through continuously developing lineages into the Permian, and possibly beyond, in Texas and elsewhere, I believe they may represent three basic evolutionary stocks among the Sphinctozoa. They appear full-blown in the record and there is no trace of their prior divergence from a common ancestor. It is worth noting, however, that *Maeandrostia* has an overall structure very much like that of the early Cambrian *Archaeocyathus* Billings, 1861, and *Cystauletes* one not too distant from that of the Silurian *Aphrosalpinx* Miagkova, 1955.

The earliest Pharetronida Zittel, 1878 appear to be two genera that enter the record in the early Permian (Lenox Hills/Sakmarian) of Texas. Other genera appear later in the Permian in several parts of the world and the group expands into the Mesozoic. The Calcarea are active in reef-building in Permian times and continue so into the Triassic.

The late Palaeozoic Calcarea (Fig. 12) have a more concentrated geographic distribution than the siliceous sponges. One might also call it a Tethyan distribution. This may reflect a climatic control and perhaps also a center of origin.

All the classes of sponges have had at least some representatives that lived in shallow water during the Palaeozoic. This can be seen from a list of those groups that have been associated with algae (that is, with the euphotic zone), or with reefs that reveal evidence (accumulations of talus) of having grown into rough water (Fig. 13). However, we can demonstrate certain persistent environmental preferences of particular sponge groups by noting their association with certain sediment types. We can also trace through time particular associations of sponges characteristic of certain environments and note changes in the composition of these associations (Fig. 14).

One such environment is that of quiet water, identifiable by

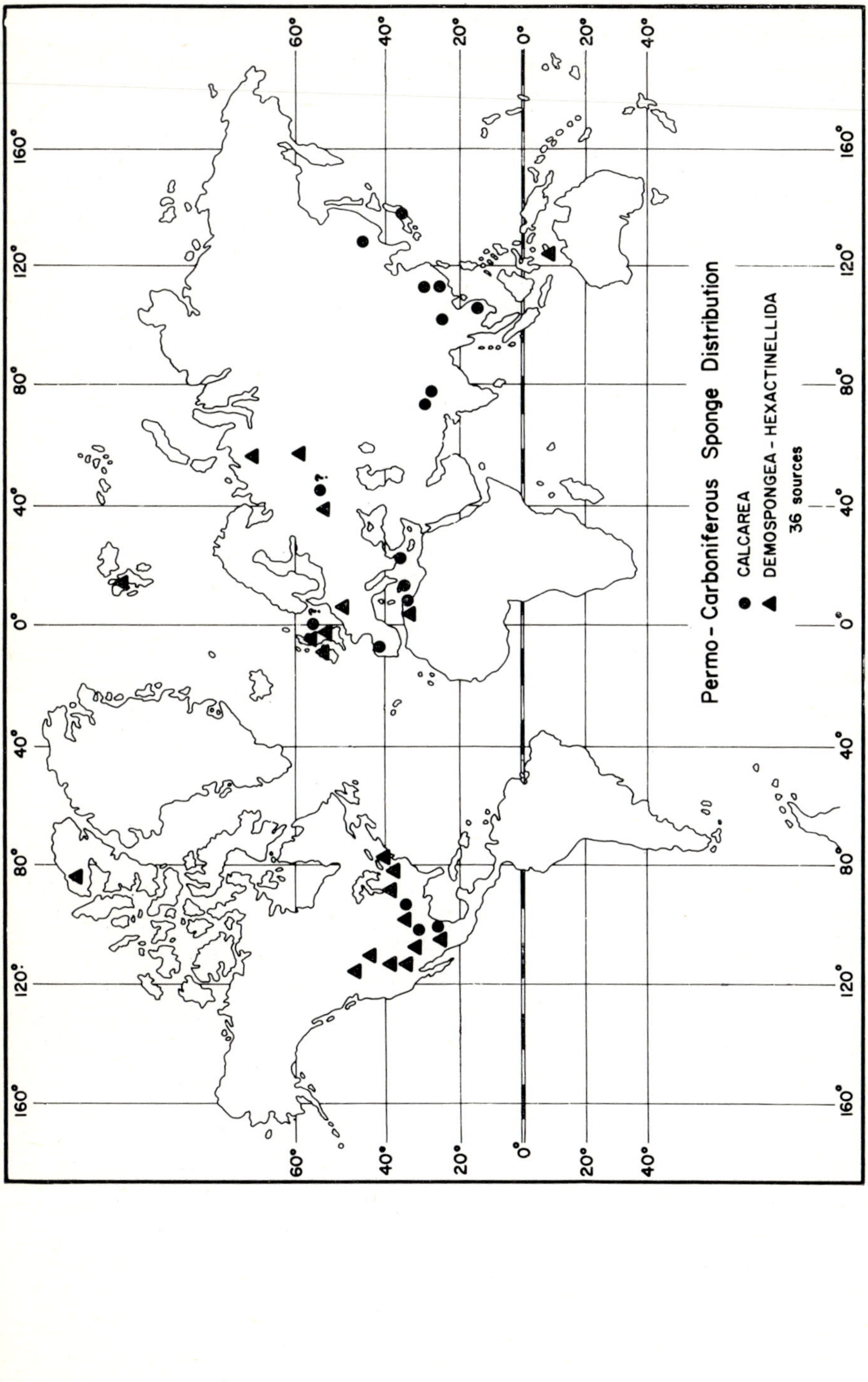

FIG. 12. Geographic distribution of all the occurrences of Calcarea, Hexactinellida and Demospongea from Carboniferous and Permian rocks known to the writer. Some localities close to each other have been combined.

PERIOD	OCCURRING WITH ALGAE	OCCURRING WITH REEFS
Permian	Hexactinellida Demospongea Heteractinida Calcarea	Hexactinellida Demospongea Calcarea
Carboniferous	Hexactinellida Demospongea Heteractinida Calcarea	Hexactinellida Demospongea
Devonian	Hexactinellida Demospongea ? Receptaculitida	Hexactinellida Demospongea ? Heteractinida ? Receptaculitida
Silurian	Hexactinellida ? Demospongea ? Heteractinida Receptaculitida ?	Demospongea Heteractinida Receptaculitida
Ordovician	Hexactinellida ? Demospongea Receptaculitida	Hexactinellida ? Demospongea Receptaculitida
Cambrian	Archaeocyatha	Archaeocyatha

FIG. 13. Shallow-water sponges of each period as shown by occurrences with algae or reefs. Question marks indicate uncertainty as to the reality of the association in life.

fine-grained sediment and other geologic criteria. This environment has been the special province of the siliceous sponges (Hexactinellida and Demospongea). Abundant occurrences of Palaeozoic hexactinellids are almost always in such sediments, although sporadic occurrences in rougher waters, and even in reefs, are known. Quiet waters are not necessarily deep; where depths can be estimated, they range from the 500-m-deep basin facies of the Texas Permian (Newell, 1957) to the 1-m-deep late Carboniferous coastal lagoon represented by a black shale above a coal bed (Zangerl and Richardson, 1963). Black shales form a distinct subgroup within this environment and are the special home of thin-walled, delicate hexactinellids of persistently primitive type.

A quite different environment, of agitated, shallow, water bearing relatively little suspended sediment, identifiable by deposits built largely of whole or fragmented, coarse, skeletal debris, is commonly designated the "shelly-facies". It has a characteristic sponge fauna from

PERIOD	QUIET WATER	SHELLY FACIES	REEF BUILDERS
Permian	DEMOSPONGES HEXACTINELLIDS	Sphinctozoa Pharetronida Chiastoclonellidae Heliospongiidae † Wewokellidae † Rhizomorina Hindiidae (Hexactinellida)	Sphinctozoa Pharetronida
Carboniferous	DEMOSPONGES HEXACTINELLIDS	Sphinctozoa Wewokellidae Heliospongiidae (Rhizomorina) (Hexactinellida)	
Devonian	DEMOSPONGES HEXACTINELLIDS	Stromatoporoidea Hindiidae Astraeospongiidae Receptaculitida (Hexactinellida)	Stromatoporoidea
Silurian	DEMOSPONGES HEXACTINELLIDS	Stromatoporoidea Astylospongiidae Astraeospongiidae Hindiidae (Hexactinellida ?)	Stromatoporoidea
Ordovician	DEMOSPONGES HEXACTINELLIDS	Stromatoporoidea Anthaspidellidae Hindiidae Dystactospongiidae Astylospongiidae Receptaculitida (Hexactinellida ?)	Stromatoporoidea Anthaspidellidae Receptaculitida
Cambrian	DEMOSPONGES HEXACTINELLIDS HETERACTINIDS	Archaeocyatha Stromatoporoidea	Archaeocyatha

FIG. 14. Sponge taxa associated with three types of environments in different periods. Under "Shelly Facies" and "Reef Builders" the taxa are arranged in approximate order of abundance, the most abundant at the top; those in parentheses are minor elements. Daggers indicate extinction before the end of the period; question-marks indicate uncertainty as to the reality of the association.

Ordovician times onwards. The lithistid demosponges and the heteractinids are particularly characteristic of this facies. So are the receptaculitids (an Ordovician to Devonian group variously assigned to the Porifera and to the Chlorophyceae) and the stromatoporoids. In the Ordovician the anthaspidellid lithistids are particularly dominant among the sponges of this facies. In the Silurian and Devonian it is the hindiid lithistids and the astraeospongiid heteractinids, accompanied in the Silurian by another lithistid group, the astylospongiids. In the later Palaeozoic, the receptaculitids and stromatoporoids disappear. The astraeospongiids evolve into the wewokellids. New demosponge groups become abundant (especially the Heliospongiidae Finks, 1960) and Calcarea appear in force, first the Sphinctozoa, then the Pharetronida. The rigid or heavy skeletons of all these groups are undoubtedly adapted to rough water. It should be noted that some of the lithistids of this facies may also be abundant in the quiet water facies, but not the Calcarea, Heteractinida, Stromatoporoidea and Receptaculitida Sushkin, 1962.

PERIOD	REEF – BUILDERS
Permian	Algae * Sphinctozoa * Pharetronids Bryozoa Hydrocorallines ?
Carboniferous	Corals Algae
Devonian	Corals Stromatoporoids Algae
Silurian	Corals Stromatoporoids Bryozoa Algae ?
Ordovician	Bryozoa Stromatoporoids Algae * Demosponges Corals Receptaculitida
Cambrian	Archaeocyatha Algae
Pre- Cambrian	Algae

FIG. 15. Principal reef-building organisms in each period of the Palaeozoic. Listed in approximate order of abundance under each period, the most abundant at the top. Sponges starred; question marks indicate uncertainty of identification of the organisms.

Reefs form a special environment which may be regarded as an offshoot of the shelly-facies. When sponges participate in reef-building, it is always those groups that are most abundant among the sponges of the shelly-facies. The times of greatest sponge participation in reefs are the Ordovician and Permian. If we look at a chronological list of the principal reef-building organisms (Fig. 15), we see that the corals (Tabulata and Rugosa) dominate the reefs during the intervening periods, a time that coincides with their general peak of abundance and diversity. The corals were apparently competitively superior to the sponges in the reef environment, with the exception of the stromatoporoids which appear to have been the competitive equals of the corals. Since sponges and corals do not have the same food sources (suspension feeding versus carnivorous macrophagy) the food supply was not a factor in either successful (stromatoporoids) or unsuccessful (other sponges) competition. Bryozoa appear to have been the competitive equals of the other sponges and likewise inferior to the corals. That genuine competition between these various groups existed is suggested by the tendency for approximately contemporaneous reefs in the same area to be dominated by one or the other of the reef-building groups, as described for the Ordovician by Pitcher (1964).

Acknowledgements

I thank Mr. Craig A. Munsart for drafting the illustrations after my original sketches and Mr. David Massie for the photograph of *Multivasculatus*. The remaining photographs are by the author.

References

Barrois, C. (1882). *Recherches sur les terrains anciens des Asturies et de la Galice.* Lille.

Bolkhovitinova, M. A. (1923). O kamennougol'nykh gubkakh Moskovskoi gubernii. *Vestnik Moskovskoi Gornoi Akademii*, **2**, No. 1: 61–71.

Finks, R. M. (1960). Late Paleozoic sponge faunas of the Texas region: the siliceous sponges. *Bull. Am. Mus. nat. Hist.* **120**: 1–160.

Finks, R. M. (1967). The structure of *Saccospongia laxata* Bassler (Ordovician) and the phylogeny of the Demospongea. *J. Paleont.* **41**: 1137–1149.

Flügel, E. and Flügel-Kahler, E. (1968). Stromatoporoidea (Hydrozoa Palaeozoica), *in* Westphal, F. (Ed.), *Fossilium Catalogus I* (*Animalia*), Parts 115, 116, 1–681.

Hall, J. and Clarke, J. M. (1899). A memoir on the Paleozoic reticulate sponges constituting the family Dictyospongiidae. *Mem. N.Y. St. Mus. nat. Hist.* No. 2: 1–350.

Hartman, W. D. and Goreau, T. F. (1966). *Ceratoporella*, a living sponge with stromatoporoid affinities. *Am. Zool.* **6**: 262.

Hinde, G. J. (1888). A monograph of the British Fossil sponges. *Palaeontogr. Soc.* [Monogr.] **1888** pt. 2, 93–188.

Kling, S. A. and Reif, W.-E. (1969). The Paleozoic history of amphidisc and hemidisc sponges: new evidence from the Carboniferous of Uruguay. *J. Paleont.* **43**, No. 6: 1429–1434; pl. 176, 1 fig.

Lecompte, M. (1956). Stromatoporoidea, *in* Moore, R. C. (Ed.), *Treatise on Invertebrate Paleontology*, Part F, F107–F144.

Librovitch, L. S. (1929). *Uralonema karpinskii*, nov. gen., nov. sp. i drugie kremnevye gubki iz kamennougol'nykh otlozhenii vostochnovo sklona Urala. *Trudў geol. Kom.*, n.s. **179**: 1–57.

Newell, N. D. (1957). Paleoecology of Permian reefs in the Guadalupe Mountains area, *in* Ladd, H. S. (Ed.), *Treatise on marine ecology and paleoecology* **2**, Paleoecology. *Mem. Geol. Soc. Am.* **67**: 407–436.

Ortmann, P. (1912). Die Mikroskleren der Kieselspongien in Schwammgesteinen der senonen Kreide. *Neues Jb. Miner. Geol. Paläont.* **1912**: **2**: 127–149.

Pitcher, M. (1964). Evolution of Chazyan (Ordovician) reefs of eastern United States and Canada. *Bull. Can. Petrol. Geol.* **12**: 632–691.

Reimann, I. G. (1945). New middle Devonian octactinellids. *Bull. Buffalo Soc. nat. Sci.* **19**: 16–21.

Rietschel, S. (1968). Die Octactinellida und ihnen verwandte paläozoische Kalkschwämme (Porifera, Calcarea), *Paläont. Z.* **42**: 13–32.

Schrammen, A. (1924). Die Kieselspongien der oberen Kreide von Nordwestdeutschland, 3. und letzter Teil, mit Beiträgen zur Stammesgeschichte. *Monogrn Geol. Paläont.* (1) **1**: 1–159.

Steinmann, G. (1882). Pharetronen-Studien. *Neues Jb. Miner. Geol. Paläont.* **1882 2**: 139–191.

Vandercammen, A. (1950). Contribution à l'étude des spongiaires hétéractinellides. *Bull. Inst. r. Sci. nat. Belg.* **26**: 1–23.

Walcott, C. D. (1920). Cambrian Geology and Paleontology IV, No. 6—Middle Cambrian Spongiae. *Smithson. misc. Collns.* **67**: 261–364.

Zangerl, R., and Richardson, E. S. Jr., (1963). The paleoecological history of two Pennsylvanian black shales. *Fieldiana, Geol. Mem.* **4**: 1–352.

Symp. zool. Soc. Lond. (1970) No. 25, 23–40

PHYLOGENETIC RELATIONSHIPS OF FOSSIL CALCISPONGES

BERNHARD ZIEGLER

Berlin-Charlottenburg, Institut für Geologie und Paläontologie der Technischen Universität, Berlin, West Germany

and

SIEGFRIED RIETSCHEL

Frankfurt am Main, Natur-Museum und Forschungsinstitut Senckenberg, Frankfurt am Main, West Germany

SYNOPSIS

Fossil calcareous sponges differ from the majority of modern Calcarea in one important feature; they possess firm skeletons stiffening the animals' bodies. Most modern Calcisponges possess loose spicules embedded separately in the soft tissue. Fossil representatives of apparently the same organization as these living species have been known since the Lower Carboniferous. They are, however, likely to have existed since Cambrian or perhaps even Precambrian times.

From this conservative stock several offshoots arose which had the common tendency to produce stiffened skeletons. The means, however, of realizing this were varied. In two lineages, the Palaeozoic Octactinellida and the mainly Mesozoic Pharetronida, the spicular skeleton was maintained. In the first group the firm skeleton was formed by entangling large many-rayed spicules, in the second by cementing the spicules and building up fibres. In two other lineages, the Lower Cambrian Archaeocyatha and the mainly Permian and Triassic Sphinctozoa, the spicular skeleton was replaced by non-spicular walls.

Thus, four offshoots from the main stock may be recognized, three of which are extinct, while a few genera of the fourth still persist. Unfortunately, the common phylogenetic origin of some of these offshoots cannot be proven.

INTRODUCTION

The main characteristics of the phylum Porifera are the mode of differentiation of the cells and the existence of flagellate cells, the so-called choanocytes, arranged in chambers, in tubes, or as a distinct layer. Unfortunately, the flagellate cells cannot be seen in fossils. Consequently fossil Porifera must be recognized by two features. First, by their distinctive skeleton and second by the existence of a channel system, depending upon the arrangement of the flagellate cells, leading from the outer surface through cavities within the body to an inner or upper surface. The channel system of modern sponges may be of the asconoid, syconoid, or leuconoid types. It should be mentioned that—despite some citations in the literature (e.g. Rauff, 1913)—no channel system other than that of a leuconoid type can be demonstrated unquestionably in any fossil sponge.

This contribution deals with those fossils showing the typical features mentioned above and having skeletons made up of the calcium carbonate calcite. In modern sponges calcareous skeletons seem to be restricted to the Class Calcarea, records of the association of silicious and calcareous skeletons (e.g. in *Astrosclera* Lister, 1900 and *Merlia* Kirkpatrick, 1908) still being under discussion (e.g. Vacelet, 1967a).

The skeleton of most modern calcareous sponges consists of spicules. These elements are formed by specialized cells, the scleroblasts (Woodland, 1905). Two scleroblasts together build up an individual single-rayed spicule, one being the builder, the other working as the thickener. Three-rayed or four-rayed spicules originate in three or four separate units which are each formed by a pair of scleroblasts and which join later on. Consequently sutures may often be seen at the bases of the rays.

There is still another mode of building skeletons in some modern

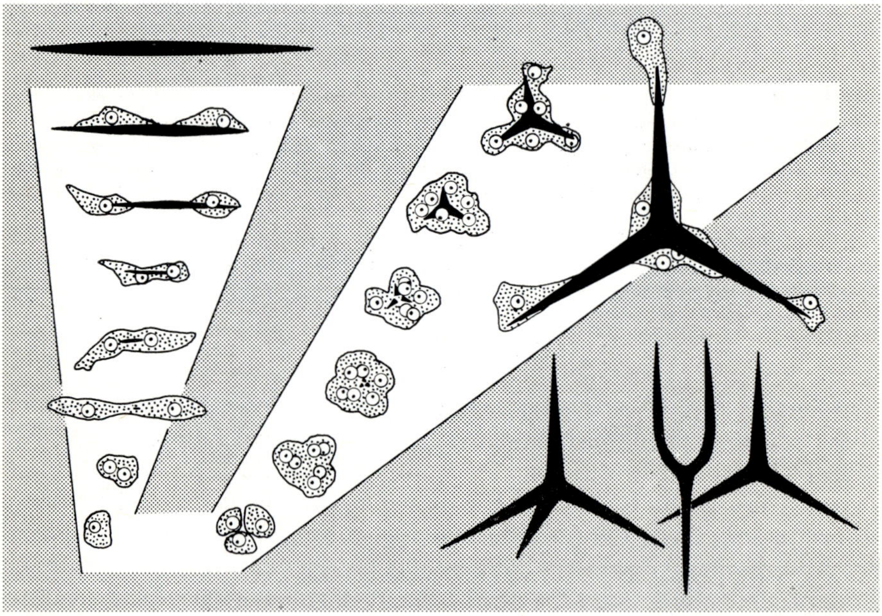

FIG. 1. Development of monaxon (left) and triaxon (middle) spicules in calcareous sponges. Types of fully grown spicules (upper left and lower right). Diagrammatic. Spicule building is begun by one or three scleroblasts. Each scleroblast divides into a builder and a thickener cell.

Calcarea. Kirkpatrick (1908) has shown that cells different from scleroblasts are able to secrete calcareous matter. These cells, cementing spicules and forming fibres in the genus *Minchinella* Kirkpatrick, 1908, he named telmatoblasts. Whether these telmatoblasts are responsible for the formation of the massive, apparently sphaeroidally structured skeletons of the genera *Murrayona* Kirkpatrick, 1910 and *Petrobiona* Vacelet and Lévi, 1958 also, or whether there are still other types of differentiated cells, does not seem to be quite clear (Vacelet, 1964). Besides, there is some doubt as to the primary nature of the sphaeroidal structure. It is possible that it may be formed by recrystallization, either during the life of the animal or soon after its death. In any case, the sphaeroidal structure requires a different primary arrangement of the crystals from that found in those cases in which laminar walls occur.

Thus, modern calcareous sponges are known to form three different types of skeleton—a. spicules, b. laminar matter, c. rigid, sphaeroidally structured masses.

TYPES OF CALCAREOUS SPONGES

Dialytina

Rauff (1893, p. 203) has proposed the order Dialytina to include the majority of modern calcareous sponges. He associated the pharetronids with this order but this is the result of his erroneous interpretation of the pharetronid fibre. According to Rauff, the main characteristic of the Dialytina is that the spicules are never amalgamated. They may be single-rayed, three-rayed or four-rayed but they are always embedded in the soft tissue in isolation.

As there is no amalgamated skeleton in the Dialytina, spicules will become isolated soon after the death of the animal. Therefore it is very unlikely that members of the Dialytina will be fossilized.

Nevertheless there are a few records of fossil detached spicules. As a rule, their affinities are doubtful. Finds of isolated polyactinal and octactinal spicules are often recorded from Palaeozoic rocks. They seem to belong partly to Octactinellids and to the Cambrian Chancelloriidae. The only spicules of Palaeozoic age which may belong to Dialytina have been described by Hinde (1887) from the British Carboniferous. Later, Hinde described isolated spicules, which possibly belong to Dialytina, from Mesozoic (1889) and Pliocene (1886) beds of Great Britain. Other finds are figured by Zankl (1965) from the Upper Triassic of the Alps. As Pharetronids and Sphinctozoans are very common in the sediments studied by him, it is possible that the isolated spicules belong to these groups and not to representatives of the Dialytina.

Moreover, some sponge-bodies apparently belonging to calcareous sponges with a skeleton consisting of individual spicules have been described. One of the best examples recorded seems to be *Leucandra walfordi* Hinde, 1889 from the Lower Jurassic of Northamptonshire. In other cases the published data are not sufficient and therefore the affinities of the genera concerned are in doubt (e.g. *Protosycon* Zittel, 1879, *Camarocladia* Miller, 1889).

The Cambrian genera *Chancelloria* Walcott, 1920 and *Allonnia* Doré and Reid, 1965 appear to be distantly related to the true Dialytina. They are now regarded as calcareous sponges because they show a similar suturing of the spicules' centre as is seen in some living Calcarea (Reid, 1967a). It is evident that in life they had calcareous spicules (Rietschel, 1968). They differ from modern Dialytina in having much larger spicules and in the fact that not only do single-rayed, three-rayed and four-rayed spicules occur but also that spicules with up to nine rays are known. If we consider the mode of formation of the spicules we may regard the Chancelloriidae as a specialized offshoot from some simpler, but still unknown, ancestors.

Thus, Calcareous sponges with isolated spicules are known since lower Cambrian times. Although there are but few records, they certainly existed in uninterrupted sequence, and in similar organization, at least from the Lower Carboniferous up to recent days. They are likely to have lived even at the base of the Palaeozoic, when they gave rise to the Chancelloriidae.

Octactinellids

The Octactinellida, Hinde, 1887 are a small but very distinctive group. They are true calcareous sponges, as is shown quite clearly by the preservation and mode of construction of their spicules (Rietschel, 1968). Their skeleton consists of 6- to 8-rayed spicules with hexagonal symmetry—the octactins. The spicules vary in size but the main network of the skeleton is built up of rather large, thick, octactins. These sometimes show (especially in the Cambrian genus *Eiffelia* Walcott, 1920) sutured centres and therefore seem to have been constructed in the way which is characteristic of recent calcisponges. The spicules are interlocked and therefore sometimes allow fossilization of the sponges' skeleton, although fusion of the spicules does not occur. Besides octactins, monaxons are present in the Devonian genus *Ensiferites* Reimann, 1945.

The typical representatives of the octactinellids lived from Cambrian to Devonian times. Best known and most frequent is the type-genus *Astraeospongium* F. Roemer, 1854 (lower Ordovician to upper Devonian); finds of isolated octactins are usually classified as belonging to this genus.

Whether this is correct for isolated octactins from the lower Carboniferous of Germany (Steinmann, 1880) is not quite clear.

The octactinellids are common and widespread fossils in palaeozoic reef- and shelf-sediments. After a rapid rise they lived on as a conservative group with few genera and little variation in skeleton and shape.

Besides the true octactinellids we know some sponges which might be related to them, such as *Wewokella* Girty, 1911 from the upper Carboniferous of North America. This is regarded by King (1943) and Croneis and Toomey (1965) as a calcareous sponge. According to other authors (Rigby and Moyle, 1959; Hoare and Sturgeon, 1968) its skeleton is built up of "sphaerasters", whereas King (1943) mentions "perfect octactins" and "polyasters". So further details about *Wewokella* are needed to determine exactly its relationship to octactinellids. In the Permian genus *Talpaspongia* King, 1943, which Finks (1960) regards as a junior synonym of *Wewokella*, Lokke (1964) described single-rayed and three-rayed spicules. This feature is quite incompatible with the characteristics of Octactinellida and even of *Wewokella*. However, the dimensions of the spicules (several millimetres) agree with these forms. Hence the affinities of *Talpaspongia* remain obscure.

Still more obscure is the systematic position of the genera *Asteractinella* Hinde, 1887 and *Tholiasterella* Hinde, 1887 from the Carboniferous. Hinde (1887) proposed his suborder Heteractinellidae for them. In these genera, as well as in the lower Cambrian *Uranosphaera* Bedford and Bedford, 1934, polyasters are the main spicules. In addition, octactins are recorded. There are serious doubts about the calcareous or siliceous nature of their skeleton. More details about these genera are required before their systematic position can be discussed.

The origin of the typical octactinellids is not proven. If, however, single-rayed spicules in calcareous sponges are held to be primitive, in accordance with the early development of spicules in modern Calcarea, the Octactinellida should have evolved from some primitive stock, perhaps through *Chancelloria* or related, still unknown, forms.

Pharetronids

The Pharetronida as described first by Zittel (1879) are characterized by fibres containing calcareous spicules cemented by a calcareous matrix. There is clear evidence to suggest that the fibres are formed by the living animal and are not, as often thought (e.g. Rauff, 1893, 1913, 1938), of secondary nature. The original formation of the fibres is known from living species (Kirkpatrick, 1908) and can be demonstrated in fossils in those cases where epizoans are fixed on the fibres and are themselves overgrown by the sponge (Ziegler, 1964), that is to say, where the epizoans have settled during the life of the animal.

In the pharetronids several different types of spicules are known. Some genera, such as the Triassic *Stellispongia* d'Orbigny, 1849, have only short, rounded, single-rayed spicules within the fibres (Steinmann, 1882). Similar uniaxial spicules with pointed extremities, however, are described from the Cretaceous genus *Pharetrospongia* Sollas, 1877 (Hinde, 1887).

Others, the Jurassic and/or Cretaceous genera *Enaulofungia* Fromentel, 1861, *Elasmostoma* Fromentel, 1859, and *Peronidella* Zittel in Hinde, 1893 for instance, are characterized by fibres containing three-rayed spicules. The shape of the three-rayed spicules may differ. In the genera mentioned above, the more or less regular type of the tripod predominates. In *Sestrostomella* Zittel, 1878 regular tripods are present in addition to three-rayed tuning-fork spicules. As well as the three-rayed spicules, single-rayed, or four-rayed, spicules may occur in most genera.

Four-rayed spicules predominate in the modern genera *Petrostroma* Döderlein, 1892, *Plectroninia* Hinde, 1900, and *Minchinella*, and in the Cretaceous genera *Porosphaera* Steinmann, 1878 and *Porosphaerella* Welter, 1910.

Cementation of the spicules may take place in different ways. In *Stellispongia* a single fibre contains bundles of more or less parallel spicules. In most other Mesozoic pharetronids, and in the modern *Minchinella*, only one or a few spicules are found in the cross-section of a single fibre. They are either merely interlocked or jointed side to side. Often the tip of a ray is connected with the side of the ray of another spicule (Wagner, 1964). There does not seem to be a fundamental difference in the mode of cementation from that of the modern genus *Plectroninia* and the Cretaceous genus *Porosphaera*, where the spicules within the fibres are said to be connected tip to tip (de Laubenfels, 1955).

The origin of the calcareous matrix of the fibres cementing the spicules was ascribed by Kirkpatrick (1908) to the action of so-called telmatoblasts—cells secreting non-spicular calcareous matter. Initially the spicules are formed by scleroblasts. Later they are cemented and incorporated into the fibres. The microscopic structure of the non-spicular parts of the fibres is not yet known sufficiently. Sometimes the cement is built up by thin lamellae. In other cases it is described as being sphaeroidally structured. It seems that the calcareous matrix of the fibres is often altered during fossilization.

A cementation of preexisting spicules similar to that in the fibres is often seen in superficial calcareous membranes, covering smaller or greater parts of the surface of the fossil pharetronids (Ziegler, 1964).

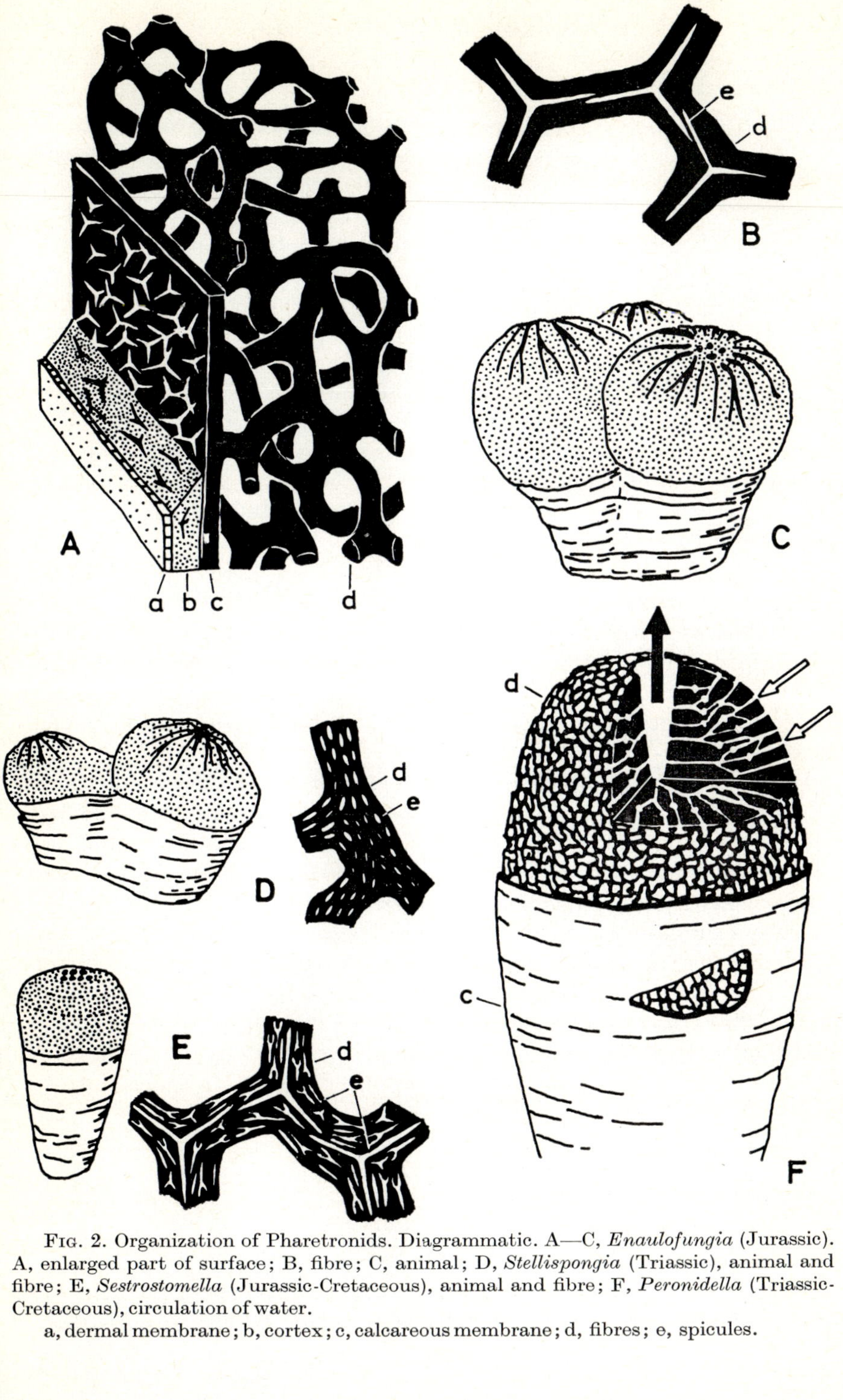

FIG. 2. Organization of Pharetronids. Diagrammatic. A—C, *Enaulofungia* (Jurassic). A, enlarged part of surface; B, fibre; C, animal; D, *Stellispongia* (Triassic), animal and fibre; E, *Sestrostomella* (Jurassic-Cretaceous), animal and fibre; F, *Peronidella* (Triassic-Cretaceous), circulation of water.

a, dermal membrane; b, cortex; c, calcareous membrane; d, fibres; e, spicules.

Thus, the skeleton of the typical Pharetronida is made up by three different elements: by spicules, by fibres cementing spicules, and by membranes cementing spicules. It is evident that isolated spicules are lost in fossils so that only fibres and membranes can be observed. In many cases, recrystallization of fibres and membranes during diagenesis has taken place. In such cases spicules can no longer be detected and hence it is rather difficult to distinguish between a pharetronid and a hydrozoan.

The Pharetronida made their first proven appearance in Permian times. Former records of Pharetronida (e.g. "Peronidella" Zittel, 1893) from Devonian beds are mostly based either on misinterpreted silicosponges or on wrongly labelled specimens. There are only a few records from Devonian beds which actually concern calcisponges of doubtful affinities. These may include ancestors of the pharetronids. In the Mesozoic pharetronids are abundant. However, most genera die out at the end of the Cretaceous and there are only a few records since the beginning of the Paleocene.

There is little doubt about the ancestry of the Pharetronida. They seem to have evolved in late Palaeozoic times from primitive Calcisponges with isolated spicules by evolving methods of cementing their spicules into fibres. Yet we have no means of ascertaining whether or not this process took place only once in the earth's history and therefore should always bear in mind the possibility of iterative evolution of the pharetronids.

Among living Calcisponges the genera *Petrobiona*, *Murrayona* and *Paramurrayona* Vacelet, 1967 are of special interest (Vacelet, 1967b). They show regular calcareous spicules within their soft tissue and are therefore true Calcisponges. However a non-spicular skeleton also occurs in these genera. In *Petrobiona* and *Murrayona* it forms basal masses or anastomosing fibres, and in *Paramurrayona* it develops as calcareous plates covering the animal (Vacelet, 1964, 1967b). As these three genera do not possess the most characteristic feature of the typical pharetronids, i.e. fibres containing spicules, their relationships to this group are not quite clear. However, a clear knowledge of their nature is very important to the understanding of another group of Calcisponges, namely the Sphinctozoa.

Sphinctozoans

The sphinctozoans were segmented animals possessing calcareous walls and channel-systems of apparently the same types as those known in sponges. In most genera of the Sphinctozoa no spicules can be observed. The main skeleton consists of walls which are either built up of thin lamellae or are sphaeroidally structured. In the Jurassic and

Cretaceous family Sphaerocoeliidae Steinmann, 1882, however, the walls quite clearly contain three- and four-rayed spicules (Hinde, 1882; Rauff, 1913; Reid, 1967b). In the Triassic genus *Thaumastocoelia*, Steinmann (1882) and Ott (1967) describe somewhat questionable irregular spicules.

In one group of sphinctozoans numerous pores are present within the wall. In another group imperforate walls are pierced by less numerous, and larger, "Ostia". Whether the wall was truly external during the life of the animal or covered by some soft tissue is not quite clear. There are, however, some facts (e.g. in the genus *Celyphia* Pomel, 1872; Ziegler, 1964) pointing to the existence of soft tissue outside the wall.

The interiors of the individual segments are either hollow—that is to say, without fossilized skeleton but, during life, certainly filled with soft tissue—or they contain different types of calcareous structures. In most genera, parts of the bodies are separated by thin diaphragms (= vesiculae). It seems that these parts no longer contained living tissue after the formation of the diaphragms. Therefore the significance of the vesiculae may be compared to that of the dissepiments or tabulae in corals or to that of the diaphragms in Bryozoa, rudist Pelecypoda, and some brachiopods (*Richthofenia* Kayser, 1881). Similar structures are known as rarities amongst pharetronids (Ziegler, 1965: p. 107; pl. 16, Fig. 6).

Other calcareous filling structures have been described by Seilacher (1961) and Ott (1967) as of reticular, trabecular or tubular types. True pharetronid fibres have never been observed, although in some genera (e.g. *Verticillites* Defrance, 1829 and on the surface of some *Celyphia*) similar vermicular elements occur. In those genera where the wall contains spicules (e.g. *Barroisia* Munier-Chalmas in Steinmann, 1882), true pharetronid fibres are entirely absent.

Apparently, it was the function of the filling tissue to support the soft tissue of the animal. The filling structures often show very clearly the course of the channel system (Seilacher, 1961).

Apart from those few sphinctozoans in which a spicular skeleton can be detected (family Sphaerocoeliidae) we have no real proof that the Sphinctozoa are sponges. Such characteristics, however, as pores, channel-system, central cavities and oscula indicate a close relationship between the two groups. Moreover, there is no feature in the organization of the sphinctozoans that could not be explained in terms of sponge morphology.

The absence of spicules in walls and filling tissue has its analogy in the non-spicular rigid skeleton of some modern Calcisponges such as *Petrobiona*, *Murrayona* and *Paramurrayona*. Isolated spicules which

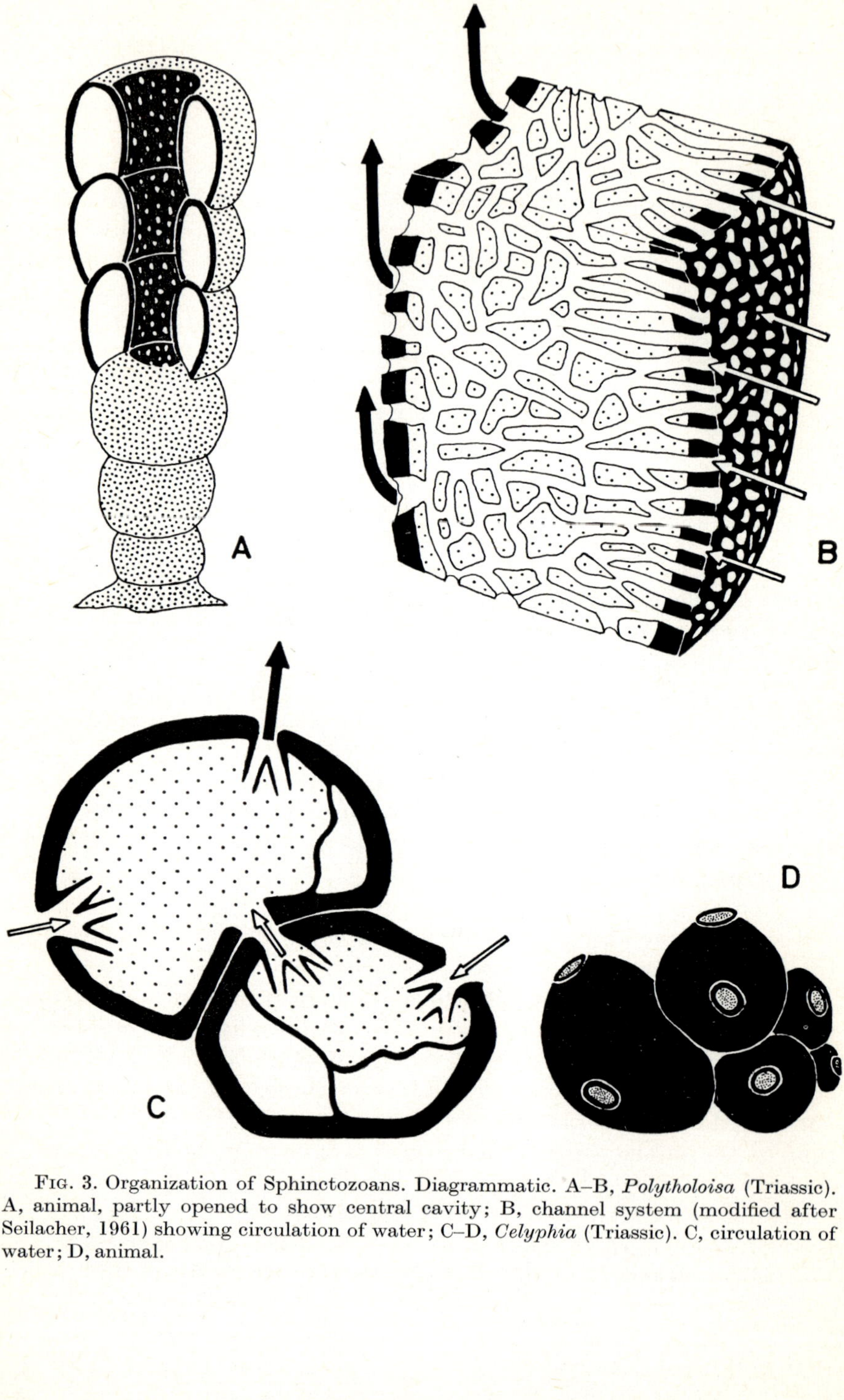

FIG. 3. Organization of Sphinctozoans. Diagrammatic. A–B, *Polytholoisa* (Triassic). A, animal, partly opened to show central cavity; B, channel system (modified after Seilacher, 1961) showing circulation of water; C–D, *Celyphia* (Triassic). C, circulation of water; D, animal.

are likely to have existed in the choanosome of the sphinctozoans have been lost during fossilization in the same way as those lying outside the fibres and membranes of the pharetronids. The formation of the non-spicular skeleton of the Sphinctozoa may be explained in the same way as in pharetronids—by the action of telmatoblasts. There is but one difference between the two groups. In pharetronids fibres and membranes are formed, as a rule, *after* the formation of the spicules. In the Sphinctozoa wall and filling structures are formed either *before* spicules develop or at places where the latter never occur. Therefore, it seems quite reasonable to us to associate the sphinctozoans with the calcisponges.

Sphinctozoans are known from the Carboniferous up to the end of the Cretaceous. They were very common animals in Permian and Triassic times. From Jurassic and Cretaceous strata only a few genera and species are described.

The most primitive Sphinctozoa are characterized by segments, with or without vesiculae, but never with any other filling tissue. The evolution of filling structures took place in Permian and—still more—in Triassic times. In the late Mesozoic sphaerocoeliids, characterized by their well developed spicules, filling tissue is once more wanting. As has been pointed out by Ott (1967), it is likely that the characteristic shape of the Sphinctozoa evolved at least twice. First, in late Palaeozoic times the bulk of the group and their Triassic descendants appear in three distinct families. Later on, in the Upper Jurassic, appears the family Sphaerocoeliidae, about whose ancestors nothing is known.

Archaeocyatha

The Archaeocyatha are a group of doubtful affinities. Most of the modern workers on the group (e.g. Okulitch, 1955; Vologdin, 1962; Zhuravleva, 1960; Hill, 1964, 1965; Debrenne, 1964) consider them as representing a totally extinct phylum close to, but different from, the Porifera.

The Archaeocyatha are represented by conical or cylindrical, sometimes irregular, cups. In most cases there are two perforated walls, an outer and an inner one, enclosing the so-called intervallum and surrounding a usually hollow central cavity. The intervallum contains skeleton of variable form—rods, radial porous parieties ("septa" = regular ones, "taenia" = irregularly bent ones) and perforated plates (tabulae). In none of the structures can spicules be detected and the whole skeleton consists of microgranular calcite.

Nearly all of the skeletal elements are perforated, with the single major exception of the dense, unperforated, "dissepiments" known in many Archaeocyatha. These are formed later than the parieties and possibly indicate that the living animal has withdrawn from parts of the skeleton. In this case the "dissepiments" are comparable to the vesiculae of the sphinctozoans.

Nothing has been preserved of the soft tissue and reconstructions of the living animals vary considerably (see Hill, 1964). Yet all agree in the assumption that the Archaeocyatha are a phylum ("type" in the Russian literature) on their own.

The reasons why the Archaeocyatha are thought not to be sponges have been summarized by Okulitch and de Laubenfels (1953). Their arguments require a short discussion.

For convenience, the counts on which archaeocyathids are thought by most authors not to be sponges may be placed in five groups.

1. Sponges are said to have a spicular skeleton, whereas archaeocyathids have granular-lamellar walls. However, some modern Calcisponges possess non-spicular skeletons. As telmatoblasts secrete granular-lamellar fibres or sheets in the Pharetronida, they, or similar cells, may also have been responsible for the formation of the granular-lamellar walls of archaeocyathids.

2. Sponges are said never to have outer and inner walls. However, some fossil pharetronids as well as some sphinctozoans have such walls. It can easily be demonstrated that these walls did not arise by post-mortem alteration, as believed by de Laubenfels, but have been formed by the living animal.

3. Sponges are said never to have parieties. Perhaps this may be true. However, there seems to be no reason why the presence of parieties is incompatible with the sponge organization. Their presence means only that the water-currents do not stream diffusely through the soft tissue but are restricted to distinct regions of the animal. Such a restriction may also be seen in the honeycombs of the hexactinellid *Aphrocallistes* Gray, 1858.

4. Sponges are said to have totally different development of the skeleton from that in archaeocyathids. Their skeleton "begins by widely-separated, needle-like spicules". Archaeocyathids, on the other hand, "start out with a temporarily imperforate sheet". However, this may depend on the nature of the cells first secreting lime. If they are scleroblasts, spicules will be formed. If they are telmatoblasts, fibres or sheets will grow. Moreover, nobody knows whether the initial stages of the calcareous cup of an archaeocyathid are formed during the early development as they now appear or whether the rather dense preserved walls are of a gerontic nature similar to that of the basal calcareous layers in some Pharetronida.

5. True calcareous sponges are said to occur first in the Devonian, the archaeocyathids becoming extinct in the Middle Cambrian. However, calcareous sponges are known now to have existed at least since the Middle Cambrian.

Thus the arguments put forward by Okulitch and de Laubenfels (1953) do not seem to be convincing. None of the features shown by archaeocyathids really conflicts with a sponge-nature of the animal. On the other hand, we admit that there is no proof that the Archaeocyatha actually were sponges, although the existence of pores connecting the outer and inner surfaces, and probably representing a channel-system for water circulation, as well as the apparent occurrence of the main tissues and functions in the intervallum of the archaeocyathids suggest strong affinities. We feel that we should not separate archaeocyathids from sponges unless and until there are less equivocal data supporting such a separation.

Fossils apparently distinct from the calcisponges

There is a group of fossils which is generally adjoined to sponges and by some authors even regarded as Calcisponges: the Receptaculitida. They are mostly spherical to conical bodies, consisting of a wall built up of particulate elements, the "meromes", and an inner cavity filled by sediment and/or calcite. The meromes bear some resemblance to sponge spicules. This has induced most authors (e.g. Hinde, 1884) to regard receptaculitids as sponge-like fossils. This resemblance, however, is due to preservation only and it can be shown that the meromes are not comparable with sponge spicules. The meromes were complex organs with a distinctive mode of growth. They were apparently calcified during the lifetime of the organism. The receptaculitid wall had neither pores nor a canal system, nor an osculum (S. Rietschel, unpublished manuscript). The systematic position of this group cannot be cleared up in this paper—receptaculitids really seem to have been calcareous algae—yet it is certain that they were not related to calcisponges.

Very obscure is the relationship of another group, the so-called Aphrosalpingoida Miagkova, 1955. They are regarded by Vologdin and Miagkova (1962) as a class closely related to archaeocyathids. Some of them, the Cambrian Syringocnemidae Taylor, 1910, resemble true archaeocyathids. The typical order Aphrosalpingida Vologdin and Miagkova, 1962 on the other hand, seems to stand in closer relationship to calcareous algae than to calcisponges and archaeocyathids. This order consists of two families, each with one monospecific genus. These fossils were described by Miagkova (1955a,b) from Upper Silurian

(Ludlow) beds of the Ural. They show features which are well known from Paleozoic codiacean algae (e.g. *Palaeoporella* Stolley, 1893), as far as can be judged from the descriptions alone. Thus we have serious doubts about the sponge-like nature of the Ludlovian "archaeocyathids".

EVOLUTION IN CALCAREOUS SPONGES

The evolutionary trends leading to the major groups of calcareous sponges cannot be observed directly because intermediate links are missing. Therefore, the phylogenetic relationships can only be inferred.

It seems that simple forms similar to those genera which predominate today (Dialytina) represent the main stock in Calcisponge evolution. From this conservative group several offshoots arose which possessed the common tendency to produce stiffened skeletons. The means, however, by which this came about have differed considerably.

In early Cambrian times the Archaeocyatha possibly evolved from the main stock by attaining a non-spicular double-walled skeleton with radial elements. Quite a different mode of stiffening the skeleton, by entangling large many-rayed spicules, is shown by the Middle Cambrian to Upper Palaeozoic Octactinellida. In Upper Palaeozoic times another offshoot from the main stock, the Sphinctozoa, once more attained a non-spicular firm skeleton. Soon after the rise of the sphinctozoans the pharetronids evolved from primitive ancestors, attaining a spicular, but solid, skeleton stabilized by the formation of the characteristic fibres.

Thus, four offshoots from the conservative stock whose existence is not, however, proven before the Carboniferous, may be recognized. Three of these are extinct, while the fourth still exists with only a few genera. As has been shown in the discussion of the pharetronids and sphinctozoans, there is, unfortunately, no possibility of ascertaining the phylogenetic coherence of each of these lineages.

The stratigraphical distribution of the four main groups of fossil Calcisponges shows that each group evolved immediately after or during the decline of its predecessors. There is but one exception; the pharetronids flourishing in the Upper Permian and the Triassic simultaneously and associated with the sphinctozoans. Although few details are known of the mode of life of these fossil animals this competition might be explained by the assumption of slightly different ecological requirements (Zankl, 1969).

As far as is known, all fossil calcareous sponges from Cambrian to Tertiary times lived in shallow waters. They are often associated with

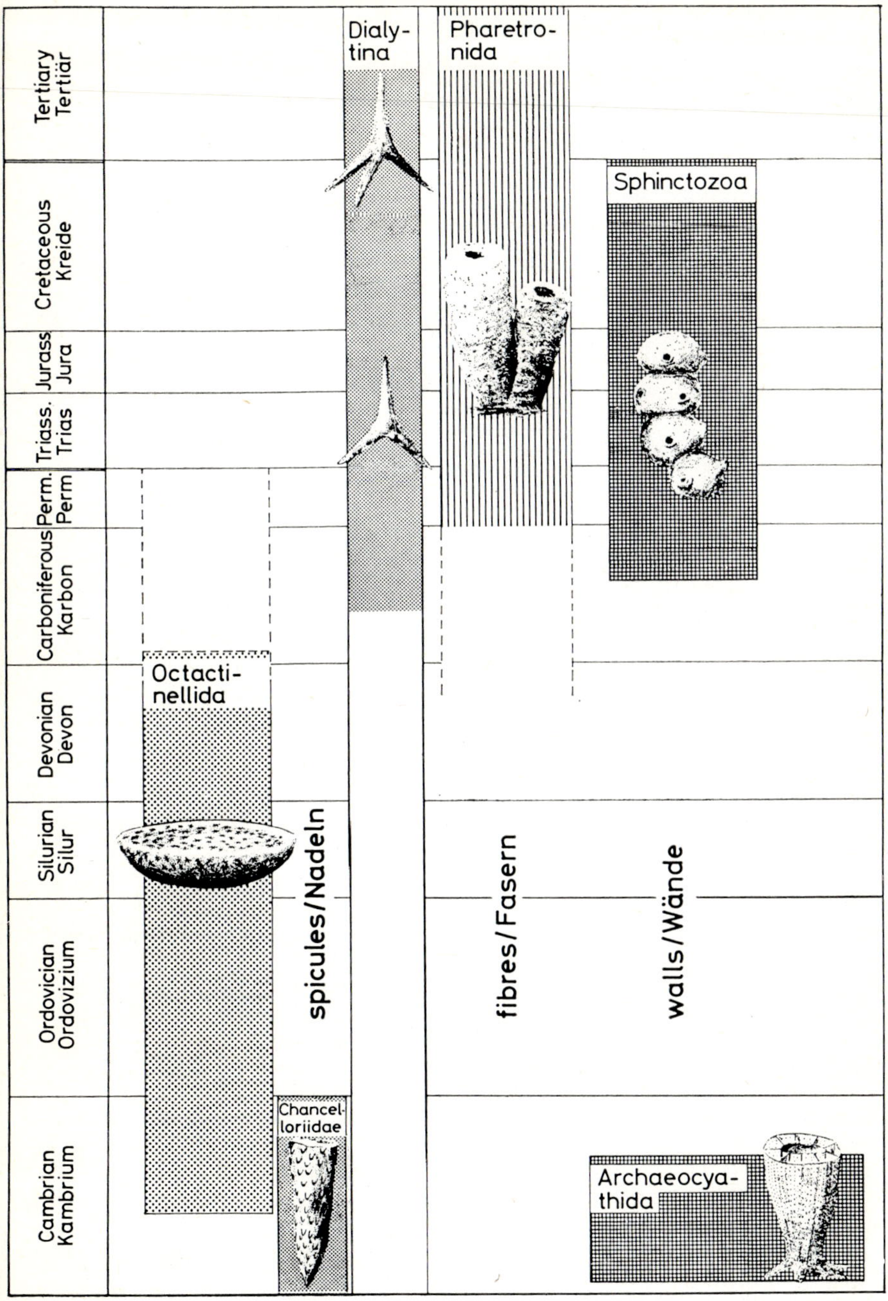

FIG. 4. Stratigraphic distribution of calcareous sponges.

reef-building organisms such as corals, or with organisms which have lived in the surroundings of reefs (calcareous algae, crinoids, brachiopods, molluscs). Apparently, the fossil calcareous sponges preferred warm waters, as the majority of them are found in calcareous sediments. In the Lower Cambrian, archaeocyathids themselves built up reefs (Zhuravleva, 1960) just as did the Sphinctozoa and Pharetronida in Triassic (Ott, 1967; Zankl, 1969), and perhaps Upper Permian, times (Newell *et al.*, 1953). Apparently, during the Upper Mesozoic and Tertiary, a change in the life habits of the pharetronids took place. They retired more and more from the shallowest zone (some genera, such as *Radicispongia* Quenstedt, 1878, *Epitheles* Fromentel, 1859 and *Porosphaera*, are associated with lithistids in somewhat deeper water) until they became restricted to waters below 80 m depth or to dark submarine caves (Vacelet, 1967b)

References

Croneis, C. and Toomey, D. F. (1965). Gunsight (Virgilian) Wewokellid sponges and their depositional environment. *J. Paleont.* **39**: 1–16.

Debrenne, F. (1964). Archaeocyatha. Contribution à l'étude des faunes cambriennes du Maroc, de Sardaigne et de France. *Notes Mém. Serv. géol. Maroc* No **179**: 1–264.

Finks, R. M. (1960). Late Paleozoic Sponge faunas of the Texas region. The siliceous sponges. *Bull. amr. Mus. nat. Hist.* **120**, 1: 1–160.

Hill, D. (1964). The Phylum Archaeocyatha. *Biol. Rev.* **39**: 232–258.

Hill, D. (1965). Archaeocyatha from Antarctica and a review of the phylum. *Scient. Rep. transantarct. Exped.*, **10**: 1–151.

Hinde, G. J. (1882). Notes on Fossil Calcispongiae, with Descriptions of new Species. *Ann. Mag. nat. Hist.* (5), **10**: 185–205.

Hinde, G. J. (1884). On the Structure and Affinities of the family of the Receptaculitidae. *Q. Jl. geol. Soc. Lond.* **40**: 795–849.

Hinde, G. J. (1886). On the Sponge-spicules from the Deposits of St. Erth. *Q. Jl. geol. Soc. Lond.* **42**: 214.

Hinde, G. J. (1887–1912). A monograph of the British fossil sponges. *Palaeontogr. Soc.* [*Monogr.*] **1887–1912**: 1–264.

Hinde, G. J. (1889). On a true Leuconid Calcisponge from the Middle Lias of Northamptonshire, and on detached Calcisponge Spicules in the Upper Chalk of Surrey. *Ann. Mag. nat. Hist.* (6) **4**: 352–358.

Hoare, R. D. and Sturgeon, M. T. (1968). The genus *Wewokella* (Porifera) in the Pennsylvanian of Ohio. *J. Paleont.* **42** (1): 81–83.

King, R. H. (1943). New Carboniferous and Permian sponges. *Bull. Kans. State geol. Surv.* **47**: 1–36.

Kirkpatrick, R. (1908). On two new Genera of Recent Pharetronid Sponges. *Ann. Mag. nat. Hist.* Ser. (8), **2**: 503–514.

de Laubenfels, M. W. (1955). Porifera—*in:* Moore, R. C. (Ed.): *Treatise on Invertebrate Paleontology*, **E**: 21–112: Lawrence, Kansas: University of Kansas Press.

Lokke, D. H. (1964). Calcareous spicules in *Talpaspongia clavata* R. H. King, lower Permian of Concho County, Texas. *J. Paleont.* **38** (4): 778–781.

Miagkova, E. J. (1955a). Kharakteristike klassa Aphrosalpingoida Miagkova, 1955. *Dokl. Akad. Nauk SSSR* **104** (4): 478–481 (Russ.).

Miagkova, E. J. (1955b). Novye predstviteli tipa Archaeocyatha. *Dokl. Akad. Nauk SSSR* **104** (4): 638–641 (Russ.).

Miller, S. A. (1889). *North American Geology and Palaeontology.* Cincinnati, Ohio: Western Methodist Book Press.

Newell, N. D., Rigby, J. K., Fischer, A. G., Whiteman, A. J., Hickox, J. E. and Bradley, J. S. (1953). *The Permian Reef Complex of the Guadaloupe Mountains Region, Texas and New Mexico.* San Francisco: Freeman & Co.

Okulitch, V. I. (1955). Archaeocyatha—*in:* Moore, R. C. (Ed.). *Treatise on Invertebrate Paleontology,* **E**: 1–20. Lawrence, Kansas: University of Kansas Press.

Okulitch, V. I. and de Laubenfels, M. W. (1953). The systematic position of Archaeocyatha (Pleosponges). *J. Paleont.* **27** (3): 481–485.

Ott, E. (1967) Segmentierte Kalkschwämme (Sphinctozoa) aus der alpinen Mitteltrias und ihre Bedeutung als Riffbildner im Wettersteinkalk. *Abh. bayer. Akad. Wiss.*, mathem.-naturwiss. Kl. *n.F.*, **131**: 1–96.

Rauff, H. (1893). Palaeospongiologie. Allgemeiner Teil. *Palaeontographica* **40**: 1–232.

Rauff, H. (1913). *Barroisia* und die Pharetronenfrage. *Paläont. Z.* **1** (1): 75–144.

Rauff, H. (1938). Über einige Kalkschwämme aus der Trias der peruanischen Kordillere nebst einem Anhang über *Stellispongia* und ihre Arten. *Paläont. Z.* **20** (2): 177–214.

Reid, R. E. H. (1967a). *Hyalostelia* Zittel and the Heteractinellida. *Geol. Mag.* **104** (4): 398–399.

Reid, R. E. H. (1967b). Spicules in the Skeleton of *Tremacystia d'orbignyi* (Hinde). *Nature, Lond.* **215** (5103): 875–876.

Rietschel, S. (1968). Die Octactinellida und ihnen verwandte paläozoische Kalkschwämme (Porifera, Calcarea). *Paläont. Z.* **42** (1/2): 13–32.

Rigby, J. K. and Moyle, R. W. (1959). Some Mississippian and Pennsylvanian Sponges from Utah. *J. Paleont.* **33** (3): 399–403.

Seilacher, A. (1961). Die Sphinctozoa, eine Gruppe fossiler Kalkschwämme. *Abh. Akad. Wiss. Lit. Mainz.* Mathem-naturwiss. Kl., 1961 (10): 721–790.

Steinmann, G. (1880). Mikroskopische Tierreste aus dem deutschen Kohlenkalke (Foraminiferen und Spongien). *Z. dt. geol. Ges.* **32**: 394–400.

Steinmann, G. (1882). Pharetronen Studien. *Neues Jb. Miner. Geol. Paläont.*, 1882 (2): 139–191.

Vacelet, J. (1964). Etude monographique de l'Eponge Calcaire Pharétronide de Méditerranée, *Petrobiona massiliana* Vacelet et Lévi. Les Pharétronides actuelles et fossiles. Thèse Doct. Sci. nat. Aix-Marseille.

Vacelet, J. (1967a). Quelques éponges pharétronides et "silico-calcaires" de grottes sous-marines obscures. *Recl. Trav. Stn. mar. Endoume* **42** (58): 121–132.

Vacelet, J. (1967b). Déscriptions d'éponges pharétronides actuelles des tunnels obscurs sous-récifaux de Tuléar (Madagascar). *Recl. Trav. Stn. mar. Endoume*, Fasc. hors sér. suppl, (6): 37–62.

Vologdin, A. G. (1962). Tip Archaeocyatha—*in:* Orlov, J. A. (Ed.), *Osnovy Paleontologii* **2**: 89–133. Izd. Akad. Nauk SSSR (Russ.).

Vologdin, A. G. and Miagkova, E. J. (1962). Klass Aphrosalpingidea—*in:* Orlov, J. A. (Ed.), *Osnovy Paleontologii* **2**: 135–137. Izd. Akad. Nauk SSSR (Russ.).

Wagner, W. (1964). Zum Skelettbau oberjurassischer Kalkschwämme. *Mitt. bayer. Staatssamml. Paläont. hist. Geol.* **4**: 13–21.

Woodland, W. (1905). Studies in spicule formation. I. The development and structure of the spicules in sycons with remarks on the conformation, modes of disposition and evolution of spicules in calcareous sponges generally. *Q. Jl. microsc. Sci.* n.s. **194** (**49**,2): 231–282.

Zankl, H. (1965). Zur mikrofaunistischen Charakteristik des Dachsteinkalkes (Nor/Rät) mit Hilfe einer Lösungstechnik. *Z. dt. geol. Ges.* **116** (2): 549–567.

Zankl, H. (1969). Der Hohe Göll. Aufbau und Lebensbild eines Dachsteinkalk-Riffes in der Obertrias der Nördlichen Kalkalpen. *Abh. senckenb. naturforsch. Ges.* **519**: 1–123.

Zhuravleva, I. T. (1960). *Arkheotziathy sibirskoi platformi.* Leningrad: Izd. Akad. Nauk SSSR [Russ.].

Ziegler, B. (1964). Die Cortex der fossilen Pharetronen (Kalkschwämme). *Eclog. geol. Helv.* **57** (2): 803–822.

Ziegler, B. (1965). Die Variabilität bei Pharetronen (Kalkschwämme). *Paläont. Z.*, **39** (1/2): 106–110.

Zittel, K. A. (1879). Studien über fossile Spongien. Dritte Abteilung. Monactinellidae, Tetractinellidae und Calcispongiae. *Abh. bayer. Akad. Wiss.* math.-phys. Kl. **13** (2): 1–48.

Symp. zool. Soc. Lond. (1970) No. 25, 41–59

PORIFERA, SPHINCTOZOA, ARCHAEOCYATHI—THEIR CONNECTIONS

I. T. Zhuravleva

Institute of Geology and Geophysics, Novosibirsk 90, U.S.S.R.

SYNOPSIS

Study of the fossil remains of Sphinctozoa and Archaeocyathi can contribute a great deal to our knowledge of the earliest stages of the evolution of multicellular animals. Initially regarded as variants of other groups of multicellular animals such as sponges, protozoa and coelenterates, the Sphinctozoa now appear to be most satisfactorily included within a group containing the calcareous sponges, while the Archaeocyathi Vologdin, 1937 are best regarded as a distinct group, more different from Porifera than from Protozoa, and sufficiently distinct to be placed in a taxon—the Archaeozoa—of equal rank with the Parazoa and Enterozoa. The Archaeozoa could well include a number of other poorly-known and enigmatic extinct multicellular animals.

INTRODUCTION

There are several groups among the fossil and modern Invertebrata an understanding of whose nature and position in the Animal Kingdom contributes much to the elucidation of some problems in general biology and historical geology. An understanding of the Porifera, Sphinctozoa and Archaeocyathi can make such a contribution. Porifera are living now, while the other two groups have died out completely—the Sphinctozoa at the beginning of the Palaeogene and the Archaeocyathi at the very beginning of the middle Cambrian.

The representatives of all three groups have some features in common, notably their sedentary life conditions, their mineral skeleton with a separate central cavity and the degree of their morphological diversification. Not surprisingly, therefore, many specialists have referred members of all three groups to a single type of animal—the Porifera.

The history of research shows that, in establishing the systematic position of each of the groups, scientists followed similar lines of reasoning. The best-known group, the Porifera, were recognized as an independent animal type in the last century (earlier they had been united with the Coelenterata and, still earlier, included in the Radiata together with a number of other polyradial marine invertebrates).

The Sphinctozoa, accepted as a separate morphological and taxonomic group only after Steinmann's (1882) researches, were compared in turn with Coelenterata, Archaeocyathi, Calcarea Porifera and other groups. Even at the present time their systematic position is uncertain. Their general neglect is one of the reasons for this situation.

The first finds of Archaeocyathi were made only between 1850 and 1861 and they were included in either the Porifera or Coelenterata until the middle of the present century.

Some attempts were made to compare the Archaeocyathi with Algae (Siphonales), Sphinctozoa, Protozoa and other organisms. Only between 1945 and 1955 were convincing reasons produced for their isolation as an independent type of animal, despite their great geological antiquity (their earliest known fossils are dated as early Cambrian 550–500 million years old).

It is necessary to outline the characteristics of the three groups of organisms before the main problem of their evolutionary status is explored further.

PORIFERA

These are lowly, multicellular animals, devoid of distinct tissues and separate organs. They possess choanocytes in special cavities in the body. The flagella of the choanocytes produce water currents. Their nutrition is intercellular and they are thus biofiltrators.

Their spicular skeleton, consisting of silica, spongin or calcite with some admixture of magnesium and strontium, is made in special cells—the scleroblasts.

The primary embryonic layers, the ectoderm and entoderm, are not distinguishable.

According to their skeletal ingredients, sponges may be classified into siliceous, cornacosiliceous, cornuous and calcareous. These divisions are considered very differently by different authors, their ranks ranging from ordinal to that of a class. However, all specialists believe that there is a single mechanism of production of spicules in all the different groups of Porifera. It is necessary to emphasize this point because especial attention will be paid to the Calcarea Porifera, while comparing the Archaeocyathi with the Porifera.

The morphological plasticity of the Porifera is very high because of the low level of their organization and their immobile life conditions. Because of this, possible pronounced polymorphism must be expected in all forms (Ziegler, 1965).

The mechanism of spicule production in Porifera has been studied in detail and all the processes in spicule development have been considered (e.g., Lévi, 1964). As has been mentioned, the spicules are produced only by special cells, the scleroblasts. Each ray, if there are several, corresponds to a single cell.

The main forms of spicules in the Porifera Calcarea are triradiates and quadriradiates. Only the Pharetronida Zittel, 1878, which are included in the Calcarea, have a skeleton which is not primarily of these main spicule types. Pharetronids have a massive skeleton and were known for a long time only as fossils. Consequently, the nature of their skeleton was in dispute for many years. Only after the recent discovery of nine living species of these sponges could Vacelet (1961) and Ziegler (1964) conclude that the pharetronid skeleton consisted of spherules and had been primarily massive. The diameter of the spherules is about 0·08 mm, while the pharetronid skeletal fibres are about 1 mm thick. It is still not possible to refute or confirm the supposition of Zittel (1878), Steinmann (1882) and others that the apparently separate fibres of the fossil Pharetronida were cemented together by spongin or other mesogloeal components.

Detailed researches into the initial stages of the development of Porifera have been carried out only in recent years (e.g. Burton, 1949; Bakus, 1964). It must be emphasized that a very young sponge possesses distinct spicules even in the earliest hours of its life (Fig. 1) and during

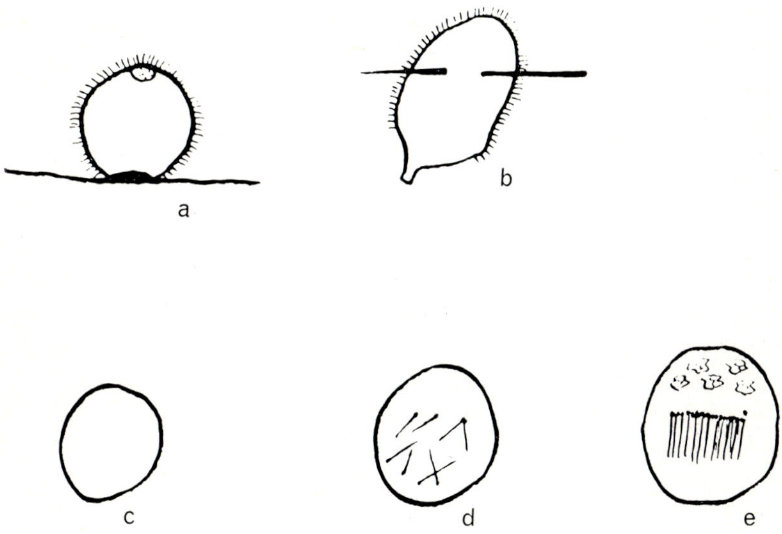

FIG. 1. Young sponges with first spicules, × 50. a, b, *Tedania gurjanovae*, a, embryo; b, parenchymella stage, containing a single spicule in the upper part (after Bakus, 1964); c, d, e, *Iophon radiatus* embryonic stages, c, without spicules; d, with first acanthostyles; e, acanthostyles and anisochelae segregated (after Burton, 1931).

growth the spicule features can change as many as three times. The sponge can attain maturity in two to three weeks.

The larvae of sponges are of two types, the amphiblastula and the parenchymella, and it seems reasonable to suppose that this distinction existed in fossil sponges.

Some authorities have concluded that the larval type and the apical or basal position of the choanocyte nucleus are important in the classification of the Porifera and in particular that of the Calcarea (Bidder, 1898; Dendy and Row, 1913; Hartman, 1958 and others). However, the researches of Vacelet (1960, 1961) show us that these characters are in some cases not consistent even within one species. Moreover, these criteria do not assist us in classifying the Pharetronida, most of the members of which group are known only from fossils.

The first records of separate siliceous spicules occur in Proterozoic rocks, the true calcareous sponges being known from Devonian times onwards. However, there is no guarantee that they will not be found in more ancient rocks, such as Cambrian deposits.

At present we recognize about 5000 species of modern Porifera, so we cannot consider the group as a failing animal type. Perhaps their more solid adaptive features have permitted them to survive throughout the Phanerozoic.

SPHINCTOZOA

This is a completely extinct group of multicellular animals (Devon-Palaeogene) and therefore we know its members only from their calcareous skeletal remnants (Fig. 2).

There has been prolonged debate on the primary structure of the sphinctozoan skeleton, centred around whether it was composed of separate spicules secondarily cemented together, or was initially deposited as a solid mass. This controversy explains why the systematic position of the group is still far from satisfactorily settled. In earlier times Sphinctozoa* were recognized as Bryozoa, Stromatoporoidea and even Nautiloidea. Sphinctozoa were also compared to pharetronid calcareous sponges (e.g. Dunikovsky, 1883; Hinde, 1884; Rauff, 1893) and Archaeocyathi (Taylor, 1910; Moret, 1952; Zhuravleva and Rezvoy, 1956). There was even the suggestion of a probable connection between Sphinctozoa and coelenterates.

Recently I have detailed additional reasons for continuing uncertainty about the systematic position of this group (Zhuravleva,

* The name for this group was devised by Steinmann (1882). de Laubenfels (1955) used the term Thalamida.

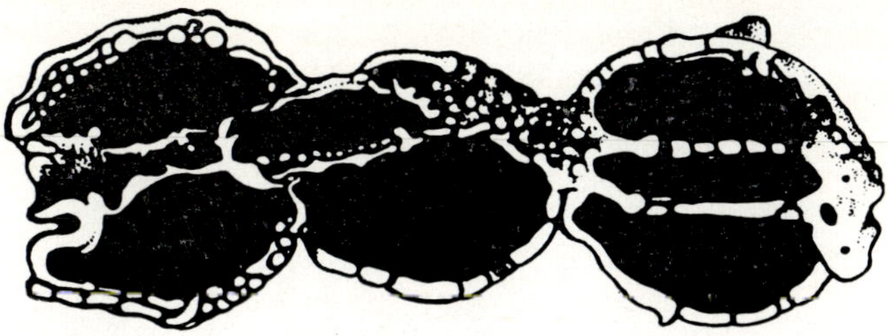

FIG. 2. Cameral structure in Sphinctozoa; a longitudinal section, × 10 (from Zhuravleva, 1959).

1965). My opinion was based on the study of some Sphinctozoa from the Permian of the Pamir, Armenia, and the northern Caucasus. Despite the good and sometimes even excellent preservation of the skeletons in the collections studied, it was not possible to discover in slides any traces of primary spicules (Fig. 3). There are conflicting data. Reid (1967) has confirmed the spicular nature of the sphinctozoan skeleton by a re-examination of those specimens in the collections of the British Museum (Natural History) which were studied by Hinde (1884). This evidence places the Sphinctozoa morphologically more distant from the Archaeocyathi and closer to the calcareous sponges.

At the same time, the cameral structure of the majority of Sphinctozoa, and their possession of vesicular "tissue", raises the questions of a possible double origin of their skeleton. This would have involved an initial intracellular formation of primary spicules (as in Porifera) followed by the secretion of laminae by a quite different mechanism, about which we are still ignorant.

As has been indicated, the sphinctozoan skeleton was calcareous, and consisted always of an "outer wall", formed by the overhanging exterior edges of the basal plates during the construction of the camera. Both the bases and the outer wall were permeated by canals. An inner wall was present in many genera. This separated the axial tube or central cavity from the cameral cavities.

The vesicular tissue and rods were sometimes present in the camerae as well as in the central cavity. The skeletal substance is, as a rule, of conjoint structure, as seen in transparent slide preparations.

The group Sphinctozoa is not large, only six families, thirty two genera and about a hundred species being known (Zhuravleva, 1965). Unfortunately, the morphology and systematics of Sphinctozoa have been little studied as yet and we lack completely any knowledge of their ontogeny.

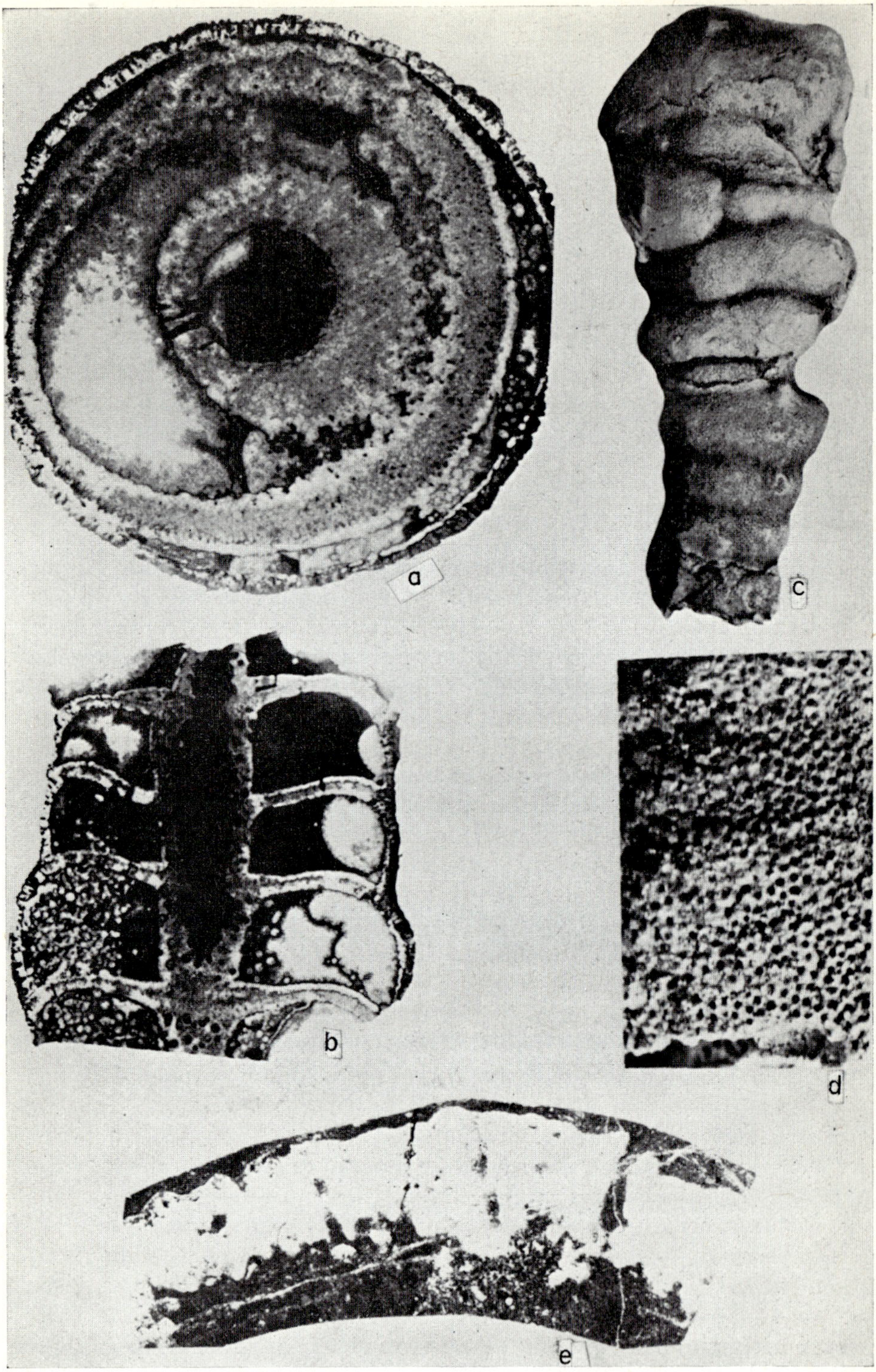

FIG. 3. a, Transverse section of *Amblysiphonella vesiculosa* (Sphinctozoa), × 3; b, longitudinal section of *Ambl. vesiculosa*, × 2; c, external view of *Amblysiphonella* sp., × 1; d, pores on the external surface of the cup of *Amblysiphonella*, × 4; e, part of a transverse section of the camera of *Amblysiphonella*, × 10.

All these uncertainties demand great care, not only in the interpretation of functional morphology, but also in the erection of hypotheses about the evolutionary position of the group. Probably it is most satisfactory at the present time to continue to regard the Sphinctozoa as lying within a group containing the calcareous Porifera, with the reservation that they may yet have to be regarded as sufficiently distinct to be placed in a separate high-ranking taxon.

ARCHAEOCYATHI

A much shorter period of study of this entirely fossil group was required before it was included in the Porifera as the Class Pleospongia Okulitch, 1943. Archaeocyathi have also been identified as representatives of Coelenterata, Protozoa and even Algae. Even now there is not unanimous acceptance of their affinities, although their independence is generally admitted. Thus Pospelov (1962) unites them with Stromatoporoidea and Yakovlev (1954) regards them as related to Rugosa, or even as having a direct affinity with Echinodermata (Yakovlev, 1956).

Only in recent years has the mode of formation of the archaeocyathian skeleton become understood. The skeleton was not spicular (Okulitch, 1943, 1955; Vologdin, 1948; Zhuravleva, 1959) but, on the contrary, was closer to the skeletal structures of foraminiferan Protozoa.

Vesicular "tissue" and cameral structure are commonplace in the Archaeocyathi. This we must regard as a convergent resemblance to the Sphinctozoan structure. Furthermore, as distinct from all known other types, the porous skeleton of Archaeocyathi could develop such complex and thin structures as outer walls with branched pores, microporous membranes, chequered structures with various forms of tumulae and so on (Fig. 4). The inner wall structure was still more variable. It could be annular, scaled, or contain interconnected porous canals and, at the very least, it was complex—that is to say, it had rings and canals combined (Fig. 5). The skeletal elements of the intervallum (the space between the outer and inner walls) were not so highly variable but here too there was a range of structure from a hollow non-skeletal intervallum in Capsulocyathina Zhuravleva, 1964 to an intervallum with completely unporous vertical parities in *Compositocyathus* Zhuravleva, 1960 and some other Archaeocyathi. In addition, mention must be made of the development of different types of rods, up to a complete imitation of the dictyonal structure of Hexactinellida (e.g. *Dictyocyathus* Bornemann,1891).

The form of the cups of the Archaeocyathi was also variable.

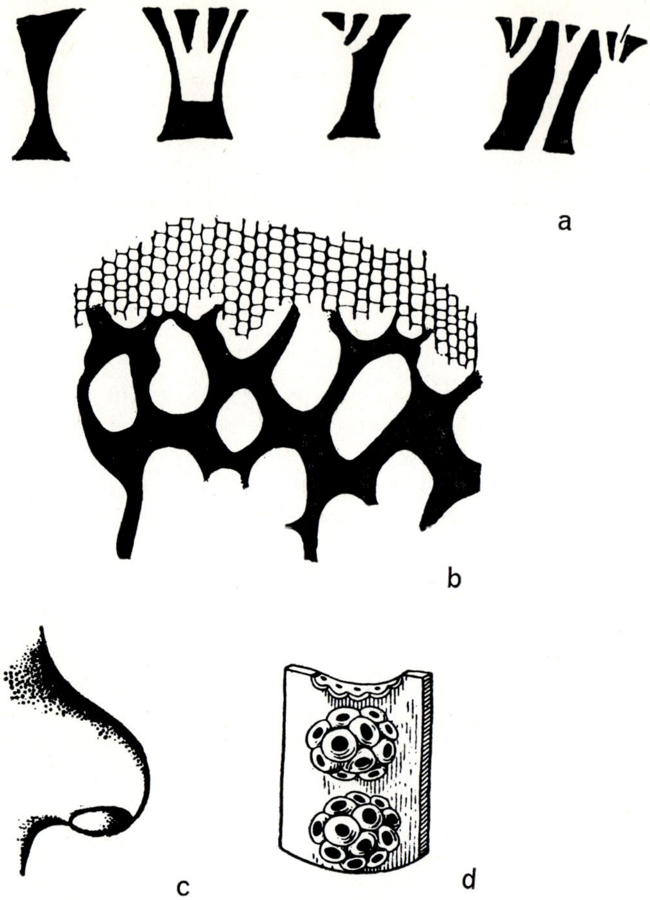

FIG. 4. Different types of porosity of the complex outer walls of Archaeocyathi Regulares: a, branched outer wall in *Cryptaporocyathus*, part of transverse section of cupl, × 60 (from Zhuravleva, 1960); b, an outer wall with microporous membrane, tangentia section of cup (after Rozanov, 1964); c, simple tumulus of the outer wall, reconstruction after Rozanov, 1964; d, raspberry-like tumuli of the outer wall, reconstruction after Rozanov (1964).

Besides the more common conical or cylindrical cups, hemispherical and sac-like discoidal and irregular cup forms are known.

Like the Porifera, Archaeocyathi were organisms with very poorly expressed individuality and we find among them branched, massive and serial colonies.

The archaeocyathian skeleton structure, as known from the excellently preserved samples taken from early Cambrian layers of the

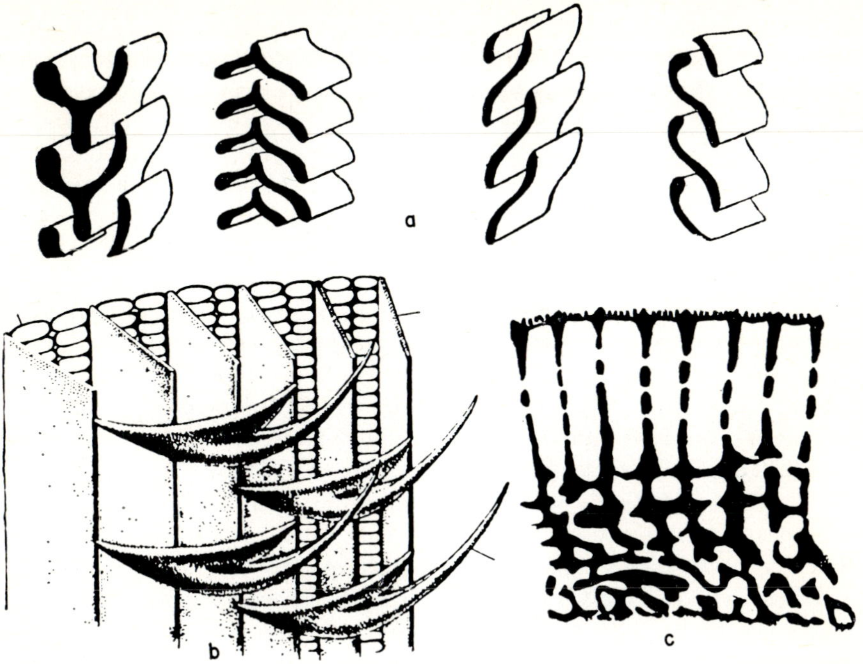

FIG. 5. Various structures of the porosity of the complex inner wall in Archaeocyathi Regulares: a, various forms of the rings of the inner wall, reconstruction after Rozanov 1964; b, view from the side of the central cavity of an inner wall with scales, × 60 (from Zhuravleva, 1960); c, part of a transverse section of an inner wall with connected curved canals (after Rozanov, 1964).

Siberian Platform, was micrograined, with considerable admixture of dispersed organic material. The outlines of the archaeocyathian skeleton in slide preparations are very sharp. All the data confirm the opinion of Vologdin (1962a,b) that the skeleton of Archaeocyathi was originally organo-calcitic.

The skeletal pores of Archaeocyathi were radically different from those of the "walls" of Sphinctozoa, or from those occurring in the cortical layer of Porifera. The pores of the representatives of the two latter groups appeared as holes or intervals between the spicules or elements of the skeleton. The archaeocyathian pores appeared in the very first stages of the beginning of the skeleton development as a basic element in the skeletal layers themselves.

The archaeocyathian skeleton has been well studied from the very early stages of ontogeny—even at cup diameters of as little as 0·02 to 0·08 mm. The first essential element of the skeleton was the small imperforate cup, laid down within the single outer wall of the animal.

The wall microstructure was the same as in adult forms, so we can be certain that there were no spicules in the early stages of the development of Archaeocyathi. A little later, at the next stage of development of the cup skeleton, the cup wall became porous and the inner wall and the first intervallum skeletal elements appeared. Simultaneously, the wall pore-system became more complicated, although such additional complexity is not characteristic of all genera of Archaeocyathi. The outer wall always achieved complex porosity earlier in ontogeny than the inner wall.

The origin of the skeletal inner wall was considered initially to be the same, without exception, in all Archaeocyathi. Later, two different types of development, and therefore of the appearance of the inner wall, among Regulares Vologdin, 1937, and Irregulares Vologdin, 1937 were detected (Bedford and Bedford, 1939; Okulitch, 1943; Zhuravleva, 1960). Today we recognize three main types of development (Fig. 6).

Type 1. The inner wall of Monocyathida Okulitch, 1935, Subclass Regulares, was directly joined with the outer wall by the pushing of the overlying skeletal elements of the cup into the inner cavity. This process varied slightly in its details. Thus, the outer wall of Monocyathina Okulitch, 1935 was connected at a sharp angle with the covering pelta and membrane, but remained clearly distinct from the latter. In the Capsulocyathina the outer wall curves smoothly inwards to become the inner wall. In both cases the type of in-pushing is very distinctive and thus the inner walls must be regarded as analogous rather than homologous. This general type of inner wall construction may be termed invaginal (Fig. 6a,b).

Type 2. Among the other Regulares (those with parities or radial rods in the intervallum), the inner wall appeared simultaneously with the first radial rods, the delimitation of the central cavity being synchronous with the formation of the hollow, porous cone of the inner wall. It is not yet finally established whether this cone was open at its top (i.e. the lower part of the cup) or remained closed. This inner wall, recognizable by the characteristic growth of skeletal elements towards the longitudinal axis of the cup, can be termed the centripetal, or immigrational, type (Fig. 6d).

Type 3. Lastly, all Irregulares, from the single-walled Rhizacyathida Zhuravleva, 1955 to those forms with a very complex inner wall (Protocyclocyathidae Vologdin, 1962) had certain features in common during the very early stages of cup formation. Following the stage of the small, imperforate cup (common to all Archaeocyathi Irregulares) occurred a stage during which the whole inner cavity of the cup became filled with vesicular tissue and variously oriented rods. At this stage

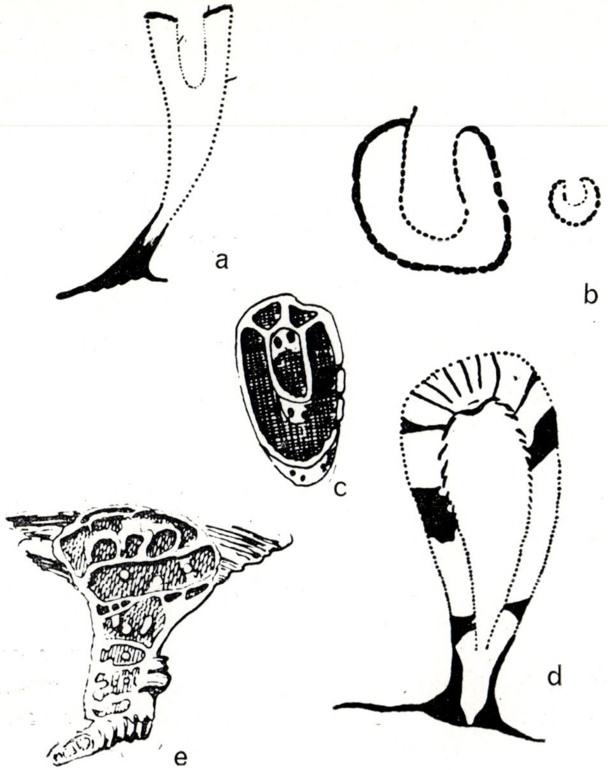

FIG. 6. Different types of appearance and development of the inner wall of Archaeocyathi. a, b, inner wall of the invaginal type; a, *Archaeolynthus*, with pelta (from Zhuravleva, 1963); b, *Capsulocyathus*, showing merging of inner and outer walls (from Zhuravleva *et al.* 1964); c, d, inner walls of the immigrational (centripetal) type; c, in *Ajacicyathus*, × 12 (after Bedford, 1939); d, in *Leptosocyathus*, × 12 (from Zhuravleva, 1960); e, inner wall of the centrifugal type in Metacyathus, × 8 (after Bedford, 1937).

there was no indication of a division of this cavity into a central space and an intervallum. Only much later did the free space in the centre of the cup appear, analogous to the central cavity of most Regulares (Fig. 6e). All the skeletal elements, except the sparse vesicular tissue, were situated in the peripheral space comparable to the intervallum of Regulares. The inner wall skeletal structures were completely absent in some Irregulares. Other irregular Archaeocyathi had, in place of the skeletal inner wall, thickened ridges on the rods (e.g. *Dictyocyathus*) and only a minority of Archaeocyathi Irregulares (*Bicyathus* Vologdin, 1939, *Anthomorpha* Bornemann, 1891) had a special skeletal inner wall with pores and other elements.

This variation in the structure of the inner wall of Irregulares does not override the essential characteristic of the group—the formation of the central cavity by the centrifugal growth of all skeletal elements. The three main types of development of the inner wall reflect fundamental phylogenetic differences between the three main groups of Archaeocyathi. These differences are reflected in the present state of archaeocyathian classification, which involves subdivision of the group into the Monocyathida Okulitch,1935, Ajacicyathida Okulitch,1943, and Irregulares. It seems quite clear that the inner walls of these three groups are not homologous.

However, such subdivision does not take into account any possible differences in patterns of larval development (cf. amphiblastula or parenchymella in Porifera). It is indeed unfortunate that there will never occur opportunities for studying the soft tissues of larval Archaeocyathi.

Other variations in the skeletal development of Archaeocyathi, such as outgrowth patterns (Fig. 7) have been known for many years (Taylor, 1910; Okulitch, 1935). For a long time such outgrowths were considered sufficiently distinctive to be diagnostic of genera (e.g. *Somphocyathus* Taylor,1910) or were regarded as parasitic growths on the main cup (e.g. *Tersia* Vologdin,1931). The outgrowths of the inner wall into the central cavity were believed to represent a special "internal organ" comparable with the gastral cavity of Coelenterata (Vologdin, 1962b).

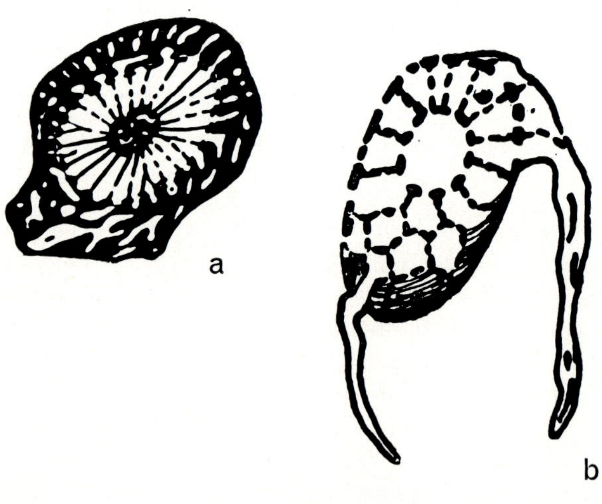

FIG. 7. Protuberances in the archaeocyathian *Cambrocyathellus tschuranicus*; a, secondary thickening of the outer wall and infilling of the central cavity, × 6 (from Zhuravleva, 1960); b, tersia-like protuberance, × 30 (from Zhuravleva, 1960).

Nowadays we possess many data which enable us to regard these structures (named "protuberances" by Hill (1964) and "stereoplasmas" by Voronin (1963)) as intrinsic parts of the main skeleton and as characters inherent in Archaeocyathi from the very beginning (Figs 7, 8). Whether or not the appearance of such structures was related only to specific environmental conditions (biohermal facies) alone or also to certain stages of development is still difficult to establish. Probably it was caused by the morphological plasticity of Archaeocyathi, which was so pronounced that even now we distinguish between forms with and without such protuberances at the generic level.

The establishment of the primary nature of the protuberances is of prime importance in that it allows us to regard the origin of the organomineral skeleton of Archaeocyathi as similar to that occurring in some skeletal Protozoa (Zhuravleva, 1959, 1963).

It seems very possible that the archaeocyathian skeleton was purely organic in its first ontogenetic stage. Only at a slightly later stage of development was this skeleton "impregnated" with calcite. Therefore we cannot regard the archaeocyathian skeleton as either purely external or purely internal in its origin.

The evidence given above confirms convincingly the conclusion of many workers that the Archaeocyathi are a discrete and evolutionarily independent animal type (e.g. Vologdin and Zhuravleva, 1947; Beklemishev, 1952, 1964; Okulitch and de Laubenfels, 1953).

At present we recognize about 700 species and more than 140 genera of Archaeocyathi which are arranged in two large categories of either Class or Subclass rank, the Regulares (with Monocyathida) and the Irregulares (Zhuravleva, 1968).

Through the use of the law of homologous series, researches with Archaeocyathi have reached the stage at which we can now predict the presence of some forms in early Cambrian deposits and then find them there (e.g. *Tumulifungia* Zhuravleva,1968 and others, Rozanov and Missarzhevsky, 1966).

The skeletal structures of some Archaeocyathi have replicated convergently, in even the most minute details, those of other animal types, such at Protozoa, Porifera (calcareous and non-calcareous), Sphinctozoa, Stromatoporoidea, Rugosa, Tabulata, Aphrosalpingoidea, and Soanitidae; in other words, representatives of virtually all groups of the lower Invertebrata and of Protozoa known at the present. Table I shows some of the most interesting examples of these convergences. There is no need to dwell upon the convergent similarity of the external form of the cup in Archaeocyathi and other animals. The cup form was so plastic and variable that examples would be too abundant.

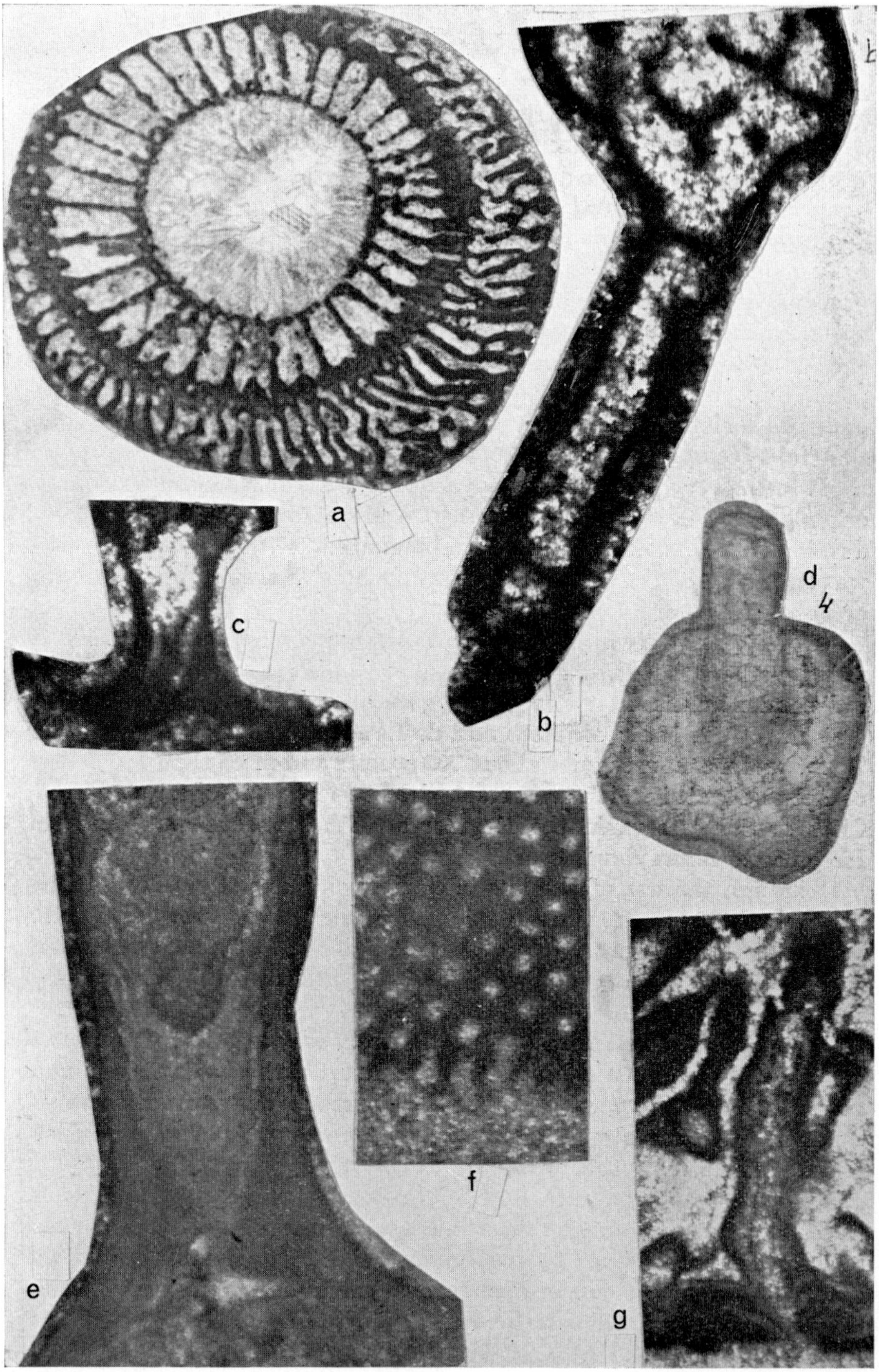
a
b
c
d
e
f
g

Thus it can be seen that the Archaeocyathi were one of the earliest groups to become established as a distinct morphological type. Unfortunately, most published schemes of the animal kingdom ignore the Archaeocyathi entirely.

DISCUSSION

Basing our arguments on the data given in this paper, we can make a very strong case for the independence of Archaeocyathi and therefore draw a more precise picture of the evolutionary radiation of the animal kingdom in its very early stages (Zhuravleva, 1959; Beklemishev, 1964).

After dividing the Animal Kingdom into the subkingdoms Protozoa and Metazoa, we consider among the latter two subdivisions—the Parazoa and the Enterozoa. The first is characterized by the absence of distinct larval layers in its embryonic stages, while the latter is characterized by their presence.

According to the mechanisms of skeletal construction during their early developmental stages, Archaeocyathi were representatives of a group evolving independently during the evolutionary development of multicellular animals. From this point of view, Archaeocyathi cannot be included in either the Parazoa or Enterozoa without changing the accepted scope of these groups. According to the manner of their skeletal construction (multicamerality, skeletal microstructure, protuberances, porosity) they were closer to the Protozoa. Simultaneously, the Archaeocyathi had all the features characteristic of typically multicellular animals. All this allows us to distinguish within the Metazoa yet another superdivision, which has been named Archaeozoa (Zhuravleva, Konyushkov and Rozanov, 1964).

We know that there existed a suborder of double-walled sac-like Archaeocyathi (Capsulocyathina Zhuravleva,1964), fully corresponding in their appearance to the hypothetical hemispherical blastaea (Metchnikov, 1955; Zachvatkin, 1949; Livanov, 1955). From this we might

---

FIG. 8 (facing page). a, Protuberances at the external surface of *Cambrocyathellus* sp., transverse section of cup, × 10, Lower Cambrian, East Sajan; b, early stage of development of the cup in *Dictyocyathus salairicus*, longitudinal section of cup, × 40, Lower Cambrian, Kuznetsky Alatau; c, early stage in the development of the cup in *Archaeolynthus* sp., longitudinal section of cup, × 60, Lower Cambrian, Kuznetsky Alatau; d, longitudinal section of a sac-like cup of *Capsulocyathus* sp., × 10, Lower Cambrian, Kuznetsky Alatau, e, early stage of cup development in *Ajacicyathus* sp., longitudinal section of cup, × 60, Lower Cambrian, Jakutija; f, a micrograined structure in an archaeocyathian part of a tangential section of the cup parities, × 60, Lower Cambrian, Jakutija; g, early stage of cup development in *Archaeolynthus* sp. produced by budding, × 40, Lower Cambrian, Kuznetsky Alatau.

TABLE I

The convergent similarities of some Archaeocyathi and representatives of other marine groups

Character of similarity	Archaeocyathi (genus, family)	Other animals
Skeletal elements		
Outer wall with microporous membrane	*Rhabdocyathella* Vologdin,1937 *Erbocyathidae,* Vologdin and Zhuravleva,1956	Aperture of some Protozoa
Unporous small cup with the basal plate at the very beginning of development	All Archaeocyathi	Coelenterata
Porous parities	Ajacicyathina Zhuravleva,1960 and others	*Calostylis* Lindström, 1869 (Rugosa, Coelenterata)
Common similarity in skeleton structure		
Cup with many camerae but without central cavity	*Kameshkovia* Vologdin, 1956	Some Protozoa
Cup with central cavity and base, but without parities or rods	*Aptocyathus, Putapacyathus*	Some Sphinctozoa
The same but with rods in the intervallum	*Tabulacyathus* *Archaeosycon* Taylor, 1910	Some Sphinctozoa
Cup with radial rods only in the intervallum	Dokidocyathidae	Receptaculitidae, Soanitidae
Cup with radial tubes in the intervallum	Syringocnematidae Taylor,1910	Aphrosalpingoida
Cup with rods in the intervallum and with indistinct outer and inner walls	*Dictyocyathus*	Dictyonina Zittel,1878 (Triaxonida, Porifera)
Cup with numerous porous bases and vertical skeletal elements between them in the intervallum	*Claruscyathus* Vologdin 1932 Metacoscinidae Bedford,1936	*Clathrodictyon* Nicolson and Murie,1878 (Stromatoporoidea)
Cup with unporous parities and sometimes with a secondarily filled central cavity	*Anthomorpha*	Rugosa (Coelenterata)

suppose that the very first multicellular animals belonged to the Archaeocyathi. In short, it is the Archaeozoa which must be considered the most ancient Metazoan group and not either of the other two branches of this subkingdom.

Thus, among the three groups that we have been considering—the Porifera, Sphinctozoa and Archaeocyathi—two types may be recognized. These are the Porifera, which may be attributed to the Parazoa, and the Archaeocyathi, which can be included in the Archaeozoa. The Sphinctozoa had some features of skeletal structure which permit us to suppose their closer affinity to Porifera than to Archaeocyathi. But this is not the final solution to the problem. It is quite probable that Sphinctozoa were also an independent group of a higher rank than we now believe them to be.

The Archaeozoa could also include such enigmatic animals as Aphrosalpingoidea (Miagkova, 1955a,b), Soanitidae (Miagkova, 1965), probably Receptaculitida (= Squamiferida Sushkin,1962) and Cribricyathi (Radugin 1962; Vologdin, 1966; Yankauskas, 1965), as well as other problematic groups of "archaeocyathian-like" construction. For the most part, representatives of these groups are very rare and so the nature of the calcareous skeleton can be studied with little profit as yet. However, it is possible that the evolutionary independence of some of these perplexing organisms was considerable. The very great diversity among these groups (sometimes even greater than that between Porifera and Sphinctozoa) and the great range of time from the Cambrian to the middle Palaeozoic favour such a hypothesis.

In conclusion, it can be stated that that part of the animal kingdom containing the lower multicellular organisms has been sadly neglected. We may infer, despite the scarcity of data, the representatives of distinctive animal types among the many rapidly extinguished forms. So far, only the Poriferan type has been studied in sufficient detail for us to identify it, without hesitation, as a clearly definable superdivision (Ivanov, 1968).

References

Bakus, G. J. (1964). Morphogenesis of *Tedania gurjanovae* Koltun. *Pacif. Sci.* **18**: 58–63.

Bedford, R. and Bedford, J. (1939). Development and classification of Archaeos (Pleospongia). *Mem. Kyancutta Mus.* **6**: 67–82.

Beklemishev, B. N. (1952). Osnovi sravitnelnoy anatomii bespozvonochnych, *Sov. Nauka*

Beklemishev, B. N. (1964). *Osnovi sravitnelnoy anatomii bespozvonochnych.* **1**. Moscow: Akad. Nauk. SSSR.

Bidder, G. P. (1898). The skeleton and classification of calcareous sponges. *Proc. R. Soc.* (B) **64**: 61–76.
Burton, M. (1931). The interpretation of the embryonic and post-larval characters of certain tetraxonid sponges, with observations on post-larval growth stages in some species. *Proc. Zool. Soc. Lond.* **1931**: 511–525.
Burton, M. (1949). Non-sexual reproduction in sponges, with special reference to a collection of young *Geodia*. *Proc. linn. Soc. Lond.* **160**: 163–178.
Dendy, A. and Row, R. W. H. (1913). The classification and phylogeny of the calcareous sponges, with a reference list of all the described species systematically arranged. *Proc. zool. Soc. Lond.* **1913**: 704–813.
Dunikovsky, E. (1883). Die Pharetronen aus den Cenoman von Essen und die systematische Stellung der Pharetronen. *Palaentographica* **29**: 281–324.
Hartman, W. D. (1958). A re-examination of Bidder's classification of the Calcarea. *Syst. Zool.* **7**: 97–110.
Hill, D. (1964). The Phylum Archaeocyatha. *Biol. Revs.* **39**: 232–258.
Hinde, G. J. (1884). *Catalogue of the fossil sponges of the British Museum*. London: Trustees of the British Museum.
Ivanov, A. V. (1968). Proischozhdenise mnogokletochnich zhivotnich. *Nauka*.
de Laubenfels, M. W. (1955). Porifera, in *Treatise on invertebrate palaeontology* Moore, R. C. (Ed.), **E**: 21–112. Lawrence, Kansas: University of Kansas.
Lévi, C. (1964). Ultrastructure de la larve parenchymella de Démosponge. I. *Mycale contarenii* (Martens). *Cah. Biol. Mar.* **5**: 97–104.
Livanov, N. A. (1955). Puty razvitiya zhivotnovo mira. *Sov. Nauka*.
Metchnikov, I. I. (1955). *Sobraniye Sotchinenii* **1–3** [collected works].
Miagkova, E. J. (1955a). K. charakteriske klassa Aphrosalpingoida Miagkova, 1955. *Dokl. Akad. Nauk SSSR* **104**: 478–481.
Miagkova, E. J. (1955b). Noviye predstaviteli tipa Archaeocyatha. *Dokl. Akad. Nauk SSSR* **104**: 638–641.
Miagkova, E. J. (1965). Soanitinovaya gruppa organismov. *Paleont. Zh.* **3**: 16–22.
Moret, L. (1952). Embranchement des Spongiaires. *Traité de Palaéontologie*, **1**: 333–374. Piveteau, I. (ed.). Paris: Masson.
Okulitch, V. I. (1935). Cyathospongia, a new Class of Porifera to include the Archaeocyathinae. *Trans. R. Soc. Can.* (3) (IV) **29**: 75–106.
Okulitch, V. I. (1943). North American Pleospongia. *Spec. Pap. geol. Soc. Am.* **48**: 1–112.
Okulitch, V. I. (1955). Archaeocyatha. In *Treatise on invertebrate palaeontology* **E**: 1–20. Moore, R. C. (ed.). Lawrence, Kansas: University of Kansas.
Okulitch, V. I. and de Laubenfels, M. W. (1953). The systematic position of the Archaeocyatha (Pleosponges). *J. Paleont.* **27**: 481–485.
Pospelov, A. G. (1962). K voprosu o sistematicheskom polozhenii archaeocyat. *Mater. geol. zapad. sib. Kraya* **63**: 11–12.
Radugin, K. V. (1962). O rannich formach archaeocyat. *Mater. geol. zapad. sib. Kraya* **63**:
Rauff, H. (1893). Palaeospongologie, Allgemeiner Teil. *Palaeontographica* **40**: 1–232.
Reid, R. E. H. (1967). Spicules in the skeleton of *Tremacystia d'Orbignyi* (Hinde). *Nature, Lond.* **215**: 875–876.
Rozanov, A. U. and Missarzhevski, V. V. (1966). Biostratigrafia i fauna nizhnich gorizontov Kembriya. *Nauka*.
Steinmann, G. (1882). Pharatronen-Studien. *N. Jb. Min. Geol. Pal.* **1882**: 139–191.

Sushkin, M. A. (1962). Squamiferida. In *Osnovy palaeontolgii* **2**. Sokolov, B.S. (Ed.) Moscow: Akad. Nauka SSSR.

Taylor, G. (1910). The archaeocyathinae from the Cambrian of South Australia. *Mem. R. Soc. S. Austr.* **2**(2): 1–188.

Vacelet, J. (1960). Les choanocytes et la larve de *Petrobiona massiliana* Vacelet and Lévi, 1958—Eponge Pharétronide mediterranée. *C.r. hebd. Séanc. Acad. sci. Paris* **251**: 2405–2407.

Vacelet, J. (1961). The order Pharetronida in Hartman's classification of the Calcarea. *Syst. Zool.* 10: 45–47.

Vologdin, A. G. (1948). K stroyeniyu tela pravilnych archaeocyat (po dannym izuchenniya *Archaeocyathus demboi*). *Izv. Akad. Nauk SSSR* (biol.) No. 1: 93–99.

Vologdin, A. G. (1962a). Archaeocyatha. In *Osnovy paleontologii* 2: 87–142. Sokolov, B. S. (Ed.). Moscow: Akad. Nauk SSSR.

Vologdin, A. G. (1962b). K anatomii archaeocyat. *Paleont. Zh.* **1962**(2): 9–20.

Vologdin, A. G. (1966). Kribrocyaty kembriya SSSR. *Trudÿ Paleont. Inst.* **109**: 1–58.

Vologdin, A. G. and Zhuravleva, I. T. (1947). *Morfologiya pravilnych archaeocyat.* Referat rab. Otdel. Biol. Nauk., Akad. Nauk, SSSR.

Voronin, J. I. (1963). O stereoplasmaticheskich obrasovaniyach septalnych archaeocyat. *Byull. mosk. Obschh. Ispyt. Prir.* (geol.) **38**: 149.

Yakovlev, V. N. (1954). K voprosu o svyazi mezhdu archaeocyatami i korallami. *Dokl. Akad. Nauk SSSR* **94**: 771–773.

Yakovlev, V. N. (1956). O nekotorych nepodtcherknutych osobyenostyach stroyeniya *Archaeocyathus* Taylor i evo vozmozhnoi rodstvyennoi svyazi siglokozhimi. *Dokl. Akad. Nauk SSSR* **109**: 855–857.

Yankauskas, T. V. (1965). K paleontologii drevnich tolsh Severozapadnoi tchasti Vostochnovo Sajana. *Izv. Tomsk Polytechn. Inst.* **135**:

Zachvatkin, A. A. (1949). Sravitelnaya embryologia nizhnich bespozvonotnych. *Sov. Nauk.*

Zhuravleva, I. T. (1959). O polozhenii archaeocyat v filogenetichekoi systeme. *Paleont. Zh.* **1959** (4): 30–40.

Zhuravleva, I. T. (1960). *Archaeocyathi sibirskoi platformi.* Moscow: Akad. Nauk SSSR.

Zhuravleva, I. T. (1963). *Archaeocyathi Sibiri.* Odnostenniye archaeocyati (otrad Monocyathida i Rhizocyathida). Moscow: Akad. Nauk SSSR.

Zhuravleva, I. T. (1965). Sphinctozoa. V svodke 'Razvitiye i samena morskych organismov na rubezhe paleozoiya i mezozoiya'. *Trudÿ Paleont. Inst.* **108**: 147–149.

Zhuravleva, I. T. (1968). *Morskiye fauny i problemy stratigraphii nizhnevo Kembria.* Moscow: Akad. Nauk SSSR.

Zhuravleva, I. T., Konyushkov, K. K. and Rozanov, A. U. (1964). *Archeocyathi Sibiri. Dvustenniye Archaeocyathi.* Moscow: Akad. Nauk SSSR.

Zhuravleva, I. T. and Rezhvoy, P. D. (1956). K sistematike iskopayemych gubok i archaeocyat. *Dokl. Akad. Nauk. SSSR* **111**: 449–451.

Ziegler, B. (1964). Die Cortex der fossilen Pharetronen (Kalkschwämme). *Eclog. geol. Helv.* **57**: 803–822.

Ziegler, B. (1965). Die Variabilität bei Pharetronen (Kalkschwämme). *Paläont. Z.* **39**: 106–110.

Zittel, K. A. (1878). Zur Stammesgeschichte der Spongien. In *Festschrift Bayer. Akad. Wiss. München* 50 *Jahr. Prof. von. Siebold.*

SPICULES AND EVOLUTION

Symp. zool. Soc. Lond. (1970) No. 25, 63–89

TETRAXONS AND DEMOSPONGE PHYLOGENY

R. E. H. REID

Department of Geology, The Queen's University, Belfast, Northern Ireland

SYNOPSIS

According to Schulze (1887) and Dendy (1921) the regular tetraxon, or calthrops, is the central type of demosponge spicule and all monaxonid demosponges have been derived from choristids in phylogeny. Dendy's (1921) detailed development of this concept can provide a complete evolutionary explanation of demosponge spiculation. On the other hand, rejection of some of Dendy's concepts by later authors, including the writer, leaves many monaxonid sponges with no sure relationship to choristids. Such monaxonids may never have had tetraxons as megascleres or even at any stage of phylogeny. Lévi (1957) has claimed that some are of altogether independent origin. In addition, some evidence can be interpreted as suggesting that the primitive demosponge spicules were monaxons, not tetraxons. This evidence includes (1) the origin of some triaene tetraxons from monaxons in ontogeny, (2) the occurrence of monaxonid sponges in the Cambrian period, long before the oldest known (Lower Carboniferous) choristids, and (3) the occurrence of tetraxon-like desmas in some Palaeozoic lithistids, which are probably of monaxonid origin.

The writer agrees with previous authors who have rejected Dendy's concept of sigmatose microscleres as a homologous series but supports Dendy's treatment of the choristids against that of de Laubenfels (1936). The development of some triaenes from monaxons could be palingenetic but it could also be caenogenetic. Other cases of caenogenetic change in ontogenetic prototypes can be demonstrated in spicules of the lithistid Tetracladina. The Palaeozoic record probably records only a very small fraction of the sponges which are likely to have existed in that period; the Lower Carboniferous choristids must also have already existed for long enough to evolve all the main types of tetraxon magascleres found in living choristids. Further, nothing is known of the microscleres of the Palaeozoic sponges, with which the early monaxonids and later choristids might, in fact, have no relationship. Four lines of evidence suggest early Palaeozoic existence of sigmatomonaxonid sponges, i.e. the group regarded by Lévi as of independent origin. Last, variation in the desmas of Ordovician anthaspidellids suggests that their tetraxon-like desmas were monaxial or anaxial.

INTRODUCTION

This paper deals with some problems in the interpretation of demosponge phylogeny, from a viewpoint of spicular morphology. It is concerned chiefly with the supposed status of the regular tetraxon, or calthrops, as the central type of demosponge spicule. This concept was chosen for special discussion as being the most fundamental in assessment of the overall pattern of phylogeny and because evidence which could be cited against it is partly palaeontological.

TETRAXONID SPICULATION: THE HYPOTHESIS OF SCHULZE AND DENDY

The notion that the calthrops is the central type of demosponge spicule arose from Schulze's (1880) work on plakinids. The spicules of the genera he studied are mainly tetraxons, developed either as calthrops or as lophose variants of this type, but they also include forms showing passage from triactins to diactins in a manner implying reduction. This evidence, the absence of triaenes in these sponges, and the simple anatomy of *Plakina* Schulze, 1880, led Schulze (1887) to interpret the calthrops as the primitive demosponge spicule. He assumed that all monaxons are derived from tetraxons and interpreted all monaxonid demosponges as derived from forms with tetraxons.

The development of this concept by Dendy (1921, 1924) can provide a complete explanation of skeletal evolution in Demospongia with spicules. According to Dendy (1921), small spicules like those of the plakinid *Dercitopsis* Dendy, 1905 are the prototypes of both the

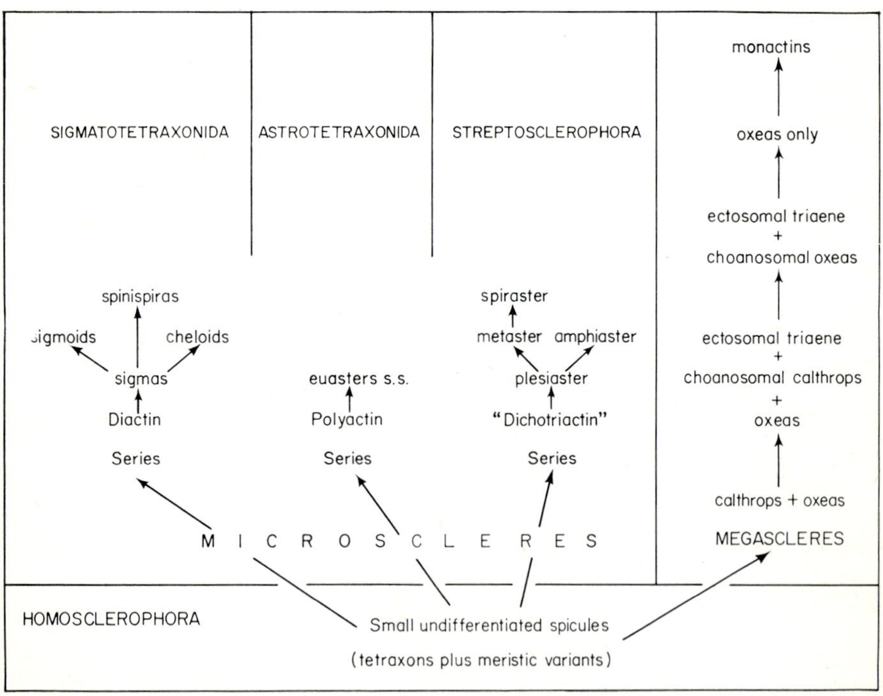

FIG. 1. Skeletal evolution in non-lithistid Demospongia ("Tetraxonida") according to Dendy (1921, 1924).

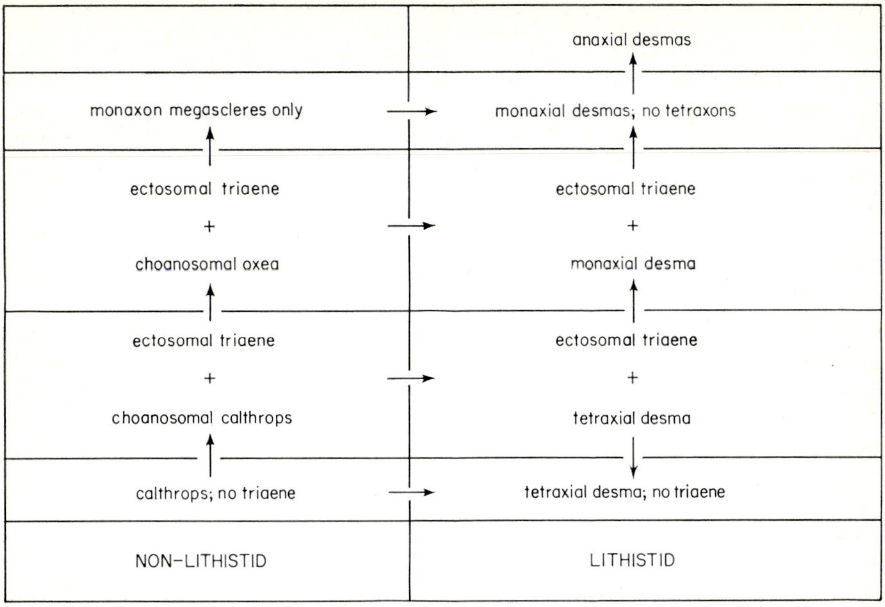

FIG. 2. Lithistid megaspiculation interpreted in terms of Dendy's picture of spicular phylogeny and additional secondary evolution in lithistid sponges.

megascleres and the microscleres of higher forms. The spicules of *Dercitopsis* vary meristically from diactins to pentactins although tetractins predominate; they can thus be regarded as showing both positive and negative meristic variation, away from a central tetractin. The higher Demospongia have evolved in three main stocks, distinguished by different types of characteristic microscleres: (1) polyactinal microscleres, or asters s.s., comprising the euasters of ancorinids, geodiids, coppatiids and tethyids and interpreted as true polyactins arising by positive meristic variation; (2) diactinal microscleres, also called sigmatose, comprising sigmas *sensu* Dendy (i.e. including sigmaspires), spinispiras and other sigmoids and cheloids; and (3) "dichotriactinal" microscleres, comprising the intergrading plesiasters, amphiasters, metasters and spirasters of pachastrellids and theneids, which are interpreted (Dendy, 1924) as derived from triactins by actinal branching. The corresponding taxa in Dendy's final (1924) usage are Astrotetraxonida Hentschel, 1909, *sensu* Dendy, Sigmatotetraxonida Hentschel, 1909, *sensu* Dendy 1916 and Streptosclerophora Dendy, 1924. The megascleric evolution of these sponges began with the megascleres consisting of calthrops and oxeas. After differentiation of triaenes as an ectosomal specialization, restriction of tetraxons to triaenes has sometimes been followed by their loss, leaving purely monaxonid sponges. Genera in all stages of this sequence are represented among the Astrotetraxonida and Sigmatotetraxonida, and forms up to triaenose choristids in the Streptosclerophora. Loss of tetraxons before evolution of triaenes could also be envisaged, although this was not emphasized by Dendy.

This picture of megascleric evolution can also explain the spiculations found in lithistid sponges (Fig. 2). First, lithistids derived from non-lithistids at various stages of skeletal evolution could include forms (a) with tetraxial desmas but no triaenes, (b) with tetraxial desmas and ectosomal triaenes (c) with monaxial desmas and ectosomal triaenes and (d) with monaxial megascleres only. Sponges whose spiculation is explicable in this manner are, in order, (a) the Hindiidae Rauff, 1895, if their desmas are really tetraxial (see p. 73), (b) the typical Tetracladina Zittel, 1878, having tetraclone desmas and triaene dermalia, (c) the Megamorina Zittel, 1878, and probably the sublististid *Helminthophyllum* Schrammen, 1936, and (d) some or all Rhizomorina Zittel, 1878, *sensu* Schrammen, 1910 and the sublithistid desmacidontids (e.g. *Helophloeina* Topsent, 1928). Some types are, however, more likely to represent further secondary evolution in lithistid sponges themselves. For instance, both (a) the occurrence of occasional tetraxial desmas among rhizoclones of *Macandrewia* Gray, 1859 and (b) the resemblance of corallistid dicranoclones to tetraclones and triders of the tetracladine Acrochordoniidae Schrammen, 1910, and Discodermiidae Schrammen, 1910, suggest that the Dicranocladina Schrammen, 1924 (= Corallistidae Sollas, 1888 plus Macandrewiidae, Schrammen, 1924), have been derived from the Tetracladina by reduction of the crepides of the desmas from tetraxons to monaxons. The anaxial types of desmas (sphaeroclones and variants) have no evident non-lithistid prototypes and are explicable (Sollas, 1888) as representing a final stage in reduction of the crepis.

SUBSEQUENT DEVELOPMENTS

Unfortunately some of Dendy's views on the relationships of microscleres, and consequently parts of his picture of phylogeny, have not been accepted by later authors. This is especially the case with his concept (Dendy, 1921) of diactinal or "sigmatose" microscleres. In addition, some evidence suggests that monaxons could be primitive in some or even all Demospongia, in which case Dendy's picture of phylogeny would be partly or completely inverted. On the other hand, some of Dendy's views have been criticized in ways which do not appear justified.

First, Topsent (1928) rejected the identity of sigmaspires with sigmas s.s. following Sollas (1888), and with it the supposed derivation of "Sigmatomonaxonida" (= Halichondrina Vosmaer, 1887, *sensu* Topsent, 1898, Haplosclerina Topsent, 1898 and Poecillosclerina Topsent, 1898) from choristids with sigmaspires (tetillids, Dendy; craniellids, de Laubenfels). Topsent (1928) also rejected the supposed derivation of spinispiras from a sigma, claimed by Dendy (1921) to have occurred by way of "discorhabds", and grouped the monaxonid sponges with spinispiras and those with euasters in his Hadromerina Topsent, 1898. Topsent's rejection of these parts of Dendy's classification, and the picture of phylogeny implied by it, has been followed by de Laubenfels (1936), Burton (e.g. 1956) and the writer (Reid, 1963a) although with different opinions on how the sponges concerned should be classified. In phylogenetic speculation, the same rejection opens the way for departure from the ideas of Schulze and Dendy. Monaxonid sponges with spinispiras and those with sigmas or allied types of microscleres can still be envisaged (cf. Topsent, 1928) as derived in some way from the choristids but, once it is admitted that they have no sure choristid relatives, they can also be pictured as possibly never having had tetraxons, either as megascleres or at any stage of phylogeny. This applies especially to the sponges with sigmas and allied types of microscleres, as interpreted by Lévi (e.g. 1957) and the writer (Reid, 1963a). Lévi in particular interprets them as primitively monaxonid and as having evolved their spicules independently of forms with tetraxons.

Second, even greater changes in classification were made by de Laubenfels (1936), in two main directions. de Laubenfels rejected all taxa based solely on microscleres, preferring to regard the megascleres and general structure as more important. The choristids in particular were arranged as Choristida Sollas, 1888 (*sensu* de Laubenfels, 1936, *non* Sollas, 1888) and Carnosa Carter, 1875 (*sensu* de Laubenfels, 1936, *non* Carter, 1875) on the basis of their megascleres, using microscleres only to distinguish some families. Genera with spinispiras were referred mainly to a restricted Hadromerina but in part to his Epipolasida de Laubenfels, 1936,* in the main on a basis of the presence or absence of tylostyles. Topsent's Haplosclerina and Poecillosclerina were also altered so that both include sponges with chelas. Reliance on the microscleres as primary criteria of relationship was thus generally abandoned,

* This taxon was attributed (de Laubenfels, 1936) to Sollas (1888) but Sollas' (1888) "Epipolasidae" is a family group taxon and the name is also invalid because it is not based on that of a genus.

although some use was still made of them. In addition, de Laubenfels inverted the arrangement of taxa used by Dendy and Topsent so that keratose sponges are placed first and sponges with calthrops last. He thought that the calthrops is derived usually from a triaene, and that plakinids are possibly reduced forms. de Laubenfels' classification has been criticized by Burton (1956) as giving an "incorrect" picture of phylogeny, presumably meaning an inverted one compared with that given by Dendy, although this was not emphasized by de Laubenfels.

While some of the relationships between microscleres pictured by

Microscleres	Choristids	Monaxonids	BOTH
Euasters s. s.	yes	yes	YES
Streptoscleres ("dichotriact" streptasters)	yes	no	NO
Sigmaspires	yes	yes*	YES
Spinispiras	no	yes	NO
Sigmatoscleres (sigmas s. s. + allied sigmoids and cheloids)	no	yes	NO

FIG. 3. Distribution of choristid and monaxonid Demospongia of the five main types of microscleres which are characteristic of different groups of sponges. Sigmaspires and sigmas s.s. are treated as distinct, following Topsent (1928). (* Present in a few species only.)

Dendy are unlikely to be genuine, it is difficult to agree with de Laubenfels' treatment of the choristids. So far as this treatment was explained, it depended chiefly on the notions (de Laubenfels, 1936, pp. 167, 173) that *Thenea* Gray, 1867 " . . . differs from *Ancorina* Schmidt only in that the streptasters of the latter are much less bent than those of the former", and that the fine spinulation of many sigmaspires " . . . very probably proves them to be reduced spirasters with spines only vestigial". These notions do not seem correct to the writer. The "streptasters" of *Ancorina* are spinulated microrhabds but those of *Thenea* are intergrading plesiasters, metasters, amphiasters and spirasters, of the types comprising Dendy's (1924) "dichotriacts", and otherwise found only in pachastrellids and some lithistids. The sigmaspire has no known connection with the spirasters of this series; it is never accompanied by metasters or plesiasters and seems to be simply a monaxon. If taken as monaxial it is comparable with members of the spinispira series which vary from spinulate strongylospires to coarse spirasters but these microscleres are found in monaxonid sponges which de Laubenfels (1936) classified as Hadromerina or Epipolasida. Further, fine spinulation is more probably an incipient condition when fine spinules are more numerous than coarse ones. In forms with spinispiras, a fine spinulation of the microscleres may also occur in a megascleric oxea (in e.g. *Cliona vastifica* Hancock, 1849), of which a coarse spiraster is not a likely phylogenetic prototype. The most strongly ornamented sigmaspires are also forms resembling chelas (in e.g. *Chrotella amphiacantha* Topsent, 1904), not spirasters. Last, *Acanthotetilla* Burton, 1959 lacks sigmaspires but these may be represented by finely spined oxeas which are submegascleric in size; this again suggests a relationship to a megascleric oxea.

In this light, Burton's (1956) preference for the microscleres as indices of natural relationships clearly seems to be well justified. Both of the orders (Choristida Sollas and Carnosa Carter, *sensu* de Laubenfels, 1936) to which de Laubenfels referred choristids *sensu* Sollas' 1888 include sponges whose principal microscleres are (1) euasters (2) plesiasters, metasters and spirasters and (3) sigmaspires, in different cases. Nothing distinguishes de Laubenfels' orders but the absence of calthrops as megascleres in his Choristida, since typical triaenes occur in some forms placed as Carnosa, and the calthrops and triaene intergrade in some forms called Carnosa (e.g. *Poecillastra* Sollas, 1887). The choristids *sensu* Sollas thus seem clearly to include three groups of sponges, distinguished by different types of microscleres, which the orders distinguished by de Laubenfels cut across artificially. De Laubenfels also made his arrangement seem to be based on that of Carter by use

4

of his order Carnosa (see de Laubenfels, 1936, p. 177. In fact Carter placed all choristids but plakinids in his Holoraphidota Carter, 1875.

On the other hand, the possible inversion of Dendy's whole picture of phylogeny must be taken more seriously because several lines of evidence suggest a picture which is opposite to Dendy's. In particular, (1) some long-shafted triaenes develop from monaxons in ontogeny (which was emphasized by Dendy (1921) himself as the key to spicular phylogeny); (2) the oldest known choristid sponges are Lower Carboniferous although monaxonids are known from the Cambrian and (3) apparently tetraxial desmas occur in some Palaeozoic lithistids which are probably derived from monaxonids. This evidence, however, needs critical discussion.

THE ONTOGENY OF TRIAENES

If the development of some triaenes from initial monaxons in ontogeny is palingenetic, it should imply the origin of triaenes from monaxons in phylogeny, and hence that of all choristid sponges from monaxonids unless other tetraxons are supposed to have had a different origin. The first implication was recognized by Sollas (1888) but, as Sollas also noted, other triaenes and all calthrops are tetraxons from their earliest known stages. Therefore, unless tetraxon megascleres of choristid sponges have had two different origins—which does not seem likely since all types intergrade morphologically—there must have been caenogenetic change in the nature of the crepidal body. This could have been either from monaxon to tetraxon or from tetraxon to monaxon.

Caenogenetic alteration of crepidal bodies in phylogeny has not been recognized widely but clearly has occurred in some cases. Two good examples are provided by spicules of tetracladine lithistids. First, in various fossil members of the Phymatellidae Schrammen, 1910 and Phymaraphiniidae Schrammen, 1910, which are supported by stalks or root processes, the normal tetraxial choanosomal desmas grade into elongate monaxial radical desmas in the basal parts. The intergradation may occur through forms in which the crepis was triaenose or triactinal, implying that the monaxon crepides may be either monactinal or diactinal homologically. Second, tetracladine dermalia show gradation from normal dichotriaenes to discotriaenes and monaxial variants of the latter, by way of intermediate phyllotriaenes. In some phyllotriaenes, the crepis has deuterocladi but in others, and in most discotriaenes, it is a simple orthotriaene, with the length of the cladi less than half the radius of the fully grown cladome. In discotriaenes of e.g. *Discodermia* du Bocage, 1869, the crepidal cladi may be rudimentary. The spicules may then grade into forms in which the disc is entirely anaxial and develops from one end of a strongyle which becomes the rhabdome. Whether the phylogeny of these desmas and dermalia is supposed to have proceeded from tetraxon to monaxon, or in the opposite direction, the ontogenetic prototypes of the supposed phylogenetic end-forms do not correspond with their phylogenetic prototypes and must have changed caenogenetically.

Moreover, unless it is mistaken to picture (a) the tetracladine radical desmas as functional variants of the normal tetraxial desmas, and (b) an anaxial disc as an unlikely prototype of a normal actinal cladome; the probable direction of phylogeny is tetraxon to monaxon in both cases, not the opposite.

The ontogenetic development of some triaenes from monaxons could have a similar basis. If triaenes have arisen from regular tetraxons in phylogeny, by reduction of the cladi but hypertrophy of the rhabdome, this might also lead to delay in formation of the cladi when the rhabdome is large by comparison. This would produce a situation in which triaenes arose from monaxons in ontogeny, although not derived from monaxons in phylogeny, and could be the reason why all spicules which are known to show this type of development are long-shafted triaenes. Moreover, a proterogenetic prolongation of delay in formation of the cladi, in forms with a hypertrophied rhabdome, might culminate in failure to form cladi and thus in replacement of triaenes by monaxons. This is a means by which "epipolasid" monaxonids could originate from chloristids, as envisaged by Sollas (1888) and Dendy (1921).

The development of some triaenes from monaxons in ontogeny is, therefore, not critical evidence for either their origin from monaxons in phylogeny or the origins of choristid sponges from monaxonids. On the contrary, it could have an altogether opposite significance.

In passing, the concept of "epipolasids" (Sollas, 1888) as monaxonids derived from choristids by loss of triaenes in phylogeny is not the general basis of the "Epipolasida Sollas" of de Laubenfels (1936). This order is of far larger scope than Sollas' Epipolasidae, and comprises chiefly forms removed from Topsent's Hadromerina on a basis of the absence of tylostyles.

THE PALAEOZOIC RECORD

Turning to the fossils, the apparent existence of monaxonids long before the first known choristids could be cited as supporting the origin of the latter from the former although, in practice, the early monaxonids are not even certainly demosponges. Critical discussion again shows this evidence to be dubious on four distinct counts: (1) the sparsity of the Palaeozoic record, (2) the diversity of the earliest known choristids, (3) the absence of microscleres in Palaeozoic material and (4) a possibility that the early monaxonids were sigma-bearing sponges, unrelated to any choristids.

First, the Palaeozoic record is so sparse that little trust can be placed in it. Taking genera listed by de Laubenfels (1955) and excluding forms based only on burrows, the number of non-lithistid genera recorded from the pre-Carboniferous systems are: Cambrian—8; Ordovician—4; Silurian—2; Devonian—2. Even counting genera which may (de Laubenfels, 1955) be either monaxonids or hexactinellids, and others based on burrows which may not be those of sponges, the totals are only increased by two genera each for the Cambrian and Ordovician, and by five genera for the Devonian. The overall totals are 13 to 21 genera, from systems representing a period of over 300 million years. The current number of modern nominal non-lithistid genera is over 800. By comparison, and even if this total exaggerates the number of true biological genera, we probably know only a very small fraction of the Palaeozoic fauna, unless Demospongia were then far less numerous than at present.

Second, and supporting this conclusion, the Lower Carboniferous choristids show far too great a diversity in their spicules to be the true first choristids. Tetraxon spicules from the Lower Carboniferous D_1 of Co. Fermanagh, Northern Ireland, and D_2 of Ayrshire, Scotland, include the following types: (1) minute calthrops and lophose calthrops, comparable with those of plakinids; (2) large calthrops with simple or branching rays, like those of e.g. *Pachastrella abyssi* Schmidt, 1870; (3) spiny forms grading from calthrops to short-shafted triaenes, or developed as cladocalthrops; (4) spiny short-shafted triaenes grading from sigmatriaene to anatriaene, the former like those found in *Thrombus* Sollas, 1888 but much larger, and (5) typical long-shafted trianes, developed as plagiotriaenes, orthotriaenes, protriaenes, anatriaenes, dichotriaenes and trachelotriaenes. There are also other spicules of types having five or six rays which appear to belong to Demospongia; examples are those of *Spiractinella* Hinde, 1888 which has been taken to be a hexactinellid because some of the spicules are hexactinal but which also had tetractins which are tetraxons (calthrops and short-shafted triaenes) (Reid, 1963b). Comparable spicules occur in a few modern choristids, e.g. *Yodomia* Lebwohl, 1914, in which a pentactinal centrotriaene occurs and may have hexactinal variants (in *Y. perfecta* Dendy, 1916). In other words, whatever direction is ascribed to the phylogeny of the spicules, in Lower Carboniferous times the choristids must already have existed for long enough to have evolved every major type of tetraxon megasclere seen in modern forms, plus plakinid-like spicules and other aberrant types. The only type lacking is the rare amphitriaene. Further, if the almost total lack of innovation of types of megascleres in post-Carboniferous choristids is an index of the rate of choristid evolution before the Carboniferous period, the pre-Carboniferous history of chorisids should be pictured as having taken several times longer than the time which has elapsed from the Lower Carboniferous to the present.

Third, the existence of unknown choristid sponges in the early Palaeozoic is suggested by the form of the desmas of the lithistid Hindiidae (= Eutaxicladina *sensu* Schrammen, 1924) which are first known from the Ordovician system. The desmas of these sponges are characteristically tripodal, with little variation from this pattern, and are most comparable with tetracladine triders in which a brachyome either bears no zygome or is lacking altogether. They can also be compared with tripodal variants of other types of desmas which are monaxial (dicranoclones, megaclones) or anaxial (sphaeroclones), but these types show far greater variability in form and none is characteristically tripodal. The desmas of the late (Carboniferous–Permian) *Scheiia* Tschernyschew and Stepanov, 1916 have no brachyome-like feature but those of the earlier *Hindia* Duncan, 1879 have a prominent capstan-like outgrowth, like that seen in basal triders of the living tetracladine genus *Kaliapsis* Bowerbank, 1869.* The skeletal architecture of the hindiids is also most nearly matched elsewhere in the calcareous sponge *Porosphaera* Steinmann, 1878, in which spicules forming the skeletal framework are undoubted tetractins. If hindiid desmas were tetraxons and arose from a non-lithistid tetraxon, their character must imply the existence of an earlier choristid prototype.

Fourth, nothing is known of the microscleres of any of the Palaeozoic sponges, except that loose supposed sterrasters are recorded from the Lower Carboniferous. Presumably, a choristid derived from a monaxonid would inherit the same type of microscleres, assuming that the monaxonid had microscleres. In modern forms, however, not all types of microscleres are common to both choristids and monaxonids. Of five main types of microscleres which are characteristic of different groups of families (Fig. 3), those common to monaxonids and choristids comprise only euasters and sigmaspires. Neither spinispiras nor the restricted sigmoid-cheloid series (sigmatoscleres, Reid, 1963a) are found in choristids, if we agree with Topsent (1928) that the sigma s.s. is not to be equated with the sigmaspire. On the other hand, Dendy's (1924)

* The inclusion of this genus in the Rhizomorina by de Laubenfels (1955) is unfounded; the desmas are not rhizoclones but tetraxial tetraclones and triders.

"dichotriacts" (streptoscleres Reid, 1963a), found in choristid sponges, are not known from monaxonids. This is plainly a warning that pre-Carboniferous monaxonids might in fact have no connection with the ancestry of Carboniferous choristids. One may add that, while the sigmaspire occurs in both choristids and monaxonids, its occurrence in the latter is restricted to a very few genera (e.g. *Raphidotethya* Burton, 1934, *Tentorina* Burton, 1959). In contrast, the monaxonids with spinispiras or sigmatoscleres, with their allies, comprise more than 70 % of the class Demospongia at present. It is very desirable to ask whether the early monaxonids may represent either of the two latter groups of modern monaxonid sponges. The possible representation of sigmatomonaxonids (Sigmatosclerophora Burton, 1929, *sensu* Burton, 1937; Haplosclerina and Poecillosclerina Topsent (e.g. 1928); spiculate Ceractinomorpha Lévi, 1953) is especially important if these forms are correctly regarded (Lévi, 1957; Reid, 1963a) as descended independently of choristid sponges.

A further point here is that, if it is correct to regard (a) primitive Demospongia as having had spicules of microscleric size, like living plakinids, and also (b) sigmatomonaxonid sponges as having descended independently of choristids, then restriction of the pre-Carboniferous record to megascleric material means that even a complete stratigraphical record would show no more than which of these groups evolved megascleres first.

Four lines of indirect evidence point to the early Palaeozoic existence of sigmatomonaxonid sponges.

(1) The characteristic skeletal architecture of the orchocladine lithistids of the family Anthaspidellidae Ulrich, 1890 (Ordovician–Permian) is strikingly similar to that of the living sublithistid genus *Lithochela* Burton, 1929, especially in anthaspidellids in which the skeletal fibres are cored by oxeas (e.g. *Climacospongia* Hinde, 1883). This need not imply direct relationship but points to the origin of anthaspidellids from monaxonids with the same type of skeleton as *Lithochela*, now found only in sigmatomonaxonids.

(2) The echinated skeletal fibres of the Ordovician *Saccospongia* Bassler, 1889, figured by Finks (1967), suggests a sigmatomonaxonid.

(3) As recognized by Lévi (1960), the "asteroid" desmas of *Crambe* Vosmaer, 1886 are strikingly like those of the Palaeozoic Astylospongiidae Rauff, 1893, which also possessed monaxon spicules.

(4) The sponges which Hartman and Goreau (see pp. 205–240) compare with the Palaeozoic stromatoporoids are sigmatomonaxonids (e.g. *Merlia* Kirkpatrick, 1911).

This evidence again shows a clear possibility that the early Palaeozoic monaxonids have no bearing on the origin of choristids.

One may add here that, if Dendy (1921) was right in regarding a plakinid-like ancestor as the starting point of choristid phylogeny, the stratigraphic record is unlikely to provide supporting evidence because of the microscleric size of plakinid spicules. Microscleres are generally rare in the whole fossil record and most spicules attributable to plakinids are from rocks which also preserve microscleres. But, as noted, small spicules which may be plakinid occur among the oldest choristid spicules known.

THE DESMAS OF ORCHOCLADINE LITHISTIDS

The last palaeontological question is the nature of tetraxon-like desmas seen in the Orchocladina Rauff, 1895, including the anthaspidellids. These forms were regarded as Tetracladina by Rauff (1893–1895), because some of the desmas resemble tetraclones or amphitriders, or are comparable with an amphitrider with the central shaft shortened to a point. Other orchocladine desmas are, however, more similar to various types of rhizoclones and Hinde (1887–1912) thought that the desmas in general were more probably related to the bipolar type of rhizoclone seen in the rhizomorine genus *Cnemidiastrum* Zittel, 1878.

Evidence of the nature of orchocladine desmas is provided by their form in the Ordovician anthaspidellids *Calycocoelia* Bassler, 1927 and *Nevadocoelia* Bassler, 1927 (Fig. 4). In both, the predominant desmas have the form of smooth-shafted bipolar rhizoclones with the zygomes extending up and down the skeletal columns (not trifid, as in dendroclones s.s., cf. Rauff, 1893–1895). This type of desma passes into simple linear rhizoclones with lateral zygomes, including the curved type of rhizoclone which is usual around skeletal pores or canals in rhizomorine sponges, and it also passes into triradiate desmas (Fig. 4, top left) through intermediates with one zygome deeply bifid. These triradiate desmas are comparable with triradiate rhizoclones found in numerous typical Rhizomorina, except that the zygomes are restricted to the ends. In *Nevadocoelia*, a third group of variants of the central type begins morphologically with bipolar desmas in which both zygomes are deeply bifid, in the same plane or different planes; these forms grade in turn into four-armed desmas, with the arms arranged in one plane, or more or less in the manner of the arms of a tetraclone. In the skeleton, the flat cross-like forms have two arms directed to each of two skeletal columns but the triradiate and tetraclone-like forms have the arms directed to more than two columns. The variation in forms of the four-armed

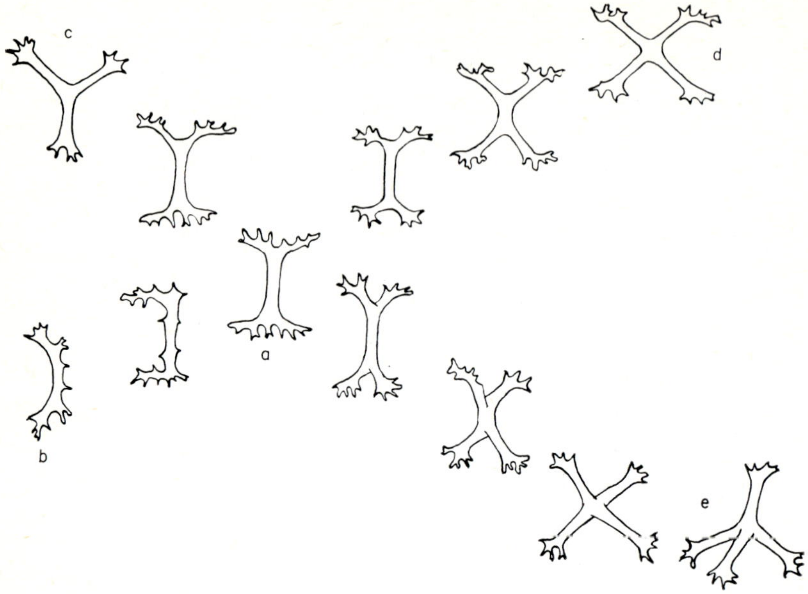

FIG. 4. The suggested relationship of anthaspidellid "tetraclones" to other desmas, based chiefly on spicules of *Calycocoelia* Bassler, 1927 and *Nevadocoelia* Bassler, 1927, a, Desma with the form of a bipolar rhizoclone; b, desma with the form of a simple curved rhizoclone; c, three-armed desma; d, flat (cross-shaped) four-armed desma; e. tetraclone-like desma.

desmas thus apparently depended on how they grew into the skeleton and also on the skeleton having anthaspidellid structure. In contrast, the tetraxon form of true tetraclones depends on the possession of a tetraxon crepis and is characteristically maintained by four-armed examples irrespective of skeletal architecture.

The anthaspidellid tetraclone therefore seems to be a desma derived from a bipolar rhizoclone (or a "dendroclone" with the ends bifid instead of trifid), by dichotomous division of the zygomes and shortening of the epirhabd. By comparison with typical dendroclones and chiastoclones *sensu* Rauff (1893–1895), they were four-armed chiastoclones homologically, unless any which occur in other genera had a different origin. The typical dendroclone and chiastoclone represent a parallel series of forms, derived by division of the primary zygomes into three or sometimes more branches. The forms with a distinct central shaft were presumably monaxial; by comparison, the chiastoclones, including the supposed "tetraclones", would then be either monaxial with a short crepidal body or simply anaxial.

SUMMARY AND VERDICT

The above discussion of the status of tetraxons in phylogeny may now be summarized. First, the derivation of tetraxons from monaxons, and of choristid sponges from monaxonids, is undoubtedly possible. A great deal of modern comparative evidence suggests the opposite conclusion but all modern sponges are phylogenetic end-forms, whose characters give no objective indication of phylogenetic direction. There are several lines of evidence which suggest that monaxons were the primitive spicules. On the other hand, this evidence is far from conclusive, for several reasons.

(1) The ontogenetic development of some triaenes from monaxons is genuine and could be palingenetic but it could also be caenogenetic, and could then represent an incipient stage in the evolution of "epipolasid" monaxonids (*sensu* Dendy, 1921). Moreover, the sponges with triaenes which develop in this manner are not specially simple in other ways but are, on the contrary, mainly radiate and corticate sponges.

(2) Where lithistid specializations make it possible to ascribe a probable direction to caenogenetic evolution in intergrading tetraxons and monaxons, this direction, in the examples cited, is from tetraxon to monaxon.

(3) The Palaeozoic record of non-lithistid sponges probably shows only a small fraction of the sponges which existed. The first known choristids had already existed for long enough to evolve every major type of tetraxon megasclere found in modern species, as well as some aberrant types. If the lack of innovation of new types in later choristids is a measure of the rate of evolution before the Carboniferous period, the group is probably several times older.

(4) If the desmas of *Hindia* were tetraxial, they imply the earlier existence of a choristid ancestor.

(5) Pre-Carboniferous monaxonids could be members of groups unconnected with choristid ancestry. This applies especially to the sigmatomonaxonid sponges, which have no sure relationship to any choristids, and four lines of indirect evidence point to the early Palaeozoic existence of sigmatomonaxonids.

(6) The tetraclone-like desmas of some Orchocladina were probably not tetraxial but monaxial or anaxial, and appear to have no likely relationship to the tetracladine tetraclone.

Therefore, I see no substantial evidence that choristid tetraxons have arisen from monaxons in phylogeny, at least at a megascleric level. Their origin in this manner is possible but, in terms of present knowledge, only possible and not probable.

Moreover, it seems to me more likely that any derivation of tetraxons from monaxons would have happened at an earlier stage of demosponge phylogeny. One needs to ask how the tetraxon arose in the first case. If it is believed that some modern monaxonids have never had tetraxons during their phylogeny, at least some primitive Demospongia must have had monaxial spicules only. The primitive monaxonids could be forms derived from plakinid-like sponges, by reduction of tetractins or triactins to diactins. But phylogeny could also have had an opposite direction if the primitive spicules were monaxons, and radiate spicules arose from assembled monaxons which became their rays. This might happen, for instance, if a number of monaxon units were formed within one scleroblast, like trichites, but united in radial groups. This could in turn be the basis of the type of meristic variation seen in plakinid spicules, choristid euasters and some choristid megascleres, in which the number of rays is varied in a 3, 4, 5, . . . , etc. sequence. I think this at least as likely as the origin of radiates as such, assumed by Schulze (1887).

SUGGESTED PHYLOGENY OF THE DEMOSPONGIA

A picture of demosponge phylogeny (Fig. 5) may first indicate primitive Demospongia as microspiculate sponges, with radiate spicules (including tetraxons) but monaxons only in others. Some monaxonid sponges can then have evolved without possession of tetraxon megascleres at any stage of phylogeny. I think this more likely than the independent origin of the spicules of the sigmatomonaxonids. Except in their range of shapes, the spicules of sigmatomonaxonids show no apparent difference from those of forms possessing tetraxons. On the contrary, the specialization of choanosomal megascleres to form desmas is identical in character in forms as apparently distant as tetraxon-bearing lithistids (Tetracladina, Dicranocladina, Megamorina) and sublithistid desmacidontids. This is more consistent with common inherited potentialities than independent origins. In addition, a lack of transitions, or morphological intermediates, between modern phylogenetic end-forms is not proof of independent origins; it is just as consistent with long established independent histories. One can only expect transitions or intermediates between groups which are not yet fully differentiated.

The suggested relationships of modern Demospongia are as follows.

(1) The plakinids (or Homosclerophora Dendy, 1905) are probably descended from a primitive tetraxonid stock, another part of which was ancestral to most higher choristids and some monaxonids. They

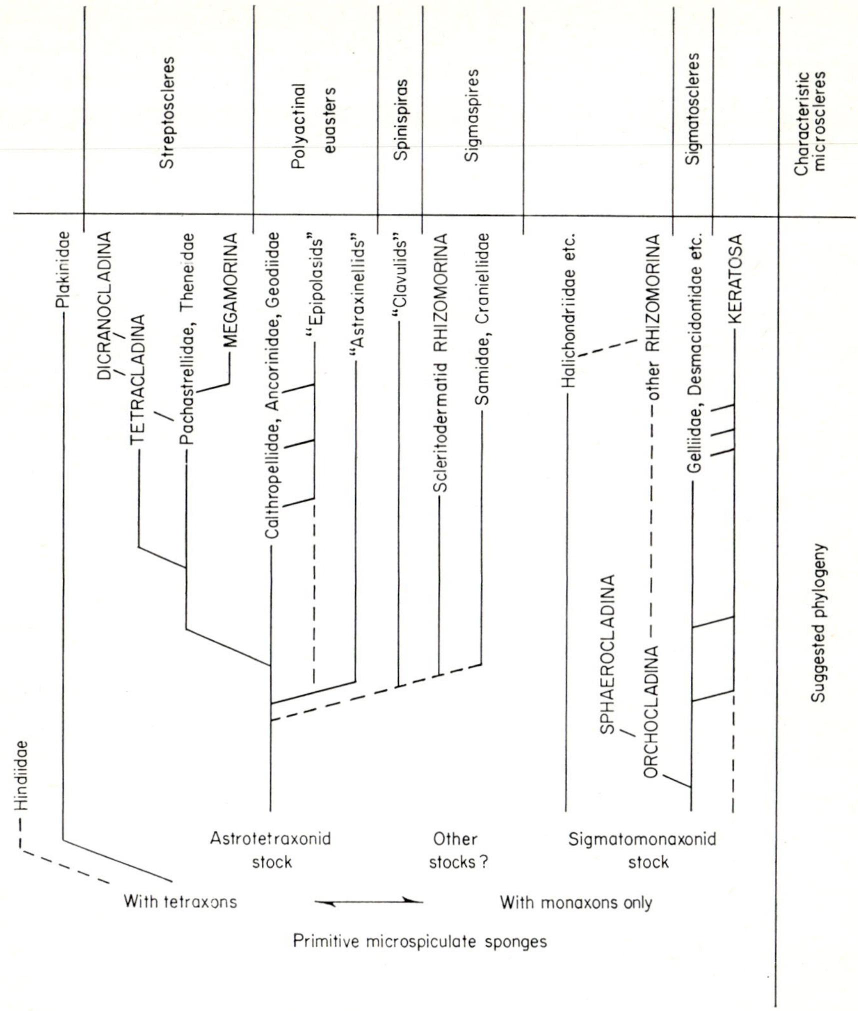

FIG. 5. Suggested phylogeny of the Demospongia, based chiefly on spicular criteria. Taxa given in capitals are orders or suborders. Family names in vernacular and parentheses are invalid taxonomically, although in current use. "Epipolasids" here does not include sponges with sigmaspire or spinispira microscleres, included as Epipolasida by de Laubenfels (1936), which are "clavulids" or rarely craniellids.

must have had a very long independent history and show non-ancestral specializations including candelabra spicules and the *Plakina* amphiblastula. The group is, however, diverse and may include forms which are closer to typical choristids than *Plakina*.

(2) Two groups of modern choristids can be interpreted as parallel stocks descended from an astrotetraxonid radical, with similar evolution in their megaspiculation but divergent evolution of different characteristic microscleres. The latter are "dichotriact" streptasters (Dendy, 1924; streptoscleres, Reid, 1963a) in a stock comprising pachastrellids (*sensu* von Lendenfeld, 1907) and theneids, but polyactinal euasters in a second stock comprising calthropellids, ancorinids and geodiids. The radical stock may be reconstructed as comprising sponges with oxeas and calthrops or short-shafted triaenes as megascleres, and simple (e.g. diactinal to pentactinal) oxyasters as microscleres. The difference in size between the prototypic megascleres and microscleres may also have been less than that between typical examples and comparable, for example, with that between the large and small spicules of *Plakinastrella* Schulze, 1880. The characteristic microscleres of the two descendant stocks will represent different modifications of the primitive oxyasters.

The monaxonid coppatiids and tethyids are probably at least partly epipolasids *sensu* Sollas (1888), of ancorinid origin, although some are not close to any living ancorinid.

(3) The "astraxinellids" are acceptable as allied to the astrotetraxonids, because some contain genuine meristically varying euasters. Dendy (1921) was mistaken in regarding the euasters of *Hemiasterella* Carter, 1879 as pseudasters, derived from spinispiras. They vary meristically, in some forms to triactin and diactin, in the same way as ancorinid oxyasters. The group, however, has no evident connection with any living choristids and might represent a stock in which megascleres were developed from monaxons only. It may also be the source of the other axinellids, if these do not belong with the sigmatomonaxonids.

(4) The "clavulid" sponges may be of similar origin because very simple oxyasters occur in a few species (e.g. slender examples with two to four rays, in *Thoosa fischeri* Topsent, 1891). The characteristic microscleres, however, when any are present, are spinispiras and variants of this type. These spicules are probably derived from a megascleric oxea, with which, for instance, finely spinulated examples may share both fine spinulation and a central annulation in *Cliona vastifica*. Other *Cliona* spinispiras vary from this type to massive spirasters and even **C**-shaped sigmasters (in e.g. *C. aethiopica* Burton, 1932) which are not likely prototypes for megascleric oxeas. If the so-called sigmodiscorhabd, relied on by Dendy (1921) as relating these microscleres to sigmas, has any connection with spinispiras, it will probably be a modified form of a **C**-shaped spinispira (or, sigmaster), as developed in, for instance *C. aethiopica*, or *Spirastrella tristellata* Topsent, 1918.

Dendy's observation that the microscleres are sigmasters in *Timea curvistellifera* Dendy, 1905 supports his (1921) view that the "euasters" of other species of *Timea* Gray, 1867 are pseudasters, unless the genus is composite.

(5) The craniellid choristids appear to have no sure relationship to other choristid sponges although possessing indistinguishable triaenes. They are specially distinguished by sigmaspire microscleres and by occurrence in some species of incubated embryos which are not amphiblastulas. The sigmaspire might somehow represent a reduced euaster but is comparable with the "clavulid" spinispira morphologically and could be derived from the type of acanthoxea which may represent it in *Acanthotetilla*. The monaxonid *Raphidotethya* is probably an "epipolasid" craniellid but is unlikely to represent the origin of true sigmatomonaxonids.

(6) The halichondriids and the sigmatomonaxonids proper (i.e. the Haplosclerina and Poecillosclerina of Topsent) appear to have no close relatives but the keratose sponges. The keratose sponges are probably best pictured as iterative derivatives of the spiculate sigmatomonaxonids at various stages of their history, the oldest being those which are least like any living monaxonids and retain some probably primitive characters (e.g. large eurypylous chambers). If this group is not derived from an astrotetraxonid source through axinellids (cf. Topsent, 1928), it may represent descent from primitive Demospongia without tetraxons.

(7) The lithistids are certainly polyphyletic. The Tetracladina, Dicranocladina and Megamorina all have microscleres of streptosclere types in a few living genera (e.g. *Racodiscula* Zittel, 1878; *Corallistes* Schmidt, 1870; *Costifer* Wilson, 1925). The Tetracladina are probably of pachastrellid origin although possibly not all from one source; but the desmas of the Megamorina appear to be derived from an ophirhabd, through heloclones to megaclones, so that this group seems nearer to theneids. The Dicranocladina are more probably derivatives of acrochordoniid and theonellid Tetracladina, with the crepides of the desmas reduced from tetraxons to monaxons. The Orchocladina, in contrast, were probably sigmatomonaxonids derived from a stock with chalinid skeletal architecture. The Sphaerocladina Schrammen, 1924, whose desmas resemble those of *Crambe*, may be forms derived from Orchocladina by development of sphaeroclones from chiastoclones. This could occur through adaptation to growth of the skeleton in layers, with all arms of the desmas facing inwards. The rhizomorine Scleritodermatidae Sollas, 1888, with sigmaspire microscleres, should be allied to the craniellids or possibly the "clavulids", but other Rhizomorina may be derived from anthaspidellid Orchocladina or other sources. The

hindiids, if their desmas were tetraxial, could be derivatives of a primitive tetraxonid although perhaps not near the ancestry of any modern sponges.

Last, while these views are based on the spicules, it is not claimed that these provide either the whole or the only key to phylogeny. Divergent evolution has presumably affected every character which varies in demosponges. Furthermore, if one trusts analogy with the evidently independent histories of megascleres and microscleres, at least some major changes in various other attributes were probably uncorrelated. Hence reliance on any single character will always be to some extent misleading. The principal divisions of the class suggested by the spicules, however, are probably so ancient that any other more plastic attribute will probably be also less reliable.

APPENDIX: SOME ASPECTS OF SPICULAR NOMENCLATURE

Students of demosponge spicules are compelled to use a complex nomenclature by the wide range of morphological types which they need to distinguish. Most of the general basis of modern zoological nomenclature will be found in the introductory part of Sollas' (1888) "Challenger" monograph although various additions and changes have been made in later literature. For details of Dendy's usages, see his papers of 1921 and 1924. The palaentological nomenclature has been generally based on Rauff's (1893–1895) version of the "Challenger" nomenclature, with additions made by Rauff (*ibid*) and by Schrammen in various later papers. The best English summary of the palaeontological nomenclature is by O'Connell (1919, pp. 34–44). The glossary of de Laubenfels (1955) is incomplete and some of his usage is badly confused (see below).

The nomenclature used above has three aspects in which readers may be helped by the following explanations. These aspects are (1) the usage of "euaster" and "streptaster" as names for microscleres, (2) the palaeontological nomenclature of desmas and certain other spicules, which could be adopted to advantage by zoologists, and (3) the badly confused usage of certain terms by de Laubenfels (1955), in terms of which the work of previous authors may seem incomprehensible if the errors which were made by de Laubenfels are not understood.

Euasters and streptasters

The terms euaster and streptaster (Sollas, 1888) have both been used in various ways for different types of spicules which are not homologous.

Morphologically (Sollas, 1888), a euaster is a microsclere in which rays (or apparent rays) radiate from a centre, and it is distinguished by this character from streptasters *sensu* Sollas (1888), in which ray-like spines are emitted from a linear axis which is sometimes straight but typically spiral. The choristid microscleres to which Sollas (1888) applied the name euaster are forms in which simple examples (oxyasters, etc.) have the number of rays varied in the sequence 2, 3, 4, 5, 6, 7, etc., up to over a hundred in some advanced types. Some monaxonid sponges (e.g. *Tethya* Lamarck, 1814) have similar microscleres but other monaxonid microscleres, which are euasters morphologically, do not vary in this manner. Instead they grade into monaxial megascleres (e.g. acanthoxeas, acanthostyles) or spinispira microscleres. Euasters s.s. in this paper means the meristically varying kind, which include the asters s.s. of Dendy (1921). The others, grading into monaxial megascleres of microscleres, are pseudoeuasters, for details of which see Dendy (1921).

Second, the streptasters of Sollas (1888) include two kinds of microscleres which again are apparently not homologous: (a) the intergrading plesiasters, amphiasters, metasters and spirasters of pachastrellids and theneids, which are Dendy's (1924) "dichotriacts" and my streptoscleres (Reid, 1963a), and (b) ancorinid sanidaster which may also pass into amphiasters but otherwise grades into spinulated microhabds like those of *Ecionemia* Bowerbank, 1862. There are also two later restricted usages. Streptasters *sensu* Burton (e.g. 1956), which are the characteristic microscleres of his taxon Streptastrosclerophora Burton, 1929 (= Streptosclerophora Dendy, 1924), are "dichotriacts", or streptoscleres; but a streptaster as defined by de Laubenfels (1955) is a sanidaster although he also applied the same name to streptoscleres which he regarded as bent sanidasters. In addition, if "streptaster" is extended to the nomenclature of monaxonid microscleres, it also takes in spinispiras and various discasters with no known relationship to either streptoscleres or sanidasters.

My own term "streptosclere" was introduced because of this confusion and because most of the streptosclere series, called "dichotriacts" by Dendy (1924), are not dichotriactins morphologically.

Palaeontological nomenclature

The nomenclature explained here is mostly used in this paper but a number of additional terms are included for completeness. Where possible, examples from modern sponges are cited.

General nomenclature of desmas

The articulating megascleres of lithistid sponges are called **desmas** (Sollas, 1888) as in zoology or sometimes **desmomes** (Rauff, 1893). The inceptional body of a desma is its **crepis** (Sollas, 1888) or **crepidome** (Rauff, 1893) and the articulating parts are **zygomes**. Ray-like arms which are partly or entirely anaxial are called **clones** (Rauff, 1893); they were also called "clads" by de Laubenfels (1955) but are not true rays or cladi because of the absence of a protorhabd.

Individual types of desmas are given names which end "-clone" with some prefix; it ought to be noted that such prefixes often do not refer to the form or other attributes of clones, which may be lacking (in e.g. heloclones and some rhizoclones—see below).

Tetraclones and variants (Rauff, 1893)

Tetraclones are tetraxial (or, tetracrepid) desmas which do not show a triaenose symmetry. The four arms may be similarly or differently developed and the zygomes are typically terminal. The crepis may be a calthrops or a short-shafted triaene whose form need not be reflected in the form of the adult desma. The desma may be smooth (as in *Theonella* Gray, 1868) or tuberculate (as in *Discodermia* du Bocage, 1869).

Triders are variants of tetraclones having a triaenose symmetry, with three similar arms which comprises a **clonome**. The commonest types are desmas with one arm either shorter or longer than the others, or lacking altogether so that the desma takes on a tripodal form. A shortened arm is a **brachyome**; it may possess or lack a zygome. The crepis may be a calthrops or a tripod.

An **amphitrider** is a tetraclone variant resembling an amphitriaene and sometimes also having an amphitraene crepis (as in "*Jereopsis*" *fruticosa* Wilson, 1925). A **mesotrider** has a mesotriaene (centrotriaene) crepis, as in some desmas of *Brachiaster* Wilson, 1925.

Rhizoclones (Rauff, 1893, restr. Schrammen, 1910)

Rhizoclones are non-tuberculate monaxial desmas with spine- to root-like zygomes which are often (but not always) mainly lateral. They vary extensively in shape but are divisible into four main intergrading types: (1) **linear**, with mainly lateral zygomes along a straight or curved shaft; (2) **bipolar**, with root-like zygomes at the ends of a smooth or spiny shaft; (3) **pseudoradiate**, with three or more spiny arms radiating from a centre and (4) intermediate and irregular types. Any of these types may predominate in given cases but all may occur together. Those of modern sponges (e.g. *Azorica* Carter, 1873; *Scleritoderma* Sollas, 1888; *Plakidium* von Lendenfeld, 1907) are mainly linear and irregular.

Rhizoclones *sensu* Rauff (1893) also include the corallistid dicranoclones distinguished by Schrammen (1910).

Dicranoclones (*Schrammen*, 1910)

Dicranoclones are tuberculate monaxial desmas which are characteristically arch-shaped, tripodal, tetrapodal or rarely polypodal, and have terminal zygomes like those of tetraclones. The tubercles are typically strongest on the convex sides of the arching arms. The central part may bear a specially prominent tubercle or a short brachyome-like outgrowth with no zygome, called the **"basaltheil"** by Schrammen, but this may be lacking or sometimes double. The typical dicranoclones grade into desmas of irregular shapes and also into non-tuberculate forms which Schrammen (1910) called **megarhizoclonids.** In irregular and megarhizoclonid forms, the crepidal canal is usually in a central shaft but in typical dicranoclones it is sometimes in the "basaltheil" (hence this name).

Dicranoclones are the characteristic desmas of the Corallistidae Sollas, 1888, although generally subordinate to irregular and megarhizoclonid desmas in modern genera (e.g. *Corallistes* Schmidt, 1870). The name dicranoclone was used by de Laubenfels (1955) for a variety of unrelated desma types (see below).

Heloclones and megaclones (*Schrammen*, 1910)

These are related types of desmas, called *rhabdoclones* by Rauff (1893) but distinguished by Schrammen (1910). A **heloclone** has an elongate and usually sinuous shaft with no branches or with small digitations at the ends. The zygomes are lateral notches which are usually subterminal. An axial canal usually traverses most of the length of the desma. A **megaclone** is an arch-shaped desma with a short central crepidal canal, or a comparable form in which the two primary arms are branched, or extra arms grow out laterally from the epirhabd. The zygomes vary from oblique lateral facets to tongue- or hand-like terminal expansions which grip other desmas. These two types of desmas are usually separate but occasionally intergrade in one skeleton. They are usually non-tuberculate but slight tuberculation may occur in some megaclones.

Heloclones and megaclones are, respectively, the characteristic desmas of the triaene-bearing Isoraphiniidae Schrammen, 1924 and Pleromatidae Sollas, 1888, of which *Costifer* Wilson, 1925 and *Pleroma* Sollas, 1888 are modern examples; but desmas like megaclones also occur in some probably unrelated sponges, e.g. *Desmatiderma* Topsent, 1929 (a sublithistid desmacidontid).

Sphaeroclones (*Schrammen*, 1910) *and variants*

A **sphaeroclone** s.s. is an anaxial desma in which several (usually 4–6) ray-like arms are emitted from one side of a globular centrum which may be spined on the other side. In the living *Vetulina* Schmidt, 1870, the sphaeroclones develop from an anaxial corpuscle and their centra contain a granular **"nucleus"** which originates as an esternal pit. Some of the fossils show what seems to be a sinuous canal in the centrum but this is probably secondary. The typical sphaeroclones grade into forms with simply radiating arms which I call **astroclones**; these were called anomoclones by Schrammen (1936) but this name (Rauff, 1895) belongs properly to a different kind of desma (see below).

Desmas similar to some sphaeroclones also occur in *Crambe* Vosmaer, 1886 but differ from those of *Vetulina* in showing a number of granular inclusions corresponding with the number of arms. The desmas called sphaeroclones could thus have two origins.

Didymoclones (*Rauff*, 1893)

These are monaxial desmas resembling two sphaeroclones linked by a shaft between their centra and are sometimes wrongly said to consist of two linked sphaeroclones. They are true single desmas which developed from a crepidal strongyle and also grade into rhizoclones. There are no modern examples.

Ennomoclones (*Rauff*, 1893)

This name was applied by Rauff to the trider-like desmas of hindiids and the sphaeroclones of the Astylospongiidae Rauff, 1893, on the assumption that the astylospongiid sphaeroclone is a **"dichotrider"** form of the hindiid trider, derived by branching of three primary arms (clones, Rauff). This relationship is not demonstrable and Schrammen interpreted the hindiid desmas as tetraxons.

Dendroclones and chiastoclones (*Rauff*, 1894)

These names were used by Rauff for desmas of the Anthaspidellidae Ulrich, 1890 (= Archaeoscyphiidae Rauff, 1894 + Aulocopiidae Rauff, 1895) and Chiastoclonellidae Rauff, 1895, which he placed in a special section of the Tetracladina Zittel, 1878 under the name Orchocladinidae. A **dendroclone** has the form of a tetracladine amphitrider but typically occurs set transversely to skeletal fibres as do the desmas of *Lithochela* Burton, 1929 (hence Orchocladinidae, from orchos, Gr., row of trees). A **chiastoclone** is a comparable desma with a very short central shaft or with the arms radiating from a centre. As shown in the paper above, these desmas were probably monaxial, or possibly anaxial in the case of chiastoclones.

Anomoclones (*Rauff*, 1895)

These are irregularly varying desmas of the Anomoclonellidae Rauff, 1895, which may more of less resemble chiastoclones, didymoclones or thick rhizoclones. According to Rauff (1895) they are irregular forms of his "ennomoclones" but they are more probably related to chiastoclones. The name anomoclone was also used by Schrammen (1936) for irregular sphaeroclonar desmas of the Lecanellidae Schrammen, 1936, which are better called astroclones, and resemble some *Crambe* desmas.

Palaeontological usage of de Laubenfels

Some of the terms defined above were used by de Laubenfels (1955) in ways which are not their normal usage and which can be confusing if one does not know the normal usage and the individual genera concerned.

(1) The first family of the lithistid suborder Rhizomorina Zittel, 1878, whose desmas are called rhizoclones, is the Kaliapsidae de Laubenfels, 1936. *Kaliapsis* Bowerbank, 1869 is in fact a tetracladine lithistid, whose desmas are tetraclones and triders. In two other genera (*Corallistes* Schmidt, 1870, *Procorallistes* Schrammen, 1910) the desmas are mainly dicranoclones or megarhizoclonids.

(2) The Dicranocladina Schrammen, 1924 are divided between the Eutaxicladina Rauff, 1895 and Rhizomorina Zittel, 1878, largely on the basis of whether or not their dicranoclones are predominantly tripodal to polypodal. Dicranoclones replace Rauff's ennomoclones as the nominally characteristic desmas of Eutaxicladina, although Schrammen's (1910) dicranoclones are forms of Rauff's rhizoclones, not his ennomoclones. The desmas of genera included are sometimes true dicranoclones (in e.g. *Gignouxia* Moret, 1926), sometimes hindiid triders (in *Hindia* Duncan, 1879), sphaeroclones (in e.g. *Astylospongia* Roemer, 1860), or dendroclones and chiastoclones (in *Chiastoclonella* Rauff, 1894). The sphaeroclones of *Ozotrachelus expectatus* (Schrammen) (de Laubenfels, 1955, Figs 44 5a–c) are called dicranoclones or sphaeroclones according to whether they are viewed from the side or from above (cf. Schrammen, 1910, textpl. 8, Figs 1a–g).

(3) Sphaeroclones replace didymoclones as the nominally characteristic desmas of the Anomocladina Zittel, 1878 *sensu* Rauff (1895) but only one genus included (*Mastosia* Zittel, 1878) has sphaeroclones; the desmas of the others are didymoclones (in e.g. *Cylindrophyma* Zittel,

1878), dendroclones or bipolar rhizoclones (in *Anthaspidella* Ulrich, 1889), or chiastoclones and anomoclones (in *Anomoclonella* Rauff, 1895). All other lithistids with true sphaeroclones are in the Eutaxicladina, where their desmas are called dicranoclones, sphaeroclones or both dicranoclones and sphaeroclones. One genus included (*Heliospongia* Girty, 1908) has no desmas of any sort and no megascleres but simple oxeas.

References

Burton, M. (1956). The Sponges of West Africa. *Atlantide Rep.* **4**: 111–147.

Dendy, A. (1921). The tetraxonid sponge spicule: a study in evolution. *Acta zool. Stockh.* **2**: 95–152.

Dendy, A. (1924). On an orthogenetic series of growth forms in certain tetraxonid sponge spicules. *Proc. R. Soc.* (B) **97**: 243–250.

Finks, R. M. (1967). The structure of *Saccospongia laxata* Bassler (Ordovician) and the phylogeny of the Demospongea. *J. Paleont.* **41**: 1137–1149.

Hinde, G. J. (1887–1912). A monograph of the British fossil sponges. *Palaeontogr. Soc.* [Monogr.]: **1887–1912**: 1–264.

de Laubenfels, M. W. (1936). A discussion of the Sponge fauna of the Dry Tortugas in particular and the West Indies in general, with material for a revision of the families and orders of Porifera. *Papers Tortugas Lab.* **30**: iii + 1–225.

de Laubenfels, M. W. (1955). Porifera. In *Treatise on Invertebrate Palaeontology* **E**: E22–E122, Ed. Moore, R. C. Geological Society of America and University of Kansas Press.

Lévi, C. (1957). Ontogeny and systematics in Sponges. *Syst. Zool.* **6** (4): 174–183.

Lévi, C. (1960). Les Spongiaires à desmes astéroids. *Bull. Inst. océanogr. Monaco* No. **1179**: 1–7.

O'Connell, M. (1919). The Schrammen collection of Cretaceous siliceous Sponges in the American Museum of Natural History. *Bull. Am. Mus. nat. Hist.* **41**: 1–261.

Rauff, H. (1893–1895). Palaeospongiologie. *Palaeontographica* **40**: 1–346.

Reid, R. E. H. (1963a). A classification of the Demospongia. *Neues Jahrb. Geol. Paläont.* Monatshefte **4**: 196–207.

Reid, R. E. H. (1963b). *Spiractinella* Hinde is a Demosponge. *Ir. Nat. J.* **14**: 129–131.

Schrammen, A. (1910). Die Kieselspongien der oberen Kreide von Nordwestdeutschland: I, Tetraxonia, Monaxonia, und Silicea incertae sedis. *Palaeontographica*, suppt. 5, 1–175.

Schrammen, A. (1936). Die Kieselspongien des oberen Jura von Süddeutschland. *Palaeontographica* **84**: 149–194, and **85** (A): 1–114.

Schulze, F. E. (1880). Untersuchungen über den Bau und die Entwicklung der Spongien: 9, Die Plakinidae. *Z. wiss. Zool.* **34**: 407–451.

Schulze, F. E. (1887). Report on the Hexactinellida collected by H.M.S. "Challenger" during the years 1873–1876. *Rep. Sci. Res. H.M.S. "Challenger"* **21**: 1–513.

Sollas, W. J. (1888). Report on the Tetractinellida collected by H.M.S. "Challenger" during the years 1873–1876. *Rep. Sci. Res. H.M.S. "Challenger"* **25**: i–clxvi + 1–458.

Topsent, E. (1928). Spongiaires de l'Atlantique et de la Méditerranée, provenant des croisières du Prince Albert 1er de Monaco. *Résult. Camp. sci. Prince Albert I* **74**: 1–376.

Symp. zool. Soc. Lond. (1970) No. 25, 91–123

THE COMPOSITION, DEVELOPMENT, FORM AND ORIENTATION OF CALCAREOUS SPONGE SPICULES

W. CLIFFORD JONES

Department of Zoology, University College of North Wales, Bangor, Wales

SYNOPSIS

Calcareous sponge spicules are biocrystals composed of impure calcite. The rays of large examples exhibit a concentric lamination and an axial region. Each spicule is enveloped by an organic sheath but no organic residue has been detected within the spicule. The spicule rays are characteristically oriented in relation to the crystallographic axes of their calcite. Magnesium, strontium and traces of lithium, barium, sodium and manganese have been detected in certain species.

A pair of sclerocytes usually participates in the formation of each ray of the smaller spicules but rays bearing one and four cells have also been described. Doubtless more cells contribute to the production of colossal rays. Little is known about the early stages of development. The growing rays are enveloped in a thin membrane, which is surrounded by the sclerocytes. The rays grow in length at rates of the order 1·0–2·5 μ/h. Calcium and carbonate ions must be dissolved in the external environment of the juvenile sponge for spicules to form. The rate of production of spicules is not uniform in either time or space.

The spicules and the spicule calcite are oriented in characteristic ways in the sponges. The evidence indicates that mechanical factors play an important part in the mechanism of orientation of the spicules but as yet there is no satisfactory explanation for the orientation of the spicule calcite.

In certain sponges the shape of the spicules varies progressively with the position of the spicule in the sponge. Both intrinsic and extrinsic factors affect the spicule form. Thus, whereas the spicule rays tend to grow in planes containing the optic axis of the calcite, the directions taken within the planes are guided by the bounding surfaces of the mesohyl.

TYPES OF CALCAREOUS SPONGE SPICULES

Calcareous sponge spicules are classified according to the number of rays, one to four, which they possess. The types are designated: monact (monactine) or uniradiate; diact or biradiate; triact or triradiate and tetract or quadriradiate. Some sponges only have one type, whereas others have several, subserving different functions. Thus *Clathrina coriacea* (Montagu, 1812) only has triacts, while *Leucosolenia complicata* (Montagu, 1812) has four types, namely triacts, tetracts, slender monacts and curved, often lanceolate, monacts. The tetracts are essentially similar to the triacts with an additional ray (the apical ray) projecting from the plane of the other three (the basal triact system). In *Leucandra nivea* (Grant, 1825) a considerable range of size is exhibited by the spicules, from the tiny micromonaxons of 40 μ length by

2 μ width to the colossal triacts with rays of up to 1·2 mm length and 80 μ width. The term "monaxon" is used when it is not clear whether a uniaxial spicule should be regarded as a monact or a diact. Since such spicules are pointed at each end (oxeote), they have also been called "oxea". True diacts occur in *Clathrina lacunosa* (Johnston, 1842). Special types, such as the "tuning fork", the "cruciform", the "dagger", the "anchor", the "needle-eye", the "tripod" and the "nail-head", also occur but these appear to be characteristic variants of the basic types and need not concern us further. The rays of calcareous spicules are generally smooth but occasionally rows of spines may be borne in the apical ray of tetracts.

PROPERTIES OF CALCAREOUS SPONGE SPICULES

Structure

The structure of the spicules is best investigated using giant specimens but care must be exercised in attempting to extrapolate the results so obtained to the smaller varieties of spicule. The colossal monaxons of *Leucandra aspera* (O. Schmidt, 1862) and *L. alcicornis* Gray, 1858 and the huge triacts of *Leucaltis solida* O.S., 1862 were mainly used by earlier researchers (von Ebner, 1887; Bütschli, 1901). Figure 1A shows the main features exhibited by the huge triacts of *Leucandra nivea*. Faint longitudinal striations can be seen in the rays, which betray a concentric lamination that is better seen on transverse sections after corrosion by acetic acid (Fig. 1B). The rays appear to be joined together along distinct lines of union at the central junction, while along the axis of the rays is visible a distinct line, the so-called axial filament. According to Minchin and Reid (1908) this is composed of organic matter but these authors undoubtedly confused this filament with the retractile organic sheath which envelopes the spicule rays (Jones, 1967). There is no detectable organic residue within the sheath following corrosion by dilute acids (Kölliker, 1864; von Ebner, 1887; Jones, 1955b). However, the sheath is certainly organic. Around developing rays there is a delicate, contractile membrane, outside which connective tissue fibres are laid down (Jones, 1967). The sheath stains well with orcein (Jones, 1956) and shrinks when the spicules are decalcified (Jones, 1955b). Two kinds of collagen fibres have been detected by Travis, François, Bonar and Glimcher (1967) in the decalcified remnants of *Sycon* (= *Scypha*) spicules. Travis *et al.* also detected quartz lines on X-ray diffraction patterns of demineralized spicules of *Sycon* The significance of this finding is obscure.

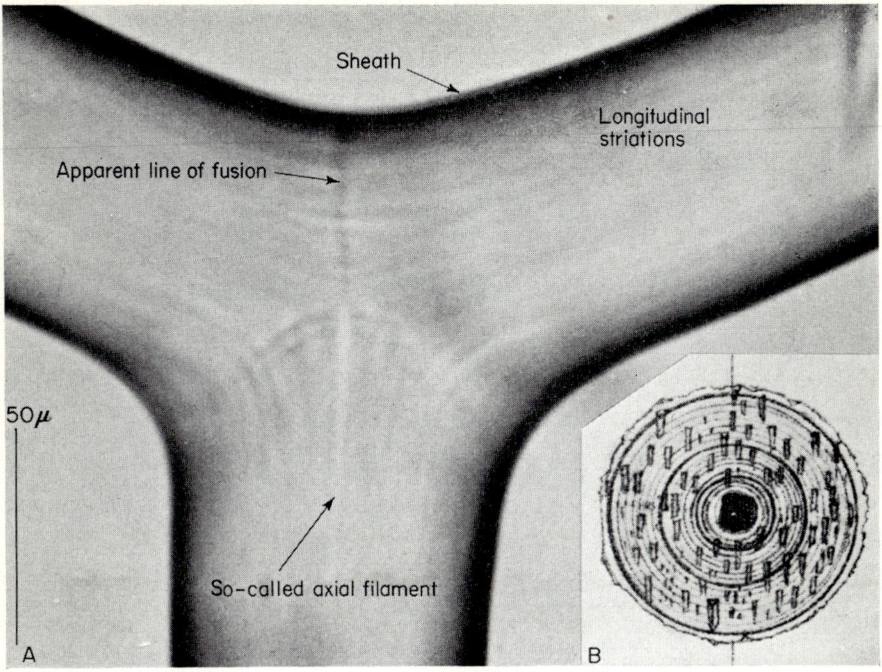

FIG. 1. A, Large triact of *Leucandra nivea* showing the structural details which can be seen with the optical microscope; B, transverse section of an unpaired ray of a colossal triact of *Leucaltis solida* after brief treatment with concentrated acetic acid. Note the concentric lamination, corrosion figures and central cavity (after von Ebner, 1887).

It has long been known that the spicules consist very largely of calcium carbonate and that they are crystalline and birefringent (Haeckel, 1872; Minchin, 1909) but it was Sollas (1885) who first demonstrated that they were a form of calcite. Their properties are compared with those of calcite in Table I. It can be seen that while in some important respects (refractive indices, for example) the spicules behave like calcite, in others they are clearly different. Thus their specific gravity is somewhat less and their behaviour on heating is different. They appear to contain more water than calcite, losing more weight when heated. They decrepitate with strong heating and become brown when viewed by transmitted light, blue-white by reflected light, owing to the production of innumerable small vacuoles in their substance. Their lower specific gravity is due, presumably, to the organic matter composing the sheath and to the probable occurrence of submicroscopic vacuoles containing water or trapped remnants of the crystallizing fluid. Such vacuoles could coalesce to form the larger cavities

TABLE I

Comparison between calcareous sponge spicules and calcite

Reference	Test	Spicules	Calcite
von Ebner (1887)	Refractive index (sodium line D)	Birefringent $\omega = 1{\cdot}658$ $\epsilon = 1{\cdot}486$	Birefringent $\omega = 1{\cdot}658$ $\epsilon = 1{\cdot}486$
Sollas (1885)	Specific gravity	2·61–2·63	2·72
Bütschli (1907)		2·639	
von Ebner (1887) Bütschli (1907)	Heat	Become brown by transmitted light, blue-white by reflected light due to appearance of vesicles	No vesicles become visible
Bütschli (1908)	Heat at 400°C	Loss in weight = 4·64%	Loss in weight = 0·028% (Iceland spa)
Paine (1964)	Heat above 140°C	Water progressively lost	
von Ebner (1887)	Strong heat	Decrepitate into leaf-like fragments	Does not decrepitate
Carter (1869)	Mechanical breakage	Conchoidal fracture	Cleavage along cleavage plane (sometimes conchoidal fracture)
von Ebner (1887)		Cleavage along cleavage plane occasionally	
von Ebner (1887)	Crushing	No shear lamellae produced	Shear lamellae become visible
von Ebner (1887)	Corrosion by 10–15% KOH	Rapid. Central axial canal appears	None after 24 hours, even with some heating
Bütschli (1907)	Corrosion by 35% KOH	Corrosion rapid	Corrosion rapid
von Ebner (1887)	Corrosion by formic acid	Triangular corrosion figures appear and small solution bodies	Larger solution bodies

when the spicules are heated (Bütschli, 1901). Moreover, since the vacuoles have an uneven distribution, appearing in layers, they could well be the explanation of the slight differences in refractive index in the spicule substance, which give rise to the concentric lamination of the spicule rays seen under the optical microscope (Bütschli, 1901).

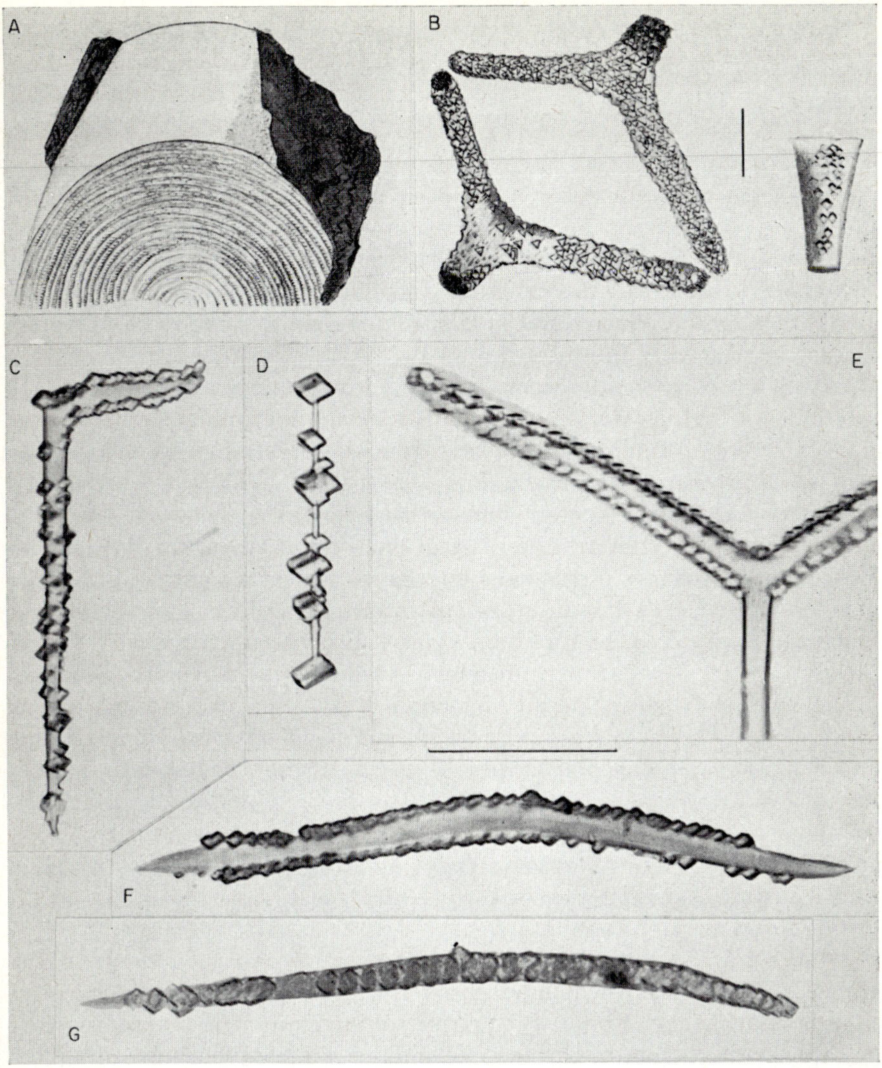

FIG. 2. A, Fragment of a paired ray of a colossal triact of *Leucaltis solida* after macerating for 24 h with 10% potash. Differential corrosion has revealed the concentric lamination (after von Ebner, 1887). B, corrosion figures produced by formic acid on the surfaces of paired rays (seen from above and below) and an unpaired ray of a colossal triact of *Leucaltis solida* (after von Ebner, 1887). The line indicates 100 μ. C–G, spicules of *Leucosolenia complicata* bearing crystals (mainly cleavage rhombohedra) of calcite. C, tetract in side view; D, slender monact; E, triact; F, lance-headed monaxon; G, curved monaxon. Potash was used to isolate the spicules in C, D, E and F from the sponge. Potassium nitrate solution was used for the spicule in G. Note that whereas in G the crystals occur on the surfaces parallel to the optic axis (which lies in the plane of the paper), in F they are on the surfaces at right angles. C–G all to the same scale, the line indicating 50 μ (after Jones, 1955a).

Differences are also found in respect of corrosion by acids and alkalis (Fig. 2A). The axial region is particularly susceptible to corrosion (Fig. 1B), a fact which has given rise to the concept of an axial canal (Haeckel, 1872) along the axis of each ray. However, variations in the rate of corrosion of the various parts could well be a reflection of the presence of impurities, not necessarily organic, in the spicule calcite and von Ebner (1887) has explained the so-called axial filament and concentric lamination as being the result of an alternate secretion of less pure and more pure calcite as the spicule grows. The first formed part, at the union of the axial canals, is also readily corroded and would consist of relatively impure calcite.

From his corrosion experiments on giant spicules von Ebner (1887) was able to deduce that each spicule is crystallographically homogeneous; the crystalline substance maintains its crystallographic orientation without disruption throughout the entire spicule. Consequently the corrosion figures spread across the boundaries between the concentric layers seen on transverse sections of the rays (Fig. 1B), while on the outer surfaces of the rays triangular and other corrosion figures develop (Fig. 2B). These indicate the orientation of faces of the calcite crystal in relation to the direction of growth of the rays. Figure 3 shows those parts of a calcite crystal which would be needed if the spicules shown were to be chiselled out of a single crystal. The calcite crystal is a combination of the corrosion rhombohedron, which von Ebner designated $-\mathbf{2R}$, and the deuteroprism, $\infty\mathbf{P2}$. Using the modern Miller-Bravais notation (see Phillips, 1963) the faces would be respectively $\{\mathbf{02\bar{2}1}\}$ and $\{\mathbf{1\bar{2}10}\}$. Since the optic axis (the principal symmetry axis) of the calcite is a trigonal axis, there are three equivalent settings of each spicule when it is one way up and three more when it is turned upside down.

Apart from a large ectosomal tetract of *Sycandra bucchichii* (von Ebner, 1887), a ray of which had broken to reveal a face of the cleavage rhombohedron, von Ebner was unable to determine the precise crystallographic orientation of the rays of smaller spicules, because corrosion produced no more than a series of stripes on the ray surfaces, instead of recognizable crystal faces or edges. He was able to determine the direction of the optic axis, however, using the polarizing microscope and the birefringence of the spicules and this enabled him to classify the spicules in terms of the relation of the spicule rays to the principal planes of the calcite crystal, that is, the three vertical planes which meet at 120° and include the optic axis. He deduced from the spicule symmetry and the orientation of the optic axis that one of the rays at least must lie in a principal plane but which way up the spicule would have to be

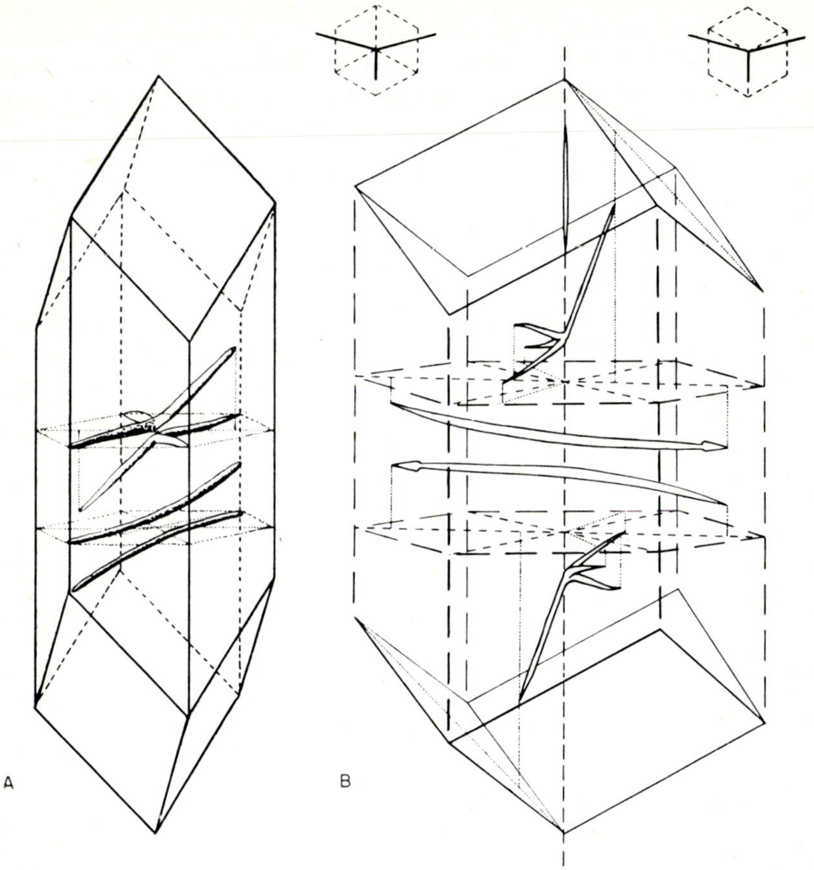

FIG. 3. A, Triact of *Leucaltis solida* and monaxon of *Leucandra aspera* oriented crystallographically within the combination of the corrosion rhombohedron {$\mathbf{0\bar{2}21}$} and the prism {$\mathbf{1\bar{2}10}$} (after von Ebner, 1887); B, the directions of the rays of spicules of *Leucosolenia complicata* in relation to the faces of the cleavage rhombohedron {$\mathbf{10\bar{1}1}$}. A tetract, a lance-headed monaxon and a slender monact are shown. The triact has essentially the same crystallographic orientation as the basal triact system of the tetract, as can be seen in the two small diagrams above. These show the spicule viewed along its optic axis in relation to the three upper faces of the cleavage rhombohedron. On the left the canal side of the unpaired ray is nearer the eye; on the right, the exopinacodermal side is nearer (after Jones, 1955a).

set he could not determine. This has, however, now been done for a few species, using a crystallization technique (Jones, 1955a).

If the spicules are spread over a slide and left in a thin film of saturated calcium bicarbonate solution, prepared by bubbling carbon dioxide through a suspension of calcium carbonate for a few minutes and then filtering, crystals form on the surfaces of the rays and, provided the spicules have not become too corroded prior to crystallization and crystallization is allowed to proceed undisturbed, only a few scattered and well-formed crystals develop on each spicule (Fig. 2C–G). The crystals usually have the form of the cleavage rhombohedron (Miller-Bravais $\{\mathbf{10\bar{1}1}\}$, and since there is crystalline continuity from crystal to spicule calcite, one can easily determine the relationship between the direction of the rays and the faces of this rhombohedron. Crystalline continuity is shown by the parallel orientation of the crystals and by the simultaneous darkening of the whole spicule together with its overgrown crystals when they are rotated between crossed nicol prisms. The sheath apparently does not prevent the oriented overgrowth of the crystals, which suggests that the sheath is perforated. Certainly, oriented crystals cannot be obtained by crystallization experiments using isolated decalcified sheaths (Jones, 1955b).

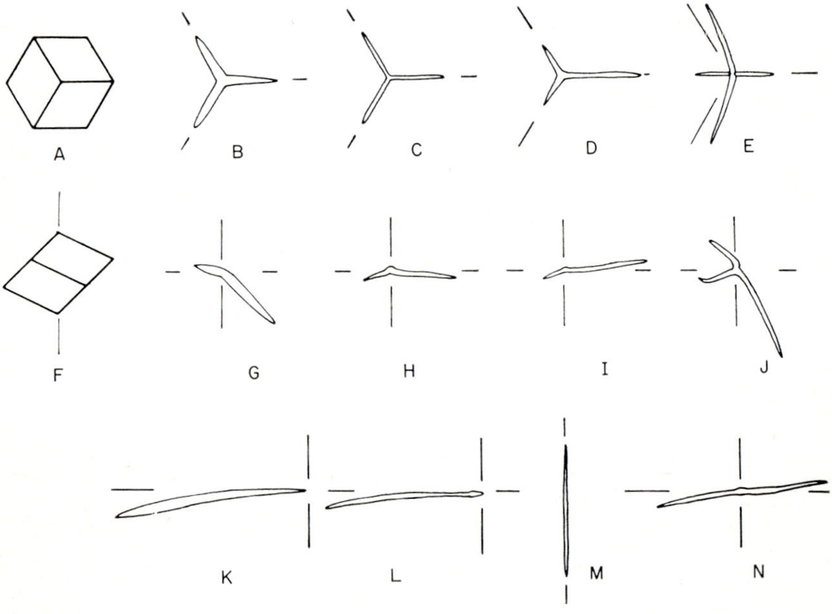

FIG. 4. The directions of spicule rays in relation to the faces of the cleavage rhombohedron of calcite (Miller-Bravais: $\{\mathbf{10\bar{1}1}\}$). A–E, viewed from above; F–N, viewed from the side. A and F, the cleavage rhombohedron used as an indicatrix; B and G, giant triact of *Leucaltis solida* (after von Ebner, 1887); C and H, triact of *Clathrina coriacea* (perregular); D and I, triact of *Clathrina lacunosa* (parasagittal); E and J, tetract of *Leucosolenia complicata* (sagittal); K, giant monaxon of *Leucandra aspera* (after von Ebner, 1887); L, lance-headed monaxon of *L. complicata*; M, slender monact of *L. complicata*; N, diact of *C. lacunosa*.

Figure 4 shows the crystallographic orientation of the spicules for which this has been determined so far. The cleavage rhombohedron is used as an indicatrix. The examples depicted illustrate von Ebner's classification of triact calcareous spicules. He distinguished four groups. First, the regular spicules, exemplified by *Clathrina coriacea.* These spicules have rays of equal length and the rays lie in the three principal planes, each ray making the same angle with the optic axis (88° in *C. clathrus* (O.S., 1864)). The optic axis is thus oriented perpendicular to the facial plane of the spicule (the plane upon which the ray tips lie). The remaining spicules are either sagittal or irregular. The irregular spicules have no ray lying in a plane containing the optic axis and bisecting the angle between the other two rays. The sagittal spicules have at least one such ray and there are two sub-categories of these spicules. First the primary sagittal spicules, which have all three rays in principal planes so that the angles between the rays are 120° when these are projected into a plane at right angles to the optic axis. Second, the secondary sagittal spicules, in which the paired rays meet at an angle which is not 120° when so projected. In the latter spicules the angle is commonly 150–180°, according to von Ebner. Borojević *et al.* (1968) have proposed the terms "parasagittal" and "sagittal" to replace the primary and secondary sagittal nomenclature of von Ebner. The true angle between any two rays is the angle measured when the plane of the two rays is perpendicular to the line of sight. Sagittal spicules have one such angle differing from the other two, which are the same. Thus one can speak of unpaired and paired angles and also of unpaired and paired rays. The tetracts are essentially similar to the triacts, apart from the addition of the apical ray. The monaxons appear to be of two kinds at least. The slender monacts of *L. complicata* have their long axis parallel to the optic axis, while the curved monacts start approximately at right angles to the optic axis but then curve in a principal plane as they grow.

The composition of some calcareous sponge spicules

Qualitative analysis by von Ebner (1887) showed that spicules of *Leucandra aspera* contained calcium carbonate, magnesium, sulphate, sodium and possibly water. Phosphate, chloride, aluminium, iron and potassium were all absent. Bütschli (1908), however, found that spicules of the same species contained 0·26% iron oxide, besides 86·76% calcium carbonate, 8·84% magnesium carbonate, 1·97% calcium sulphate dihydrate and 3·14% water. In addition he found 0·42% organic matter on average but was inclined to think that this was derived from foreign matter contaminating the sponge specimens used, because the amount varied widely in the three samples analysed.

Using the technique of arc spectrography, Jones and Jenkins (1969, in press) have recently found that spicules of *Leucandra nivea* contain calcium, magnesium and strontium. Traces of lithium, sodium, barium and manganese were also detected. We have also estimated quantitatively the calcium, magnesium and strontium contents of spicules from a number of species of calcareous sponges by means of atomic absorptiometry and flame photometry. The results, which will be published elsewhere, indicate that the percentage magnesium content varies from species to species, even when these are growing in the same littoral pool at the same time of year. The values obtained ($\pm 0{\cdot}05\%$) range from $1{\cdot}30\%$ for *Sycon ciliatum* (O. Fabricius, 1780) to $3{\cdot}15\%$ for *Clathrina coriacea*. For all the samples tested, however, the strontium contents were much the same, and in the range $0{\cdot}20$–$0{\cdot}25\%$ ($\pm 0{\cdot}02\%$). The spicules of some species (for example, *Leucandra nivea*) were sorted into different size categories and it was found that the smaller spicules tended to have slightly less magnesium than the larger, the amount increasing progressively up to the maximum for the largest fraction. Thus for the four fractions of *L. nivea*, values of 2·55, 2·60, 2·65 and $2{\cdot}80\%$ magnesium were obtained for the smallest to the largest spicules. It appeared from this work that the composition was not correlated with the spicule form, but merely with spicule size, although further tests are needed for this to be confirmed. It is perhaps unfortunate that the magnesium content of the various spicule fractions in the same sponge species varies, for this will hinder the possible use of chemical analyses as a means for clarifying the phylogenetic relationships of the Calcarea.

The spicules are clearly magnesian calcites and Jones and Jenkins determined the spacings of the **112** planes of the crystalline substance, using an X-ray diffractometer. They found a correlation between the molar percentage of magnesium carbonate and the difference between these spacings and the spacing for pure calcite, in accordance with what is already known about natural and artificial magnesian calcites in general. It would appear from this that the magnesium is uniformly distributed throughout the spicule substance. On the other hand, the smaller magnesium content of smaller spicules suggests that the material first secreted may contain less magnesium than the later crystallizing mineral. In addition, the concentric lamination exhibited by giant rays may be caused by varying magnesium content. An attempt, therefore, has been made to investigate the distribution of magnesium, strontium, sulphur and calcium within a section of a single spicule by means of an electron probe micro-analyser (Jones and James 1969, in press). No significant variation in the composition was detected on sections through the central junction and rays of large triacts of *L. nivea*. Either the resolution (1 μ) or sensitivity of the microanalyser, when used for X-ray analysis, is inadequate to reveal the reasons for the appearance of axial region and concentric lamination.

THE DEVELOPMENT OF SOME TYPES OF SPICULES

Description

The spicules are secreted by cells called sclerocytes of which there are usually two per spicule ray. A scleroblast is a cell which divides to form sclerocytes. According to Minchin (1896, 1898) such cells are derived from the exopinacoderm. The pupa formed immediately after the sponge larva has attached itself to the substratum exhibits developing spicules in the flat, squamous cells forming the external surface (Minchin, 1896; Duboscq and Tuzet, 1936, 1937) but at a later stage (the pre-olynthus) the surface layer has differentiated into two, an outer pinacoderm and an inner mesohyl, the latter containing immigrated scleroblasts in a hyaline matrix. The immigration of cells from the exopinacoderm continues even in the adult sponge, according to Minchin, and certainly flask-shaped pinacocytes, protruding into the mesohyl, are commonly in evidence in *Leucosolenia* (Urban, 1906; Jones, 1966) and other calcareous sponges. Nevertheless, Pavans de Ceccatty (1955) has asserted that the scleroblasts are autochthonous elements, derived by the transformation of internal blastomeres.

If the former description of their derivation is correct, the scleroblasts would be expected to have a cytoplasm and nucleus similar to those of the pinacocytes. This is so; they are also compact and irregularly rounded cells. In *Clathrina* the scleroblasts have a granular cytoplasm like the pinacocytes, the granules tending to disappear as spicule growth proceeds. The cells first associate to form trios (Fig. 5A and D). Each of the three cells then divides, so as to form a sextet of three outer cells superimposed upon three inner ones. The latter maintain close contact with the bases of the choanocytes or endopinacocytes lining the oscular tube. Between each of the three pairs of cells a spicule ray is secreted and each inner sclerocyte moves progressively outwards from the centre, lengthening the ray as it goes, while the outer sclerocytes remain at the central junction, usually in the angles between the rays. The same sequence occurs in *Leucosolenia*, (Minchin, 1908), *Sycon* and *Grantia* (Woodland, 1905) with the difference that in these forms the outer three cells are not superimposed exactly on the inner three. Minchin described the arrangement as irregular but Woodland stated

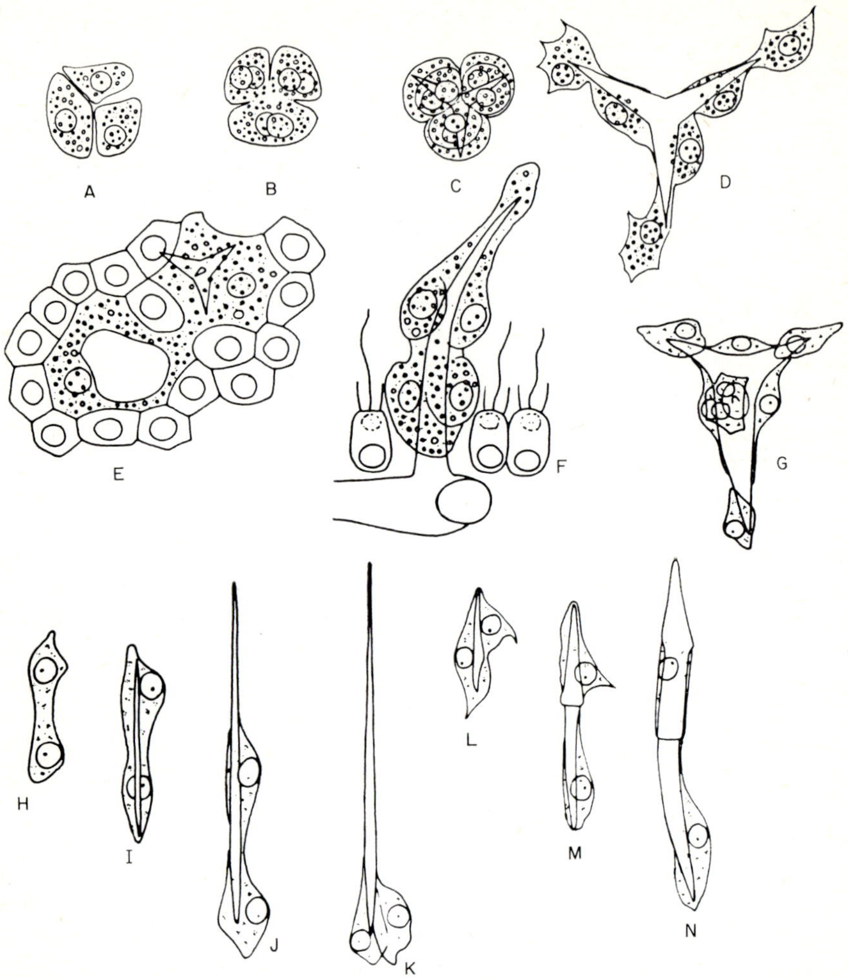

FIG. 5. The development of calcareous sponge spicules. A–D, triact of *Clathrina coriacea*; E and F, apical ray of tetract of *Clathrina contorta* (E, surface view, F, side view); G–N, spicules of *Leucosolenia complicata*; G, tetract. H–K, slender monact; L–N, lanceolate monaxon (all re-drawn after Minchin, 1898, 1908).

that the scleroblasts cleave in the centripetal and approximately radial directions so that the inner three daughter nuclei tend to be farther apart than the outer three. A possibly related arrangement with the outer three nuclei each lying in the space between two of the inner nuclei has been seen fairly frequently in *L. complicata* (Jones, 1957) and can be termed a "rosette". Whether or not the outer trio rotates immediately after cleavage until its constituent cells nestle in the gaps between the cells of the inner trio has not yet been determined. Possibly in *Clathrina* such a rotation is prevented by the stiffness of the surrounding matrix. The hypothetical tilting of the sextet in *Leucosolenia*, suggested by Jones (1954a) to explain the orientation of the optic axis in the radiate spicules, has not been confirmed by histological investigation.

It has been stated above that the three cells of the trio divide prior to the formation of the sextet, but no mitoses have yet been observed. However, cell division is most probable, for cell clusters with less than six nuclei, but with some nuclei paired and smaller than the rest, have been observed (Woodland, 1905; Minchin, 1898). The smaller paired nuclei could well be daughter nuclei and, if so, this would indicate that scleroblast division need not be synchronous. Woodland never saw two of the three nuclei of a trio smaller than the third. Minchin (1908) never observed any trios in *Leucosolenia* but they undoubtedly occur.

It is not certain whether the scleroblasts cleave completely at first or the cytoplasm of the two daughter cells remains confluent. Boundaries between the cells of the sextet cannot be detected with the light microscope and the sextet may well be a syncytium. However, at later stages of spicule formation the two cells per spicule ray are quite distinct from one another (Woodland, 1905; Jones, 1967) and spicule formation appears to be an extracellular process. The spicule primordium is first produced at the centre of the sextet and the rays then grow outwards in the direction of the enveloping inner sclerocyte. The observation of Minchin and Woodland that the primordium first consists of three separate rays which later fuse together centrally is probably erroneous and based on the examination of corroded preparations (Maas, 1906). As mentioned above, the central junction consists of relatively less pure, more easily corroded calcite. In *Leucosolenia* and *Sycon* the unpaired ray develops precociously.

Minchin called the inner three cells "founder cells", because these were responsible for the lengthening of the spicule ray, while the outer three cells he regarded as "thickener cells". These remain near the central junction for a time, closely applied to the proximal ends of the rays, but eventually they move outwards along their respective rays, apparently thickening them as they go. Woodland, however, disagreed that they had a thickener function. The founder cell leaves the tip of the fully-formed spicule ray just before the so-called thickener cell arrives (Minchin), or just as it starts to move (Woodland). In *Clathrina coriacea* the latter cell remains permanently on the ray tip but in *Sycon* both cells eventually leave the spicule ray. The clubbed tips of the rays of *C. clathrus* were attributed by Minchin (1908) to the persistence of the founder cell on them.

The development of the basal triact system of tetracts is the same as that of triacts, so that one need only consider the mode of formation of the apical ray. In *Clathrina contorta* Bowerbank, 1866 this grows from the central junction but in *L. complicata* it is joined to the unpaired ray near the central junction (Minchin, 1908). In *Sycon* both types of union are found (Woodland, 1905). In the Clathrinidae Minchin, 1898 the apical ray is secreted by a sclerocyte which is derived from a nearby porocyte (Fig. 5E) or, in the region of the oscular rim, from an endopinacocyte. Porocytes and endopinacocytes are basically similar in that, as the oscular tube grows in length, endopinacocytes become surrounded by choanocytes at the inner boundary of the oscular rim and then transform into porocytes. When producing a sclerocyte the porocyte divides, even while its pore is open, giving off a cell which resembles the porocyte in the details of its nucleus and cytoplasm. This sclerocyte applies itself to the central junction of the triact system, with its nucleus lying over the centre. The apical ray then grows out, displacing the nucleus to one side. Later the sclerocyte separates from the porocyte and its nucleus divides into two but the cytoplasm does not appear to divide. In *C. contorta* each nucleus divides once again so that the apical ray, now very long, exhibits four nuclei distributed along its length in the enveloping cytoplasm (Fig. 5F). In *C. cerebrum* (Haeckel, 1872), however, the nucleus of the original sclerocyte does not divide at all but granules, staining with picrocarmine, appear scattered about in the region of the "thorns" of calcite which beset the apical ray in this species. In *C. reticulum* (O.S., 1862) only one nucleus is found on the short apical rays, whereas two occur around the long apical rays on some tetracts (Minchin, 1898). In *Leucosolenia* the apical ray is likewise formed by an additional cell which can remain single or divide to form a pair. In *Sycon* the apical sclerocyte divides when the ray has reached two-thirds of its full length and the two cells thus produced envelope the whole ray throughout the life of the sponge (Woodland, 1905).

The monact in *L. complicata* is formed by a scleroblast which has not left the pinacoderm covering the external surface of the oscular tube or even, in some slender monacts, the endopinacoderm lining the oscular rim. According to Minchin (1908) the nucleus first divides but the division of the cytoplasm is delayed until after the start of spicule secretion (Fig. 5H–N). Woodland (1905), Hammer (1908) and Tuzet and Connes (1963) have also observed the development of monacts within binucleated cells in *Sycon* but Woodland thought that two distinct cells could participate in the production of the thick monacts in this genus. Hammer also stated that for some monacts two cells could coalesce, rather than one scleroblast divide. Developing monacts of *Grantia compressa* (O. Fabricius, 1780) (Bidder, 1898, 1937) and *Leucandra apicalis* (Urban, 1906) also bear two sclerocytes.

The long axis of the binucleated cell extends inwards into the mesohyl and the monact first appears as an internal rod parallel with this axis. Both Minchin and Woodland state that the spicule is secreted in a mould, seen as a pale streak which stretches at first from nucleus to nucleus, but the possibility of corrosion of the primordium during staining, etc. must not be overlooked. At a later stage the two cells become separate. The inner one is amoeboid in appearance and moves towards the choanoderm. This sclerocyte is responsible for the growth in length of the monact. It lies on the concave side of the developing curved monacts. Again it was termed the "founder" cell by Minchin. The other cell generally maintains contact with the pinacoderm and it thickens the monact as this is protruded through the external surface of the sponge. This "thickener" cell also forms the distal "lance-head" of the lanceolate curved monacts. The distal end of the monact is thus formed first of all. Once the lance-head has reached full thickness, the protrusion of the spicule begins. How this is achieved is unknown but the founder cell apparently ceases to move once it has reached the choanoderm. When the spicule is almost fully formed, the thickener cell moves down the shaft and both cells then leave the monact together, the thickener rounding off its end somewhat before so doing. In *Sycon coronatum* (Ellis and Solander, 1786) the founder cell leaves the monact before the thickener (Woodland) but eventually the thickener goes also. In *Leucandra phillipensis* Dendy, 1892 each monaxon has a single cell on its inner end (Dendy, 1893). Sometimes monacts may be formed by a pair of sclerocytes at sites far removed from the exopinacoderm and choanoderm in *Sycon*, in parts where the mesohyl is extensive, such as in the region of attachment of the sponge to the substratum.

Almost nothing is known of the development of the huge monaxons, triacts and tetracts of the syconoid and leuconoid species. Dendy (1893) observed very large sclerocytes attached to the rays of a large triact in the cortex of *Grantiopsis cylindrica* Dendy, 1892, and he believed that the concentric layers of the huge spicules were secreted by secondary cells (Dendy, 1891). Maas (1900) also believed that the larger monaxons of *Sycon* were completed by a number of adventitious cells. Minchin has observed two fusiform cells on the shaft of a monaxon of a species of *Clathrina*, associated with a branched cell at the apex, and another example from the same species with four shaft cells and one apical cell.

Such large monaxons, however, may have been nothing more than aberrant triacts. Urban (1906) also depicted a triact with four cells on the unpaired ray and two on each of the two paired rays. It is clear from the development of the apical rays of tetracts that it is not essential to have two sclerocytes per ray; there may be one, two or four. Maas in 1906 reiterated his view that small monacts of *Sycon* develop in a single cell.

The fate of the sclerocytes which leave the spicule rays is uncertain. Presumably they move away to start the process of spicule formation over again. Korotkova (1962) found that scleroblasts in *L. complicata* migrate between the two layers of pinacoderm constituting a healing membrane (which spreads across a gap in the wall) and produce spicules there. The pinacocytes are in an extended, flattened condition in these healing membranes, being under tension, so presumably recruitment of scleroblasts from the pinacoderm does not occur.

It is interesting that during the ontogeny of the sponge the monacts are the first type of spicules to be developed (Minchin, 1896; Maas, 1900). This is not necessarily an indication of a more primitive status, for monacts are also the first to appear on healing membranes which develop across wounds in the oscular tube of *Leucosolenia* (Jones, 1957). Since only one scleroblast is required for the formation of the monact, whereas three must meet for that of the triact, one would expect the monacts to appear earlier, once spicule formation begins. Indeed, the monact can develop *in situ* in the pinacoderm of the metamorphosed larva, as mentioned above.

Growth rates in Leucosolenia

Spicule growth rates have been determined by extending pieces of the wall of oscular tubes of *Leucosolenia variabilis* (Haeckel, 1870) in seawater beneath a 2/3″ objective and photographing the wall at intervals of about 3 h (Jones, 1959). It was found that the rays of triacts and the basal triact systems of tetracts grow slowly in length at first but that the rate rises to a steady value when the rays are about 5 μ long and then remains more or less constant up to lengths of about 70 μ at least. The unpaired ray grows precociously but the paired rays tend to catch up so that the two types of ray are growing at roughly the same rate when longer than about 20 μ. As measured on one piece of wall, the average rates for the unpaired ray were 0·34, 1·72 and 2·44 μ/h when the rays were 1·5–5, 5–25 and over 25 μ long, respectively. For the paired rays the corresponding mean rates were 0·98, 2·47 and 2·53 μ/h. The rays of aberrant triacts and T-shaped spicules grow at similar rates to those of the Y-shaped triacts on the same piece of wall.

However, the growth rates of all the spicule rays were influenced by the conditions of the experiment. Mechanical factors, for example the degree of stretching of the wall, probably affected the growth rate. Thus on a severely stretched piece the paired rays grew more slowly than the unpaired ray in the range 5–25 μ whereas another piece derived from the same oscular tube, but not so severely stretched, yielded higher growth rates on average. As similar rates were measured for a variety of rays having markedly different orientations with respect to the crystallographic axes of the spicule calcite, it can be concluded that the growth rate is independent of the type and orientation of the crystal faces at the growing points. The growth rate is probably determined by the speed at which the founder sclerocytes move over the base of the choanoderm, assuming an ample supply of material for crystallization.

It might be argued that spicule growth would always be inhibited by the technique of extending pieces of the wall which would normally be curved, but photographic records of entire, growing oscular tubes gave growth rates of the order 1·0–2·25 μ/h (Jones, 1964b). A factor which could well influence the growth rate, however, is the relative stiffness of the mesohylial matrix. There is some evidence that the matrix softens, or becomes firmer, as the physiological condition of the oscular tube alters (Jones, 1956). Even in the same tube at the same time, there is great variation in the rate of growth of spicule rays (for example, 0·42 to over 2 μ/h).

Less is known about the growth rates of monacts, but rates of 1·59 μ/h (from 40 to 78 μ) and 1·53 (from 35 to 70 μ), both at 16°C, have been measured (Jones, 1964b), while three other monacts grew in length at the apparent rates of 3·53–3·78 μ/h.

The mechanism of spicule secretion

The developing spicule rays of *L. complicata* are enclosed in a thin, retractile membrane (Jones, 1967) which presumably functions as a sort of calcium pump. The larger spicule rays are enclosed by a thicker spicule sheath, presumably consisting of the initial membrane with an outer covering of connective tissue fibrils. It was known to Minchin (1898) that the sheath becomes more distinct as the rays of triacts of *Clathrina* grow in length, suggesting that the sclerocytes also secrete the material which condenses to form the fibrils. The sheath stains with picronigrosin in the same way as the mesohylial matrix but more intensely. There seems no reason why the sclerocytes should not secrete mesohylial matrix for they, like the inwardly projecting pinacocytes from which they were derived, contain coarse-surfaced endoplasmic reticulum and a Golgi apparatus (Jones, 1966, 1967) and hence are capable of manufacturing and secreting protein. However, with the electron microscope no fibrils have been seen between a sclerocyte and the spicule ray membrane which it encloses.

Little has been discovered about the mechanism of secretion, although Maas (1904a, 1904b, 1906) long ago gained evidence for this by rearing settled larvae of *Sycandra* in artificial media. In normal seawater the larvae settled and completed their metamorphosis, producing numerous spicules on the same or following day, but in artificial media containing no carbonate ions no spicules developed. At the most, shadowy granular rods, quite free of calcite, were produced by the "pupae" instead of monaxons. Maas hence concluded that the carbon dioxide produced by the developing spongelets was either not utilizable, or insufficient, for calcite production. When calcium ions also were absent, the larval cells tended to separate and fall out before the larvae attempted to attach themselves.

The calcium and carbonate ions required for spicule production presumably penetrate to the sclerocytes via the pinacocytes covering the surface of the metamorphosed larvae, for these produce spicules before they have acquired a functional water conducting system (Maas, 1906). Recently Glover and Sippel (1967) have shown that magnesian calcites can be crystallized from seawater at temperatures down to 0°C by the addition of sodium bicarbonate. They suggest that the following reaction would account for the formation of magnesian calcites in marine organisms:

$$m\ Ca^{++} + n\ Mg^{++} + (m+n)\ HCO_3^{-} = Ca_mMg_n(CO_3)_{m+n} + (m+n)\ H^{+}$$

The magnesium molar percentage of spicules of *Leucandra nivea* is 11·2 (Jones and Jenkins, 1969, in press). From the graph given by Glover and Sippel this magnesian calcite would crystallize from a solution containing Mg : Ca in the molar ratio of about 4 : 1. The corresponding ratio for seawater is approximately 5 : 1, and the difference indicates a selective process producing the spicule calcite. Indeed this must be so, because spicules of different species growing in the same rock pool at the same time contain different amounts of magnesium.

Production rate in Leucosolenia

Some information on the factors controlling spicule production have been gained from photographic records of living oscular tubes of *L. variabilis* (Jones, 1965). The number of growing spicules present varies greatly from time to time in a particular oscular tube and at any instant the production per unit area varies with the level along a tube. The average number of new radiate spicules per 250^2 μ^2 area of the wall per day varied from 0·04 to 2·24, the maximum being four. In excised oscular tubes mounted on fine glass jets through which seawater was flowing, production rates of 95/mm^2/day, or about 5 per 250^2 μ^2, were obtained (Jones, 1958). The oscular tube grows most rapidly some distance behind the oscular rim, the growth rate diminishing towards each end of the tube. The production of new spicules per unit area tends likewise to decline towards each end of the tube. Within each area, however, the production is quite at random, both in time and space, as would be expected from a chance meeting of scleroblasts.

SPICULE ORIENTATION IN ASCONOID SPONGES

In *L. complicata* the triacts and tetracts are embedded in the mesohyl of the wall of the oscular tube, with the apical rays of the latter projecting into the cavity. The spicules all tend to be oriented in the same way, the unpaired ray pointing away from the osculum and the paired rays partially embracing the tube. The monacts project from the outer surface of the tube, the slender monacts somewhat haphazardly over most of the tube although tending to be roughly at right angles to the surface. Near the osculum they become inclined more and more in the direction of the exhalant stream, until finally, at the oscular edge itself, they lie parallel to the axis of the tube and form an oscular crown. Slender monacts can arise also from the inner lining of the oscular rim, just within the oscular border. The curved monacts tend to lie in an axial plane of the oscular tube, their distal ends tilted in the direction of the osculum and their proximal ends curving into the mesohyl (Fig. 6).

Because the form of all the spicule types bears a definite relation to the crystallographic axes of the spicule calcite, it follows from the orientation of the spicules that their crystalline substance must also tend to be oriented with respect to the morphology of the oscular tube. Thus the optic axis of the spicules tends to lie in an axial plane of the tube. In the triacts and tetracts it is inclined towards the osculum, the angle of inclination being 30–35° at the oscular edge and diminishing with increasing distance of the spicule from the edge (Fig. 7). As can be seen from the graph, the angles measured on different tubes need not be the same at corresponding levels. Those shown on the lower graph were measured on a thinner tube. For the curved monacts, in which the optic axis is approximately perpendicular to the spicule axis at the distal edge, there is no clear pattern of orientation of the optic axis along the tube, other than the tendency for it to lie in an axial plane,

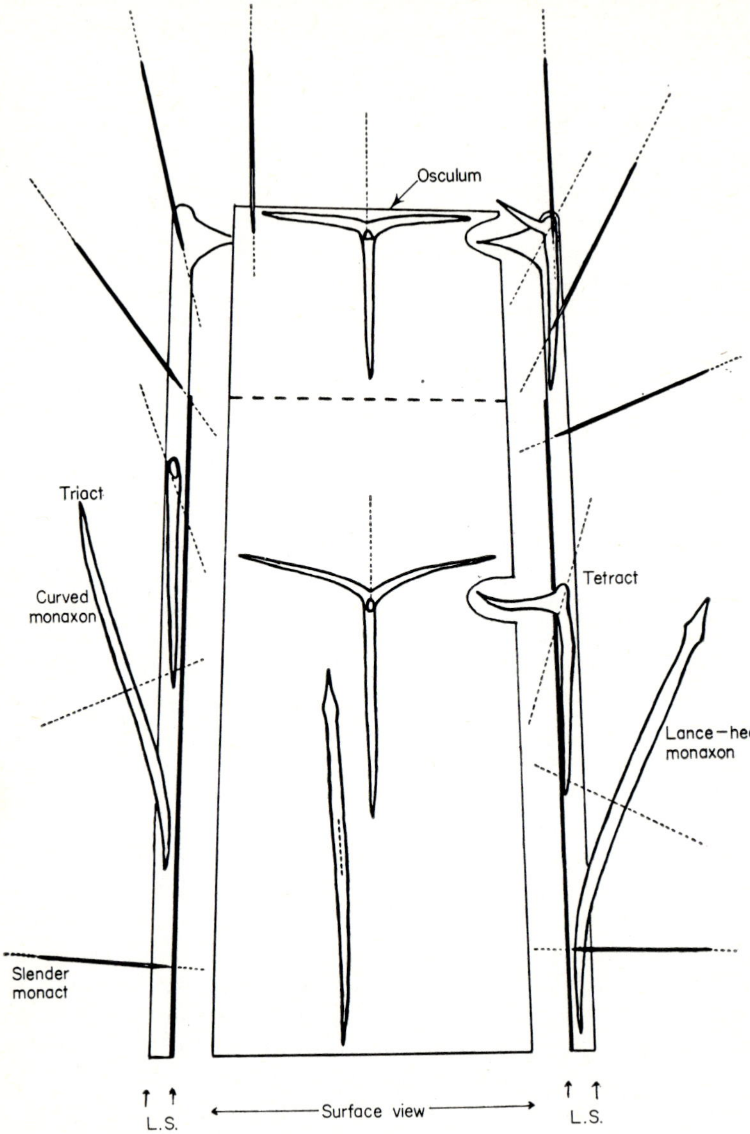

FIG. 6. Diagram summarizing the orientation of the spicules and their optic axes (shown as dotted lines) in an oscular tube of *Leucosolenia complicata*. L.S. = longitudinal section (after Jones, 1954a).

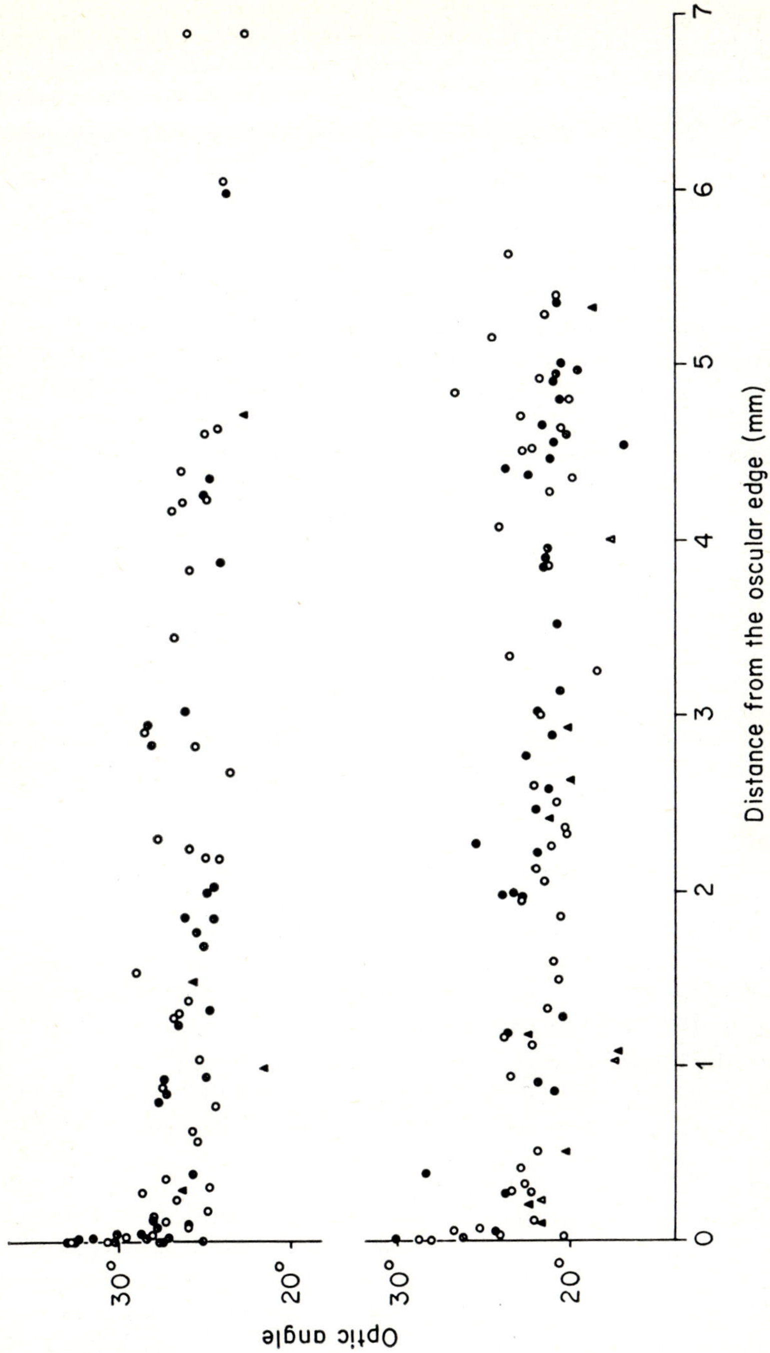

FIG. 7. The angle of inclination of the optic axis (optic angle) of triacts and tetracts in two oscular tubes of *Leucosolenia complicata*, plotted against the distance of the spicules from the edge of the osculum (after Jones, 1954a).

because the curvature of these spicules is variable. With the slender monacts the orientation of the optic axis is the same as that of the spicules themselves. As well as the optic axis, the horizontal axes of the spicule calcite must also be oriented in a definite way with respect to the oscular tube, excepting possibly the slender monacts for which there is as yet no evidence.

In the diverticula which grow from the basal parts of each oscular tube the orientation of the newly forming spicules tends to be the same as in the main tube—the triacts having their unpaired rays pointing distally, at least while the distal end of the diverticulum remains sealed. When a new osculum breaks through here the newly forming spicules in the diverticulum tend to be oriented in the opposite direction, their unpaired rays now pointing away from the new osculum. Thus a relationship appears to exist between the orientation of the radiate spicules and the direction of the water current in the cavity of the tube. However, this relationship is not consistently enforced, as Minchin (1908) discovered, for in *L. lieberkühnii* (O.S., 1866) the newly forming spicules in the long, blindly ending diverticula may exhibit the reversed orientation before the osculum has broken through.

The apparent relationship between the direction of the internal current and the spicule orientation has been tested experimentally by excising oscular tubes and mounting them on fine glass jets through which water was allowed to siphon (Jones, 1958). The results are summarized in Fig. 8. It was found that the oscular rim itself was of no direct significance in spicule orientation, for when the oscular rim had been excised and water was passed through the tube in the normal direction, the newly forming spicules exhibited the normal orientation, apart from the one or two at the cut end which had obviously been rotated by the mechanical act of cutting. When one remembers that the spicule orientation is exhibited right up to the oscular edge, at which no special glandular cells occur, one must admit that the manufacture of a chemical substance at the oscular edge and its diffusion down the tube in order to guide the groups of cells forming the spicules is unlikely.

When oscular tubes, with or without their osculum, were mounted so that the water current passed through them in the opposite direction to that before the tube was excised, a pattern of reversed orientation was obtained at the distal end of the tube. From a knowledge of the growth rates of the spicules and their sizes on these tubes, it was clear that only newly forming triacts and tetracts were concerned in the reversed arrangement, even though it is known, from photographic records, that tiny developing spicules can rotate through more than 180° (Jones, 1961a,b, 1964b). A time interval of several hours always

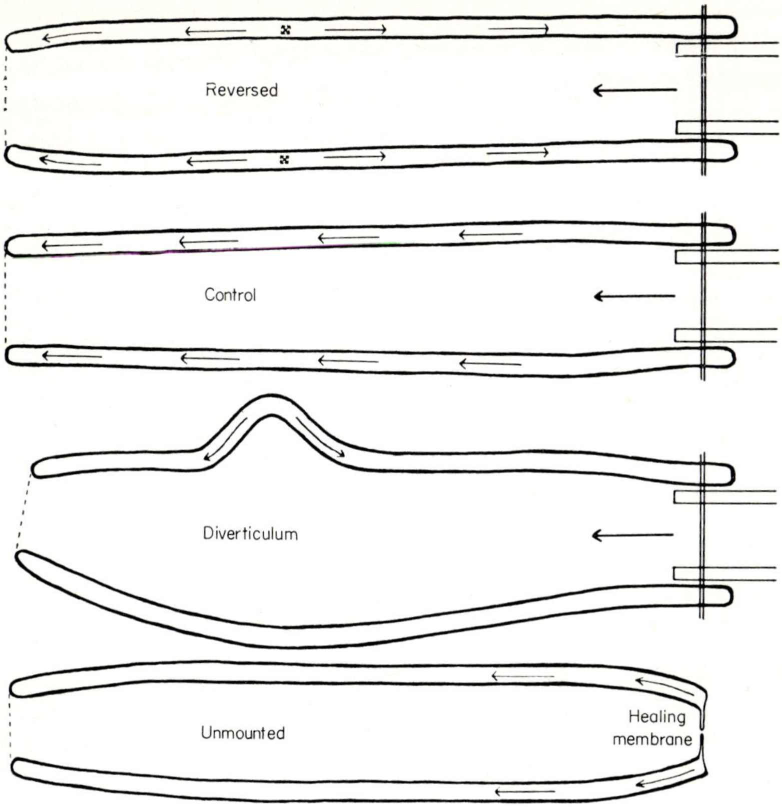

Fig. 8. Diagram summarizing the results of the capillary jet experiments. The reversed tubes were mounted with their oscular ends over the jet, the control tubes with their basal ends. In both cases diverticula occasionally developed on the tubes. The unmounted tubes were left on the bottom of the experimental dish. The prevailing orientation shown by the tri- and tetracts which developed during the experiment is indicated by the arrows in the walls of the tubes; a triact is oriented in the direction of the arrow when its unpaired ray points in the opposite direction.

occurred before tiny reversed spicules became visible in the mounted tubes. In a few cases it was also possible to deduce that a zone of reversed orientation spread back from the distal end of the tube until it reached almost to the halfway level along the tube. Thus, whereas both large and small developing, reversed, spicules could be found close to the distal end of the tube, only small ones occurred some distance from this end after the same period of time on the jet.

Simultaneous with the establishment of a zone of reversed orientation, a new oscular rim developed, apparently by a withdrawal of the choanoderm boundary from the cut edge and a lengthening of the tube. Concomitantly the spicules near to, and at, the cut edge tended to become crowded together. Undoubtedly these changes involved a displacement of the three layers of the wall (choanoderm, pinacoderm and mesohyl) relative to each other.

At the proximal end of the tubes mounted in reverse the orientation of the small, developing spicules was the same as that of the spicules existing before the tube was excised. Thus, the reversed tubes exhibited two "orientation fields" after some time on the jets, one for each end of the tube, the developing spicules in each having their unpaired rays pointing away from the nearer end of the tube. In between these two "fields" there lay a thin zone in which the orientation of the developing spicules was at random.

It is clear from the orientation in the proximal halves of the reversed tubes that there is no strict correlation between the orientation and the direction of the internal current. Moreover, from the reversed orientation in the distal halves of these tubes, it is clear that the mechanism of orientation is an easily reversible one. Thus one can rule out the possibility that static structural features, already laid down in the wall, guide the formative cells. On the other hand, there is a strong suggestion that mechanical factors play a leading role. The mounted tubes exhibited changed in shape. They tended to constrict at each end and sometimes they produced diverticula. The orientation of the developing spicules always seemed to be related to these mechanical changes, the unpaired rays of the spicules pointing up the slopes of the bulges. A hypothesis of "jelly-shear" was hence formulated (Jones, 1954a, 1965), according to which the spicules are oriented by a movement of the mesohylial matrix relative to the sextets, which are in close contact with the cell layer lining the tube. The evidence indicated that the choanocytes rested upon an endopinacoderm, to the other side of which the sextets were attached (Jones, 1961b, 1964a). The movement of the matrix could thus push the cells into an arrangement offering the least resistance to the shear. The most probable direction for the shear during the normal growth of the tube was considered to be towards the osculum and it was thought that, besides orienting the inner trio of founder cells, the shear could also tilt the axis of the sextet by displacing the outer three cells in the direction of the osculum relative to the inner three and thereby account for the tilting of the optic axis of the spicule calcite. This requires the assumption that the first calcite rudiment crystallizes by oriented overgrowth upon an organic precursor in such a way that the optic axis coincides with the direction of the mitotic spindles in the three cleaving scleroblasts. However, this hypothesis must now be abandoned. Histological studies have failed to confirm that the sextet is tilted in the manner assumed. Furthermore, an examination of the spicules of *Clathrina coriacea* and *C. lacunosa* has revealed that in these forms, despite the fact that the spicules are oriented in the same way as in *Leucosolenia*, with one ray (unpaired in *C. lacunosa*) pointing away from the osculum, the optic axis is perpendicular to the tube wall in *C. coriacea* and tilted backwards by a few degrees in *C. lacunosa*. Clearly, the same mechanism cannot orient the inner trios of the sextets and, at the same time, tilt the optic axis forwards, or backwards, or not at all. It is also significant that in the zone of confused orientation between oppositely oriented fields referred to above, the spicule form was normal. If the haphazard orientations were due to the absence of "jelly-shear" and the "jelly-shear" was the cause of the inclination of the optic axis, the optic axis here should have been perpendicular to the wall. Yet, in fact, it was tilted in the normal direction relative to the unpaired ray.

Other objections can be raised to the "jelly-shear" hypothesis. It is less probable now that the matrix shears in the direction of the osculum because it appears that the choanocytes are not the sole source of the jelly as was earlier assumed. The inwardly projecting exopinacocytes exhibit a well-developed coarse-surfaced endoplasmic reticulum and Golgi apparatus suggesting secretory activity. Moreover, it is known that a thin layer of endopinacoderm does not lie beneath the choanoderm (Jones, 1966). The changes occurring as the tube grows in length (Jones, 1965) thus require reinterpretation. However, it is still possible that displacements of the matrix determine the orientation of the sextets. The developing spicules usually move towards the osculum, if they move at all (Jones, 1964b), and the production of aberrant spicules by dislodgement or deflection of spicule primordia also indicates the occurrence of at least local displacements. Spicule rotation plays no part in the normal mechanism of orientation because the smallest spicule primordia usually exhibit the same orientation as the fully-grown spicules in their vicinity (Minchin, 1908). The mechanism thus concerns the trio or sextet and not the later stages of spicule development. Whether the cell-clusters are pushed into position, or merely guided by some influence, cannot be decided at present. A serious difficulty here is the triradiate symmetry of the trio. If one assumes that the three cells are equivalent, it becomes difficult to envisage how the trio can be guided by an axial field in the oscular tube wall, for two of the cells would have to behave in a different manner from the third. For a mechanical hypothesis, much would depend on how the cells were anchored.

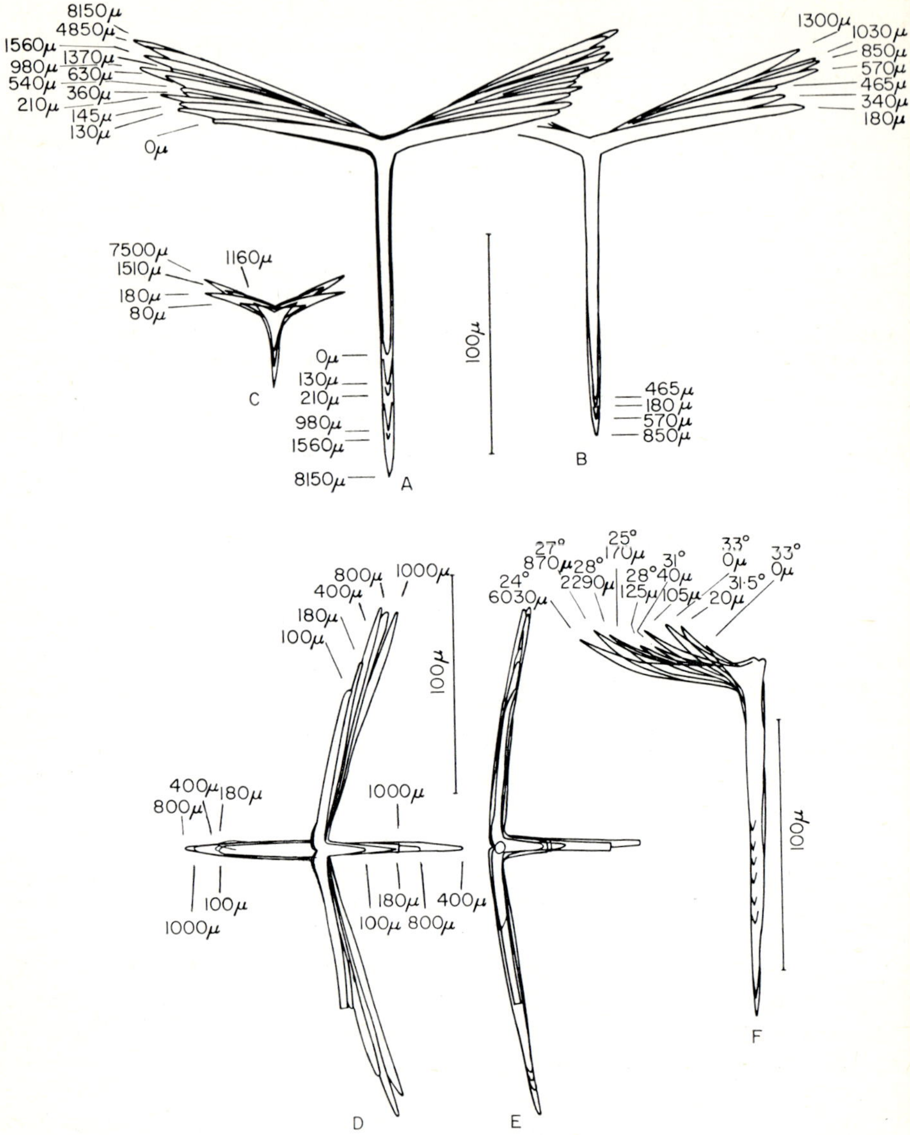

FIG. 9. Superimposed drawings showing the variation of spicule form with distance from the oscular edge in *Leucosolenia complicata*. A, B, tri- and tetracts in contact with the lining epithelium in two oscular tubes; C, developing tri- and tetracts from the tube concerned in A; D, isolated tetracts viewed along their optic axes; E, the same, in transverse view; F, tetracts seen in side view. In this case the optic angle is indicated as well as the distance from the oscular edge (after Jones, 1954b).

FACTORS CONTROLLING SPICULE FORM

The skeleton in *L. complicata* is even more complicated than the orientation of the spicules and the calcite would suggest, for the shape of the triacts and tetracts varies according to the level at which the spicule was developed along the oscular tube. Moreover, the shape varies in several different aspects. Figure 9 shows superimposed drawings of spicules from different parts of the tube, as seen in surface, transverse, side and planar views (Jones, 1954a,b). In surface view the line joining the spicule centre to the tip of the unpaired ray of the basal triact system is perpendicular to the line of sight, whereas in transverse view the two lines coincide. The side view drawings require no further explanation, while planar view is obtained by looking along the optic axis of the spicule. The radiate spicules of *Leucosolenia* belong to the secondary sagittal class of von Ebner; their paired rays do not lie in principal planes of the calcite crystal. Nevertheless they tend to appear straight in planar view, with the angle of 140–160° between the paired rays projected into a plane perpendicular to the optic axis. The angle decreases with distance from the oscular edge but no close correlation was found between this "planar angle" and the angle of inclination of the optic axis. However, the measurements were made on isolated spicules which could have been formed at different times, when the diameter of the tube was different. The form of the monacts does not vary consistently with position along the oscular tube. The curved monacts are variable but are probably not all formed at the same time, while the slender monacts offer little scope for variation, other than in size.

In *Clathrina coriacea* all the spicules are of the same type (aberrants excepted) and there is little scope for variation with position. The skeleton can be regarded as more primitive than that of *Leucosolenia*, in that there is no differentiation of T-spicules at the oscular edge. Recently I have investigated the form of the triacts of *C. coriacea* by embedding single spicules in Gurr's laktoseal and placing the laktoseal block in immersion oil between the glass hemispheres of a Vickers universal stage. Additional information on the factors controlling spicule form has thereby been obtained. In the first place, it is clear that there is only a tendency for uniplanar growth to occur. The angles between the rays in planar view need not be exactly 120° and the rays may curve gently in this aspect (Fig. 10). Previously it had been observed that the paired rays of, in particular, the triacts of *Leucosolenia* sometimes curve gently in planar view (Jones, 1954b), while von Ebner (1887) depicted some radiate spicules with paired rays curved markedly in the plane perpendicular to the optic axis. The tendency for planar growth in

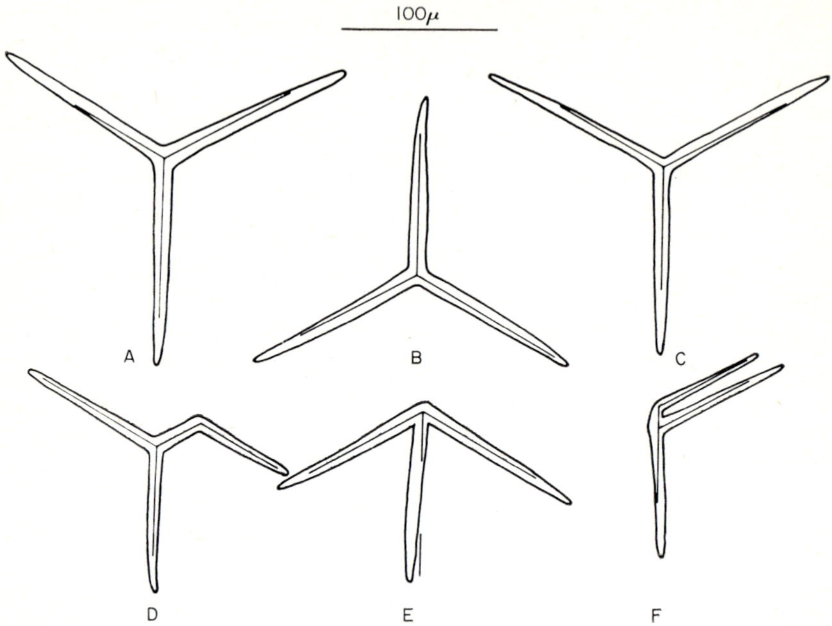

FIG. 10. Triacts of *Clathrina coriacea* viewed along their optic axis. The angles between the lines drawn along the spicule rays are 120°. Note that the angles between the rays need not be precisely 120° in this aspect (planar view) and that the rays can curve. D–F, aberrant spicules indicating the tendency for the rays to grow in planes that are at approximately 120° to each other.

both *Leucosolenia* and *Clathrina*, however, is revealed by the directions of growth of the rays of aberrant spicules (Fig. 10D–F). Aberrants with kinked rays tend to have sections of their aberrant rays parallel to one of the other rays of the spicule. In these cases the directions of growth could not have been determined merely by the arrangement of the cells of the trio at the onset of spicule secretion. Nor can extrinsic factors (e.g. structural features of the wall) be alone in guiding the cells, because photographs of developing spicules have shown that spicules with normal form, but different orientation, can develop at the same time close together in the wall and that developing spicules can rotate without there being any detectable effect on the subsequent form (Jones, 1964b). It would appear, then, that the planar growth of the rays must be intrinsically determined. To some extent the spicules with kinked rays remind one of crystals (e.g. snowflakes) bearing dendritic offshoots on the main rays, except that the number of dendrites and rays is greatly restricted for the spicules by the limited number of founder cells (one per ray). However, it is difficult to envisage how a physical process of crystallization can determine uniplanar growth, when the planar angle and the directions of the rays within the planes vary so much and by such small gradations from one to another. The spicules are not true crystals, however, in that they cleave by conchoidal fracture, and the possible occurrence of organic matter in their substance may provide the missing factor required for the explanation of planar growth. Preferential growth of crystal faces could well account for the coincidence of morphological and optic axes in the slender monacts and also for the initial growth at right angles to the optic axis in the curved monaxons (Fig. 4 K,L). It could also account for the elliptical cross-section of the paired and apical rays in tetracts, and for the growth perpendicular to the optic axis at the tips of the apical rays in *Leucosolenia* (Jones, 1955a). The calcite is certainly of significance; when calcium and carbonate ions are lacking in the medium surrounding recently attached "pupal" sycons, organic "spiculoids" of irregular form are produced instead of normal spicules (Maas, 1904a).

Granted that the rays grow in planes whose intersection coincides with the inclined optic axis, and granted that the founder sclerocytes tend to move in contact with the cells lining the oscular tube, the shapes of the paired rays of the radiate spicules of *Leucosolenia* in surface view become intelligible as the intersections of the planes with the curved wall of the tube. However, the paired rays sometimes exhibit a slight double curvature in surface view, possibly because they do not follow the curvature of the choanoderm closely near the spicule centre. In the triacts of *Clathrina coriacea* the proximal portions of the rays unite to form a shallow tripod and it seems likely that growth starts at the centre of the sextet, the rays then growing down within the founder cells towards the choanoderm before curving into a direction parallel with the choanoderm. In the arrow-shaped aberrants in this species, the middle ray grows distally in parallel with the other two rays when the spicules are observed in side view, indicating that its founder sclerocyte has moved in contact with the exopinacoderm while the other two progressed along the base of the choanoderm. The growth of the rays is clearly guided by one or both boundary surfaces of the mesohyl and the aberrant spicules of *Leucosolenia* in general are produced by abnormal orientations of the spicule primordia in relation to the curvature of the wall. Woodland (1905) has already commented on the irregularity of form of larval spicules which develop in a thick mass of sponge "jelly". Besides being guided by cell layers, the sclerocytes may also be deflected by, for example, currents. Bidder (1898) observed that the sclerocyte lies in the concavity of the curved apical ray and the same seems true for curved monaxons also, so that it is possible that the curvature is the result of mechanical deflection of the founder cell, in the former case by the water current in the oscular tube. On the other hand, the surface tension of the cell membrane might draw the cell into the concavity, if the curvature were produced by intrinsic factors, e.g., the apical rays tending distally to grow into a direction perpendicular to the optic axis. In the vicinity of the choanoderm they grow more or less at right angles to the lining before curving into this direction (Fig. 9E), indicating once again that the founder sclerocyte may be directed by the current in its vicinity because the choanocytes drive water radially inwards.

Besides the tendency for uniplanar growth of the rays and the extrinsic mechanical influences which tend to deflect the sclerocytes, the spicule form is determined by the initial setting of the calcite primordium. The inclination of the optic axis undoubtedly is a factor in the production of Y-spicules with curved paired rays in *Leucosolenia* although the form in surface view also depends on the planar angle being greater than 120°, particularly with the T-spicules. Precisely what mechanism determines the orientation of the primordium is completely unknown. Possibly the initial crystal nucleus develops a characteristic shape and is forced into the appropriate orientation by the shape of the space available at the centre of the sextet. Perhaps the calcite crystallizes epitaxially upon an organic precursor. But if so, by what mechanism or mechanisms are the space or precursor oriented?

One final factor which is obviously of importance is the number of sclerocytes concerned, for the number of founder sclerocytes controls the number of rays. The precise orientation of the sextet seems to be of relatively minor importance, at least in tubes of large diameter, for the curvature in these is so gradual that apparently symmetrical spicules can develop even in oblique orientations. It is only when the spicule primordium rotates through some degrees about the axis of the unpaired ray that markedly aberrant spicules arise. The factors which control the association and division of scleroblasts, the establishment of sites of spicule secretion and the production of special spicules in particular parts of the sponge (for example, T-spicules at the osculum, anchor spicules at the base of *Sycon*) are at present completely unknown.

Acknowledgements

For permission to reproduce figures I am indebted to *Die Österreichische Akademie der Wissenschaften* (for Figs 1B,2A,2B and 3A) and the Clarendon Press, Oxford (for Figs 2C–G,3B, 6, 7 and 9).

References

Bidder, G. P. (1898). The skeleton and classification of calcareous sponges. *Proc. R. Soc.* (B) **64**: 61–76.

Bidder, G. P. (1937). The perfection of sponges. *Proc. Linn. Soc. Lond.* 149th Session: 119–145.

Borojević, R., Fry, W. G., Jones, W. C., Lévi, C., Rasmont, R., Sarà, M. and Vacelet. J. (1968). Mise au point actuelle de la terminologie des Éponges. *Bull. Mus. natn. Hist. nat. Paris* (2) **39** (6).

Bütschli, O. (1901). Beobachtungen über Kiesel-und Kalknadeln von Spongien. *Z. wiss. Zool.* **69**: 235–286.

Bütschli, O. (1907). Ueber die Einwirkung von konzentrierter Kalilauge und konzentrierter Lösung von kohlensaurem Kali auf kohlensaurem Kalk, sowie über zwei dabei entstehende Doppelsalze von kohlensaurem Kali und kohlensaurem Kalk. *Verh. naturh. -med. Ver. Heidelb.* **8**: 277–330.

Bütschli, O. (1908). Untersuchungen über organische Kalkgebilde, nebst Bemerkungen über organische Kieselgebilde. *Abh. K. Ges. Wiss. Göttingen* **60** (3): 1–177.

Carter, H. J. (1869). Calcareous sponges. *Ann. Mag. nat. Hist.* (4) **3**: 466.

Dendy, A. (1891) A monograph of the Victorian sponges. I. The organization and classification of the Calcarea Homocoela, with descriptions of the Victorian species. *Trans. R. Soc. Vict.* **3**: 1–81.

Dendy, A. (1893). Studies on the comparative anatomy of sponges. V. Observations on the structure and classification of the Calcarea Heterocoela. *Q. Jl microsc. Sci.* **35**: 159–257.

Dendy, A and Row, R. W. H. (1913). The classification and phylogeny of the calcareous sponges, with a reference list of all the described species, systematically arranged. *Proc. zool. Soc. Lond. 1913*: 704–813.

Duboscq, O. and Tuzet, O. (1936). Les amobocytes et les cellules germinales des Éponges calcaires. *Mém. Mus. r. Hist. nat. Belg.* Sé 2: 209–226.

Duboscq, O. and Tuzet, O. (1937). L'ovogénèse, la fécondation et les premiers stades du développement des Éponges calcaires. *Archs Zool. exp. gén.* **79** (2): 157–316.

von Ebner, V. (1887). Über den feineren Bau der Skelettheile der Kalkschwämme nebst Bemerkung über Kalkskelete überhaupt. *Sber. Akad. Wiss. Wien* Abt.I. **95**: 55–149.

Glover, E. D. and Sippel, R. F. (1967) Synthesis of magnesium calcites. *Geochim. cosmochim. Acta* **31**: 603–613.

Haeckel, E. (1872). *Die Kalkschwämme.* Berlin: Georg Reimer.

Hammer, E. (1908). Neue Beiträge zur Kenntnis der Histologie und Entwicklung von *Sycon raphanus*. *Arch. Biontol.* **2**: 289–334.

Jones, W. C. (1954a). The orientation of the optic axis of spicules of *Leucosolenia complicata*. *Q. Jl microsc. Sci.* **95**(1): 33–48.

Jones, W. C. (1954b). Spicule form in *Leucosolenia complicata*. *Q. Jl microsc. Sci.* **95** (2): 191–203.

Jones, W. C. (1955a). Crystalline properties of spicules of *Leucosolenia complicata*. *Q. Jl microsc. Sci.* **96** (2): 129–149.

Jones, W. C. (1955b). The sheath of spicules of *Leucosolenia complicata*. *Q. Jl microsc. Sci.* **96** (4): 411–421.

Jones, W. C. (1956). Colloidal properties of the mesogloea in species of *Leucosolenia* *Q. Jl microsc. Sci.* **97** (2): 269–285.

Jones, W. C. (1957). The contractility and healing behaviour of pieces of *Leucosolenia complicata*. *Q. Jl microsc. Sci.* **98** (2): 203–217.

Jones, W. C. (1958). The effect of reversing the internal water-current on the spicule orientation in *Leucosolenia variabilis* and *L. complicata*. *Q. Jl microsc. Sci.* **99** (2): 263–278.

Jones, W. C. (1959). Spicule growth rates in *Leucosolenia variabilis*. *Q. Jl microsc. Sci.* **100**(4): 557–570.

Jones, W. C. (1961a). Properties of the wall of *Leucosolenia variabilis*. I. The skeletal layer. *Q. Jl microsc. Sci.* **102**(4): 531–542.

Jones, W. C. (1961b). Properties of the wall of *Leucosolenia variabilis*. II The choanoderm and the porocyte epithelium. *Q. Jl microsc. Sci.* **102**(4): 543–550.

Jones, W. C. (1964a). Photographic records of living oscular tubes of *Leucosolenia variabilis*. I. The choanoderm boundary, the choanocytes and the pore arrangement. *J. mar. biol. Ass. U.K.* **44**: 67–85.

Jones, W. C. (1964b). Photographic records of living oscular tubes of *Leucosolenia variabilis*. II. Spicule growth, form and displacement. *J. mar. biol. Ass. U.K.* **44**: 311–331.

Jones, W. C. (1965). Photographic records of living oscular tubes of *Leucosolenia variabilis*. III. Irregular growth of the oscular tube. *J. mar. biol. Ass. U.K.* **45**: 1–28.

Jones, W. C. (1966). The structure of the porocytes in the calcareous sponge *Leucosolenia complicata* (Montagu). *Jl. R. microsc. Soc.* **85**: 53–62.

Jones, W. C. (1967). Sheath and axial filament of calcareous sponge spicules. *Nature, Lond.* **214** (5086): 365–368.

Jones, W. C. and James, D. W. F. (1969). An investigation of some calcareous sponge spicules by means of electron probe micro-analysis. *Micron* (in press).

Jones, W. C. and Jenkins, D. A. (1969). Calcareous sponge spicules: a study of magnesian calcites. *Calc. Tiss. Res.* (in press).

Kölliker, A. (1864). *Icones histologicae oder Atlas der vergleichendren Gewebelehre*. I. *Der feinere Bau der Protozoen*. Leipzig: Engelmann.

Korotkova, G. P. (1962). Behaviour of the cellular elements in the calcareous sponge *Leucosolenia complicata* Mont. during regeneration. *Acta. biol. hung.* **13**: 1–30; 24 figs.

Maas, O. (1900). Die Weiterentwicklung der Syconen nach der Metamorphose. *Z. wiss. Zool.* **67**: 215–240.

Maas, O. (1904a). Ueber die Wirkung der Kalkentziehung auf die Entwicklung der Kalkschwämme. *Sber. Ges. Morph. Physiol. Münch.* **50**: 4–21.

Maas, O. (1904b). Über den Aufbau des Kalkskelette der Spongien in normalem und in $CaCO_3$ freiem Seewasser. *Verh. dt. zool. Ges.* **14**: 190–201.

Maas, O. (1906). Über die Einwirkung Karbonatfreier und Kalkfreier Salzlösungen auf erwachsene Kalkschwämme und auf Entwicklungsstadien derselbens. *Arch. Entw Mech. Org.* **22**: 581–602.

Minchin, E. A. (1896). Note on the larva and postlarval development of *Leucosolenia variabilis* with remarks on the development of other Asconidae. *Proc. R. Soc.* (B) **60**: 42–52.

Minchin, E. A. (1898). Materials for a monograph of the ascons. I. On the origin and growth of the triradiate and quadradiate spicules in the family Clathrinidae. *Q. Jl microsc. Sci.* **40**: 469–588.

Minchin, E. A. (1908). Materials for a monograph of the ascons. II. The formation of spicules in the genus *Leucosolenia*, with some notes on the histology of the sponges. *Q. Jl microsc. Sci.* **52**: 301–335.

Minchin, E. A. (1909). Sponge-spicules. A summary of present knowledge. *Ergebn. Fortschr. Zool.* **2**: 171–274.

Minchin, E. A. and Reid, D. J. (1908). Observations on the minute structure of the spicules of calcareous sponges. *Proc. zool. Soc. Lond. 1908*, **2**: 661–676.

Paine, R. T. (1964). Ash and calorie determinations of sponge and opisthobranch tissues. *Ecology* **45**: 384–387.

Pavans de Ceccatty, M. (1955). Le système nerveux des Éponges calcaires et siliceuses. *Annls Sci. nat.* (Zool) 11e série **17**: 203–298.

Phillips, F. C. (1963). *An introduction to crystallography.* 3rd Edn.: 340 pp. London: Longmans, Green & Co. Ltd.

Sollas, W. J. (1885). On the physical characters of calcareous and siliceous sponge-spicules and other structures. *Scient. Proc. R. Dubl. Soc.* **4**: 374–392.

Travis, D. F., François, C. J., Bonar, L. C. and Glimcher, M. J. (1967). Comparative studies of the organic matrices of invertebrate mineralized tissues. *J. Ultrastruct. Res.* **18**: 519–550.

Tuzet, O. and Connes, R. (1963). Recherches histologiques sur la reconstitution de *Sycon raphanus* O.S. à partir des cellules dissociées. *Vie Milieu.* **13**: 703–710; 4 figs.

Urban, F. (1906). Kalifornische Kalkschwämme. *Arch. Naturgesch.*, **72**J. (**1**): 33–76.

Woodland, W. (1905). Studies in spicule formation. I. The development and structure of the spicules in Sycons: with remarks on the conformation, modes of disposition and evolution of spicules in calcareous sponges generally. *Q. Jl microsc. Sci.* **49**: 231–282; 3 pl., 11 figs.

Symp. zool. Soc. Lond. (1970) No. 25, 125–133

ULTRASTRUCTURE OF THE SPICULES OF *SPONGILLA LACUSTRIS* (L.)

EJNAR J. FJERDINGSTAD

*Baylor College of Medicine, Texas Medical Center Houston, Texas, U.S.A.**

SYNOPSIS

The ultrastructure of the silica skeleton needles (spicules) of the freshwater sponge *Spongilla lacustris* (L., 1759) has been investigated using electron microscopy of ultrathin sections of osmium tetroxide fixed, Epon embedded material. In contrast to earlier conclusions based on conventional and polarized light microscopy, it appears that the spicules are formed of submicroscopic quartz crystallites arranged in a plane perpendicular to the longitudinal axis of the spicule. Treatment of the sections with hydrofluoric acid removes the crystallites completely, leaving other spicule structures intact. The implications of these results and the conflicting earlier evidence are discussed, together with some contradictory results obtained with electron diffraction methods.

INTRODUCTION

The freshwater sponge *Spongilla lacustris* (L.) (Class *Demospongiae* Sollas, 1886, order *Haplosclerinae* Topsent, 1928, family *Spongillidae* Gray, 1848) belongs, as do all other freshwater sponges, to a group of Porifera characterized by the possession of silica skeleton needles or spicules. *S. lacustris* possesses two types of spicules, the megascleres, 200 μ long and 6–8·5 μ wide, and the microscleres, 60–100 μ by 3·5–6 μ. The first of these possess a smooth surface, the latter are provided with numerous small spines.

As established by a number of authors, for example Carter (1874a, 1874b), Woodland (1908) and Schulze (1925), these spicules are formed by a cell type called the scleroblast. The scleroblast exhibits the same features as those sponge cells generally termed archaeocytes, i.e. a large nucleus with a very prominent nucleolus and pronounced cytoplasmic basophilia (Wintermann-Kilian, 1951). Such cells arise without much further differentiation from the binucleate gemmule cells.

Spicule formation has been described by several authors such as Carter (1874a, 1874b), Woodland (1908) and Minchin (1909) and more recently in its very first stages by Lévi (1963). The initial step is the

* The investigations reported here were carried out at the Institute of General Zoology, University of Copenhagen, Copenhagen, Denmark.

formation of an organic "axial thread" upon which silica is then deposited. The complete formation of a spicule takes about 100 h, during which time it grows both in length and in width (Jørgensen, 1944).

Chemically it has been established that siliceous spicules consist of 92% silicon dioxide, 7% water, and trace quantities of magnesium, potassium and sodium (Bütschli, 1901). It has been shown (Jørgensen, 1947) that the sponge can use only dissolved silicic acid and not solid quartz.

When viewed through the light microscope the spicule exhibits very little structural organization. It appears homogeneously transparent, the axial thread being the only clearly visible structural element. From this appearance, supported by the absence of birefringence under the polarization microscope, a general conclusion has been drawn that the silica is present in the spicule in an amorphous state, as is the case in the mineral flint, and possibly is laid down in concentric layers separated by small amounts of organic matter (see reviews by Minchin, 1909 and by Schulze, 1925). Minchin expressed this very firmly, "... all investigators are unanimous in stating that the substance of the spicule is perfectly amorphous and non-crystalline". Lévi (1963), who did not, however, continue investigation of spicule formation as far as the deposition of silica, affirmed his belief in this concept. Still more recently the results of an electron microscope study of the silica spicules of the freshwater sponge *Heteromyenia* were described by Drum (1968). He was unable to obtain thin sections of the spicules, however, and his micrographs of spicule replicas do not show any structures that are not known from light microscope observations. He apparently believed the silica to be present as opaline deposits, i.e. in the amorphous state. As will be shown in the next sections, this is very far from being the case, at least as far as *Spongilla lacustris* is concerned.

MATERIALS AND METHODS

The sponges used in this study were obtained from the river Mølleå, north of Copenhagen, and were fixed immediately after arrival at the laboratory. Small pieces were immersed in 2% osmium tetroxide buffered to pH = 7·2 with collidin-hydrochloric acid according to Bennett and Luft (1959) for 1 h and subsequently embedded in Epon (Luft, 1961). Ultrathin sections were cut with an LKB Ultrotome, using a diamond knife. Even so, very few useful sections were obtained, since the brittle spicules tended to shatter completely rather than section. Some slight shattering, however, did prove very advantageous in the elucidation of the spicule structure. Five hundred Å sections were found to be most suitable, for in thicker sections the spicules appeared nearly opaque. In order to make cytoplasmic structure stand out with sufficient detail it was found necessary to use a double contrasting method, combining uranyl acetate and lead hydroxide staining.

After fixation, some pieces of sponge were treated with hydrofluoric acid in ethanol to dissolve silica (a few drops of 30% hydrofluoric acid were added to the 70% dehydrating ethanol) before embedding. Such tissue was extremely easy to section, even when a glass knife was used. However, this treatment resulted in rather extensive membrane damage.

The sections were picked up on carbon stabilized Parlodion films and observed in a Siemens Elmiskop Ib. For the intact spicules the high accelerating voltages 80 kV and 100 kV were employed to give better penetration of the siliceous matter. This resulted, as might have been expected, in a lower contrast of cytoplasmic structures. Forty kV were used for the material treated with hydrofluoric acid.

OBSERVATIONS

The structure of the intact spicule

Figure 1 shows a cross section of a partially shattered spicule. The axial thread appears as a finely granulated, not very dense, structure about 3300 Å wide. The micrographs do not give much indication of its composition but it is certainly very different from the cytoplasm of the surrounding scleroblast.

Around the axial thread the siliceous matter of the spicule is arranged as elements 9500 Å long and 1500 Å wide on average, with pointed ends. These units all lie parallel to one another, with their longitudinal axes parallel to the smaller diameter of the somewhat oval cross-section of the spicule. The silica units form two rows from the periphery of the spicule to the surface of the axial thread. At the common edge of the rows, the triangularly pointed ends of the units are seen to interlock tightly in the undamaged parts of the spicule. In Fig. 1, the outer border of the outer layer of units is not sharply defined and does not show the pointed ends. This may represent a stage of incomplete formation of the outer layer. In other preparations the outer edge of the peripheral row is sharply demarcated, though not triangular.

Very little material seems to be present between the individual units. Although in some preparations there is evidence of a slightly granular, less electron dense, substance occurring in a thin layer (Fig. 2).

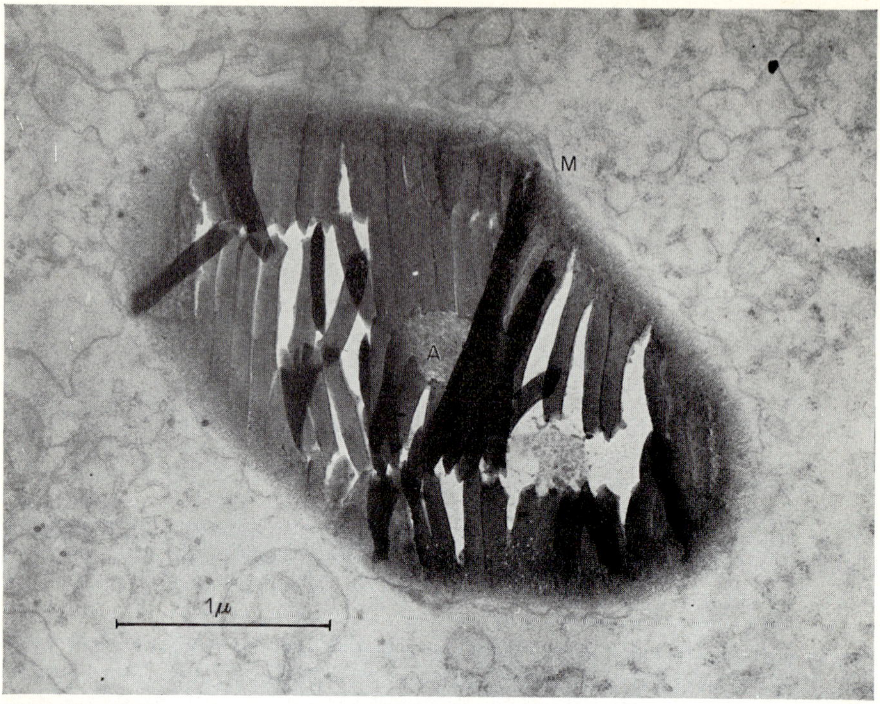
M
A
1μ

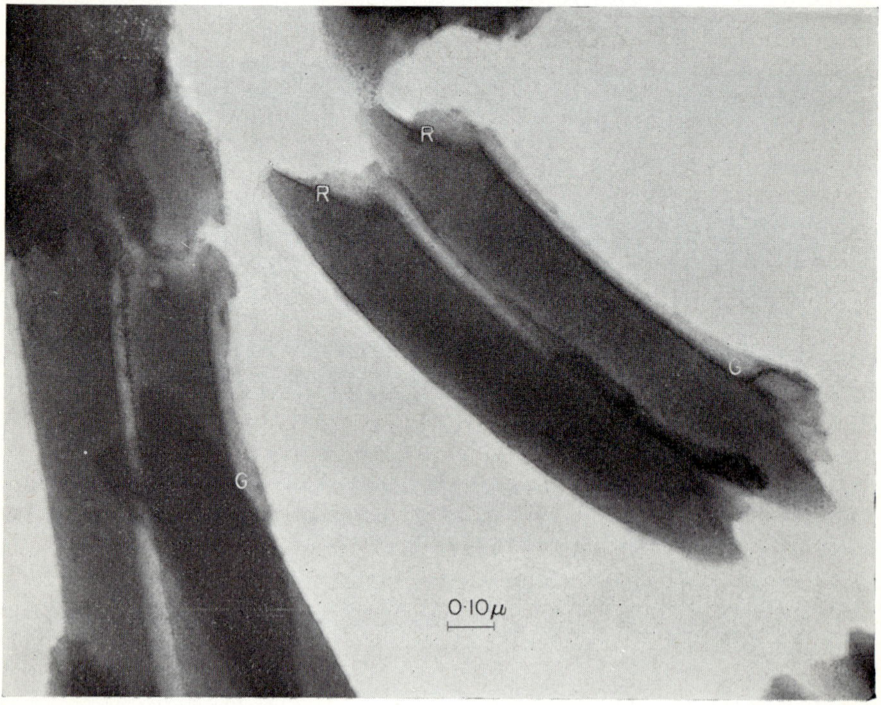
R
R
G
G
0·10μ

The spicule is enveloped in a membrane of the common unit membrane type (Fig. 1).

Figure 2 shows individual units at a higher magnification. No internal structure is evident. However, the pointed ends of the units appear to be pyramidal, with sharp ridges between the faces of the pyramids. The slightly curved shape of the units is probably due to deformation during sectioning, which is possible even in very brittle materials at the dimensions in question.

The structure of the spicule after treatment with hydrofluoric acid

Figure 3 shows what remains of the spicule after treatment with hydrofluoric acid. There is no visible trace of the units, confirming that these must consist of very pure silica and contain, at most, minute amounts of organic material.

In the centre of the cavity left by the removal of the siliceous material the axial thread may still be seen, showing no apparent damage as a result of the treatment.

The spicule cavity in Fig. 3 is surrounded by a large number of rather small fibrillae, which exhibit a definite periodicity (Fig. 4). These are parts of the spongiolin which, by binding the individual spicules together, participates in the formation of the skeleton of the sponge. The striking similarity of this material to collagen has been pointed out by Gross, Sokal and Rougvie (1956). However, its periodicity of about 150 Å in *S. lacustris* is much shorter than might be expected (but see e.g. Picken, 1960 and Borojević and Lévi, 1967).

The spongiolin fibrillae form a dense sheath around the spicule for about 3000 Å. There is no evidence of any unit membrane having surrounded the spicule at this stage, which is probably a much later stage than that shown in Fig. 1. The latter probably depicts a spicule still in process of formation. This is indicated by the smaller diameter of the spicule in Fig. 1 — 3 μ compared to a diameter of 4 μ for the spicule in Fig. 4. Unfortunately, it was not possible to obtain sections of similar areas in the untreated sponge because of the high density of spicules in these areas. This high density presented insuperable difficulties to the sectioning techniques.

FIG. 1 (Upper facing page). Slightly oblique transverse section of a spicule at intermediate magnification. The spicule is somewhat damaged by the sectioning. This makes even more obvious its composition of regular "units". A, axial thread; M, membrane limiting the spicule from the cytoplasm of the scleroblast. × 28 400.

FIG. 2 (Lower facing page). Individual spicule units at high magnification. R, ridge between faces of the pyramidal ends of the units; G, granular, less electron dense material, possibly cementing together the units of the intact spicule. × 62 000.

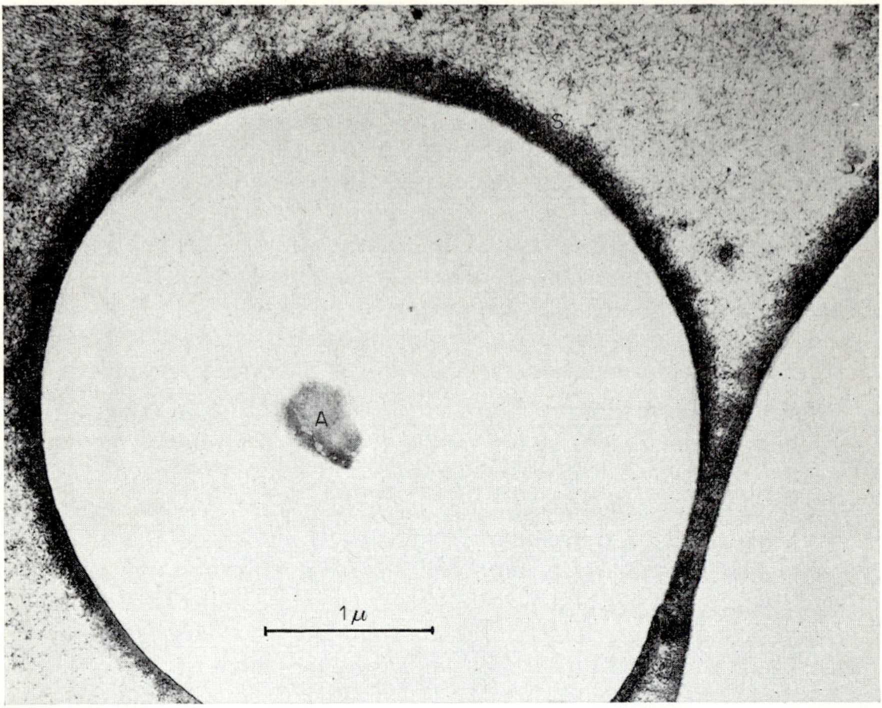
S
A
1 μ

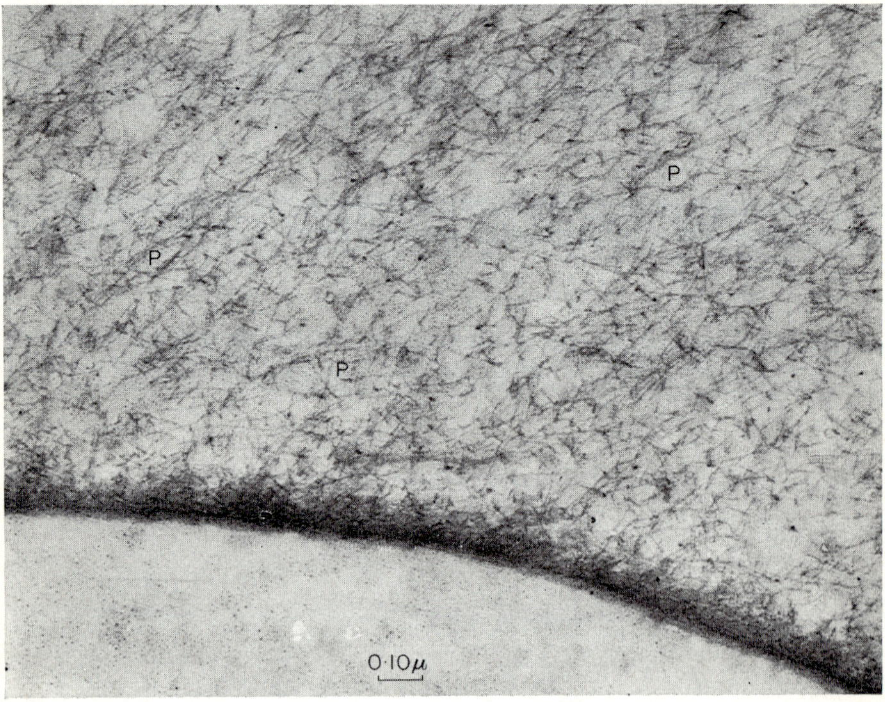
P
P
P
0·10μ

DISCUSSION

The immediate conclusion to be drawn from these observations is that the structure of *S. lacustris* spicules deviates markedly from the description given by earlier investigators. The siliceous substance is certainly not perfectly homogeneous, nor is it deposited in any pattern of homogeneous layers surrounding the axial thread. Instead, we find a bilaterally symmetrical arrangement of parallel "silica units" of extremely regular structure. The nature of these units remains to be elucidated.

An explanation suggests itself if the trigonal trapezohedric crystal structure of common quartz is considered. It is readily apparent that a thin section of such a crystal would possess the shape of the individual silica units seen in the electron micrographs. Considering only morphological evidence, it would therefore seem most likely that the spicules are composed of regularly arranged, submicroscopic, quartz crystallites. Thus, they would be similar to the spicules of calcareous sponges, the crystalline nature of which has long been recognized (see Jones, 1967). The complete solubility of the units in hydrofluoric acid would, at least, confirm that they are made of pure silica.

In order to confirm the possibility of the crystallinity of these units, some investigations with electron diffraction techniques were carried out. However, it was not possible to obtain any consistent diffraction patterns, neither when using "selected area diffraction" of a single sectioned unit, nor when applying the "powder diagram" method to crushed spicules which had been isolated from fresh, or fixed, sponge tissue by treatment with sodium hypochlorite. A few faint diagrams were obtained but these were far from being comparable in clarity with the very clear and consistent patterns obtained with similar preparations of pure quartz (Merck, analytical grade). Thus, the diffraction method can hardly be stated to confirm the concept of crystallinity of the units. This is in good agreement with the failure of polarized light to indicate crystallinity. There remains the puzzling question of the real nature of these structures, a question to which no answer can be given in the present paper.

FIG. 3 (Upper facing page). Section of spicule after treatment with hydrofluoric acid. No trace is left of the silica units. The axial thread (A) is unchanged. The spicule is surrounded by a sheath (S) of spongiolin gradually merging with the looser mesh of similar fibrillae binding the spicules together. × 21 400.

FIG. 4 (Lower facing page). Part of the same section as in Fig. 3 seen at higher magnification. The period-city of spongiolin fibrillae is evident (P). × 56 400.

It has already been stressed that the "units" or "crystallites" are arranged in parallel, closely packed rows instead of being deposited radially, in contrast to the description of "concentrous layers" by earlier authors. Such a parallel arrangement is relevant to the mechanical strength of the spicule, as it is the only manner of arrangement in which spaces between the units can be eliminated. That the units are bound together firmly is indicated by the considerable force required to grind isolated spicules in a mortar and also by the fact that, although spicules will break if ordinary paraffin sections are made, this is generally a clean break with very little shattering or fragmentation. Possibly the small amounts of less electron dense material between the units serve to cement them together. This material may also be the site of the trace amounts of magnesium, potassium and sodium.

The formation and structure of the axial thread has been described in detail by Lévi (1963), who found that in *Mycale contarenii* (Martens, 1824) it arises at the contact area between two large flattened vacuoles with a thickened limiting membrane. The newly deposited material was observed to consist of 70–100 Å helical fibrillae. The present observations have not added anything to this description. However, an interesting similarity to some electron microscope observations on calcareous sponge spicules by Jones (1967) may be noted. Observing material decalcified by acid ethanol during embedding, he stated that "the cavities formerly occupied by the spicule rays contained no trace of structure whatsoever". Even so, two cavities in his Fig. 5 clearly show structures very similar to the axial thread as observed in siliceous spicules in this study and of about the same diameter (about $0 \cdot 3\,\mu$). Consequently, investigation of the ultrastructure of the intact calcareous spicule appears to be highly desirable. Considering the much lower degree of hardness of this material, such an investigation should not pose too many problems.

Finally, it may be noted that the transport of silicate through the membrane surrounding the spicule and the mechanism of the regular deposition of the units pose some interesting questions.

Acknowledgements

The author would like to thank Professor D. v. Wettstein for lending him the diamond knife used in the investigations.

Thanks are also due to Professor H. V. Brønsted for helpful discussion and suggestions.

References

Bennett, H. S. and Luft, J. H. (1959). S-collidine as a basis for buffering fixatives. *J. biophys. biochem. Cytol.* **6**: 113–114.

Borojević, R. and Lévi, P. (1967). Le basopinacoderme de l'éponge *Mycale contarenii* (Martens). Technique d'étude des fibres extracellulaires basales. *J. Microscopie*, **6** (6): 857–862.

Bütschli, O. (1901). Einige Beobachtungen über Kiesel- und Kalknadeln von Spongien. *Z. wiss. Zool.* **69**: 235–286.

Carter, H. J. (1874a). On the nature of the seed-like body of *Spongilla*; on the origin or mother cell of the spicule; and on the presence of spermatozoa in the spongida. *Ann. Mag. nat. Hist.* (4) **14**: 97–111.

Carter, H. J. (1874b). Further instances of the sponge spicule in its mother cell. *Ann. Mag. nat. Hist.* (4) **14**: 456–458.

Drum, R. W. (1968). Electron microscopy of siliceous spicules from the freshwater sponge *Heteromyenia*. *J. Ultrastr. Res.* **22**: 12–21.

Gross, J., Sokal, Z. and Rougvie, M. (1956). Structural and chemical studies on the connective tissue of marine sponges. *J. Histochem. Cytochem.* **4** (3): 227–246.

Jones, W. C. (1967). Sheath and axial filament of calcareous sponge spicules. *Nature, Lond.* **214**: 365–368.

Jørgensen, C. B. (1944). On the spicule formation of *Spongilla lacustris* (L.) I. The dependence of the spicule formation on the content of dissolved and solid silicic acid of the milieu. *Biol. Medd. Kgl. Dansk. Vidensk. Selsk.* **19**: 1–45.

Jørgensen, C. B. (1947). On the spicule formation of *Spongilla lacustris* (L.) and *Ephydatia fluviatilis* (L.). II. The rate of growth of the spicule. *Biol. Medd. Kgl. Dansk. Vidensk. Selsk.* **20**: 1–21.

Lévi, C. (1963). Scléroblastes et spiculogénèse chez une éponge silicieuse. *C. r. hebd. Séanc. Acad. Sci., Paris* **256**: 497–498.

Luft, J. H. (1961) Improvement in epoxy resin embedding methods. *J. biophys. biochem. Cytol.* **9**: 409–410.

Minchin, E. A. (1909). Sponge spicules, a summary of present knowledge. *Ergebn. Fortschr. Zool.* **2**: 171–274.

Picken, L. E. R. (1960). *The Organisation of Cells and other Organisms.* 629 pp.; 34 pl., 113 figs. Oxford: The Clarendon Press.

Schulze, P. (1925). Zum morphologischen Feinbau der Kieselschwammnadeln. *Z. Morph. Ökol. Tiere* **4**: 615–625.

Wintermann-Kilian, G. (1951). Entwicklungsphysiologische Untersuchungen an Süsswasserschwämmen. *Zool. Jb.* **71**: 427–486.

Woodland, W. (1908). Studies in spicule formation VIII. Some observations on the scleroblastic development of hexactinellid and other siliceous sponge spicules. *Q. Jl microsc. Sci.* **52**: 139–157.

Symp. zool. Soc. Lond. (1970) No. 25, 135–162

THE SPONGE AS A POPULATION: A BIOMETRIC APPROACH

WILLIAM G. FRY

Marine Science Laboratories, Menai Bridge, Anglesey, Wales

SYNOPSIS

A review of the literature reveals that data on sponge spicule complements have not been exploited to the full for taxonomic purposes. The reasons for this lack of exploitation are discussed and it is concluded that more whole-hearted acceptance of the concept of sponges as mixed populations of cells and cell-products permits logical development of powerful taxometric procedures. The logic of one such procedure, which involves the Kolmogorov–Smirnov test, and which has been programmed for electronic computation, is demonstrated with data from four populations of *Ophlitaspongia seriata*. The procedure not only discriminates between populations which might be expected, from their spatial and ecological distinctions, to differ morphologically, but also isolates key characters together with their powers of discrimination.

INTRODUCTION

Accuracy of descriptive terminology is a limiting factor in any taxonomic work, for the more imprecise the terminology the fewer are the differences which can be recorded. However, descriptive terminology must not only be precise; it should also be of such a form that it can be used to record degrees of difference between individual specimens, populations or taxa.

Degrees of difference between such items are usually measured either as the proportion of differences in a large number of characters, or as degrees of difference in single characters. The first type of measurement —frequently referred to as a measure of "overall similarity"— is that normally associated with numerical taxonomy. The second type is that normally associated with the construction of taxonomic keys and the writing of taxonomic descriptions. However, a measure of "overall similarity" involves a large number of decisions about degrees of similarity in single characters. Each of these decisions is made in terms of the probability that two character states are the same, although the fact that such decisions are statements of probability is either ignored or obscured by preoccupation with concepts of homology and analogy.

There is a considerable wealth of literature dealing with methods of assessing measures of overall similarity but in recent years relatively little has been written about the difficulties of assessing degrees of difference between individual characters. Some exceptions to this generalization are the works of Cracraft (1967), Davis and Heywood (1963), Fry (1964), Goodall (1966), Inglis (1966), Kendrick (1964, 1965), Kendrick and Weresub (1966), Proctor and Kendrick (1963), and Sneath (1967), which point to the need for a re-examination of the real meaning of statements about differences between character states.

There is a general opinion amongst taxonomists, recently most clearly stated by Sokal (1965), that statements about degrees of taxonomic differences should not be couched in terms of probability—and yet every statement about a difference between the character states of two items can be regarded as a cryptic statement of probability.

In this paper it will be shown that not only can probability statements be used to assess differences between sponges but also that the organization of sponges is such that we must regard them for taxonomic purposes as statistical populations of cells and cell products. In so doing we must describe and compare these populations in ways which leave us no alternative other than to make such assessments of difference in terms of probability. Unless sponge taxonomists are prepared to recognize this, it seems unlikely that sponge taxonomy will ever become a truly precise and coherent body of knowledge.

Within the framework of this attitude there is little room for discussion of the significance of the outward appearance of sponges. It will suffice to say that there exist descriptive vocabularies for these features which, by virtue of their adjectival verbal nature, are extremely imprecise and therefore of only limited usefulness in making taxonomic assessments. Instead, discussion will be limited in this paper to spicule frequencies and dimensions and also to one particular aspect of the aquiferous system. This is not to say that all features of sponge cells and other ultrastructure are excluded from consideration. On the contrary, the conclusions of this discussion undoubtedly apply with equal force to cell characters and, while I would contend that the electron microscope is not an essential tool for routine taxonomic work, there can be no doubt that it is certain recent studies of sponges with the electron microscope which should compel us to pay more attention to the concept of sponges as populations of cells and their products. I refer in particular to recent work by Borojević and Lévi (1964, 1965, 1966) which has confirmed in the most striking way the inference of other writers (e.g. Brien 1943, 1967; Brien and Meewis, 1938; Maas, 1893; Meewis, 1940) that it is the large numbers of interactions between individual cells which controls the form and successful functioning of the sponge.

Such a picture is highly reminiscent of mixed populations of whole organisms and, as rigorous statistical procedures have proved so fruitful in ecology in the comparison of ecosystems, there seems to be no reason why some of the ecologists' statistical techniques should not prove fruitful in comparing populations of sponge cells and their products.

MATERIAL STUDIED

The data on which this study is based were obtained from the encrusting sponge *Ophlitaspongia seriata* (Grant, 1826). The species was investigated because it is generally accepted as being the only species of its genus in European waters; the study is thus of intra-specific variation. In addition, it is a relatively abundant form, growing

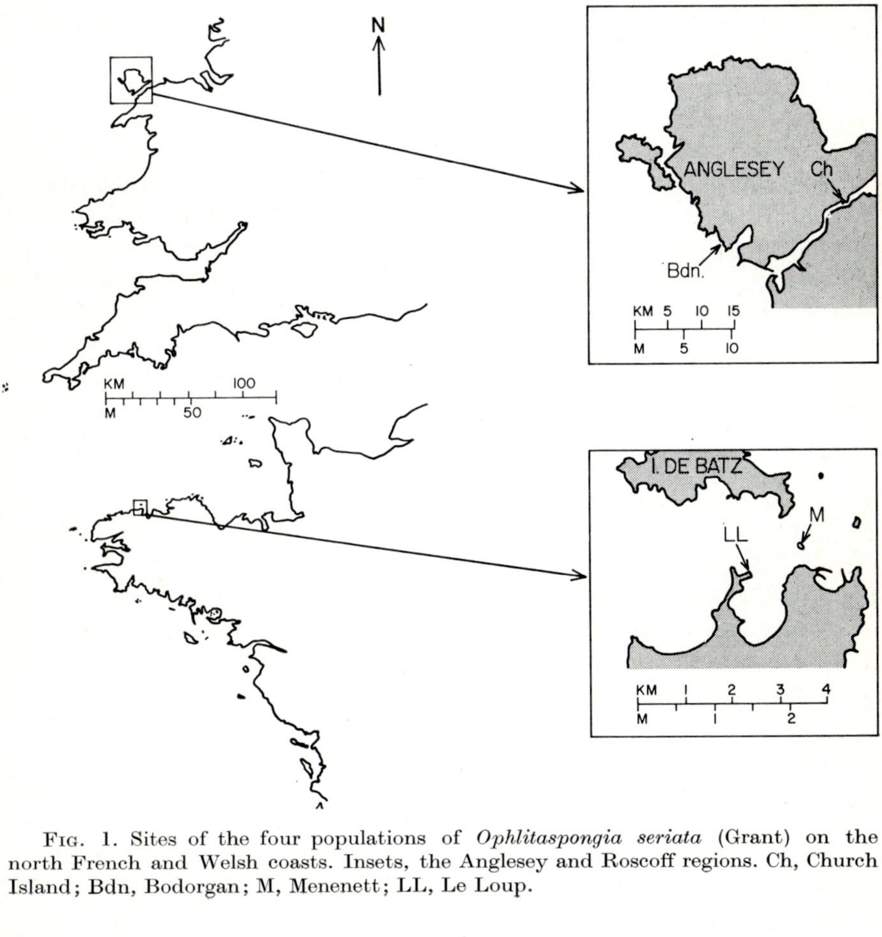

FIG. 1. Sites of the four populations of *Ophlitaspongia seriata* (Grant) on the north French and Welsh coasts. Insets, the Anglesey and Roscoff regions. Ch, Church Island; Bdn, Bodorgan; M, Menenett; LL, Le Loup.

in the readily accessible environment of the lower littoral and upper sublittoral. The material was collected from four regions (Fig. 1); two on the North coast of Wales, at Church Island and Bodorgan, and two in the vicinity of Roscoff, at the rock outcrops Menenett and Le Loup.

The Menai Straits, in which Church Island is situated, are traversed by strong tidal currents producing a residual flow through the south-west exit of the Straits and along the Anglesey coast towards Bodorgan (Admiralty Chart 1464; Buchan, Floodgate and Crisp, 1967; Harvey, 1965, 1968; Harvey and Spencer, 1962). The physical regimes at Church Island and Bodorgan differ in that the water of the Menai Straits has lower salinity, higher mean surface temperatures (Hughes, 1966) and much higher silt content in the larval liberation period than has the open water of the Irish Sea. At Church Island the sponges occur, in thick *Laminaria* beds, on large boulders which are never exposed to violent wave action. The Bodorgan collecting site is fully exposed to the open Irish Sea.

At Le Loup the sponges occur on large tumbled boulders, amongst *Laminaria* holdfasts, while Menenett is a relatively smoothly eroded rock outcrop. Both Roscoff sites are very exposed.

The average swimming period of the larvae of *Oph. seriata* from North Wales specimens is approximately 24 h at the mean water surface temperature for the larval season and it seems reasonable to suppose that Church Island specimens' larvae can reach Bodorgan, and vice versa, and that similarly there can be an exchange of larvae between Menenett and Le Loup. On the other hand, a direct exchange of larvae between North Wales and the Roscoff region cannot be expected.

If there is geographical variation in the species it would be reasonable, therefore, to expect to find the two Roscoff populations highly similar overall and the two North Wales populations highly similar to each other but both markedly dissimilar to both the Roscoff populations. In addition, it might be expected that the Church Island population would differ more from the Bodorgan population than the Menenett from the Le Loup, either because of greater environmental differences, or partial genetic isolation, or both.

METHODS

Analysis of the aquiferous system

The first application of the concept of sponges as populations was to the oscula. Use of Fisher's Coefficient of Dispersion according to the methods of Blackman (1942) and Salt and Hollick (1946) revealed that in 30 specimens of *Oph. seriata* the oscula were overdispersed. The nature of dispersion of organisms or other items on plane surfaces is not very readily susceptible to numerical analysis (see Johnson, 1958), especially when—as is the case with the oscula of *Oph. seriata*—the items are few in number. An analogue of the method employed by Crisp (1961), by which the frequency distribution of all distances between oscula is compared with the theoretical distribution of such distances computed from the mean of the observed minimum distances between nearest-neighbour oscula ($\bar{x}io$), revealed that in all the specimens studied the frequency distributions of *io* are positively skew. Figure 2 shows an example of the way in which only a minor departure from normality in the distribution of minimum interoscular distances reduces the power of Crisp's analytical method when applied to a small number of data.

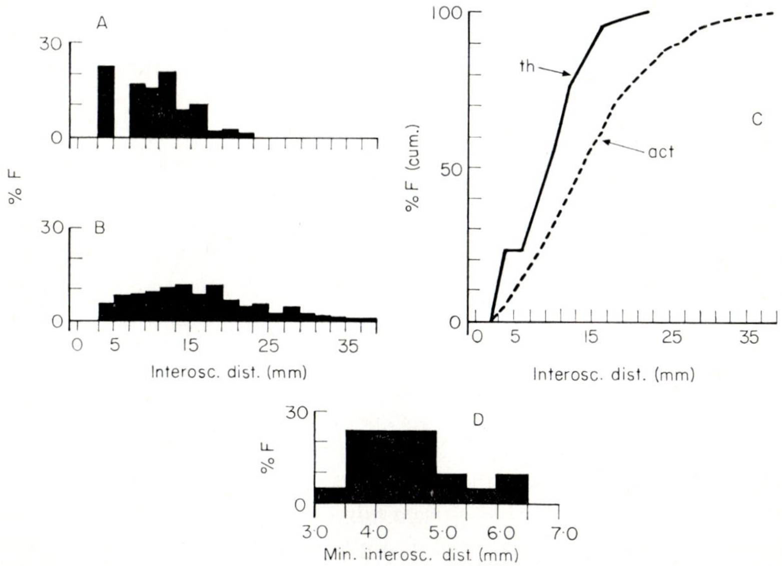

FIG. 2. The dispersion of oscula in a specimen of *Oph. seriata*. A, theoretical percentage frequency distribution of all interoscular distances based on the mean minimum interoscular distance ($\bar{x}io$ = 4·61 mm) of Specimen 16; B, actual percentage frequency distribution of all interoscular distances on Specimen 16; C, cumulative percentage frequency distribution curves based on A and B. *th* represents distribution A, *act* represents B. *th* and *act* can be considered to be derived from the same population only at $P < 0{\cdot}01$; D, percentage frequency distribution of *io* values from Specimen 16.

Such positive skewness is readily explicable in terms of observed phenomena. Most specimens of this sponge are found growing around and over objects such as barnacles, hydroids, tube-worms and algae.

Unless the sponge can overgrow such projections, any regular arrangement of oscula is bound to be disturbed and there will be frequent need for the sponge to increase *io* values. If the proper functioning of the sponge requires that increase in some interoscular distances requires compensating decreases in others, then we should expect to find markedly kurtose, but symmetrical, frequency distributions of *io* in every sponge. Such kurtose frequency distributions of *io* have not been observed, while distributions approaching the normal have been found only in those very few specimens growing on clean rock surfaces. From these observations it appears reasonable to regard specimens of *Oph. seriata* as populations of oscula and to consider the formation of an osculum as the response to a need for an exit for water drawn in and processed by a certain volume of sponge. Such volumes can be referred to as aquiferous modules (*am*). Considering the sponge as a population of aquiferous modules overcomes the difficulties of trying to trace the boundaries of influence of individual oscula and of describing plasticity in the aquiferous system—difficulties which have greatly inhibited the use of data on oscula of encrusting sponges.

The process of growth at the sponge edge is another reason why we should expect the frequency distributions of *io* values to be positively skew. Bearing in mind the functional definition of aquiferous modules, and assuming for convenience that such modules have circular perimeters, then, in an encrusting sponge like *Oph. seriata*, each module will be a cylinder with radius of $\frac{1}{2}io$. If *io* is totally invariable the extension of the sponge could not occur beyond a distance equal to $\frac{1}{2}io$ from the peripheral oscula (Fig. 3) and thus the sponge could not grow in the horizontal plane. On the other hand, additional oscula would not be required as exits for exhalant currents until the edge of the sponge had advanced from the peripheral oscula to a distance equal to $3 \times \frac{1}{2}io$, i.e. until new aquiferous modules had been formed. If this hypothesis is correct, the distance of the sponge edge from the peripheral oscula at all times will be related to the amount of variation in *io* values. If, for the purposes of convenience, frequencies of *io* values are assumed to be normally distributed, and if the statistic $\bar{x}io + 2\sigma$ is accepted as delimiting approximately 95% of the values of *io*, then we should expect to find only very few instances in which the sponge edge is further than $3(\frac{1}{2}\bar{x}io + \sigma)$ or closer than $\frac{1}{2}\bar{x}io - \sigma$. Of 45 specimens of *Oph. seriata* in which the oscular distributions have been analysed, only three have portions of the sponge edge further from an oscule than $3(\frac{1}{2}\bar{x}io + \sigma)$ and in each case the area of sponge lying outside this perimeter is 2% or less of the total sponge area. In no sponge examined was any portion of undamaged edge found to lie closer to an osculum than $\frac{1}{2}\bar{x}io - \sigma$.

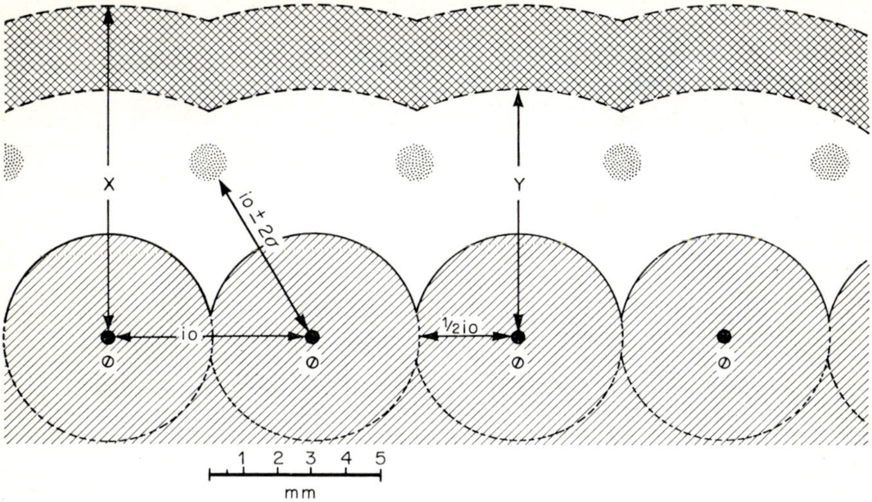

FIG. 3. Theoretical model of the horizontal growth of *Oph. seriata* with $\bar{x}io \pm \sigma$ equal to $6{\cdot}0 \pm 0{\cdot}5$ mm, *io* values being normally distributed. Ø, osculum; *io*, distance between two nearest-neighbour oscula; stippled circles, most probable sites of next new oscula; hatched area, existing sponge tissue; cross-hatched area, most probable region in which new sponge edge will lie at the appearance of the next new oscula; $X = 3(\frac{1}{2}io + \sigma)$; $Y = 3(\frac{1}{2}io - \sigma)$.

Aquiferous modules have been defined in terms of volume and, therefore, the sponge thickness must be taken into account. While values of $\bar{x}io$ (Fig. 4) suggest that the Church Island population stands somewhat separate from the other three populations, data on thickness ($\bar{x}h$) within the areas bounded by the peripheral oscula do not give any such clear picture (Fig. 5). However, if the volume of the aquiferous module (v_{am}) is to be regarded as a simple expression of the hydromechanics of the aquiferous system then *io* and *h* should be related as $\frac{1}{2}io = \sqrt{v_{am}/\pi h}$. Table I shows the mean values of v_{am} for the four populations based on the $\bar{x}io$ and $\bar{x}h$ values for the populations and for this relationship. The mean values of $\bar{x}v_{am}$ indicate a marked difference between the North Wales and Roscoff populations.

Figures 6 and 7 show the distribution of values of $\bar{x}io$ and $\bar{x}h$ about the theoretical curves based on $\frac{1}{2}\bar{x}io = \sqrt{v_{am}/\pi h}$ for each population. Over the ranges of *io* and *h* in which the data occur, the theoretical curves approximate to a straight line not far from the horizontal and so the fit of the data to the regression equation $\frac{1}{2}io = 0h + a$, where $a = \bar{x}h$, has been tested as well as to the regression equation $\frac{1}{2}io = \sqrt{v_{am}/\pi h}$. Table II shows the goodness of fit of the data to these pairs of null hypotheses. For every population, the relationship $\frac{1}{2}\bar{x}io = \sqrt{v_{am}/\pi h}$ is

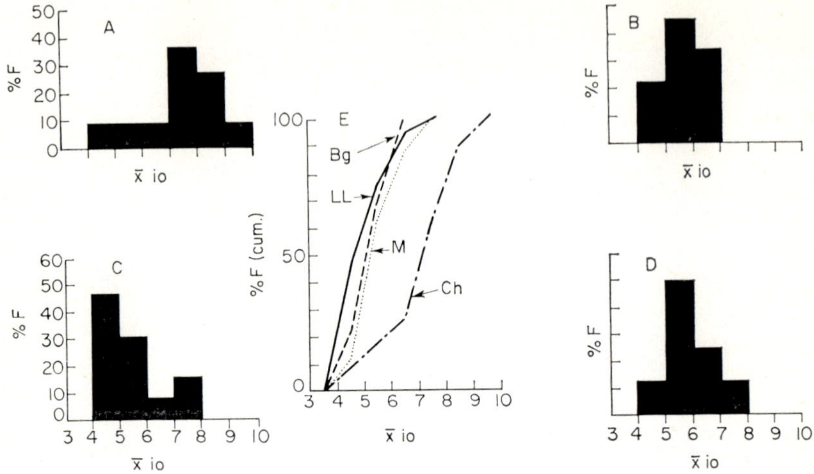

Fig. 4. A–D, percentage frequency distributions of $\bar{x}io$ values in the Church Island (A), Bodorgan (B), Le Loup (C) and Menenett (D) samples of specimens; E, cumulative percentage frequency curves of $\bar{x}io$ for Church Island (Ch), Bodorgan (Bg), Le Loup (LL) and Menenett (M) samples. $\bar{x}io$ values in mm.

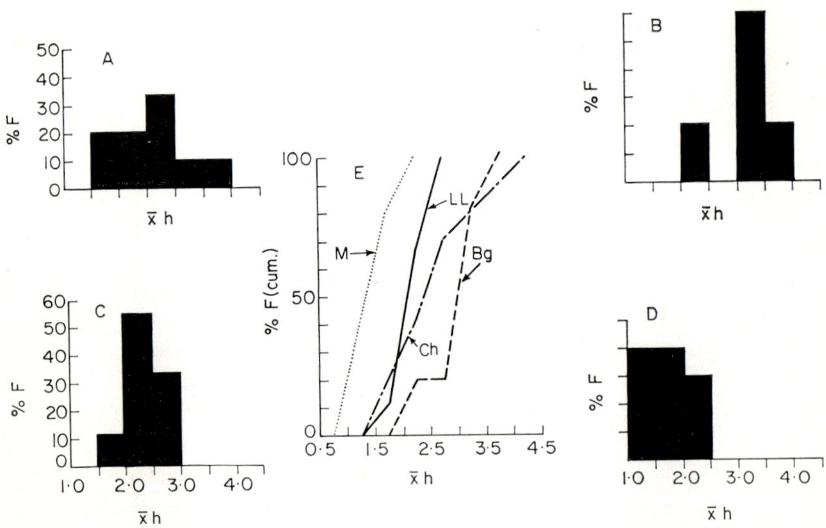

Fig. 5. A–D, percentage frequency distributions of $\bar{x}h$ (thickness) values in the Church Island (A); Bodorgan (B), Le Loup (C) and Menenett (D) samples of specimens; E, cumulative percentage frequency curves of $\bar{x}h$ for the Church Island (Ch), Bodorgan (Bg), Le Loup (LL) and Menenett (M) samples. $\bar{x}h$ values in mm.

TABLE I

Mean dimensions of aquiferous modules

	$\bar{x}io$	$\bar{x}h$	$\bar{x}v_{am}$
Church Is.	7·46 mm	2·77 mm	121·01 cu mm
Bodorgan	5·79	3·16	83·16
Menenett	5·36	2·28	51·38
Le Loup	5·83	1·71	45·64

more acceptable than the more simple regression equation. In addition, the difference between the two North Wales populations is shown to be less than that indicated by their $\bar{x}v_{am}$ values.

TABLE II

Test of regression equations $\frac{1}{2}io = \sqrt{v_{am}}/\pi.h$ *and* $\frac{1}{2}io = 0h + a$

Data	Null hypotheses	P	
Church Is	121·01	>0·95	
	a = 3·73	>0·25	<0·50
	83·16	>0·10	<0·25
	51·38	<0·005	
	45·64	<0·005	
Bodorgan	83·16	>0·10	<0·25
	a = 2·89	>0·05	<0·10
	121·01	>0·90	<0·95
	51·38	>0·05	<0·10
	45·64	<0·005	
Menenett	51·38	>0·50	<0·75
	a = 2·68	<0·005	
	45·64	>0·50	<0·75
	83·16	>0·05	<0·10
	121·01	>0·01	<0·02
Le Loup	45·64	>0·02	<0·05
	a = 2·92	>0·05	<0·10
	51·38	>0·10	<0·20
	83·16	>0·05	<0·10
	121·01	>0·05	<0·10

Much more detailed analysis of this hypothesis is required and it is introduced here only because the sampling technique used in the assessment of spicule populations is based upon acceptance of the concept of aquiferous modules.

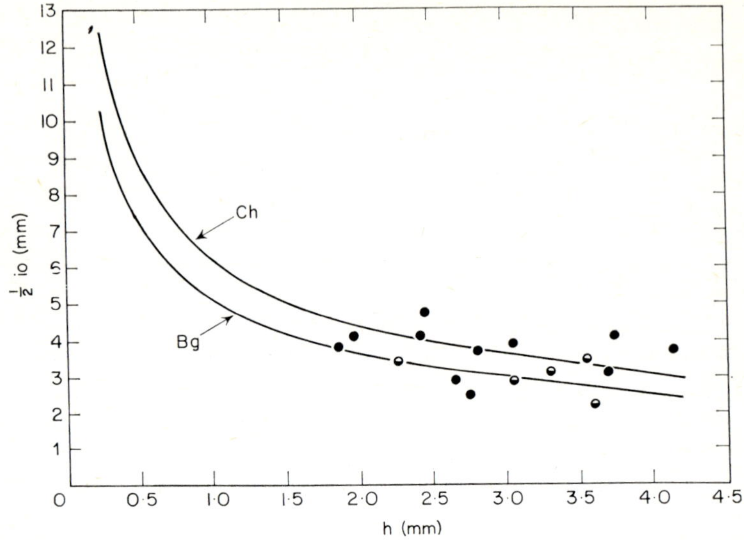

FIG. 6. Plot of $\bar{x}\frac{1}{2}io$ against $\bar{x}h$ for Church Island (●) and Bodorgan (◒) specimens. Ch represents the regression equation $\bar{x}\frac{1}{2}io = \sqrt{121{\cdot}01/\pi} \, . \, \bar{x}h$; Bg represents the regression equation $\bar{x}\frac{1}{2}io = \sqrt{83{\cdot}16/\pi} \, . \, \bar{x}h$.

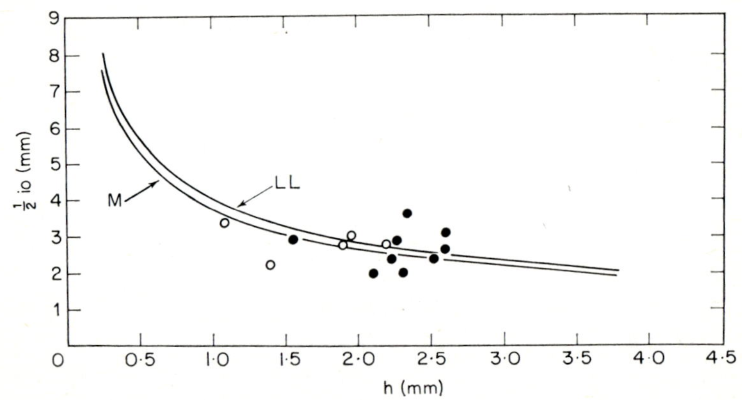

FIG. 7. Plot of $\bar{x}\frac{1}{2}io$ against $\bar{x}h$ for the Le Loup (●) and Menenett (○) specimens. LL represents the regression equation $\bar{x}\frac{1}{2}io = \sqrt{51{\cdot}28/\pi} \, . \, \bar{x}h$; M represents the regression equation $\bar{x}\frac{1}{2}io = \sqrt{45{\cdot}64/\pi} \, . \, \bar{x}h$.

Analysis of spicule populations

I do not wish to discuss the classification of spicules according to form. There already exist several such classifications (e.g. Bowerbank, 1864; Ridley and Dendy, 1887; Sollas, 1888; Dendy, 1921; Arndt, 1935;

de Laubenfels 1953; Koltun 1959, 1967), most of which are adequate for descriptive purposes although, as Finks (1967) has so clearly demonstrated, they do in some measure imply unsubstantiated homologies. Instead, I wish to discuss the relative frequencies and size frequency distributions of the three spicule types occurring in *Oph. seriata*. These are toxa, tylostyles and subtylostyles. They have been referred to by some authors as toxa, subtylostyles, and styles respectively (e.g. Lévi, 1960).

Five specimens from each of the four populations were selected for the study of spicule analysis techniques. All the specimens were collected during the same month so as to reduce the chances of variation being excessively complicated by seasonal growth patterns.

Five samples were taken from each specimen. Each sample consisted of a disc cut and removed with fragments of the underlying rock substrate. Two discs from each specimen had an osculum at their centre—these are referred to as oscular, or **Ø**, samples, two had their centres situated midway between two oscula—these are referred to as interoscular, or **IO**, samples, and one was taken so that it included a portion of the periphery. The last, referred to as edge, or **E**, samples, were found to be highly similar either to **Ø** or to **IO** samples and their data were later omitted from the analysis. The diameter of each disc was given by $\bar{x}io-2\sigma$, $\bar{x}io$ and σ being calculated separately for each specimen.

Every disc was bisected, the **E** discs being bisected at right angles to the sponge periphery, and one half-disc was used to make a spicule preparation, while the other half-disc was used for subsequent histological studies.

Spicule preparations were made by a carefully standardized technique, involving final suspension of the spicules in hot Euparal and their mounting under a 19 mm × 52 mm coverglass.

Accepting a simple classification of the spicules into the three form categories of tylostyle, subtylostyle and toxa, two features of the spicule populations immediately suggest themselves for descriptive and comparative purposes. These are the size frequency distributions and the relative frequencies of each spicule type.

It would be unthinkable at the present time to omit from a taxonomic description a mensural description of spicules encountered. Yet several taxonomists have, at some time or other, expressed misgivings about such descriptions (e.g. Carter, 1874; Burton, 1932; Hartman, 1958; Connes, 1963; Simpson, 1965). In addition, misgivings are apparent in the mensural descriptions themselves. For example, a statement such as "... The obliquely placed dermal oxeas are somewhat larger, say about $8\mu \times 420\,\mu$ long" (de Laubenfels, 1961) or the presentation of raw mensural data (e.g. Lambe, 1896; Traxler, 1896; Weltner, 1910; Burton, 1926; Cordero, 1928; Lévi, 1952; Hartman, 1955; Wells, Wells and Gray, 1960) reflect considerable uncertainty about the significance of the data collected. Another, frequently employed, and equally uncertain, descriptive method involves drawings of a selection of spicules together with scale indicators.

A number of attempts have been made to produce more meaningful statistical descriptions of spicule populations. Perhaps the most frequently encountered formula is the *Minimum Extreme–Maximum Extreme*, although this is becoming replaced by the *Minimum Extreme*–Mean–*Maximum Extreme* formula (Hartman, 1958; Little 1963; Bakus, 1966). Other workers have plotted size-frequency distribution curves (Martinez, 1940; Rioja, 1940) while a very few have used parameters such as mean, standard deviation, e.m., standard error, or variance (e.g. Brønsted, 1930; Sasaki, 1941; Sarà, 1951, 1953a, 1953b; Hartman, 1958).

Implicit in the use of such statistical descriptions are the assumptions that they are true indicators of the nature of the spicule populations and that they are to be used for comparing and distinguishing between separate spicule populations. Yet in apparently only three instances have such statistical comparisons been made (Hartman, 1958; Sara, 1953a; Little, 1966).

The relative frequencies of different types of spicules have been used to describe and evaluate specimens (e.g. Bakus, 1966; Bergquist, 1961a,b; Hartman, 1955, 1958; Jørgensen, 1944; Sara, 1953b) although the precision of such descriptions ranges from the verbal such as "appear to be absent", "occasional", "rare" or "abundant", to actual percentage frequency values. There do not seem to be any published statistical comparisons of such data.

It would appear that this lack of exploitation of statistical techniques is based upon two uncertainties. The first is uncertainty as to the statistical validity of the samples of data collected, i.e. how representative are the data of the parent population. The second is uncertainty about the taxonomic meaning of values of such indices of difference as Probability or Coefficient of Difference.

Possibly because of these uncertainties, considerable attention has been paid to minor variations in spicule form such as patterns of microspination. Undoubtedly, such minor variations can provide most valuable taxonomic information, but we know from the writings of Hentschel (1929), Jewell (1935, 1939), Jørgensen (1944, 1946, 1947), Hartman (1958), Rasmont (1963), and others, that spicule form may be influenced by environmental factors and the physiological state of the animal, as well as by the genotype. Thus we cannot regard the presence or absence of minor modifications of form as providing unequivocal taxonomic information. Such modifications should be regarded in the same way as other characters, and their description and use in comparison should be based on precise sampling and analytical techniques.

Useful although such parameters as mean, standard deviation, variance, etc., may be, if calculated from sufficient data, they have descriptive power only if the frequency distribution which they describe is unimodal and symmetrical. If size-frequency distributions of spicule populations are bi- or polymodal, or markedly skew or kurtose, then potentially extremely valuable biological information is lost by restricting description of the distributions to values of their first two moments. In addition, the statistical methods of comparing values of these parameters depend for their power to a large extent upon the approximation to normality of the frequency distributions which they purport to describe.

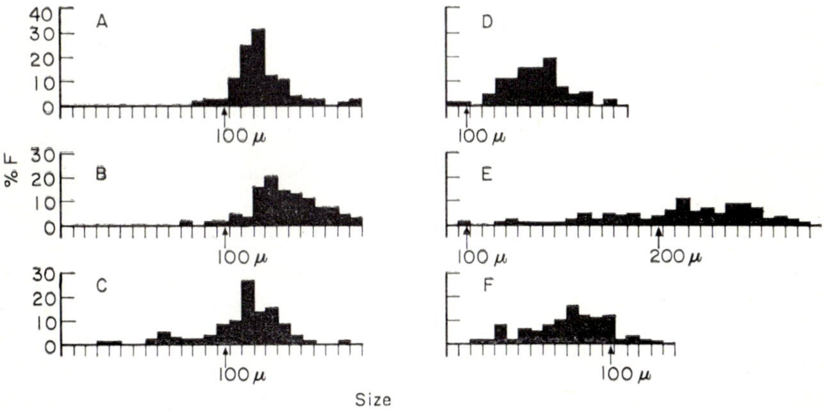

FIG. 8. Size frequency distributions of samples of spicules. A and B, subtylostyle lengths in specimens of *Oph. seriata* from Church Island; C, subtylostyle lengths of a specimen of *Oph. seriata* from Bodorgan; D, tylostyle lengths from a Church Island specimen of *Oph. seriata*; E, tylostyle lengths from a thinly encrusting specimen of *Suberites sp.* from the Irish Sea; F, toxa lengths (chord) from a Church Island specimen of *Oph. seriata*. Each histogram represents more than 100 data from two oscular and two interoscular samples. Size class intervals 6·75 μ.

That we have no right to assume normal distributions of size frequency of sponge spicules is demonstrated clearly in Fig. 8. Adequate parametric descriptions of these frequency distributions should include means, variances, measures of skewness and kurtosis and measures of polymodality. This would involve extremely laborious calculation and, for measures of skewness and kurtosis, enormously large samples of data. Measures of polymodality require graphical analysis of the type evolved by Harding (1949), which provides only non-numerical hypotheses for further testing (Sokal, 1965), or more elaborate analytical methods such as Tchebychef orthogonal polynomials (Aitken, 1957; Snedecor and Cochran, 1967), or moving means (Taylor, 1965).

On the other hand, because of the availability of computers and advances in non-parametric statistical methods, description of frequency distributions of spicule size can be given quite adequately by histograms such as those in Fig. 8, or by lists of percentage frequencies for each size class. Providing that the difficulties of translating probability values of difference into measures of taxonomic difference can be overcome, it would appear to be more fruitful to describe size-frequency data in this manner, together with sample size, than by values of parameters such as means, variances, etc.

A non-parametric method of comparing frequency distributions, the Kolmogorov-Smirnov test, has been evolved and developed by Smirnov (1948) and Massey (1950a,b, 1951a,b). This test can be used to assess the probability of two samples, or a sample and a population, being derived from the same parent population, by relating the maximum difference between the two cumulative step functions to the sizes of the samples or populations being compared. The Kolmogorov–Smirnov test statistic, D, unlike other test statistics such as g_1, g_2 or Student's t, does not rely for its power upon the form of, or the nature of some parameters in, the frequency distribution of the samples (Fraser, 1966). Massey (1951b) has put forward evidence to show that the Kolmogorov-Smirnov test is a better all-round test than the χ^2 test and the nature of the test obviates the need to group together classes wherever the total of data in any class is less than five—a procedure, necessary in the χ^2 test, which can result in the loss of biological information while increasing the power of the test.

Together with a modified d test for comparing the relative frequencies of spicules of different form, the Kolmogorov–Smirnov test has been incorporated in a programme written for the U.C.N.W. Elliott 4130 computer and designed for comparing samples from several populations. The programme has been written so that the input is of raw data. The d test for comparisons of relative frequency is based upon

$$d = \sqrt{\frac{k_1 - k_2}{k(1-k)(1/n_1 + 1/n_2)}},$$

where

$$k_1 = \frac{a_1}{n_1},\ k_2 = \frac{a_2}{n_2},\ k = \frac{a_1+a_2}{n_1+n_2},$$

and d is treated as a normal variable (Bailey, 1959). In the case of *Oph. seriata*, in which there are three types of spicule, it seems unsatisfactory to include the results of comparisons of frequencies of all three spicule types in the final assessment of overall spicule similarity, as this would, in effect, merely increase the importance of the results of comparisons of relative frequencies of the other two spicule types. For this reason the results of comparisons of relative frequencies of tylostyles were excluded from the final assessment of overall spicule similarity. For reasons which are given below it proved unwise, however, to reject the tylostyle frequency data from all the computations.

The confidence which can be ascribed to the results of such computations is limited by the degree to which the data represent the populations of spicules in the spicule preparations. Use of Fisher's Coefficient of Dispersion revealed that, in a random selection of 30 spicule preparations of *Oph. seriata*, all three spicule types are always clearly overdispersed. More detailed analysis of these slides showed that there existed no simple pattern of symmetry of dispersion, except possibly in the case of the tylostyles. The degree of departure from a

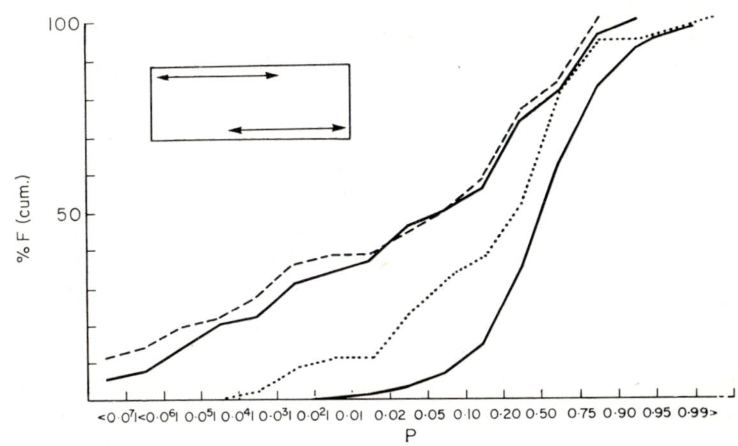

FIG. 9. Departure of spatial distribution of spicules under a 19×52 mm coverglass from a simple symmetrical pattern. Cumulative percentage frequency curves of P values derived by comparing relative frequency data from two longitudinal edge transects (positions of transects shown in inset) on 34 spicule preparations. Lower unbroken line, random variation about perfect simple symmetry; thin unbroken line, data for toxa; broken line, data for subtylostyles; dotted line, data for tylostyles. Mean P values: toxa 0·27, subtylostyles 0·25, tylostyles 0·41.

simple symmetry dispersion is remarkably similar in the subtylostyles and toxa (Fig. 9).

Two spicule preparations, selected so that the estimated total density of spicules in one (Slide I : 139) was approximately one and a half times that in the other (Slide I : 136), were analysed in further detail (Fig. 10). The object of the analysis was to find the location of optical

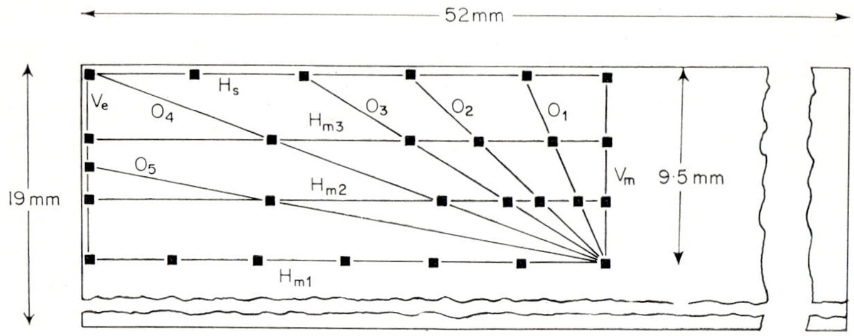

FIG. 10. Spatial distribution of areas studied for the analysis of variation in relative frequency and size frequency distribution of three spicule types in relation to their dispersion under 19 × 52 mm coverglasses. The areas, represented by black rectangles, are distributed so as to lie along 12 transects. V, vertical; O, oblique; H, horizontal; e, end; s, side; m, median.

transects which would provide data representing most faithfully the relative frequencies and size-frequency distributions estimated to represent the total populations under the coverglasses. Table III shows that the data from the transects Horizontal Median 2 (Fig. 10, H_{m2} were consistently highly representative. Results for the Oblique transects are not shown in Table III as they not only proved to be less consistently representative but data are difficult to collect from them. As a result of these tests all the data on frequency and size used in subsequent computations were obtained from Horizontal Median 2 transects.

The results obtained by comparing relative frequency and size frequency data from **Ø** samples of every region with the equivalent data from **Ø** samples of every other region, and similarly **IO** sample data with other **IO** sample data, are represented by values of Probability that the observed differences between the data compared are due to chance errors induced by sampling factors. By definition, such probability values cannot be regarded as direct measures of taxonomic difference. Even if obtained in very large numbers they can only be used as evidence for rejecting null hypotheses if their values are modified by some factor such as

$$\frac{2}{k(k-1)}$$

(Tate and Clelland, 1959). Even then, resultant accepted hypotheses must be regarded as extremely tentative.

TABLE III

Individual transacts × total data on rest of slide

		Relative frequencies			Size frequency distributions		
		Subtylos.	Tylos.	Toxa	Subtylos.	Tylos.	Toxa
Slide I :136	Vert m	P >0·37	P >0·50	P >0·14	P >0·008	P >0·50	P >0·50
	Vert e	0·42	0·50	0·50	0·025	0·50	0·50
	Horiz s	0·09	0·40	0·40	0·004	0·50	0·50
	Horiz m_1	0·02	0·08	0·16	0·95	0·50	0·50
	Horiz m_2	0·95	0·48	0·48	0·31	0·50	0·50
	Horiz m_3	0·39	0·29	0·29	0·95	0·50	0·50
Slide I:139	Vert m	$<0{\cdot}0^{8}1$	$0{\cdot}0^{4}1$	$0{\cdot}0^{5}1$	0·11	0·50	0·50
	Vert e	0·50	0·11	0·42	0·10	0·40	0·50
	Horiz s	$0{\cdot}0^{3}1$	$0{\cdot}0^{3}1$	0·28	0·50	0·40	0·50
	Horiz m_1	0·24	0·29	0·50	$0{\cdot}0^{3}2$	0·40	0·50
	Horiz m_2	0·41	0·29	0·95	0·50	0·40	0·50
	Horiz m_3	0·02	0·09	0·16	0·95	0·50	0·50

Incorporated into the computer programme is a series of steps which, while not overcoming these difficulties, does appear to produce less tentative hypotheses. The logic of this procedure is based upon the assumption that if three populations of spicules *a*, *b* and *c* have the frequency distributions $\bar{p}_a$, $\bar{p}_b$ and $\bar{p}_c$ of values of a character such that $\bar{p}_a$ is more similar to $\bar{p}_b$ than $\bar{p}_c$ is to either $\bar{p}_a$ or $\bar{p}_b$, then comparisons of all samples in *a* with all samples in *b* and *c*, and all samples of *b* with all samples of *c*, will produce frequency distributions of Probability values $\bar{p}'_{a:b}$, $\bar{p}'_{a:c}$ and $\bar{p}'_{b:c}$ such that not only will $\bar{p}'_{a:c}$ and $\bar{p}'_{b:c}$ be more similar to each other than to $\bar{p}'_{a:b}$, but also that $\bar{p}'_{a:b}$ will contain a higher proportion of high Probability values than will $\bar{p}'_{a:c}$ and $\bar{p}'_{b:c}$.

The proportion of high probability values in such secondary distributions ($\bar{p}'$) will be reflected by the mean value of $\bar{p}'$. If the primary distributions $\bar{p}$ are normal then the ranking of degrees of similarity between $\bar{p}_a$, $\bar{p}_b$ and $\bar{p}_c$ could be given by $\bar{x}\bar{p}'_{a:b}$, $\bar{x}\bar{p}'_{b:c}$, and $\bar{x}\bar{p}'_{a:c}$. However, it has been shown above (Fig. 8) that there are no grounds for assuming normality of distribution of the primary distributions of character values and so another measure of differences between $\bar{p}'$ distributions is needed.

The non-parametric Kolmogorov–Smirnov test was used to compare the secondary distributions ($\bar{p}'$). The sampling pattern used for *Oph. seriata* resulted in the production of secondary distributions of 100 Probability values and this was felt to be a sufficiently large sample for the subsequent analysis to be sound. Fewer samples from more specimens, or more samples from fewer specimens, would produce the same large samples of results of comparisons.

The secondary distributions ($\bar{p}'$) were compared with each other to give single probability values p″ for each comparison. Critical values of p″ (Cp″) were selected, at or above which secondary distributions ($\bar{p}'$) are considered to be identical, and differences between their values of $\bar{x}\bar{p}'$ are ignored.

TABLE IV

$\bar{x}\bar{p}'$ Relative frequencies of spicules

	Subtylostyles		Tylostyles		Toxa	
	Ø × **Ø**	**IO** × **IO**	**Ø** × **Ø**	**IO** × **IO**	**Ø** × **Ø**	**IO** × **IO**
Chch Is × Bdn	0·102	0·110	0·242	0·274	0·127	0·133
Chch Is × Mntt	0·097	0·099	0·274	0·175	0·096	0·114
Chch Is × LeLp	0·097	0·128	0·264	0·180	0·101	0·156
Bdn × Mntt	0·109	0·114	0·274	0·261	0·134	0·112
Bdn × LeLp	0·111	0·118	0·205	0·246	0·137	0·119
Mntt × LeLp	0·217	0·133	0·337	0·452	0·195	0·133

The relative frequencies of subtylostyles in **Ø** samples from the four populations may be used to illustrate the procedure. Table IV shows the $\bar{x}\bar{p}'$ values for this character as well as the other five spicule frequency characters of **Ø** samples used in the analysis. Table V shows the p″ values obtained by comparing secondary distributions of the relative frequencies of subtylostyles in **Ø** samples. With **C**p′ at 0·90, $\bar{p}'_{C:M}$, $\bar{p}'_{C:LL}$ are considered identical as are $\bar{p}'_{C:B}$, $\bar{p}'_{B:M}$, and $\bar{p}'_{B:LL}$. The mean of the three latter secondary distributions is higher than the mean of the two former secondary distributions, and thus the similarity between Church Island and Bodorgan, Bodorgan and Menenett, and Bodorgan and Le Loup is held to be greater than that between Church Island and Menenett and Church Island and Le Loup. The similarity between Menenett and Le Loup is much higher, as expressed by the mean of the secondary distribution $\bar{x}\bar{p}'_{M:LL}$. Thus at **C**p″ = 0·90 there are considered to be three ranks of similarity (r = 3) for this character and each comparison is accorded a rank of similarity (r') (Table VI). Similar rankings

TABLE V

p″ *Relative frequency subtylostyles* **Ø** × **Ø**

	C × B					
C × B	1·00	C × M				
C × M	0·73	1·00	C × LL			
C × LL	0·46	0·91	1·00	B × M		
B × M	>0·99	0·40	0·30	1·00	B × LL	
B × LL	>0·99	0·81	0·38	0·46	1·00	M × LL
M × LL	0·13	0·30	0·16	0·13	0·53	1·00

TABLE VI

Relative frequency subtylostyles **Ø** × **Ø**.
Final ranks (r') *at critical values of* p″

	0·90	0·80	0·50
C × B	2	1	1
C × M	1	1	1
C × LL	1	1	1
B × M	2	1	1
B × LL	2	1	1
M × LL	3	2	1
r =	3	2	1

TABLE VII

Half-matrix of values of a coefficient of similarity based on **C**p″ = 0·90 *and all spicule characteristics considered except tylostyle relative frequencies*

	C			
C	100·00	B		
B	62·41	100·00	M	
M	44·94	55·65	100·00	LL
LL	49·05	55·36	93·45	100·00

for all the other characters are obtained at **C**p″ = 0·90 and a measure or coefficient of similarity is given by $\Sigma r'/\Sigma r \times 100$. Table VII shows values of this coefficient at **C**p″ = 0·90 for the four populations of *Oph. seriata*. Such a half-matrix can be analysed by some of the multivariate techniques which have been evolved to depict disparity between taxonomic units (see Sokal and Sneath, 1963 or Gower, 1966, 1967, for

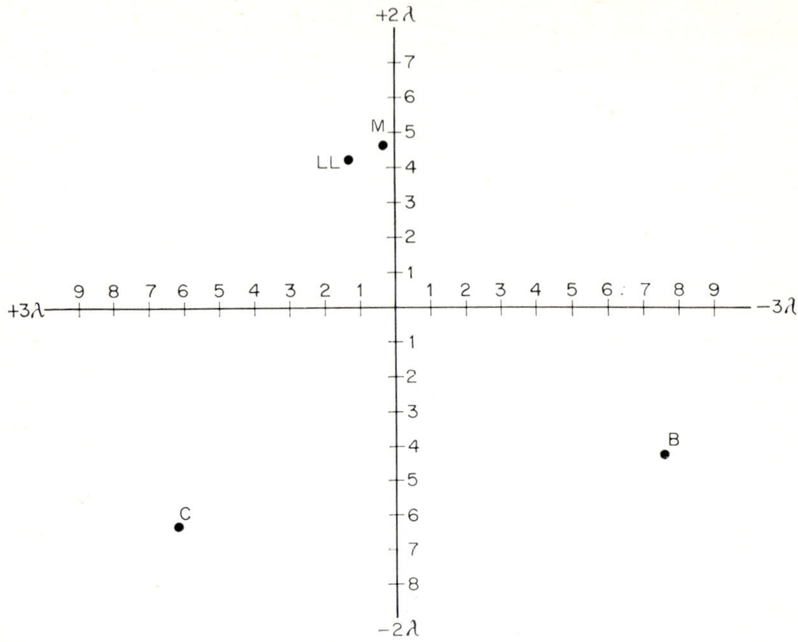

FIG. 11. Taxometric dispersion of the four populations of *Oph. seriata* depicted by plots of the second and third vectors of the latent roots of values of similarity at Cp″ = 0·90 based on spicule characteristics of **Ø** and **IO** samples, excluding tylostyle relative frequencies C, Church Island; B, Bodorgan; M, Menenett; LL, Le Loup. The figure is based on the results of computations performed at Rothamsted Experimental Station by courtesy of Mr. J. C. Gower.

recent reviews of these techniques). Figure 11 shows the results of such an analysis. Even without such refined analysis it is clear that, in terms of the spicule characters employed in the analysis, the two Roscoff populations are very different from the two North Wales populations. What is perhaps surprising is the degree of difference between the Church Island and Bodorgan populations.

Systematic reduction of **Cp″** reduces the number of ranks of similarity (r) and, when all similarity ranks are unity, the range of variation of the populations has been encompassed. In the present study it is not until **Cp″** = 0·01 that all ranks are unity; in other studies different final values of **Cp″** can be expected and this could be used as a measure of variation within the units studied. But perhaps more immediately useful in taxonomic or ecological studies is the sequence in which r becomes unity for different characters. In the present study tylostyle relative frequency is the last character for which r becomes unity as **Cp″** is decreased (Table VIII). Subtylostyle size frequency

TABLE VIII

Values of Cp″ *at which* $r = 1$ *for 6 spicule characters*

Cp″	**Ø** × **Ø**	**IO** × **IO**
0·90	—	Toxa RF
0·80	Tylostyle RF	—
0·70	—	Subtylostyle RF
0·60	Toxa RF	—
0·50	Subtylostyle RF	—
0·40	Tylostyle SF	Subtylos, Tylos, Toxa SF
0·20	Toxa SF	—
0·10	Subtylostyle SF	—
0·01	—	Tylostyle RF

distribution is the next to last in which r becomes unity. It is possible that this information could provide us with a method for finding key characters. In fact we find that the unmodified percentage frequency data for the four population (Table IX) suggest that, in **IO** samples, low percentage frequency of tylostyles is a character which separates the Roscoff populations from the North Wales populations. That r becomes unity for tylostyle relative frequency in **Ø** samples at a very much higher value of **Cp″** underlines the need for careful sampling techniques.

The total data on subtylostyle size frequency (Fig. 12) again confirm the usefulness of the analytical method in producing key characters. The pronounced bimodality of the two Roscoff populations' frequency distributions is almost absent from that of the two North Wales populations.

A key character for distinguishing between the two North Wales populations is equally easily found. Throughout the lowering of **Cp″** values it is found that for toxa size frequency distribution in **Ø** samples $\bar{p}'_{C:B}$ is consistently accorded the lowest rank r', until r becomes unity at **Cp″** = 0·20. The total data on toxa size frequency distribution shown in Fig. 12 confirm the difference between the Church Island and Bodorgan populations.

Included in Fig. 12 are the data taken from one of two slide preparations (Norman Colln No. 2773) of *Ophlitaspongia papilla* Bwk., 1866, in the collections of the British Museum (Natural History).* The specimens were collected from Guernsey—considerably closer to Roscoff than to North Wales—and while the data are taken from a

* Probably from the holotype, B.M.(N.H.) Regn. No. 1910: 1: 1: 395.

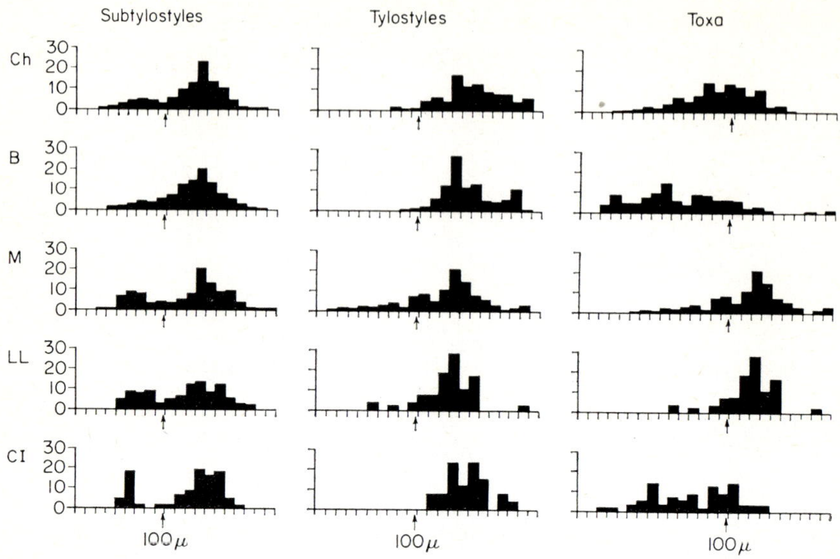

FIG. 12. Percentage size frequency distributions of three spicule types in oscular samples from four populations of *Oph. seriata*. Ch, Church Island; B, Bodorgan; M, Menenett; LL, Le Loup. CI, data for slide preparation which is probably from the holotype of *Oph. papilla* Bwk, 1866, collected from Guernsey. With the exception of the CI data, the histograms are based on data from two oscular samples from five specimens from each population. Size represents greatest chord length, and is shown in class intervals of 6·75 μ.

TABLE IX

Mean percentage frequencies of spicules

	Ø	IO	Ø + IO	E
Subtylostyles				
Church Is	65·88 %	69·63 %	67·75 %	77·33 %
Bodorgan	72·80	62·01	67·40	65·76
Menenett	55·47	63·36	59·41	63·48
Le Loup	58·85	61·00	59·92	57·04
Tylostyles				
Chch Is	6·73 %	7·22 %	6·98 %	4·40 %
Bdn	3·95	8·39	6·18	9·62
Mntt	4·22	3·20	3·72	3·73
Le Lp	2·37	2·91	2·65	2·26
Toxa				
Chch Is	27·39 %	23·15 %	25·27 %	18·27 %
Bdn	23·25	29·60	26·42	24·61
Mntt	40·31	33·44	36·87	32·79
Le Lp	38·78	36·09	37·43	40·70

single preparation, they do suggest that Bowerbank may have been correct in singling out this form as different from *Oph. seriata* (Grant, 1826). Of considerable interest is the record of a third specimen of *Oph. papilla* from Westport Bay (Bowerbank, 1882). Unfortunately, this specimen cannot be traced.

Insofar as spicule characters are concerned, a precise and useful description of the range and nature of the variation encountered in the four populations of *Oph. seriata* can be given, as a result of this analysis by Table VIII, Fig. 12 and Table IX. In addition a key for distinguishing the four populations can be given in the following form:

1. Interoscular Sample Tylostyle Relative Frequency
   Roscoff $\neq$ N. Wales at **Cp**$''$ $>$ 0·01
2. Oscular Sample Subtylostyle Size Frequency Distribution
   Roscoff $\neq$ N. Wales at **Cp**$''$ $>$ 0·10
3. Oscular Sample Toxa Size Frequency Distribution
   Church Is $\neq$ Bodorgan, Menenett and Le Loup at **Cp**$''$ $>$ 0·20

Whether or not these differences represent phenotypic or genotypic variation, or some combination of the two, cannot be determined without prolonged laboratory culture. It has been mentioned above that more samples from fewer specimens might be used to produce large samples of secondary distributions (p'). Larvae of *Oph. seriata* can readily be induced to swarm under laboratory conditions and the members of such swarms fuse after metamorphosis, whether the whole swarm be liberated from either Church Island or Bodorgan specimens or from both. Geographical variation appears to occur in this species, and doubtless in others, and it seems highly likely that on occasion sponges arise from the fusion of post-larvae of different genotypes. Until such a notion is disproved, or complete and permanent intermixture of cells of fused postlarvae is shown to be the rule, then several samples from each specimen would appear to be a necessity for reliable and truly predictive taxonomy.

Acknowledgements

The research for this contribution was carried out during the tenure of the Red Hand Compositions Research Fellowship. Besides thanking the sponsors of this Fellowship, I wish to thank also Mr. I. G. Jones for his invaluable work of writing the programme for the computations, Miss S. M. K. Stone for much help with the literature, Mrs. Patricia Yorke and Mr. John Leighton Jones for their patient and painstaking assistance in collecting data, and Dr. J. H. Simpson for his timely reminders of the distinction between sophistry and precision.

References

Admiralty Chart. (1963). No. 1464. Menai Strait.

Aitken, A. C. (1957). *Statistical Mathematics.* 153 pp. Edinburgh and London: Oliver and Boyd.

Arndt, W. (1935). Porifera. *Tierwelt N.u- Ostee* **3a** (27): 140 pp.; 239 figs.

Bailey, N. T. J. (1959). *Statistical Methods in Biology.* 200 pp. London: English University Press.

Bakus, G. J. (1966). Marine poecilosceleridan sponges of the San Juan Archipelago, Washington. *J. zool., Lond.* **149** (4): 415–531; 2 pls., 27 figs.

Bergquist, P. R. (1961a). Porifera: Demospongiae. Biological Results of the Chatham Islands 1954 Expedition Pt 5. *Bull. N.Z. Dept. Sci. industr. Res.* **139** (N.Z. Oceanogr. Inst. Mem. No. 13): 169–206; 6 figs.

Bergquist, P. R. (1961b). A collection of Porifera from northern New Zealand, with descriptions of seventeen new species. *Pacif. Sci.* **15** (1): 33–48; 18 figs.

Blackman, G. E. (1942). Statistical and ecological studies in the distribution of species in plant communities. I. Dispersion. *Ann. Bot.* N.S. **6**: 351–370.

Borojević, R. (1966). Étude éxpérimentale de la différenciation des cellules de l'éponge au cours de son dévéloppement. *Devl. Biol.* **14** (1): 130–153; 15 figs.

Borojević, R. and Lévi, C. (1964). Étude au microscope éléctronique des cellules de l'éponge *Ophlitaspongia seriata* (Grant), au cours de la réorganisation après dissociation. *Z. Zellforsch. mikrosk. Anat.* **64**: 708–725; 15 figs.

Borojević, R. and Lévi, C. (1965). Morphogénèse éxpérimentale d'une éponge à partir de cellules de la larve nageante dissociée. *Z. Zellforsch. mikrosk. Anat.* **68**: 57–69; 9 figs.

Bowerbank, J. S. (1864). *A Monograph of the British Spongiadae* **1**: 290 pp.; 37 pls., Ray Society, London.

Bowerbank, J. S. (1866). *A Monograph of the British Spongiadae* **2**: 388 pp., Ray Society, London.

Bowerbank, J. S. (1882). *A Monograph of the British Spongiadae*, **4**, (Supplementary), Ed. A. M. Norman: 250 pp.; 17 pls., Ray Society, London.

Brien, P. (1943). L'embryologie des éponges. *Bull. Mus. r. Hist. nat.* Belg., **19** (16): 1–20; 3 figs.

Brien, P. (1967). Les Éponges. Leur nature Metazoaire—Leur Gastrulation—Leur État Colonial. *Ann. Soc. r. Zool. Belg.* **97** (4): 197–235; 12 figs.

Brien, P. and Meewis, H. (1938). Contribution à l'étude de l'embryogénèse des Spongillidae. *Archs Biol.*, (*Paris*) **49**: 177–250; 2 pls.

Brønsted, H. V. (1930). Biometrische Studien über Schwämme. I. Die Nadellange von einem Individuum von *Halichondria panicea* Pall. aus 'Strömmarne' an der Skagerackskuste von Schweden. *Vidensk. Meddr. dansk naturh. Foren.* **88**: 1–13; 3 figs.

Buchan, S., Floodgate, G. D. and Crisp, D. J. (1967). Studies on the seasonal variation of the suspended matter in the Menai Straits, I. The inorganic fraction. *Limn. Oceanogr.* **12** (3): 419–431: 10 figs.

Burton, M. (1926). Observations on some British species belonging to the genus *Reniera. Ann. Mag. nat. Hist.* (9) **17**: 415–424.

Burton, M. (1932). Sponges. '*Discovery*' *Rep. 6:* 237–392; 10 pls.

Carter, H. J. (1874). On the nature of the seed-like body of *Spongilla*; on the origin or mother cell of the spicule; and on the presence of spermatozoa in the Spongida. *Ann. Mag. nat. Hist.* (4) **14**: 97–111; 1 pl., 27 figs.

Connes, R. (1963). La spiculation des Spongillidae, est-elle un critère valable de systématique. *Bull. soc. zool. Fr.* **88**: 388–392.

Cordero, E. H. (1928). La variabilidad de la longitud de las espiculas esqueleticos de dos especies de esponjas de agua dulce del genera Uraguaya (Communicacion Preliminar). *Physis*. **9**: 259.

Cracraft, J. (1967). Comments on Homology and Analogy. *Syst. Zool.* **16** (4): 356–359; 1 fig.

Crisp. D. J. (1961). Territorial behaviour in barnacle settlement. *J. exp. Biol.* **38**: 429–446; 1 pl., 8 figs.

Davis, P. H. and Heywood, V. H. (1963). *Principles of angiosperm taxonomy*: 556 pp.; 42 figs, Edinburgh and London: Oliver and Boyd.

Dendy, A. (1921) The Tetraxonid sponge-spicule: a study in evolution. *Acta zool., Stockh.* **2**: 95–152; 50 figs.

Finks, R. M. (1967). The structure of *Saccospongia laxata* Bassler (Ordovician) and the phylogeny of the Demospongia. *J. Palaeont.* **41** (5): 1137–1149; 2 pls, 5 figs.

Fisher, R. A. (1958). *Statistical methods for research workers.* 356 pp. Biological Monographs and Manuals, London and Edinburgh: Oliver and Boyd.

Fraser, D. A. S. (1966). *Nonparametric methods in statistics.* 299 pp. Wiley Publications in Statistics. New York, London and Sydney: John Wiley.

Fry, W. G. (1964). The Pycnogonida and the coding of biological information for numerical taxonomy. *Syst. zool.* **13** (1): 32–41; 4 figs.

Goodall, D. W. (1966). A new similarity index based on Probability. *Biometrics* **22**: 882-907; 1 fig.

Gower, J. C. (1966). Some distance properties of latent roots and vector methods used in multivariate analysis. *Biometrika* **53** (3 and 4); 325–338.

Gower, J. C. (1967). A comparison of some methods of cluster analysis. *Biometrics* **23**: 623–637.

Grant, R. E. (1826). Observations on the structure of some Siliceous Sponges. *Edinb. New. Phil. J.* **1**: 341–351.

Harding, J. P. (1949). The use of Probability paper for the graphical analysis of polymodal frequency distributions, *J. mar. biol. Ass. U.K.* **28**: 141–153; 6 figs.

Hartman, W. D. (1955). A collection of sponges from the West coast of the Yucatan Peninsula with descriptions of two new species. *Bull. mar. Sci. Gulf. Caribb.* **5**(3): 161–189; 10 figs.

Hartman, W. D. (1958). Natural History of the marine sponges of southern New England. *Bull. Peabody Mus. nat. Hist.* **12**: 1–155; 12 pls, 46 figs.

Harvey, J. G. (1965). Physical hydrography of the Menai Straits. Abstract, *Rep. Challenger Soc.* **3** (17): 28.

Harvey, J. G. (1968) The flow of water through the Menai Straits. *Geophys. Jl R. astr. Soc.* **15**, 517–528; 8 figs.

Harvey, J. G. and Spencer, C. P. (1962). Abnormal hydrographic conditions in the Menai Straits. *Nature Lond.* **195** (4843): 794–795; 1 fig.

Hentschel, E. (1929). Die Kiesel- und Hornschwamme des Nördlichen Eismeers. *Fauna arct.* **5** (4): 857–1042; 3 pls.

Hughes, P. (1966). The temperature and salinity of the surface waters of the Irish Sea for the period 1947–61. *Geophys. Jl R. astr. Soc.* **10** (4): 421–435; 1 fig.

Inglis, W. G. (1966). The observational basis of homology. *Syst. Zool.* **15** (3):219–228: 2 figs.

Jewell, M. E. (1935). An ecological study of the fresh-water sponges of Northern Wisconsin. *Ecol. Mongr.* **5**: 461–504; 26 figs.

Jewell, M. E. (1939). An ecological study of the freshwater sponges of Wisconsin. II. The influence of Calcium. *Ecology* 20(1): 11–28.

Johnson, R. G. (1958) Spatial distribution of *Phoronopsis viridis* Hilton. *Science N.Y.* **129** (3357): 1221.

Jørgensen, C. B. (1944). On the spicule-formation of *Spongilla lacustris* (L.). I. The dependence of the spicule-formation on the content of dissolved and solid silicic acid of the milieu. *Biol. Meddr* **19** (7):1–45; 16 figs.

Jørgensen, C. B. (1946). On the gemmules of *Spongilla lacustris* auct. together with some remarks on the taxonomy of the species. *Vidensk. Meddr dansk naturh. Foren.* **109**: 69–79: 4 figs.

Jørgensen, C. B. (1947). On the spicule-formation of *Spongilla lacustris* (L.). and *Ephydatia fluviatilis* (L.) II. The rate of growth of the spicule. *Biol. Meddr* **20** (10): 1–22; 20 figs.

Kendrick, W. B. (1964). Quantitative characters in computer taxonomy, in *Phenetic and Phylogenetic Classification*, Eds. V. H. Heywood and J. McNeill: 105–144; 2 figs. Systematics Ass. Publication No. **6**, Systematics Ass., London.

Kendrick, W. B. (1965). Complexity and dependence in computer taxonomy. *Taxon* **14** (5): 141–154.

Kendrick, W. B. and Weresub, L. B. (1966). Attempting Neo-Adansonian computer taxonomy at the ordinal level in the Basidiomycetes. *Syst. Zool.* **15** (4): 307–329; 8 figs.

Koltun, V. M. (1959). *Kremnyerogoviye gubki severnich i dal'nyevostochniech morei SSSR.* (*Otrad Cornacuspongida*). Izdat. Akad. Nauk. SSSR, 236 pp.; 43 pls., 187 figs.

Koltun, V. M. (1967). *Styeklenniye ili shestiluchiviye gubki severnich i dal'nyevostochniech morei SSSR.* (*Klass Hyalospongiae*). Izdat. Akad. Nauk. SSSR., 125 pp.; 87 figs.

Lambe, L. M. (1896). Sponges from the Atlantic coast of Canada. *Trans. R. Soc. Can.* (2) **2** (4): 181–211; 3 pls.

de Laubenfels, M. W. (1953). Sponges from the Gulf of Mexico. *Bull. mar. Sci. Gulf. Caribb.* **2** (3): 511–557; 17 figs.

de Laubenfels, M. W. (1961). Porifera of Friday Harbor and vicinity. *Pacif. Sci.* **15** (2): 192–201; 2 figs.

Lévi, C. (1952). Spongiaires de la côte de Senegal. *Bull. Inst. fr. Afr. noire* **14** (1): 34–59: 20 figs.

Lévi, C. (1960). Les Démosponges des côtes de France. I. Les Clathriidae. *Cahiers Biol. Mar.* **1**: 47–87; 25 figs.

Little, F. J. (1963). The Sponge fauna of the St. George's Sound, Apalachee Bay, and Panama City regions of the Florida Gulf Coast. *Tulane Stud. Zool.* **11** (2): 31–71.

Little, F. J. (1966). Bispecific chimerid sponges from the Antarctic. *Nature, Lond.* **211** (5047): 436–8.

Maas, O. (1893). Die Embryonalentwicklung und Metamorphose der Cornacuspongien. *Mitt. zool. Stn Neapel* **10** (3): 408–440; 2 pls., 28 figs.

Martinez, L. (1940). Nota bioestadistica acerca de los anfidiscos de las esponjas *Ephydatia fluviatilis mexicana* et *Heteromyenia repens spinulosa* del lago de Xochimilco. *An. Inst. Biol. Univ. Mex.* **11** (1): 191–196: 4 figs.

Massey, F. J. (1950a). A note on the estimation of a distribution function by confidence limits. *Ann math. Statist.* **21**: 116–119.

Massey, F. J. (1950b). A note on the power of a non-parametric test. *Ann. math. Statist.* **21**: 440–443.

Massey, F. J. (1951a). The distribution of the maximum deviation between two sample cumulative step functions. *Ann. math. Statist.* **22**: 125–128.

Massey, F. J. (1951b). The Kolmogorov–Smirnov test for Goodness of Fit. *J. Am. Statist. Ass.* 68–78; 2 figs.

Meewis, H. (1940). Contribution à l'étude de l'embryogénèse des Chalinidae: *Haliclona limbata* (Mont.). *Ann. Soc. r. Zool. Belg.* **70**: 201–243.

Proctor, J. R. and Kendrick, W. B. (1963). Unequal weighting in numerical taxonomy. *Nature Lond.* **197** (4868): 716–717.

Rasmont, R. (1963). La rolê de la taille et de la nutrition dans le détérminisme de la gemmule chez les Spongillides. *Devl. Biol.* **8** (3): 243–271; 4 figs.

Ridley, S. and Dendy, A. (1887). Report on the Monaxonida collected by H.M.S. "Challenger". *Challenger Rep.* **20** (59): 275 pp.; 51 pls.

Rioja, E. (1940). Estudios hidrobiologicos. I. Estudio critico sobre las esponjas del Lago de Xochimilco. *An Inst. Bid. Univ. Mex.* **11** (2): 173–189; 3 pls.

Salt, G. and Hollick, F. S. J. (1946). Studies of wireworm populations. *Br. J. exp. Biol.* **23** (1): 1–46; 4 figs.

Sarà, M. (1951). Sulla varabilita dei caratteri tassonomici e significato della specie nelle Calcispongie. *Annali Ist. Mus. Zool. Univ. Napoli* **3** (5): 1–6.

Sarà, M. (1953a). Variabilita di forma e di spicolazione in *Leuconia solida* (O.S.) e sua identita specifica con *L. nausicaae* (Schuffn.) e *L. pumila* (Bwk.) (Calcispongiae). *Annali Ist. Mus. Zool. Univ. Napoli* **5** (5): 1–22; 1 pl.

Sarà, M. (1953b) Variabilita delle *Leucosolenia* del Golfo di Napoli e nuove vedute sistematica del gruppa. *Annali Ist. Mus. Zool. Univ. Napoli* **5** (7): 1–109; 2 pls; 14 figs.

Sasaki, N. (1941). On the changes occurring in various parts of the body, especially in the spicules, in accordance with the increase of body length in the case of the calcareous sponge *Sycon Okadai* Hozawa. *Sci. Rep. Tôhoku Univ.* Ser. IV, Biol., **16** (4): 365–382; 12 figs.

Simpson, T. L. (1965). Sponge cells and the taxonomy of sponges. *Am. Zool.* **5** (2): 103.

Smirnov, N. (1948). Table for estimating the Goodness of Fit of Empirical Distributions. *Ann. math. Statist.* **19**: 279–281.

Sneath, P. H. A. (1967). Some statistical problems in numerical taxonomy. *Statistician* **17** (1): 1–12.

Snedecor, G. W. and Cochran, W. G. (1967). *Statistical methods.* 6th Edn.: 593 pp. Ames, Iowa: Iowa State University Press.

Sokal, R. R. (1965). Statistical methods in systematics. *Biol. Rev.* **40**: 337–391; 5 figs.

Sokal, R. R. and Sneath, P. H. A. (1963). *Principles of Numerical Taxonomy.* 359 pp. San Francisco and London: W. H. Freeman.

Sollas, W. J. (1888). Report on the Tetractinellida collected by H.M.S. "Challenger" during the years 1873–76. *Challenger Rep.* **25**: 458 pp.; 44 pls, 1 map.

Tate, M. W. and Clelland, R. C. (1959). *Nonparametric and Shortcut Statistics.* 171 pp. Danville, Illinois: Interstate Printers & Publishers.

Taylor, B. J. (1965). Analysis of Polymodal Frequency Distributions. *J. anim. Ecol.* **34** (2): 445–452; 5 figs.

Traxler, L. (1896). Uber einem neuen süsswasserschwamm aus neu-zealand (*Ephydatia kakahuensis* n. sp.). *Természetr. Füz.* **19** (1): 102–104.

Wells, H. W., Wells, M. J. and Gray, I. E. (1960). Marine sponges of North Carolina. *J. Elisha Mitchell scient. Soc.* **76**, (2): 200–245; 63 figs.

Weltner, W. (1910). Spongillidae. *Fauna Südwest-Aust.* **3** (5): 137–144.

Symp. zool. Soc. Lond. (1970) No. 25, 163–187

LES POTAMOLEPIDES AFRICAINES NOUVELLES DU LUAPULA ET DU LAC MOERO

Polyphylétisme des éponges d'eau douce

PAUL BRIEN

Institut Zoologique Torley Rousseau
Université Libre de Bruxelles, Bruxelles, Belge

SYNOPSIS

The discovery of a number of distinctive new forms of fresh-water sponges in central and west Africa points to the need for a re-examination of the hypothesis of several invasions of fresh waters by marine sponges of different families. Such a re-examination confirms the opinions of earlier authors that the fresh water sponges are a polyphyletic group. The diversity of types of asexual reproductive bodies occurring in these sponges provides very strong evidence for polyphyly. A tentative classification of fresh-water sponges, which takes into account such polyphyly and which includes the new genus *Potamophloios* and subfamily *Potamophloinae*, is proposed.

LES ESPÈCES DU BAS-CONGO, DU TANGANIKA ET DU KATANGA

En 1883, Marshall créait le genre *Potamolepis* d'après l'étude d'espèces nouvelles d'éponges d'eau douce récoltées par le Dr. Pechuel-Losche dans les régions des rapides du fleuve Congo, en amont de Matadi, aux environs d'Isangila. Selon les indications du Dr. Pechuel-Losche, reproduites par Marshall, ces éponges minces recouvraient des roches qui sont émergées aux basses eaux: juin, juillet, août et mi-septembre. Exposées au soleil, elles ont l'aspect de fientes d'oiseaux répandues sur les pierres. Aux hautes eaux, ces mêmes roches sont submergées à deux ou trois mètres de profondeur. De cette récolte, Marshall décrivait trois espèces: *P. chartaria, leubnitziae, pechuëlii.*

En 1903, Moore a fait connaître *Potamolepis weltneri* (1883) récoltée dans le lac Tanganika.

En 1938, Burton ajoutait encore deux autres espèces: *P. marshalli*, et *P. micropora* provenant des récoltes que j'avais faites, en octobre 1937, sur le bateau-drague qui, en face de Matadi, retirait du lit du fleuve, à quatre ou cinq mètres de profondeur, des pierres qui en étaient couvertes.

Dans la même étude, Burton révélait *Potamolepis schoutedeni* très abondante sur les rives rocheuses du Stanley Pool (Kinshasa) où

A. Tinan les avait observées la première fois. Toutefois, cette espèce doit être écartée du genre *Potamolepis*. Elle appartient à un autre genre, *Parametania* (Brien, 1968), de la sousfamille des *meyeniidae*, famille des *Spongillidae*. *Potamolepis leubnitziae*, *P. chartaria* (Marsh.), *pechuëlii* (Marsh.), *weltneri* (Moore), *marshalli* (Burton) *micropora* (Burton) ont une constitution très semblable. Ce sont des éponges minces, de 1 mm à 1 mm ½ d'épaisseur, revêtantes, étalées sur la roche. Elles sont dures, friables lorsqu'elles sont désséchées. Leur teinte varie du jaune pâle au gris de plomb, parfois elles sont d'un brun noisette (*P. pechuëlii*, *P. chartaria*). Elles adhèrent au support par une lame cuticulaire d'adhésion.

Le mésenchyme est soutenu par un squelette réticulé d'une texture rigide et serrée. Il est formé d'amphistrongyles incurvés, lisses ou parfois échinés (*P. micropora*), associés parallèlement par de la spongine peu abondante, en faisceaux constituant des travées délimitant les mailles dont le diamètre est variable selon les espèces: environ 300 μ. Ces mailles contiennent les éléments du mésenchyme, les canaux et les lacunes aquifères. En surface une membrane dermique, perforée de stomions inhalants, passe par dessus les mailles superficielles et les petites cavités hypodermiques qu'elles contiennent. La membrane dermique peut être soutenue par de fins spicules effilés, petits amphioxes disposés tangentiellement mais sans ordre (*B. chartaria*). La surface de l'éponge est donc unie, et dans les échantillons secs, la disposition des spicules des mailles de la surface lui donne un aspect satiné (*P. leubnitziae* et *P. pechuelii*) ou de papier buvard (*P. chartaria*). Toutefois, la surface des échantillons secs, par suite de la disparition de la membrane didermique paraît criblée de petits trous qui correspondent aux mailles du réseau squelettique. Enfin à la surface de ces éponges, les oscules sont toujours très visibles, très régulièrement répartis, presqu'également distants. Ils s'ouvrent au niveau de la surface ou sur une très légère élévation cohique (*P. leubnitziae*, *P. marshalli*, *P. micropora*). Leur orifice est régulier (*P. chartaria*, *P. marshalli*, *P. micropora*), parfois plus irréguliers (*P. leubnitziae*). Dans cette dernière espèce, on distingue très nettement les canaux aquifères exhalants superficiels, disposés en étoiles aux rayons irréguliers et bifurqués convergeants vers la base de l'oscule. Chez *P. pechuëlii*, les oscules sont au sommet de tubes osculaires légèrement côniques de plus de 5 mm de hauteur, pareils à de petites cheminées parallèles, presqu'équidistantes, perpendiculaires à la surface ou inclinées selon un même angle.

Ces éponges forment un groupe très homogène. Leurs caractères communs si distinctifs s'expliqueraient selon Marshall (1883) par leur écologie. Le courant rapide de l'eau exercerait sur leur développement une pression continue à laquelle correspondraient la minceur de l'éponge, sa surface unie, le squelette serré et rigide, au point que ces éponges, d'après cet auteur, rappelleraient les éponges Hexactinelles plutôt que les Spongillidae. Il ajoute d'ailleurs "... The skeletal elements of the sponges are, for the most part, to be referred to influence of purely mechanical nature ..." ce que l'observation révèle si souvent dans des éponges de la même espèce mais vivant dans des condition écologiques très différentes. L'influence du courant se manifeste encore, selon Marshall, sur les tubes osculaires de *P. pechuëlii* qui sont parfois inclinés parallèlement et sans doute dans le sens du courant.

Ce qui donne à ces éponges *Potamolepis* une grande originalité, c'est qu'aucun specimen ne possède de "gemmules". Cette constatation sera confirmée plus tard par Moore (1903), Burton (1938), Jaffé (1917). Marshall (1883) admet que les espèces du genre *Potamolepis* ne présentent pas de reproduction asexuée, alors qu'elles sont dans les conditions d'en produire, puisqu'elles sont exposées à la dissécation aux eaux basses. De toute façon, déclare-t-il, les mailles du squelette sont trop étroites que pour contenir des statoblastes! Les *Potamolepis* ne sont d'ailleurs pas les seules éponges d'eau douce à ne pas former des bourgeons résistants; *Lubomirskia baikalensis* du lac Baikal en est dépourvue.

Selon Marshall, l'absence de reproduction asexuée serait due à leur introduction récente en eau douce "... just like the *Lubomirskia* and probably also *Uruguaya coralloides*, they (*Potamolepis*) have adapted themselves to existence in fresh water at a recent date in comparison with the *Spongillinae* and *Meyiniinae*".

Marshall (1883) conclut au *polyphylétisme dans le temps* des Eponges d'eau douce qui toutes dériveraient, selon lui, d'une souche commune marine, peut-être les *Renieridae*, (Ridley et Dendy, 1887) mais ayant envahi les eaux douces à des moments différents ce qui expliquerait leurs caractères distinctifs acquis indépendamment et secondairement.

Cette définition et cette conception du genre *Potamolepis* doivent être précisées aujourd'hui.

En 1916, Jaffé décrit une éponge nouvelle qu'il déclarait provenir du Tanganika. En réalité elle a été récoltée en 1911 par Stappers dans le Luapula (Henri Schouteden, 1918). Cette éponge fut dénommée *Potamolepis stendelli*. L'holotype est aujourd'hui dans les collections du Musée royal d'Afrique centrale à Tervueren. Grâce à la complaisance de Mr. le Professeur J. Benoit, Chef de la section des Invertébrés, j'ai pu réexaminer ces échantillons. Ils appartiennent à une éponge mince à surface lisse, étalée sur une coquille d'*Aetheria elliptica*. Elle est jaunâtre; la texture de son squelette est celle des *Potamolepis* du Bas-Congo, décrites par Marshall (1883) et Burton (1938).

En 1966–67, le Professeur J. Symoens, directeur du Centre d'Hydrobiologie de l'Univesité de Lubumbashi, m'a donné l'occasion d'examiner un très beau spécimen d'une éponge de Luapula prélevée à Kiniama non loin de l'endroit où Stappers avait récolté *P. stendelli* Jaffé. En outre, Mr. J. J. Symoens a mis à ma disposition de remarquables exemplaires provenant cette fois du lac Moero, près de Lukonzolwa, ainsi que des fragments de la même éponge soigneusement fixés au formol picriqué (liquide de Bouin).

Tout concorde pour identifier les éponges du Luapula et du Moero à *Potamolepis stendelli* Jaffé. Il faut cependant souligner que les petits fragments étudiés par Jaffé ne peuvent donner une idée réelle et complète de cette éponge en toute sa croissance (Brien, 1967a).

Potamolepis stendelli (Jaffé, 1916) (Fig. 1b) est une espèce encroûtante, étalée sur des dalles rocheuses. Son épaisseur dépend de l'état de croissance. Dans la région marginale elle est à peine de un ou de deux millimètres; dans la région plus âgée, elle est plus considérable. La teinte varie d'un gris clair au gris de plomb. La surface de l'éponge n'est unie que dans les régions les plus jeunes. Au fur et à mesure de la croissance, la surface se hérisse d'aspérités spiculées de même texture que le reste de l'éponge. Ces aspérités, elles-mêmes, s'acroissent en se digitant en aspérités secondaires. De plus, la surface dans les régions les plus épaisses, se bosselle de protubérances à sommet arrondi d'autant plus hautes que l'éponge est plus épaisse s'élevant alors en mamelons de deux centimètres qui se lobent à leur tour devenant des boursouflures très irrégulières hérissées ellesmêmes d'aspérités.

De place en place, la surface est trouée d'oscules, peu-nombreux, très espacés, irrégulièrement répartis. Ils s'ouvrent au niveau de la

---

FIG. 1. (page ci-contre) a, Exemplaire de *Potamophloios symoensi* (Brien, 1967) récoltée par J. J. Symoens et Gaston de Witte dans le Luapula—Musée royal d'Afrique centrale Tervueren. Longueur 15 cm; largeur 10 cm; épaisseur 5 cm.
b, Exemplaire de *Potamophloios stendelli* (Jaffé, 1916) récoltée par J. J. Symoens au Luapula et lac Moero (Lukonzolwa). Musée royal de l'Afrique centrale Tervueren. Longueur 17 cm; largeur 10 cm; épaisseur de 1 mm à 1 cm ½ et 2 cm.
c, Exemplaire de *Potamophloios gilberti* (n. sp.) récoltée par R. P. Gilbert, lac Moero, embouchure de la Kasinsa. Musée royal de l'Afrique centrale Tervueren. Hauteur 50 cm; largeur 15 cm; épaisseur 5 à 7 cm.
d, Exemplaire de *Potamophloios hispida* (n. sp.) récoltée par R. P. Gilbert, lac Moero, embouchure de la Kasinsa. Musée royal de l'Afrique centrale Tervueren. Longueur 24 cm; largeur 15 cm; épaisseur de 2 mm à 3 cm.
e, Exemplaire de *Potamophloios songoloensis* (n. sp.) récolté par R. P. Gilbert, embouchure du lac Kasinsa, entre Lukonzolwa et Songolo. Musée royal de l'Afrique centrale Tervueren. Hauteur 23 cm, largeur 12 à 14 cm.

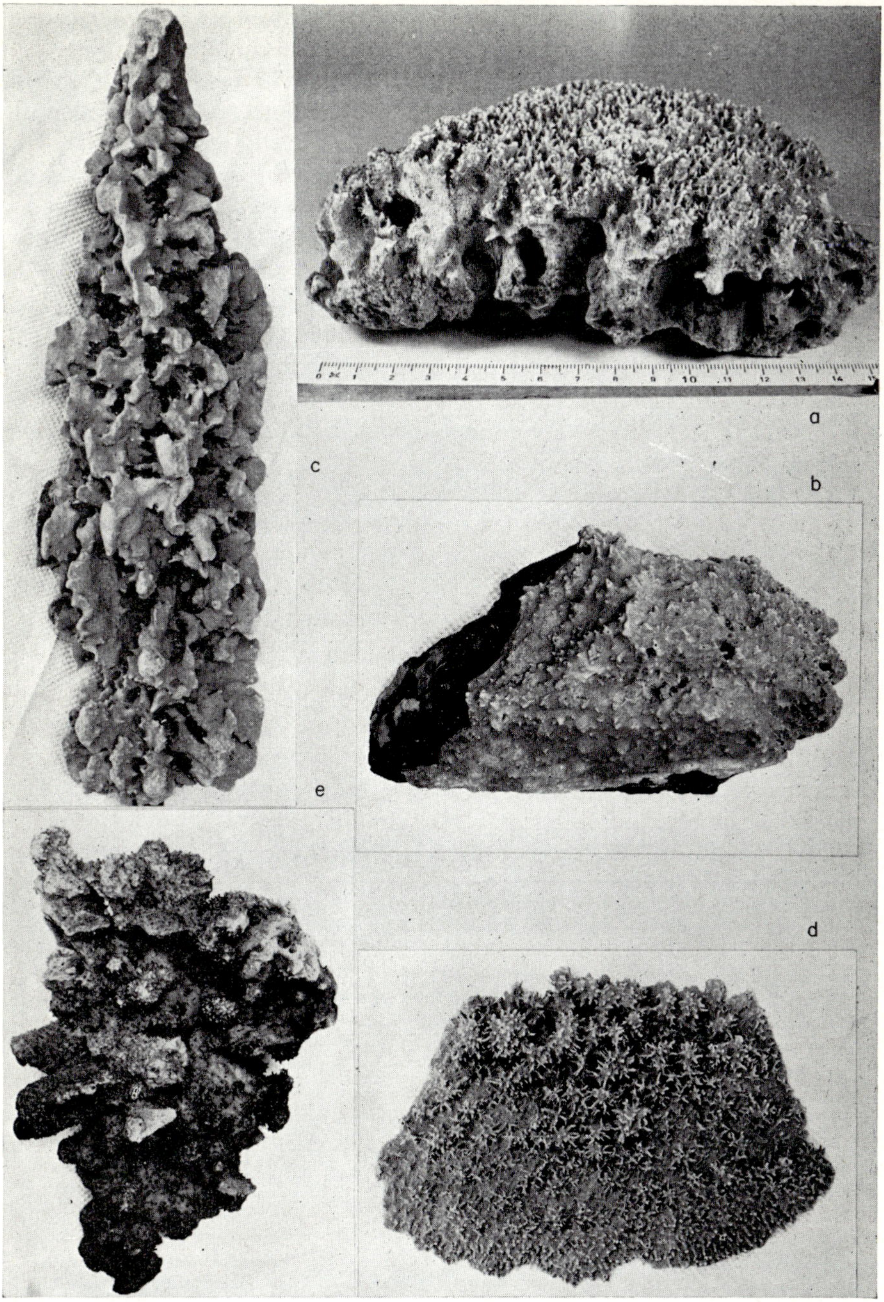
a
b
c
d
e

surface ou le plus souvent sur une forte aspérité ou au sommet d'une protubérance. Dans chaque cas, les canaux exhalants superficiels, disposés en étoile, convergent vers l'oscule et délimitent ainsi l'espace leuconoïde et son aire osculaire. Sur les échantillons secs, la surface est en outre criblée de petits trous, ce sont les mailles superficielles du réseau squelettique qui sur le vivant sont recouvertes de la membrane dermique. Le mésenchyme est soutenu par une charpente spiculée et réticulée de même type que celle des *Potamolepis* du Bas Congo. Elles consiste en amphistrongyles de 220 à 240 μ légèrement incurvés, lisses, associés parallèlement en faisceaux qui constituent les travées du réseau squelettique délimant des mailles de 300 à 400 μ. Celle-ci soutiennent les éléments du mésenchyme, les canaux et les lacunes aquifères.

Le lame dermique qui recouvre les mailles superficielles est soutenue par des amphioxes lisses, légèrement courbes, plus grêles mais de taille variable 150 à 200 μ.

L'éponge étalée sur la dalle rocheuse, y adhère par une lame cuticulaire consolidée par des amphistrongyles qui s'y déposent tangentiellement et sans ordre. Cette lame d'adhésion est unie, parcheminée sur les échantillons secs, détachés de la roche (Brien, 1967a, voir Phot. 2 Pl. I). *Habitat*: Luapula (Kiniama), Lac Moero (Lukonzolwa); *Dépôt*: Musée royal de l'Afrique central (Tervueren-Belgique).

La même année (Brien 1967 a,c), je décrivais un fragment d'une épongeinconnue, récoltée par J. J. Symoens en 1965 également à Kiniama sur le Luapula: *Potamolepis symoensi* (n.sp.) dont un échantillon remarquable et en pleine croissance était conservé à sec dans les collections du Musée de l'Afrique centrale à Tervueren. Il avait été recueilli par Gaston de Witte en 1930 à Kabunda sur le Luapula (Fig. 1a).

FIG. 2 (page ci-contre). a, Fragment de la surface de *Potamophloios hispida* (n. sp.) à un plus fort grossissement, montrant les protubérances de la surface hérissées d'aspérités. Deux d'entr' elles portent chacune un oscule.
b, Fragment de la base de *Potamophloios symoensi* (n. sp.) à un plus fort grossissement, montrant dans des anfractuosité, des "nids" de Statoblastes.
c, Fragment de la charpente réticulée rigide du squelette formé de faisceaux d'amphisstrongyles en *Potamolepis pechuelii* (Marshall, 1883); d, en *Potamophloios gilberti* (n.sp.); e, en *Potamophloios songoloensis* (n. sp.).
f, Un statoblaste de *Potamophloios symoensi* montrant la façon dont les spicules sont enrobés dans l'enveloppe cuticulaire pour consolider la coque.
g, Groupe de statoblastes de *Potamophloios stendelli* (Jaffé, 1916) incrustés dans la lame cuticulaire d'adhésion, celle-ci étant consolidée par les mêmes types de spicules que ceux de la coque des statoblastes.
h, Groupe de statoblastes a *Potamophloios hispida* (n. sp.) incrustés dans la lame cuticulaire d'adhésion. Comme dans la figure précédente, les statoblastes les plus rapprochés sont coalescent par leur coque.
i, Types de spicules de la coque des statoblastes chez toutes les espèces de *Potamophloios*: amphistrongyles pareils à ceux de la charpente squelettique, microsclères de formes et de tailles variables.

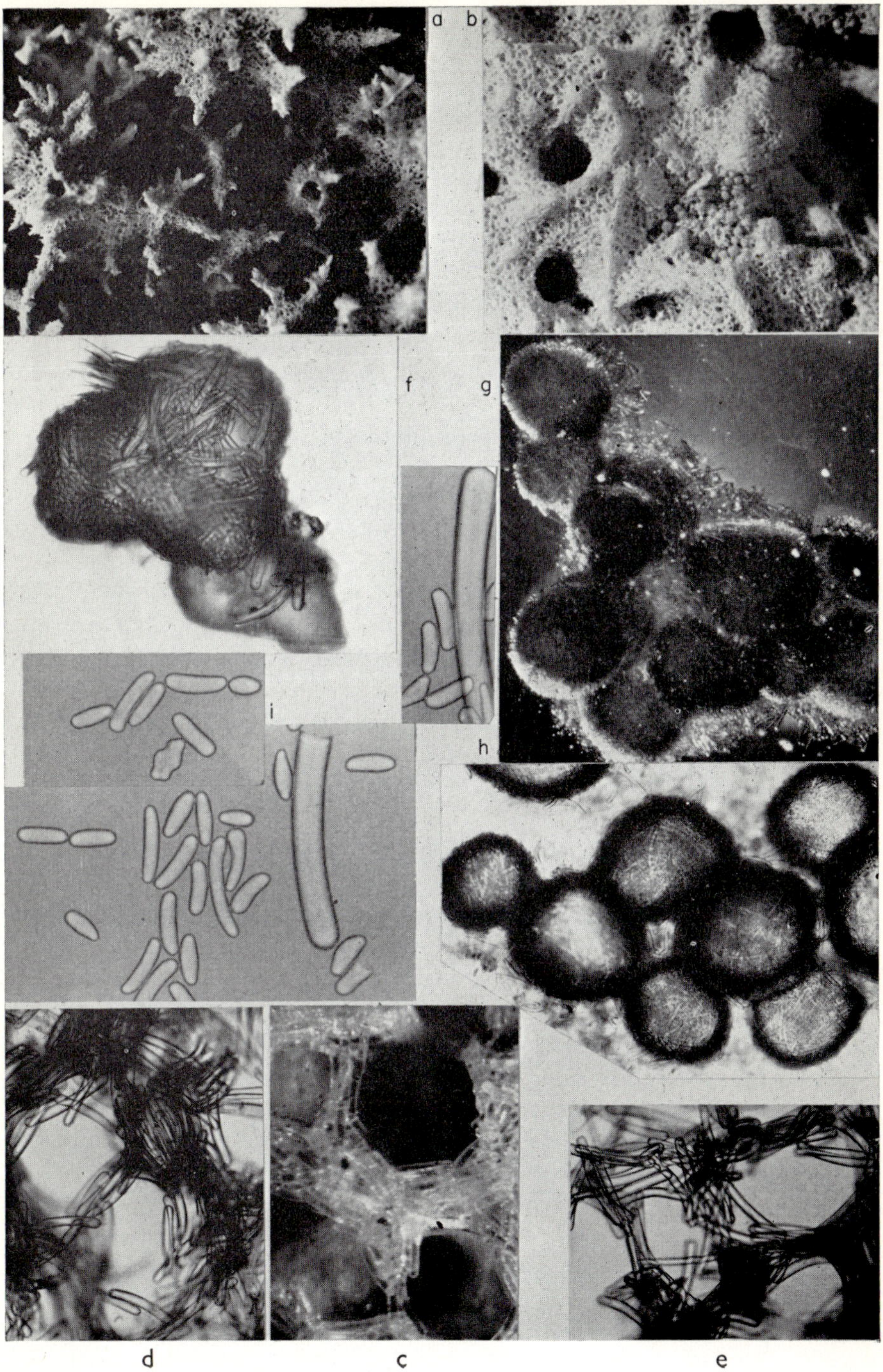
a
b
f
g
i
h
d
c
e

Potamolepis symoensi (Brien, 1967) est une éponge encroûtante, épaissie en gâteau. Elle s'étend sur 17 cm, son épaisseur est de cinq cm. Elle est d'un jaune doré. Sa face supérieure est hérissée de petites aspérités pareilles à de fines digitations irrégulières et hispides. Cette éponge est en réalité caverneuse, comme si elle était constituée de piliers irréguliers plus ou moins ramifiés et enchevêtrés, confluant par leurs base et à leur sommet. La face supérieure, hérissée d'aspérités, n'est donc pas continue, mais présente des excavations séparant les pilliers irréguliers dont est constituée l'éponge. Entre les aspérités dressées en tous sens, on distingue, quelque peu cachés sous leuren chevêtrement, les orifices circulaires des oscules peu nombreux et très irrégulièrement répartis.

La face inférieure est unie et soyeuse. Ses irrégularités sont dues aux dé nivellations du support que l'éponge épouse. Enfin on retrouve des creux laissés par la confluence incomplète des pilliers formant l'éponge (Fig. 2b).

Le squelette a une charpente réticulée. Elle est formée d'amphistrongyles légèrement incurvés, lisses, dont la corde de l'arc est environ de 200 μ. Ces amphistrongyles sont associés parallèlement en faisceaux par la spongine. Les faisceaux spiculés forment les travées du réseau squelettique délimitant des mailles qui contiennent les éléments du mésenchyme. Il s'ajoute à ces amphistrongyles, des amphioxes de tailles variables qui soutiennent surtout la lame dermique. Celles-ci recouvrent les mailles squelettiques contenant les cavités hypodermiques inhalantes. (Brien, 1967). *Habitat*: Luapula (Kiniama—Kabunda); *Dépôt*: Musée royal de l'Afrique central Tervueren-Belgique.

En 1968, le Professeur J. J. Symoens mit à ma disposition trois specimens d'éponge recueillis dans le lac Moero par le R. P. Gilbert de la mission franciscaine de Lukonzolwa. Ces éponges étaient désséchées mais avaient été prélevées en pleine croissance. Ce sont trois admirables échantillons de collections. Si distinctes qu'elles soient, par leur aspect. ces trois éponges se rapprochent des deux especes *P. stendelli*, *P. symoensi* qui viennent d'être décrites. Toutefois l'étude de ces trois espèces nouvelles, leur comparaison avec *Potamolepis stendelli* (Jaffé, 1917) et *Potamolepis symoensi* (Brien, 1967) dont il vient d'être question, m'amènent pour des raisons qui seront développées plus loin à inclure ces cinq espèces du bassin du Luapula et du lac Moero dans un genre nouveau, le genre *Potamophloios* (n. gen.). Le genre *Potamophloios* comprend donc *Potomophloios stendelli* (Jaffé = *Potamolepis stendelli* Jaffé), *Potamophloios symoensi* (Brien) (= *Potamolepis symoensi* Brien) et les 3 espèces *P. hispida* (n. sp.), *P. gilberti* (n. sp.), *P. songoloensis* (n. sp.) qui vont être décrites sucessivement.

Potamophloios hispida (Fig. 1d, Fig. 2a), est une éponge encroûtante d'un gris sale, étalée sur une dalle rocheuse plate. Son épaisseur est variable selon l'état de croissance. Dans la région marginale elle est de 2 à 3 mm. Aux endroits les plus développés, elle est d'un centimètre mais l'éponge est boursoufflée de grosses proéminences de 3 à 4 cm de hauteur.

Cette éponge rappelle quelque peu *P. stendelli*. Conservée à sec, sa surface est criblée d'orifices plus ou moins arrondis de 300 à 500 μ qui correspondent aux mailles superficielles du réseau squelettique.

La surface, ainsi perforée de ces trous, n'est pas unie mais partout fortement hérissée d'aspérités spiculées, hispides, de la même consistance que l'ensemble de l'éponge. Relativement régulières dans la région marginale, elles n'y dépassent pas 1 mm. de haut. Au fur et à mesure qu l'on se rapproche des régions plus épaisses, plus centrales, ces aspérités se surélèvent, atteignent 3 à 5 mm. En même temps, elles se compliquent d'aspérités secondaires orientées en tous sens, ce qui leur donne l'aspect de minuscules bois de cerf, très irréguliers (Fig. 2a).

Ces aspérités rappellent donc celles décrites sur les exemplaires de *P. stendelli*, récoltés en 1965, mais elles sont beaucoup plus accentuées, plus hautes, ramifiées, et aussi plus nombreuses et plus denses.

Parmi les aspérités encore petites, coniques et pointues de la région marginale, se dressent, de place en place, des tubules à peine plus grands mais au sommet de chacun desquels s'ouvre un oscule. Les tubes osculaires sont disposés selon des traînées sinueuses, parallèles au bord de l'éponge et qui marquent sans doute le processus de croissance et d'expansion de l'éponge. Mais au fur et à mesure que l'on examine les régions plus anciennes, plus épaisses de l'éponge, ils proéminent fortement entre les aspérités, elles-mêmes accrues. Ils s'élargissent, se surélèvent, deviennent des protubérances en tourelles dont le sommet est occupé par l'orifice osculaire ; leurs parois irrégulières, ont la texture de l'éponge et se hérissent de fortes aspérités digitées en tous sens (Fig. 2a). Dans les régions les plus épaisses, ces tourelles osculaires se dilatent, atteignent de 2 à 3 cm, se compliquent. Elles se dressent sur l'éponge, dardées d'aspérités épineuses, comme de jeunes cactées ou de petits échinocereus. Elles ont gardé cependant la disposition première des tubes osculaires initiaux, si bien que ces dômes osculaires, hérissés et digités, sont plus ou moins alignés en crêtes sinueuses. Ces crêtes confluent entre elles et séparent des zones de la surface déprimées en vallées profondes et étroites. Celles-ci finissent par être comblées à leur tour par des confluences transversales, de telle sorte que l'ensemble forme une masse plus ou moins caverneuse donnant à l'éponge une épaisseur de 3 à 4 cm.

Le mésenchyme est soutenu par un squelette spiculé, rigide, réticulé, constitué de faisceaux de spicules et délimitant des mailles vaguement arrondies ou polyédriques.

Les spicules sont surtout des amphistrongyles lisses, légèrement arqués, ayant tenance à se renfler un peu aux extrémités (200 à 250 μ). Il s'y mêlent, en faible abondance, des strongyles lisses plus grêles et lisses, de 150 à 200 μ, des tornotes et des amphioxes de 190 à 250 μ. Ceux-ci sont répandus dans tout le mésenchyme mais plus particulièrement dans la membrane dermique qui sur le vivant recouvre les mailles superficielles et les cavités hypodermiques qui y sont contenues (Brien, 1969). *Habitat*: Lac Moero; *Dépôt*: Musée royal de l'Afrique centrale, Tervueren–Belgique.

Potamophloios gilberti (*n. sp.*) Il s'agit d'un specimen remarquable d'une grande colonie d'éponges enveloppant la tige verticale d'un roseau de 2 cm de diamètre, lui formant un manchon d'une hauteur de 50 cm et d'une largeur de 15 cm. Elle fut recueillie en 1966 entre Lukonzolwa et Songolo, dans le lac Moero, à l'embouchure de la Kansinsa. L'éponge est d'un jaune doré. Tout en recouvrant totalement la tige d'une roseau, elle se surélève en crêtes sinueuses, quelque peu parallèles et plus ou moins longitudinales, chacune atteingnant 4 à 5 cm de hauteur. Ces crêtes sinueuses, longitudinales et irrégulières, sont elles-mêmes bosselées; elles émettent sur leurs faces des excroissances latérales qui, se rencontrant, confluent entre elles, de telle sorte que l'éponge forme autour du roseau axial une masse pyramidale caverneuse, dont la plus forte largeur est de 15 cm. Elle ferait songer à une petite termitière (Fig. 1c).

Cette puissante colonie est cependant d'une texture fine. Sa surface unie paraît criblée de petits orifices de 300 à 350 μ placés côte à côte, et qui sont les mailles superficielles du réseau squelettique contenant les cavités hypodermiques.

Quoiqu'unie, la surface est cependant légèrement bosselée et, de place en place, se hérisse d'aspérités grêles, aigües, de $\frac{1}{2}$ cm de hauteur, coniques, mais à base étroite. Elles sont de la même texture spiculée que l'ensemble. Très souvent, au sommet de ces aspérités coniques, s'ouvre un oscule de 1 mm de diamètre. Ces éminences osculaires sont irrégulièrement réparties à la surface de l'éponge. Lorsque, sur ce specimen conservé à sec, le tube osculaire est dans son intégrité, on peut voir à sa base le tracé convergeant des canaux exhalants superficiels.

Le mésenchyme est soutenu par une charpente spiculée et réticulée, rigide, constituée par des faisceaux d'amphistrongyles formant des rêts qui délimitent des mailles arrondies vaguement polyédriques.

Les spicules sont des amphistrongyles lisses, légèrement incurvés de 200 à 240 μ auxquels se mêlent, peu abondants, des strongyles lisses plus grêles de 200 à 207 μ, des tornotes et des amphioxes lisses arqués, de taille variable et répandus en désordre dans le mésenchyme (Brien, 1969). *Habitat*: Lac Moero–Haut Katanga; Musée Royal d'Afrique Centrale–Tervueren Belgique.

Potamophloios songoloensis (*n. sp.*). Cette éponge a une hauteur de 23 cm et une largeur de 12 à 14 cm. L'éponge cependant n'est pas massive. Elle est constituée de grosses excroissances digitées cylindriques, enchevêtrées de 1 à 3 cm d'épaisseur confluantes entre elles de façon à former une masse caverneuse.

L'éponge s'est développée en enveloppant un chevelu de racines ou de brindilles, si bien qu'elle devait être suspendue dans l'eau.

Toute la surface de l'éponge est légèrement ondulée. Elle paraît criblée de petits trous qui correspondent aux cavités délimitées par les mailles du réseau spiculé superficiel et sur lesquels, lorsque l'épongê est vivante s'étend le voile mince de la membrane dermique. Sur l'éponge conservée à sec, ces petits orifices placés côte à côte, plus ou moins circulaires, ont un diamètre de 300 à 500 μ.

Sur toute son étendue, la surface de l'éponge se hérisse d'aspérités coniques, aigües, irrégulièrement réparties, plus ou moins accentuées selon les endroits. Ces sortes d'épines cȏniques ont la même texture que le reste de l'éponge. Ce sont donc des surélévations du mésenchyme et de sa charpente spiculée.

De place en place, assez regulièrement espacés, on distingue les étoiles que forment les canaux exhalants convergeant vers un oscule. Lorsque la surface n'a pas été endommagée, notamment dans les concavités où elle est le mieux protégée, du centre de cette zone étoilée vers lequel convergent les canaux, s'élève un tube grêle, creux et ouvert au sommet par l'oscule. Ce petit tube peut atteindre $\frac{1}{2}$ cm. Il est très fragile. L'oscule a un diamètre de 1 mm. Le plus souvent, dans l'exemplaire conservé à sec, ces tubes osculaires sont brisés et l'on, ne distingue plus que la zone étoilée osculaire qui forme les canaux exhalants.

Le mésenchyme épais est soutenu par une charpente spiculée rigide et réticulée, constituée de faisceaux d'amphistrongyles et délimitant des mailles vaguement arrondies ou polyédriques (Fig. 2e).

Les spicules sont surtout des amphistrongyles lisses légèrement arqués, de 220 à 256 μ auxquels se mêlent très abondamment des strongyles plus grêles, arqués, de 160 à 195 μ et des amphioxes lisses arqués, de taille variable, de 90 à 175 μ (Brien, 1969). *Habitat*: Lac

Moero, près de Songolo, embouchure de la Kansinsa; *Dépôt*: Musée royal de l'Afrique centrale, Tervueren–Belgique.

LE GENRE *POTAMOLEPIS* (MARSHALL, 1883) ET LE GENRE *POTAMOPHLOIOS*

P. stendelli, *P. symoensi*, *P. gilberti*, *P. spongoloensis*, *P. hispida*, appartiennent au bassin de Luapula-Moero du Haut Katanga. De même provenance, elles ont une même structure fondamentale, une spiculation presqu'identique. La charpente squelettique réticulée, serrée donne à ces espèces leur rigidité, leur dureté, mais aussi leur friabilité. Elles se distinguent certes par leur aspect extérieur, le support sur lequel elles se développent, et quelques particularités du squelette, la densité, l'importance des aspérités, qui hérissent leur surface plus ou moins bossuées de protubérances, plus ou moins développées. Elles n'en constituent pas moins un groupe homogène. Elles ont en commun d'adhérer à leur support, dalle rocheuse, tige de roseau, chevelu de brindilles ou de radicelles, par une lame cuticulaire, consolidée par le même type de spicules que ceux de la charpente squelettique dont les amphistrongyles incurvés et lisses sont les plus importants. A la suite de Jaffé (1917) ces éponges ont été incluses dans le genre *Potamolepis*. Les fragments de *P. stendelli* décrits par Jaffé justifiaient cette opinion. Mais par ce qui vient d'être rappelé, ces éponges, en pleine croissance, n'en sont pas moins très différentes des espèces de *Potamolepis* minces à surface lisse, aux oscules régulièrement repartis, qui vivent dans des rapides du Bas-Congo, du dans le Tanganika. Celles-ci sont bien pareilles à des "écailles" fixées sur les roches du lit des rivières. Elles portent bien leur nom de *Potamolepis*. Par contre, les espèces du *Luapula–Moero* sont épaisses, elles recouvrent ou enveloppent leur support d'une écorce irrégulière, bossuée, hérissée d'aspérités souvent de plusieurs centimètres d'épaisseur. Pour cette raison, je les dénomme du nom générique: *Potamophloios*.

Le genre *Potamophloios* comprend donc des éponges qui recouvrent ou enveloppent comme d'une écorce dure et rugueuse, tout support immergé: pierres, roseaux, branches, racines. Elles sont plus ou moins épaisses, lobées et parfois caverneuses. Leur surface est toujours hérissée d'aspérites plus ou moins nombreuses et bossuées de protubérances irrégulières, elles-mêmes couvertes d'aspérités. Les oscules irrégulièrement répartis, s'ouvrent fréquemment au sommet de tubules ou de protubérances plus ou moins fortes. Le mésenchyme est soutenu par un réseau squelettique serré et rigide formé de travées d'amphistrongyles lisses, arqués, auxquels se mêlent des strongyles plus grêles, des amphioxes, parfois des tornotes, légèrement arqués et de taille variable. Tous les spicules sont lisses (Fig. 2c,d,e).

L'habitat jusqu'à présent connu est le bassin du Luapula–lac Moero dans le Haut Katanga.

LA STATOBLASTOGENÈSE DANS LE GENRE *POTAMOPHLOIOS*

Les cinq espèces du genre *Potamophloios* du Moero–Luapula se distinguent par un caractère biologique plus important encore. Toutes sont capables de former des *statoblastes.*

Marshall (1883) en créant le genre *Potamolepis,* ainsi qu'il fut rappelé précédemment était arrivé à la conviction que les éponges de ce genre ne produisaient pas de bourgeons de dissémination. Les auteurs qui l'ont suivi, Moore (1903), et Burton (1938), l'ont confirmé. Ils n'ont jamais observé de "gemmules" dans les espèces de *Potamolepis* qu'ils ont décrites.

Toutefois, dans une petite colonie de *Potamolepis pechuëlii* que j'avais récoltée dans le fleuve Congo en face de Matadi en 1937, j'ai observé des statoblastes très semblables à ceux dont il va être question à propos des éponges du Luapula et de Moero (Brien, 1967c).

D'autre part, si Jaffé lorsqu'il fit connaître *Potamolepis stendelli* déclare n'y avoir vu aucune "gemmule", c'est que celles-ci lui avaient échappé. Elles sont bien présentes, dans l'échantillon examiné; elles sont à leur place, contre le support, (Brien, 1967a,c).

Il fut dit précédemment que les éponges adhéraient à leur support, quel qu'il soit, par une lame cuticulaire. Cependant dans la région basale au contact du support, le mésenchyme subit de profondes modifications. La charpente squelettique réticulée et rigide, tend à s'y disloquer, de telle sorte que le mésenchyme devient caverneux à ce niveau et en certains endroits (Fig. 3). Tout se présente comme si les spicules amphistrongyles se dissociaient de leurs faisceaux et devenus libres, venaient se poser en désordre sur la lame cuticulaire, s'y enrober et la consolider en même temps. Il en résulte ainsi que l'éponge acquiert une base fragile, creusée de cavités; elle pourrait dès lors aisément se desquammer de son support avec cette particularité que celui-ci, décortiqué de son éponge, resterait revêtu de la lame cuticulaire basale qui ahère fortement (Fig. 3). C'est à ce niveau basal, où se produit la dislocation du squelette et contre le support, que se forment les statoblastes. Tandis qu'ils se constituent, ils trouvent ainsi l'éspace où se loger comme dans des niches appropriées à leur taille et à leur forme (Fig. 3).

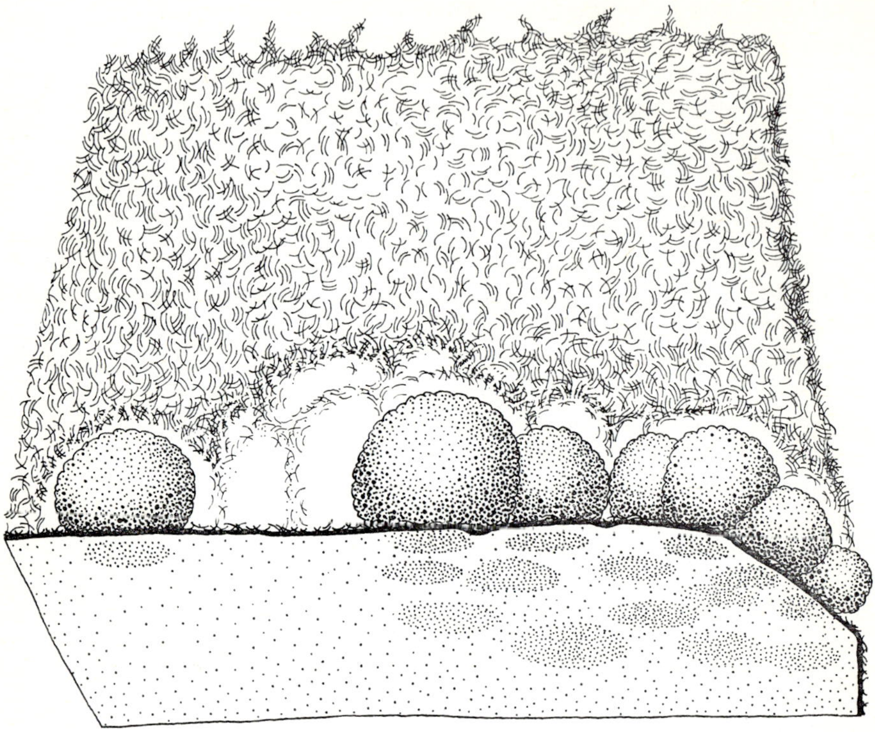

FIG. 3. Dessin semi-schématique, à la chambre claire, d'un fragment de *Potamophloios stendelli* ayant produit des statoblastes. Ceux-ci constituent un dallage de sphérules, de formes et de tailles variables, incrustées dans la lame d'adhésion, et, par l'intermédiaire de cette dernière, au support. Au niveau où se forment les statoblastes, le mésenchyme et le réseau spiculé squelettique se disloquent et se creusent de cavités dans lesquelles se logent les statoblastes. La lame d'adhésion est représentée en perspective. Elle est semitransparente et laisse voir l'ombre des statoblastes qu'elle porte.

Les statoblastes apparaissent isolément ou par petits groupes, espacés, dispersés. Ils sont disposés en une seule couche adhérant à la lame cuticulaire et au support comme des dalles convexes (Fig. 2f–h, Fig. 3).

Il en résulte que dans une éponge en place, les statoblastes ne sont jamais visibles étant recouverts par toute la masse du mésenchyme. Ils n'apparaîtront que si l'éponge se détache de son support. Dans ces conditions ils restent généralement fixés avec la lame cuticulaire au support.

De tels statoblastes n'ont jamais été signalés chez des éponges d'eau douce, car on ne peut les confondre avec les gemmules si caractéristiques des Spongillidae. Par contre, leur structure, leur forme, leur disposition, rappellent étonnamment les "gemmules" des *Suberites*, et des *Laxosuberites* qu'Annandale (1915) a décrites dans les eaux du lac Chilka en Inde et telles que nous les a fait connaître Herlant-Meewis (1948), chez les *Suberites domuncula* et *Hadriana creticula*.

Les statoblastes des *Potamophloios* sont plus gros que des gemmules. Ils ont généralement 500 μ; ils atteignent et dépassant même 1 mm. Ils sont très variables de taille et de forme. Ils sont sphéroïdes, ovoïdes, mais aussi lobés et même boudinés comme si au cours de leur formation, les amas de thésocytes destinés à donner chacun un statoblaste, avaient conflués avant d'être revêtus d'une coque commune. Ils sont convexes par toute leur face supérieure, mais la portion au contact avec la lame cuticulaire et le support est applanie. Dans l'espèce *P. stendelli*, du lac Moero et qui fut soigneusement fixée, les thésocytes qui composent les statoblastes sont de grosses cellules bourrées de plaquettes vitellines mais ils sont uninucléés alors que les archéocytes des gemmules sont binuclées. D'autre part les plaquettes vitellines y sont irrégulières de forme et de taille alors que dans les archéocytes des gemmules, elles sont remarquablement idéntiques et presque uniformément lenticulaires (Brien, 1967a). Chaque statoblaste des *Potamophloios* comme ceux de *Suberites* et *Laxosuberites* ont une enveloppe cuticulaire constituée d'au moins deux couches, l'une fine et interne, l'autre externe plus épaisse, consistante et secrétée par des cellules columellaires (Brien, 1967a). Pas plus que les *Suberites*, l'enveloppe interne du statoblaste des *Potamophloios* ne se superpose d'une couche cuticulaire épaisse spumeure, légère, alvéolaire dite "couche pneumatique" si caractérististique des gemmules des spongillides.

Enfin pas plus que chez *Suberites* l'enveloppe des statoblastes des *Potamophloios* ne se perfore d'une micropyle. Ce qui les différencie des statoblastes de *Suberites* c'est que l'enveloppe compacte des statoblastes de *Potamophloios* est très fortement armée de spicules. Ils sont de deux types. Ce sont des amphistrongyles pareils à ceux du squelette réticulé et à ceux de la lame cuticulaire (Fig. 2f–i). Tout se présente comme si au cours de la dislocation du squelette basal, les amphistrongyles, dissociés les uns des autres, tombaient en désordre non seulement sur la lame cuticulaire, mais aussi sur l'enveloppe des statoblastes. Ils s'y couchent en tous sens, comme jetés les uns sur les autres aux hasards. Englués dans l'enveloppe il lui constitue sa grosse armature. En même temps apparaissent des microsclères exclusifs à cette région basale en dislocation (Fig. 2i). Ce sont des spicules de très petite taille 20 à 50 μ, lenticulaires, ovoïdes, navicellaires, trapézoïdes ayant aussi la forme d'amphistrongyles minuscules. Ils apparaissent comme des spicules arrêtés au début de leur formation. Ils se déposent sur la lame cuticulaire à côté des amphistrongyles tombés comme de la grenaille sur l'enveloppe des statoblastes, s'immiscant entres les interstices que les amphistrongyles laissent entre eux, les recouvrant totalement et s'enrobant dans la substance de l'enveloppe comme moëllons et galets dans du mortier. Il en résulte que l'enveloppe du statoblaste épaisse devient très dure. Par sa nature, par la façon dont les spicules s'y accumulent, les statoblastes font corps avec la lame cuticulaire et s'y incrustent. Pour la même raison lorsque dans un groupe les statoblastes sont serrés les uns contre les autres, ils se soudent solidement par leur coque (Fig. 1g,h, Fig. 3).

C'est ainsi que dans les éponges revêtantes (*P. stendelli, hispida*) les statoblastes sont disséminés isolément ou par groupes espacés, adhèrant à la dalle rocheuse. Si les éponges sont enveloppantes, entourant les brindilles ou des racines (*P. songoloensis*) ou formant manchon autour d'une tige de roseau (*P. gilberti*), les statoblastes sont fixés à ce roseau et entre les brindilles par l'intermédiaire de la lame cuticulaire d'adhésion. Il est aisé d'en donner la démonstration. Des fragments de ces éponges enveloppantes ayant macéré pendant 24 h dans une solution à 5% de NaOH et de KOH, se dissocient très aisément sous la pression d'une spatule. Il ne reste plus sur l'écorce dégarnie du roseau et des brindilles, mis à nu, que la lame cuticularie sur laquelle sont incrustées les statoblasts intacts.

Si la même opération est faite sur des éponges revêtantes, préalablement détachée de la dalle rocheuse, on sépare facilement, avec une spatule, la lame cuticulaire intacte pareille à une fine feuille transparente et parcheminée sur laquelle sont disposés, à leur place naturelle, les statoblastes isolés ou en groupes dispersés.

Lorsque l'éponge encroûtante est détachée de la dalle qu'elle recouvrait, elle emporte souvent avec elle, sa lame d'adhésion et les statoblastes qui y sont fixés (Fig. 3). L'échantillon de *P. symoensi* représenté dans la Fig. 2b, a été arraché de la paroi sur laquelle elle s'était développée. On voit à sa base, ainsi découverte de ci de là, dans l'une ou l'autre dépression, de petits amas de statoblastes comme des oeufs dans leur nid.

Ni par leur mode de formation, ni leur localisation, ni par leur forme, ni leur structure, les statoblastes des *Potamophloios* ne peuvent être confondus avec les gemmules des Spongillidae (Brien, 1967–68).

La faculté de produire des bourgeons résistants est commune à bien des éponges qui n'ont aucun lien phylogénétique. Marshall (1883), Annandale (1919) et Topsent (1932) ont depuis longtemps montré que la reproduction asexuée se manifeste sporadiquement dans des éponges marines sans parenté directe : *Tethya*, *Ficulina*, *Chalina*, *Suberites* etc. . . . , sans doute aussi chez certaines Hexactinelles. Il n'y a entre ces statoblastogénèses aucune relation phylogénétique. Il en est de même pour les éponges d'eau douce. La présence de statoblastes ne signifie pas que les potamolepides aient quelques affinités avec les spongillides. Mais toutes deux sont capables de produire indépendamment des statoblastes qui ont physiologiquement la même signification adaptative mais qui sont des formations convergentes totalement étrangères les unes aux autres. S'il fallait trouver quelque similitude à la statoblastogénèse des *Potamophloios*, c'est parmi les *Suberites* qu'il faudrait la rechercher.

EMBRYOGENÈSE ET LA PARENCHYMULA DE *POTAMOPHLOIOS STENDELLI* (JAFFÉ)

La spiculation, la forme, l'aspect que peut prendre une colonie ne suffisent pas pour définir une Eponge. Il faut ajouter les caractères histologiques et biologiques notamment ceux qui concernent la reproduction et la croissance. La reproduction asexuée, la formation de statoblastes sont à cet égard des plus importants ainsi qu'il vient d'être signalé au chapitre précédent et assez facile à observer. La reproduction sexuée, l'embryogénèse, la structure de la parenchymula ne le sont pas moins, mais beaucoup plus difficiles à saisir, à moins que d'être sur place pour suivre le cycle de l'éponge pendant une année ou avoir la chance de la prélever lorsqu'elle est en période sexuelle. Ce fut le cas pour une récolte de *P. stendelli* par J. J. Symoens en mars 1966 à Lukonzolwa dont les échantillons furent soigneusement fixés. Il fut possible d'y décrire l'ovogénèse, la segmentation, la formation et la structure de la parenchymula sur le point de se libérer (Brien, 1967d, Fig. 2).

La parenchymula de *P. stendelli* est remarquable et très différente de la parenchymula connue chez *Spongilla* et *Ephydatia* (Brien, 1967a). Elle est massive. ciliée sur toute la surface, ovoïde, polarisée, l'extrémité antérieure plus large, l'extrémité postérieurs plus rétrécie. Son axe antéro-postérieur varie de 180 à 300 μ selon les individus.

Le revêtement cilié est unistratifié ; les cellules y sont trés nombreuses très comprimées, étirées en massue, la partie la plus renflée étant interne. Il est régulier, plus épais, plus serré au pôle antérieur qu'au pôle postérieur. Contre la base interne des cellules ciliées, s'étend une couche de cellules à cytoplasme clair, dont le noyau arrondi est granuleux, ce sont des collencytes, les premiers sans doute qui, à la fixation et à la métamorphose passeront à la périphérie pour constituer les éléments de l'ectomésenchyme de l'éponge définitive. Le massif cellulaire interne est

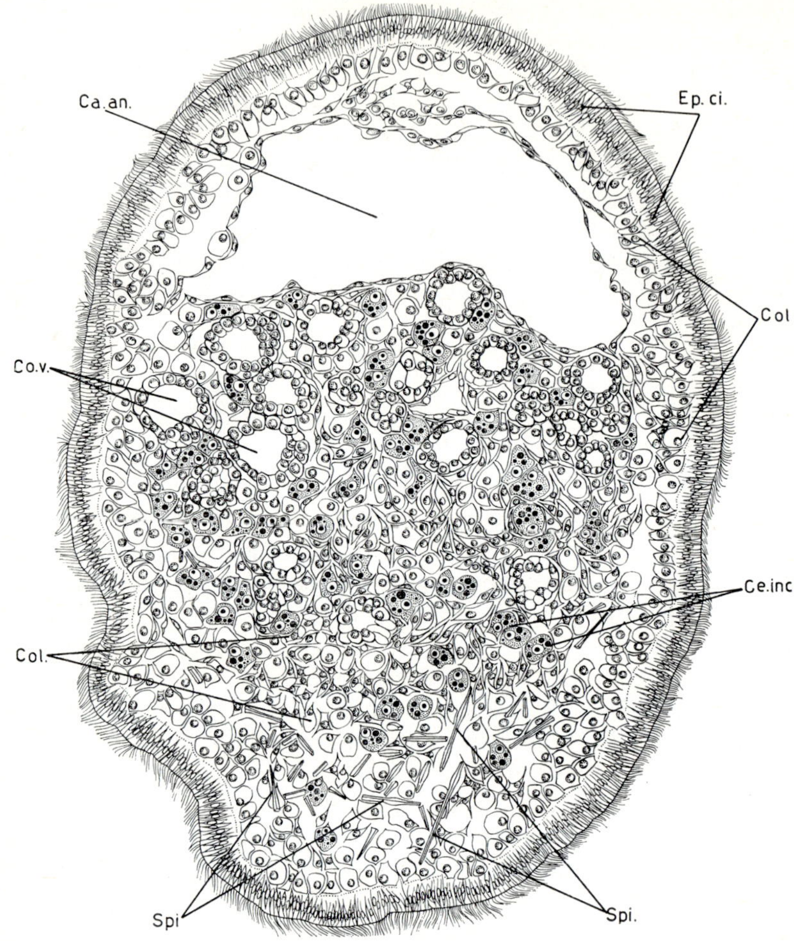

FIG. 4. Parenchymula de *Spongilla moorei* (Evans) prête à éclore mais encore contenue dans le mésenchyme de l'éponge. Ep. ci.; épithélium cilié; Ce. inc, cellules à inclusions; Spi, spicules; col collencytes; co. v, ébauches des corbeilles vibratiles; Ca. an., cavité anteriéure.

compact, il est formé de blastomères arrondis, bourrés encore d'inclusions vitellines sphéruleuses et granuleuses. Le noyau est nucléolé. Il n'est pas possible de dire qu'elle sera la destinée de ces cellules embryonnaires Seule l'étude de la métamorphose éluciderait leur histogenèse. Entre ces cellules embryonnaires, il existe des cellules différenciées. Certaines libérées de toutes inclusions cytoplasmique ont un noyau nucléolé et tout pareil à des amoebocytes; certain d'entr'eux sont vacuolaires. Des cellules étirées, étoilées à cytoplasme pâle, sans inclusions, à noyau granduleux, forment un vague réseau à travers le massif cellularie. Les scléroblastes sont le plus précocement différenciés; ils apparaissent déjà au stade de la stéréoblastula. Ce sont des cellules claires à noyau nucléolé mais qui perdent le nucléole au cours de la sécrétion du spicule Les spicules de la larve sont des amphioxes de 120 à 150 μ, parcourus par un axe organique. Légèrement incurvés, leur surface est finement épineuse. Ce sont donc des spicules exclusivement propres à la larve. L'éponge adulte ne possède pas de pareils amphioxes échinés. Dans la larve, la répartition de ces spicules est très singulière. La région antérieur massive en est privée. Les spicules appartiennent à la moitié postérieure de la parenchymula. C'est là qu'ils sont le plus volumineux, le plus nombreux. Tout à l'arrière, ils y constituent par leur ensemble une sorte de nasse évasée vers l'avant, rétrécie vers l'arrière. Les spicules se rapprochent du pôle postérieur, le perforent et font saillie à l'extérieur (voir Fig. 10, dans Brien, 1967d).

Une telle parenchymula est très différente de celle qui est connue dans les genres *Spongilla* et *Ephydatia* (Nodelke, 1893; Brien et Meewis, 1938; Brien, 1967d, Fig. 4). Dans ces genres, la parenchymula oblongue, ciliée sur toute sa surface, atteint une plus grande taille, son axe antéro-postérieur peut avoir 300 à 400 μ selon les individus. L'épithelium périphérique est également cilié sur toute son étendue et pareillement tapissé sur sa face interne d'une couche de collencytes. Mais la structure interne de la larve est très différente. Les spicules amphioxes fins et lisses y sont beaucoup moins abondants et importants. Le massif de cellules internes y est creusé d'une grande cavité antérieure qui donne à cette larve l'aspect d'un aéronef. Sous la cavité, la massif interne est composé de grosse cellules embryonnaires à noyau nucléolé, dont le cytoplasme est bourré d'inclusions vitellines. Des collencytes à noyau granuleux, à cytoplasme clair se sont différencies. Ils délimitent, sous l'aspect de pinacocytes, la cavité antérieure, et forment une trame à travers le massif cellulaire. Sous la cavité antérieure dans le $\frac{1}{3}$ antérieur du massif interne, les cellules embryonnaires sont en division active; chacune devient un amas globuleux de petites cellules cubiques à noyaux granuleux: ce sont les choanoblastes. Les sphérules qu'ils composent sont les ébauches des corbeilles vibratiles. Ainsi dans les Spongillidae, la parenchymula possède dans son massif interne tous les éléments d'une jeune éponge: collencytes, amoebocytes et choanocytes C'est une petite éponge différenciée enveloppée par un épithélium cilié, devenu un organe exclusivement vecteur, car les cellules qui le constituent ne jouent plus aucun rôle morphogénétique, puisque les corbeilles vibratiles sont déjà différenciées dans le massif intérieur. Elles seront phagocytées. Dans la parenchymula des Spongillidae, les feuillets embryonnaires ne sont plus distincts. La parenchymula équivaut à une gemmule recouverte d'un épithélium cilié, une gemmule libre et nageante.

Toutes les espéces de *Potamolepis* et de *Potamophloios* ont-elles des larves pareilles à la larve parenchymula de *P. stendelli*? Cette parenchymula peut-elle caractériser la famille des Potamolepides? Il faut attendre que l'on puisse connaître l'embryogénèse d'autres espéces, ce qui j'espère ne tardera plus, aujourd'hui que l'attention est attirée sur cette question. Elle n'en est pas moins une indication précieuse qui, s'ajoutant aux caractères morphologiques et à la statoblastogénèse, confirme la distinction profonde entre les *Spongillidae* et les *Potamolepidae*.

LA FAMILLE POTAMOLEPIDAE ET LA SYSTÉMATIQUE DES EPONGES D'EAU DOUCE

En conclusion, les éponges dont il vient d'être question ont des caractères communs. Leur spiculation, la charpente squelettique réticulée rigide y est si semblable qu'elles ont pu être réunies dans un seul genre *Potamolepis* (Jaffé 1916; Brien, 1967a,b,c,d).

Toutefois un certain nombre de particularités très importantes et qui viennent d'être rappelées les séparent néanmoins en deux groupes et entraînent la conviction qu'elles appartiennent en réalité à deux genres voisins:

le genre *Potamolepis* (Marshall, 1883) proprement dit: *P. chartaria, P. leubnitziae, P. pechuëlli, P. micropora, P. marshalli*, du Bas-Congo qui sont des éponges revêtantes, minces, à surface unie sur laquelle les oscules sont régulièrement disposés;

le genre Potamophloios (n. gen.) : *P. stendelli, P. symoensi, P. gilberti, P. songoloensis, P. hispida* du bassin du Luapula-Moero, éponges plus épaisses, à surface bossuée at hérissée d'aspérités hispides et où les oscules moins nombreux, s'ouvrant souvent au sommet d' éminences côniques ou de tubulures, sont très irrégulièrement répartis.

Ces deux genres n'en forment pas moins une unité systématique. La nature des statoblastes produits par toutes les *Potamophloios*, et peut être même par certaines *Potamolepis* (*P. pechuëlli*, Brien, 1967c), la structure de la larve connue dans une espèce, la texture morphologique du mesenchyme font de cette unité une famille phylogénétiquement distincte du reste des éponges d'eau douce, des Spongillides notamment: la famille de Potamolepididae (Brien, 1967a,b,c,d).

Il est probable que la famille des Potamolepididae englobe d'autres genres récemment décrits et sur lesquels il est prématuré de se prononcer.

Lévi a signalé en 1965, les premières éponges du Gabon parmi lesquelles *Potamolepis belingana*, éponge rameuse, d'un aspect différent de celui des espèces d'après lesquelles Marshall (1883) a créé le genre *Potamolepis*.

On ne peut oublier *Baikalolepis fungiformis* décrite en 1927, par Makuschok. Comme son nom l'indique, elle vit dans le lac Baïkal, ses amphistrongyles sont échinés. Il conviendrait de la réexaminer afin de s'assurer si elle ne se range pas parmi les Potamolepidae.

Il faut rappeler aussi le cas de *Spongilla navicella* Carter, 1881, éponge inconnue de l'Amazone, qui ne fut définie que par ses statoblastes fixés à des rameaux immergés. Or si l'on excepte le micropyle, que Carter signale (?) ces statoblastes ressemblent, par la structure de leur coque et leur position sur le support, aux statoblastes des *Potamophloios*.

Les éponges d'eau douce que Gray (1867) appelait *Potamospongida* n'ont en commun que d'être dulçaquicoles. Elles constituent un groupe hétérogène. Il est évident que les éponges des lacs, des étangs, des rivières des diverses régions de la terre, proviennent d'éponges marines qui ont pu coloniser, probablement par des estuaires, le domaine lacustre s'y adapter, s'y maintenir y proliférer et s'y différencier. Marshall (1883) était d'avis qu'elles avaient même souche, parmi les Renieridae, mais qu'elles aviaent envahi les eaux douces à des moments différents ce qui expliquerait, écrivait-il, les diversités de leurs caractères adaptifs. Il s'agirait selon lui d'un polyphylétisme dans le temps et l'espace.

Ce polyphylétisme est plus profond, il est génétique, morphologique et physiologique. Spongillidae et Potamolepidae n'ont pas la même origine phylétique. Elles se rattachment à des familles de souche marine distinctes. Enfin on doit reconnaître que les Potamospongidae de Gray comprennent d'autres familles.

Les éponges du lac Baïkal du groupe représenté par *Lubomirskia baikalensis* décrite par Dybovsky (1880), revue par Carter (1881) et Annandale (1914), ne produit pas de gemmules. C'est une éponge élastique, dressée en longs rameaux de couleur verte. Son squellette est formé d'amphioxes incurvés rassemblés en faisceaux entièrement englobés dans des fibres de spongines. Annandale (1914) qui rappelle ces faits les rapproche de *Pachychalina* (Schmidt,) et de la famille des Chalininae (Ridley et Dendy, 1881). De toute façon elle ne peut se confondre avec les Spongillinae ni les Meyeniinae. Elle appartient à une famille originellement distincte que l'on peut appeler, en attendant plus amples informations, la famille des Lubomirskiidae.

Dybovsky (1880) a décrit une autre éponge du Baïkal qui ne peut être assimilée à la première: *Lubomirskia bacilifera* (revue par Carter (1881) et Annandale (1914)). Annandale (1914) l'a rebaptisée *Baikalospongia bacilifera*. C'est une tout autre éponge. Elle est massive. Elle produit des "gemmules" qui sont basales et fixées au support et revêtues d'une simple envelloppe cuticulaire. Son squelette est formé d'amphystrongyles épineux, surtout aux extrémités, réunis bout à bout en files ou en minces faisceux mais non enrobés entièrement dans un manchon de spongine. Selon Annandale cette éponge s'écarte des Chalinidae. de Laubenfels (1936) la classe dans la famille des Adocidae.

L'immense majorité des éponges d'eau douce appartiennent, il est vrai, aux *Spongillidae*. Ce sont des éponges à grandes cavités hypodermiques dont la charpente squelettique est délicate, fragile, formée d'amphioxes ou d'amphistrongyles lisses ou épineux, associés en files ou en minces faisceaux par de al spongine, Dans leur très grande généralité, elles produisent des statoblastes, souvent innombrables, sphériques, uniformes, dont le mésenchyme peut être farci dans toute son épaisseur et qui sont appelés **gemmules**. Celles-ci possèdent une enveloppe cuticulaire interne, propre à tous les types de stratoblastes, mais surmontée d'une enveloppe secondaire plus épaisse, légère, gènéralement alvéolaire, spumeuse, dite enveloppe pneumatique consolidée par des microsclères ou gemmosclères. Ceux-ci sont souvent des amphioxes dans la sous-famille des Spongillinae. Des amphidisques, ou des spicules dérivés, se disposent dans l'enveloppe spumeuse de la sous-famille des Meyeniinae. Il existe toujours un micropyle, parfois au sommet d'un petit tube formé par l'enveloppe interne.

L'identification des nombreuses espèces de Spongillinae et de Meyeniinae, doit être refaite, non seulement par des critères morphologiques mais aussi histologiques, biologiques et physiologiques, Sans envisager cette tâche considérable, peut être même matériellement impossible, il serait nécessaire de reprendre l'examen des genres et des familles des éponges d'eau douce selon les indications qui viennent d' être rappelées. Les Potamospongida de Gray (1867) sont certainement polyphylétiques, issues de familles d'éponges marines différentes, elles ont peuplé le domaine des eaux douces à des moments différents, par des voies différentes. Provisoirement, d'une façon approximative, et à titre de simple indication, on pourrait les grouper de la façon suivante:

Famille	Sous-famille / Genre	Genres / Espèces
Les Lubomirskiidae (Brien, 1969)		*Lubomirskia baikalensis* Dybovsky, 1878.
Les Adocidae (de Laubenfels, 1936)		*Baikalospongia bacilifera* Dybovsky, 1878
Les Potamolepididae: (n.fam.) (Brien, 1967)	*Potamolepis* (Marshall, 1883) (**Potamolepidinae**	*P. chartaria* Marsh, 1883
		P. leubnitziae Marsh, 1883
		P. pechuëlii Marsh, 1883
		P. weltneri Moore, 1903
		P. micropora Burton, 1938
		P. marshalli Burton, 1938
	Potamophloios (n. gen.) **Potamophloiinae** (n. fam. 1969)	*P. stendelli* Jaffé, 1916
		P. symoensi Brien, 1967
		P. gilberti Brien, 1969
		P. *songoloensis*, Brien, 1969
		P. hispida Brien, 1969
Les Spongillidae (Gray, 1867)	**Spongillinae** Carter, 1881	*Spongilla* Lamarck, 1815
		Stratospongilla Annandale, 1909
		Corvospongilla Annandale, 1911
		etc
	Meyeniinae Vejdovsky, 1887	*Meyenia* Carter, 1881
		Heteromeyenia Potts, 1882
		Metania Gray, 1867
		Parametania (n. sp.) Brien, 1968
		Acalle Gray, 1867
		Carterius Potts, 1881
		Parmula Carter, 1881
		Trochonspongilla Vejdovsky, 1883
		Tubella Carter, 1881
		etc.

Bibliography

Annandale, N. (1914). Further notes on the sponges of Lake Baïkal. *Rec. Indian Mus.* **10**: 137–148.

Annandale, N. (1915). Fauna of the Chilka Lake (Sponges). *Mem. Indian Mus.* **5**: 23–34.

Annandale, N. (1919). Report on the aquatic fauna of Seistan, with subsidiary studies. Sponges, Hydrozoa and Polyzoa of Seistan. *Rec. Indian Mus.* **18** (2): 83–99; 1 pl., 4 figs.

Bowerbank, J. S. (1869). Monograph of Spongillidae. *Proc. zool. Soc. Lond.* **1869** 440–472.

Brien, P. (1967a). Eponges du Laupula et du lac Moëro. *Résult. Scient. Explor. hydrobiol. Bangwelo-Luapula Miss. J. J. Symoens.* **11** (58): 1–52.

Brien, P. (1967b). Un nouveau mode de statoblastogénèse chez une Eponge d'eau douce africaine: *Potamolepis Stendelli* (Jaffé). *Bull. Acad. r. Belg. Cl. Sci.* **53**: 552–570.

Brien, P. (1967c). Formation des Statoblastes dans le genre *Potamolepis*: *P. Symoensi* (Marshall), *P. Pechuelii* (Marshall) *P. Schoutedeni* (Burton). *Bull. Acad. r. Belg. Cl. Sci.* **53**: 573–590.

Brien, P. (1967d). L'embryogénèse d'une éponge d'eau douce africaine: *Potamolepis Stendelli* (Jaffé). Polyphylétisme des Eponges d'eau douce. *Bull. Acas. r. Belg. Cl. Sci.* **53**: 752–777.

Brien, P. (1969,e). Nouvelles éponges du lac Moero. *Exploration hydrobiol. lac Baugwelo et du Luapula.* (ed. I.-I. Symoens) 30 pp., 4 pl., 9 figs.

Brien, P. et Govaert-Mallebrancke, (1958). A propos de deux épongedu Tanganika. *S. Moorei* (Evans) *S. Uvirae* (n. sp.) *Mém. Acad. r. Sci. colon. Cl. Sci. nat.: med.* **8**: 3–43.

Brien, P. et Meewis, H. (1938). Contribution à l'étude de l'embryogénèse des Spongillidae. *Arch. de Biol.* **48**: 125–268 (231 pl.)

Burton, M. (1938). Some freshwater Sponges from the Belgian Congo, including new species from N. Rhodesia. *Revue Zool. Bot. afr.* **30**: 458–468.

Carter, H.S. (1881). History and classification of the known species of *Spongilla*. *Ann. Mag. Nat. Hist.* (5) **7**: 77–107.

Dybovsky, W. (1878). Mittheilung– über Spongien. *Zool. Anz.* **1**(17): 30–32.

Dybovsky, W. (1880). Structures uber die Spongien des Russichen Reiches mit besonderer Berucksichtigung der Spongien Fauna des Baikal secs. *Zap. imp. Akad. Nauk.* **7**: 3–71.

Evans, R. (1899). Description of two new species of *Spongilla* from Tanganika. *Q. J. microsc. Sci.* **41**: 471–488.

Gray, J. E. (1867). Notes on the arrangement of Sponges with description of some new genera. *Proc. zool. Soc. Lond.* **1867**: 492–558.

Herlant-Meewis, H. (1948). La gemmulation chez *Suberites domuncula* Nardo *Arch. Anat. micr. et Morph. exp.* **37** (4): 289–322.

Hilgendorff, F. (1883). Süsswasserschämme aus Central Africa. *Sber. Ges. naturb. Freunde Berl.* **1883**: 87–90.

Jaffé, G. (1916). Zwei Schämme aus dem Tanganikasee: *Spongilla Moorei* (Evans) und *Potamolepis Stendelli* (n. sp.). *Zool. Anz.* **48**; 5–14.

Jewell, M. E. (1952). The genera of North American fresh-water Sponges, *Parameyenia* (n. gen.) *Trans. Kans. Acad. Sci.* **55**, 445–457.

de Laubenfels, M. W. (1936). A discussion of the sponge fauna of the dry Tortugas in particular, and the West Indies in general, with material for a revision of the families and orders of the Forifera. *Pap. Tortugas Lab.* **30**; 34–37.

Lévi, C. (1965). Spongillides de l'Ivindo (Gabon) *Metania lissostrongyla* (Burton) et *Potamolepis belingana* (n. sp.) *Biologia Gabonica* **1**: 314–323.

Makuschok, M. E. (1927). Zur Systematik der Baikal Spongien. *Russk. zool. Zh.* **7**: 129–131.

Marshall, M. (1883). On some new siliceous Sponges collected by M. Pechuel—Lösche in the Congo. *Ann. Mag. nat. Hist.* (5) **12**: 391–412.

Moore, J. E. S. (1903). The Tanganyika Problem, Hurst and Blackett, London, 371 pp.

Nodelke, B. (1893). Die Metamorphose des Süsswasserschwämmes *Zool. Jahrb.* **8**: 152–189.

Penney, J. I. (1960). Distribution and Bibliography 1892–1957 of the fresh-water Sponges. *Univ. S. Carolina Publs.* **3** (3): 1–111.

Penney, J. I. et Racek, A. A. (1968). Comprehensive Revision of a Worldwide Collection of Freshwater Sponges. *Smithsonian Inst. United States National Museum*, 184 pp.

Potts, E. (1887). Contributions towards a Synopsis of the American forms of fresh-water sponges with descriptions of those named by other authors and from other parts of the world. *Proc. Acad. nat. Sci. Philad.* **39**: 158–279.

Topsent, E. (1932). Documents sur des *Spongillides* d'Afrique. *Bull. Mus. natn. Hist. nat.* **4** (2): 1001–1007.

Vejdovsky, V. (1883) Die Süsswasserschwämme Bohmerrs Prog. *Abh. Bohm geselsch. Wiss* **6**: folge 12.

Vejdovsky, V. (1887). Diagnosis of the European Spongillidae *Proc. Acad. Nat. Sci. Phila.* **39**: 177.

Weltner, W. (1895). Spongillidenstudien III Katal. und Verbreitung der bekennten Süsswasserschämme. *Arch. Naturgesch.* **1**: 114–144.

Weltner, W. (1897). Die Coelenteraten und Schwämme des süssen Wassers Ost-Afrikas. *Deutsch-Ost Africa* **4**: 1–8.

Symp. zool. Soc. Lond. (1970) No. 25, 189–204

LES EPONGES PHARETRONIDES ACTUELLES

JEAN VACELET

Station Marine d'Endoume, Faculté de Sciences de Marseille, Marseille, France

SYNOPSIS

While only nine extant pharetronids were known, represented by examples from the Indo-Pacific region, the discovery of *Petrobiona massiliana* in the recesses of concealed caves in the Mediterranean has revealed that the relict group, whilst no longer flourishing, is not as near extinction as was thought. The discovery of *Petrobiona* has indicated the possibility that these sponges could be modern colonists of the inner recesses of caves. Seven pharetronids have been discovered in Madagascan caves and two other species (one in course of description) in the Mediterranean. One of the latter, *Paramurrayona corticata*, characterized by a carapace of calcareous bodies, shows certain analogies to extinct Sphinctozoa. The modern ecological specialization of Pharetronida is markedly different from that of most fossil forms.

The Porifera Calcarea appear to be a heterogeneous group and the place of the pharetronids within this group is problematical. The modern pharetronids are characterized by tuning-fork triactines, by an oscular armament of tetractines and by a reinforcing of the skeleton. But the tuning-fork spicules are absent in *Lepidoleucon*, the special oscular skeleton is not yet demonstrated in all forms, and each of the five known families employs a different method of reinforcing the skeleton: tuning fork fibres (Lelapiidae), fusion of tetractines (Minchinellidae), massive aspicular skeleton (Murrayonidae), aspicular cortex (Paramurrayonidae), cortical scales (Lepidoleuconidae). This picture is probably attained by various convergent phenomena. Where known, their larval, cytological, and spicular characters allow only an uncertain position within the Calcarea.

As a result, it is proposed, provisionally, to regard the Pharetronida as a third subclass of the Porifera Calcarea and as containing the two orders Inozoa and Sphinctozoa, of which only the first has survived to present times. Certain resemblances to the other two subclasses of the Calcarea cannot be ignored. Lelapiidae, Minchinellidae, and *Petrobiona* resemble some Calcaronea, *Murrayona* and *Paramurrayona* some Calcinea. It may be supposed that these resemblances reflect evolutionary trends more clearly developed in Calcinea and Calcaronea, which may be derived from Pharetronida.

INTRODUCTION

Les Eponges Calcaires sont difficilement fossilisables, et seules se sont conservées celles dont le squelette était renforcé par des dispositifs spiculaires ou aspiculaires particuliers; ces Eponges constituent le groupe des Pharétronides Zittel, 1878, qui a eu un développement remarquable, par le nombre d'espèces et par l'abondance des individus dans certaines mers, depuis le Permien jusqu'à la fin du Crétacé. Ces Eponges montrent ensuite un déclin important: les formes segmentées (Sphinctozoaires Steinmann, 1882) disparaissent complètement, et les Inozoaires Steinmann, 1882 non segmentées ne sont plus représentées que par quelques fossiles de l'Eocène australien. Quelques formes ont survécu jusqu'à nos jours, et on comptait en 1923 neuf espèces d'Inozoaires de l'Indopacifique, chacune connue par un ou deux individus seulement. L'étude de ces reliques présentait un grand intérêt, en raison de l'importance des Pharétronides fossiles, dont elles pouvaient éclaircir la structure, très discutée, et l'écologie; les Pharétronides actuelles pouvaient aussi nous renseigner sur la phylogénie des Eponges Calcaires actuelles, pour lesquelles les documents historiques font presque totalement défaut. Mais la rareté des Pharétronides dans nos mers était un obstacle important.

Un représentant découvert plus récemment, *Petrobiona massiliana* Vacelet et Lévi (1958), présente l'avantage d'être très abondant dans une localisation écologique bien précise et particulière, les grottes sous-marines obscures méditerranéennes. Ceci a permis d'une part d'étudier plus à fond une Pharétronide actuelle, d'après plusieurs centaines de spécimens (Vacelet, 1964), et d'autre part de prévoir que les Pharétronides étaient actuellement des cavernicoles marines. Cette hypothèse s'est révélée fructueuse, et l'exploration de grottes de l'Océan Indien et de Méditerranée m'a permis de découvrir sept formes nouvelles et de retrouver une des espèces décrites par Kirkpatrick (1908) (Vacelet, 1967a et b). Il est maintenant acquis que dans les grottes les plus sombres des régions tropicales et subtropicales vivent d'autres représentants de ce groupe qui, sans présenter l'épanouissement qu'il a eu au Secondaire, est cependant bien plus répandu qu'on ne le pensait. La connaissance de leur localisation écologique entraînera certainement l'augmentation du nombre des formes connues, et nos idées sur les rapports entre Pharétronides actuelles et fossiles avec les autres Eponges Calcaires en seront certainement modifiées; ces rapports posent encore des problèmes difficiles, sur lesquels il semble cependant utile de faire un point momentané.

LES DIVERS TYPES D'ORGANISATION

On a divisé les Pharétronides en deux groupes (Steinmann, 1882): les Sphinctozoaires, segmentées, uniquement fossiles, et les Inozoaires non segmentées, fossiles et actuelles.

Les Sphinctozoaires

Depuis la fin du Crétacé on ne connaît plus aucune Sphinctozoaire, alors qu'elles étaient très répandues et ont même eu une importante activité dans la construction de récifs. Des opinions diverses ont été émises sur leur structure et sur leur position systématique, mais la plupart des auteurs s'accordent maintenant à les considérer comme des Eponges Calcaires constituant soit une division des Pharétronides, soit un groupe particulier, le terme de Pharétronide étant alors réservé aux Inozoaires. Selon Seilacher (1961), Vacelet (1964) et Ziegler (1964), elles seraient caractérisées par un "sclérosome", squelette rigide externe, dont la présence entraîne la croissance par segments successifs.

Toutefois, parmi les auteurs récents, Zhuravleva et Rezhvoy (1956) sont d'un avis différent: les spicules décrits chez certaines de ces Eponges n'existeraient pas, mais correspondraient en réalité à des espaces entre les canaux; les Sphinctozoaires, dépourvus de spicules, formeraient un groupe *incertae sedis* situé entre les Spongiaires et les Archéocyathes. Cependant, Reid (1967) confirme la réalité des spicules d'une Sphinctozoaire du Crétacé, *Tremacystia d'orbignyi* (Hinde, 1882). Selon Ott (1967), les Sphinctozoaires seraient polyphylétiques.

Les Sphinctozoaires pourvues de spicules calcaires doivent à mon sens se classer très près des Inozoaires et constituer une simple division des Pharétronides. On doit noter que *Tremacystia d'orbignyi* montre une adaptation particulière des triactines à la formation de fibres par réduction de l'actine basale (triactines filiformes) comparables aux triactines en diapason des Pharétronides Inozoaires, diapasons qui n'ont pas encore été trouvés chez des Sphinctozoaires; ces triactines filiformes sont inconnues chez les formes actuelles, mais existent chez des Inozoaires fossiles (*Corynella* Zittel, 1878 et *Raphidonema* Hinde, 1884); un autre point commun entre Sphinctozoaires et Inozoaires est l'existence chez la Pharétronide actuelle *Paramurrayona corticata* Vacelet, 1967 d'un sclérosome externe analogue à celui que l'on a décrit chez les Sphinctozoaires. Cette espèce a cependant été classée dans les Inozoaires, car elle n'est pas segmentée et les plaques calcaires de sa carapace ne sont pas soudées, mais seulement liées par de la mésoglée, formant ainsi une structure non fossilisable.

La présence de spicules ressemblant à des tylostyles d'Eponges siliceuses chez certaines Sphinctozoaires pose un problème non résolu; j'avais supposé (1964) que ces spicules indiquaient la présence de Cliones perforant le sclérosome externe; dans une lettre personnelle, Reid a bien voulu me faire part de son scepticisme; l'examen de *Barroisia anastomans* (Mantell, 1882) l'a convaincu que les spicules étaient produits par l'Eponge elle-même; cette structure a l'air constante chez *Barroisia anastomans* et *Verticillites helveticae* de Loriol, 1868. S'il s'agit bien de spicules, ceci reste mystérieux. Certaines Sphinctozoaires seraient-elles des Démosponges à squelette calcaire supplémentaire? On connaît des tylostyles chez la "Silico-Calcaire" actuelle *Merlia normani* Kirkpatrick, 1910, mais ils ne sont ni associés à des spicules calcaires, ni inclus dans le squelette principal formé de sphérolites calcaires; par contre certains acanthostyles d'*Astrosclera willeyana* Lister, 1900 sont à l'intérieur du squelette principal calcaire formé de polyèdres.

Les Inozoaires

Les Pharétronides Inozoaires actuelles comprennent cinq types d'organisation différents, correspondant chacun à une famille, dont seuls deux sont fossilisables. Une sixième famille serait uniquement fossile.

Les Lelapiidae

Les Lelapiidae sont peu différentes des autres Eponges Calcaires actuelles. Elles sont représentées par quatre espèces et deux genres (*Lelapia australis* Gray, 1867, *L. antiqua* Dendy et Frederick, 1914, *L. nipponica* Hara, 1894, *Kebira uteoides* Row, 1909) du littoral de l'Indopacifique, non localisées dans des grottes. On les classe dans les Pharétronides en raison de leurs faisceaux de spicules en diapason liés entre eux par de la mésoglée; chez *Kebira uteoides*, les diapasons sont remplacés par des triactines à actines paires vestigiales. Ces quatre espèces sont, malgré leur localisation écologique banale, de récolte exceptionnelle; leurs caractères choanocytaires et spiculaires indiquent des affinités avec la sous-classe Calcaronée: triactines en majorité sagittaux, choanocytes à noyau apical.

On a longtemps attribué à des fossiles Inozoaires une structure analogue à celle des Lelapiidae, en raison de la présence de "fibres" chez les unes et les autres; ces fibres sont en réalité fort différentes et il est évident que les faisceaux de diapasons des *Lelapia* Gray, 1867 ne constituent pas à eux seuls une structure favorable à la fossilisation.

Les Minchinellidae

Les Minchinellidae sont caractérisées par un squelette rigide obtenu par la soudure de tétractines particulières, qui ressemble assez à celui de certaines Lithistides. Elles sont maintenant connues principalement des grottes obscures. Les quatre premières espèces décrites (*Petrostroma schulzei* Döderlein, 1892, *Plectroninia hindei* Kirkpatrick, 1900, *P. deansii* Kirkpatrick, 1911, *Minchinella lamellosa* Kirkpatrick, 1908) ont été récoltées, à raison d'un ou deux exemplaires seulement, entre 90 et 195–392 m aux Iles Christmas et dans le Pacifique; *Plectroninia hindei* a été retrouvée dans une grotte méditerranéenne. Cinq autres espèces de *Plectroninia* proviennent des grottes sous-récifales de Madagascar (Vacelet, 1967a) et une nouvelle Minchinellidae, en cours de description, a été découverte récemment dans des grottes méditerranéennes.

Toutes les *Plectroninia* Hinde, 1900 actuelles sont de très petite taille (quelques mm) et encroûtantes ; leur squelette rigide est formé de quelques couches de deux catégories de tétractines solidement soudés par leurs actines basales et dont l'actine apicale, libre, est dirigée vers la couche ectosomique ; les grands tétractines forment un réseau hexagonal, à l'intérieur duquel se trouve un réseau secondaire de petits tétractines ; chez l'Eponge méditerranéenne en cours de description, qui nécessitera sans doute la distinction d'un nouveau genre, le réseau n'est formé que par ce réseau secondaire, les grands tétractines étant absents. La soudure est effectuée par un ciment calcaire, qui arrive à masquer complètement les actines soudées. *Minchinella* et *Petrostroma* atteignent des dimensions plus importantes, et dans la profondeur du squelette, les tétractines sont entièrement soudés. Les tissus vivants sont situés dans les mailles et à la surface du réseau ; ils montrent divers types de spicules, non soudés, dont des triactines en diapason groupés en faisceaux chez *Petrostroma* et chez *Plectroninia tecta* Vacelet, 1967, libres chez les autres. Chez les *Plectroninia* de Madagascar, ainsi que chez *Plectroninia hindei*, il existe des diapasons à branches épineuses ou à branches en V libres dans les mailles du réseau de tétractines. Les spicules ectosomiques de toutes les espèces trouvées dans les grottes comportent une armature particulière de tétractine autour de l'oscule ; cette particularité, qui existe aussi chez *Minchinella*, semble assez répandue chez les Pharétronides actuelles.

Dans cette famille, les définitions génériques sont peu précises, car elles sont basées sur les caractères du réseau de tétractines, caractères difficiles à apprécier et souvent mal connus, surtout chez les fossiles. *Petrostroma* se distinguerait par la soudure latérale des actines basales des tétractines du réseau, soudure qui se ferait bout à bout chez les autres genres ; *Minchinella* a une forme en lame avec face inhalante et face exhalante, disposition connue aussi chez plusieurs Eponges Calcaires fossiles et chez de nombreuses Démosponges. La disposition, radiale ou non, des grands tétractines, la présence éventuelle d'une couche basale de tétractines simplement emboîtés mais non soudés, la structure du réseau de petits tétractines constituent des différences génériques discutables qui nécessitent une révision de l'ensemble des formes actuelles et fossiles. L'attribution des espèces malgaches au genre *Plectroninia* Hinde, 1900 est donc sans doute provisoire.

Dendy et Row (1913) et Hartman (1958) ont placé les Minchinellidae dans la sous-classe Calcinea des Eponges Calcaires d'après les caractères assez incertains des choanocytes et des larves de *Minchinella lamellosa*; j'ai montré (1964) que ces caractères pouvaient aussi indiquer une affinité avec la sous-classe Calcaronea, les choanocytes décrits par Kirkpatrick (1908) correspondant à des choanocytes asphyxiques; cette affinité est confirmée par les *Plectroninia* de grottes dont les choanocytes ont le noyau nettement apical et dont les larves sont des amphiblastula chez *P. tecta*.

Le squelette rigide de cette famille a permis la fossilisation de nombreux exemplaires, connus principalement du Crétacé et de l'Eocène. Les genres *Petrostoma* et *Plectroninia* auraient des représentants actuels et fossiles.

Les Elasmostomatidae

Les Elasmostomatidae, au sens de Wagner (1964), sont uniquement fossiles. Leur squelette aurait une structure voisine de celle des Minchinellidae, mais les tétractines du réseau de ces dernières sont ici remplacés par des triactines. Ces triactines, soudés et enrobés par un ciment calcaire, restent visibles au centre de la "fibre", alors que chez les Minchinellidae, tant actuelles que fossiles, la structure spiculaire est le plus souvent complètement effacée par le processus de cimentation. Dans la fibre des Elasmostomatidae de Laubenfels, 1955, les triactines soudés sont souvent entourés par d'autres spicules, ce qui n'est pas le cas chez les Minchinellidae. L'analogie entre les deux structures n'est donc peut-être pas si grande.

Les Murrayonidae

Les Murrayonidae ne comptent encore que deux représentants actuels, *Murrayona phanolepis* Kirkpatrick, 1910 et *Petrobiona massiliana*, dont la parenté est d'ailleurs discutable. Elles sont caractérisées par un squelette principal rigide ressemblant plus ou moins à celui des Minchinellidae, mais dont l'origine est tout à fait différente: le réseau calcaire n'est pas formé ici de spicules soudés, mais de corps polygonaux de calcite, étroitement soudés.

Murrayona phanolepis n'est connue que par quelques échantillons récoltés à 83 m aux Iles Christmas; son squelette principal est réticulé, et elle est recouverte, à l'exception d'une aire inhalante spéciale, de curieuses écailles ovalaires. Elle montre un mélange des caractères des deux lignées d'Eponges Calcaires actuelles (triactines sagittaux, choanocytes à noyau basal). La structure de son oscule n'est pas connue.

Petrobiona massiliana (= *P. incrustans* Sarà, 1963), abondante dans les grottes obscures méditerranéennes entre quelques dizaines de cm et 25 m de profondeur, a ses tissus vivants situés en majorité à la surface de son squelette principal, plus massif et non réticulé; elle est dépourvue d'écailles. Ses choanocytes à noyau apical, ses triactines sagittaux et sa larve amphiblastula l'apparentent nettement à la sous-classe Calcaronée. Les phénomènes histologiques de sa reproduction ne se distinguent de ceux connus chez les Eponges Calcaires Calcaronées que par leur plus grande complexité, le schéma de base étant le même; par exemple, la nutrition de l'ovocyte de cette éponge apparaît comme le terme ultime d'une série de complexité croissante: alors que chez des *Sycon* Risso, 1826 et des *Leuconia* Grant, 1833, la cellule charriante sert principalement au transport du spermatozoïde, son rôle dans la nutrition de l'ovocyte augmente chez *Sycon elegans* (Bowerbank, 1845,) où elle devient le centre d'un complexe charriant (Duboscq et Tuzet, 1944) à rôle nutritif; ce rôle, encore accru chez *Achramorpha nivalis* Jenkins, 1908, prend une très grande importance chez la Pharétronide, où la cellule charriante, après avoir transmis le spermatozoïde à l'ovocyte, lui transmet le contenu d'une chambre choanocytaire toute entière. Plus compliqué, ce processus relève cependant d'un phénomène fondamental identique à celui observé chez les quelques *Grantia* Fleming, 1828, *Sycon* et *Leuconia* dont la reproduction est connue. L'oscule de *Petrobiona* a une armature de tétractines.

La structure des Murrayonidae est favorable à la fossilisation, en raison de leur squelette principal, et il est possible (Vacelet, 1960 et 1964) que certaines Pharétronides fossiles correspondent aux moulages d'Eponges de structure identique, la "fibre" représentant alors les cavités, occupées par les tissus et les spicules libres chez l'Eponge vivante, d'un squelette aspiculeux réticulé. Mais la répartition des spécimens fossiles entre les trois familles Minchinellidae, Elasmostomatidae et Murrayonidae est un difficile problème pour les paléontologistes.

Les Paramurrayonidae

La famille Paramurrayonidae, inconnue à l'état fossile, est représentée par *Paramurrayona corticata* Vacelet, commune dans les grottes des récifs de Madagascar. Elle possède un squelette aspiculeux cortical, sorte de carapace de plaques calcaires liées par de la mésoglée (dont j'ai déjà souligné les analogies avec le sclérosome des Sphinctozoaires), recouvert par des écailles ovales identiques à celles de *Murrayona phanolepis*. Ses triactines sont sagittaux, mais ses choanocytes à noyau basal et sa larve probablement blastula l'apparentent à la sous-classe Calcinée. Ici aussi, l'oscule possède une armature de tétractines.

Les caractères cytologiques et la spiculation, en particulier la présence d'écailles, rapprochent beaucoup cette Eponge de *Murrayona*, mais l'organisation générale est profondément différente.

Les Lepidoleuconidae

La famille Lepidoleuconidae n'est connue elle aussi que par une seule espèce des grottes des récifs malgaches, *Lepidoleucon inflatum* Vacelet, 1967. Cette minuscule Eponge est entièrement couverte d'écailles ovales, absentes seulement dans une zone inhalante latérale; les écailles ressemblent beaucoup à celles de *Murrayona* et de *Paramurrayona* Vacelet, 1967, mais elles dérivent beaucoup plus nettement de triactines, dont les actines s'épaississent considérablement. Cette carapace constitue le seul renfort du squelette.

Les choanocytes à noyau apical et les larves amphiblastula sont de type Calcaroné; par contre, les triactines dont dérivent les écailles sont équiangulaires, donc de type Calciné. L'oscule est armé de tétractines particulières.

Cette espèce, dépourvue de diapasons, n'a été placée qu'avec doute dans les Pharétronides. Ceci pose de nouveau le problème de la définition des Pharétronides et de leur unité.

L'UNITÉ DES PHARÉTRONIDES

Les descriptions précédentes montrent la grande variété des types d'organisation des Pharétronides actuelles, variété remarquable étant donné le faible nombre de formes connues. On peut même se demander si des Eponges aussi disparates représentent bien un groupe naturel, et on doit maintenant rechercher les points communs qui justifient leur regroupement et en discuter l'importance.

Le renforcement du squelette

Toutes les Pharétronides actuelles ont des structures augmentant la rigidité du squelette; des structures comparables existaient chez les fossiles, et en ont permis la fossilisation. Mais en réalité on ne peut pas accorder à ce point commun une signification phylogénique, en raison de la diversité des processus utilisés dans les différentes familles; ces structures (fibres de diapasons, soudure de tétractines ou de triactines, squelette aspiculeux interne ou externe, couverture d'écailles) peuvent bien être apparues par convergence chez des Eponges appartenant à diverses lignées; ceci est d'autant plus plausible que chacune de ces structures a son homologue chez les Eponges siliceuses: les faisceaux de spicules sont très répandus chez les Démosponges; le squelette de nombreuses Lithistides est comparable à celui des Minchinellidae; les écailles superficielles des Pharétronides de trois familles différentes aboutissent à une structure identique à celle obtenue par les discotriaenes de certaines Lithistides; le squelette aspiculeux calcaire des Démosponges *Merlia* et *Astrosclera* ressemble beaucoup à celui des Murrayonidae.

Ces renforts squelettiques rendent les Pharétronides aptes à la vie dans des eaux très agitées, comme dans les tunnels sous-récifaux de Madagascar; mais beaucoup d'entre elles vivent aussi dans des grottes calmes ou en profondeur, et il ne s'agit donc pas seulement d'une adaptation convergente.

Le spicule en diapason

Ce type de spicule, triactine dont les actines paires sont parallèles et fortement rapprochées, constitue un des plus importants points communs entre les Pharétronides actuelles. On le trouve en effet chez toutes les espèces décrites, sauf chez *Lepidoleucon inflatum* et chez *Kebira uteoides*; chez cette dernière espèce, des pseudodiactines groupés en faisceaux représentent probablement une simple modification des diapasons, par réduction des actines paires devenues vestigiales. On connaît aussi des diapasons chez plusieurs fossiles Inozoaires bien conservés. Par ailleurs les diapasons n'existent qu'exceptionnellement chez les autres Eponges Calcaires, puisqu'on ne les connaît que chez deux espèces très mal décrites, *Leucetta pandora* Haeckel, 1872 et *Leucortis pulvinar* Haeckel, 1870, qui sont peut-être d'ailleurs, comme le suggère Burton (1963), des *Lelapia*.

La forme et la disposition des diapasons sont variables. Chez les Lelapiidae, dépourvues de squelette principal, les diapasons forment des faisceaux bien définis; les actines paires sont alors longues et fortement rapprochées; cette forme a été considérée comme une adaptation à la formation de fibres. Chez les espèces à squelette solide, ces spicules ont une forme plus irrégulière et des branches généralement moins longues; ils sont alors disposés en désordre ou bien forment des fibres moins épaisses et moins longues que celles des *Lelapia*; de telles fibres existent chez *Petrostroma*, chez *Plectroninia tecta* et chez *Paramurrayona*. Enfin, chez les *Plectroninia* actuelles, on trouve en plus des triactines à actines paires en V, parfois épineuses, qui constituent une variété particulière de diapasons; ces spicules sont situés dans les mailles du squelette soudé basal.

Chez des Sphinctozoaires du genre *Tremacystia* et chez des Inozoaires fossiles (*Corynella*, *Raphidonema*), il existe une autre modification des triactines en relation avec la formation de fibres, les triactines filiformes, dont les actines latérales forment un angle très ouvert tandis que l'actine basale est vestigiale (Hinde, 1883).

Le diapason apparaît donc souvent en relation avec la présence de fibres chez des Eponges Calcaires; mais on le trouve aussi chez des espèces qui ne possèdent pas de fibres et dans ce cas il s'agit sans doute d'un caractère ancestral, rappelant l'existence de fibres chez des formes disparues. C'est surtout la présence, ancienne ou actuelle, de faisceaux de spicules qui constitue réellement un point commun. Il est évidemment possible que ce caractère soit apparu par convergence chez diverses lignées.

L'oscule armé

L'oscule des Pharétronides actuelles présente fréquemment une armature particulière; il s'agit, chez les formes sans papilles, de plusieurs rangées de tétractines serrés autour de l'oscule, dont les deux actines latérales sont parallèles au bord de l'oscule, tandis que l'actine apicale est dirigée vers le centre de l'ouverture. Cette structure a été observée chez *Petrobiona*, chez les cinq *Plectroninia* de Madagascar, chez *Paramurrayona* et chez *Lepidoleucon*; elle existe également chez la Minchinellidae méditerranéenne en cours de description. Les descriptions des diverses Lelapiidae, de *Murrayona*, de *Petrostroma*, de *Plectroninia deansii* et de *P. hindei* ne font état d'aucune structure osculaire particulière, mais ceci n'est pas très significatif en raison du mauvais état des spécimens; j'ai examiné un spécimen de *Murrayona phanolepis* conservé au British Museum (Nat. Hist.) sans trouver de tétractines autour de l'oscule, mais la plupart des spicules superficiels de cet échantillon ont disparu. Son absence est plus probable chez *Lelapia australis*: l'exemplaire étudié par Dendy, conservé au British Museum (Nat. Hist.), n'en présente pas trace malgré son bon état. Une armature osculaire n'a été signalée chez aucun fossile, mais ceci ne prouve pas son absence, et il serait intéressant de rechercher soigneusement une telle structure sur des échantillons bien conservés.

Chez *Minchinella lamellosa* et surtout chez *Plectroninia hindei mediterranea*, qui possèdent des papilles exhalantes, l'armature de tétractines est moins nette; il existe cependant, tout le long de la papille, des tétractines dont l'actine apicale est dirigée vers le centre du tube, mais cette disposition ne se distingue pas de celle qui est fréquemment réalisée, autour des cavités de beaucoup d'Eponges Calcaires, par les tétractines gastriques. Ces papilles nous donnent peut-être une interprétation de l'armature osculaire: il est possible que ces rangées de tétractines périosculaires correspondent à un stade régressif de telles papilles exhalantes, exceptionnelles chez les autres Eponges Calcaires, dont les tétractines se seraient condensées autour de l'oscule; cette armature témoignerait alors de l'existence de papilles chez les ancêtres des Pharétronides actuelles, papilles correspondant le plus souvent à la vie sur des substrats envasés.

A l'exception des Lelapiidae les Pharétronides actuelles ont donc probablement toutes une armature de tétractines autour de l'oscule. Mais ces Eponges n'ont pas l'exclusivité absolue de cette structure, puisqu'on la connaît aussi chez deux Eponges Calcaires actuellement classées dans les *Leuconia*, *L. nivea* (Grant, 1826) et *L. johnstoni* Carter, 1871.

Les caractères écologiques

Les conditions dans lesquelles vivent les Pharétronides actuelles, à l'exception encore des Lelapiidae, sont tout à fait particulières. La majorité d'entre elles ont été trouvées dans des grottes entre 0 et 25 m de profondeur en Méditerranée et en Océan Indien (donc dans des eaux dont la température, souvent supérieure à 20°C, ne descend pas en dessous de 13° en hiver), soit dans des zones très faiblement éclairées, soit à l'obscurité totale. Les autres espèces ont été récoltées dans l'Indo-pacifique entre 83 et 195–392 m, presque toujours après accrochage de l'engin de prélèvement, donc sur des substrats très accidentés et à des profondeurs où, dans les eaux tropicales, les températures sont encore relativement élevées. On ne trouve pas de Pharétronides dans les mers froides ou en profondeur en Méditerranée (où la température est aux environs de 13° toute l'année). On peut donc conclure que les Pharétronides demandent à la fois un éclairement très atténué ou nul et des eaux relativement chaudes (de température moyenne nettement supérieure à 13°), deux conditions qui ne se rencontrent simultanément que dans les grottes peu profondes des mers tempérées chaudes et tropicales, et aux environs de 100–200 m dans les régions tropicales.

Pour les Pharétronides fossiles, leur abondance remarquable dans certains gisements n'est pas conciliable avec une localisation préférentielle dans les grottes sous-marines. Les exigences écologiques qui viennent d'être définies seraient-elles particulières aux formes actuelles? Il n'est pas nécessaire d'imaginer un tel changement si l'on admet que les Pharétronides fossiles vivaient sur des surfaces subhorizontales à une profondeur suffisante pour que la lumière soit assez atténuée, mais à laquelle la température de l'eau était encore assez élevée; ces conditions, rencontrées sur de grandes surfaces dans des mers chaudes, sont compatibles avec une conservation en grand nombre. Ceci implique qu'elles aient été capables de vivre sur des substrats meubles, ou de supporter un envasement important. A ce sujet, il est intéressant de remarquer que la seule Pharétronide actuelle connue qui soit capable de vivre sur des rochers envasés, *Plectroninia hindei mediterranea* (qui, d'après de nouvelles observations, est strictement localisée aux planchers de grottes sous-marines, sous 1 à 2 mm de vase), est pourvue d'une papille exhalante; les Pharétronides fossiles possédaient-elles des papilles, dont la régression, chez les formes actuelles des parois et des plafonds de grottes, aurait donné l'oscule armé dont la signification a été discutée plus haut? Il est tentant de faire un rapprochement entre ces faits et ces suppositions, mais il est bien évident que tout ceci reste à établir.

Quoiqu'il en soit, il existe un point commun dans les exigences écologiques des Pharétronides actuelles. Il serait peut-être critiquable de lui accorder une grande valeur phylogénique, en particulier en raison de l'existence d'Eponges "silico-calcaires" dans des conditions écologiques très voisines. Ces Démosponges, dont on ne connaissait que deux représentants et dont Hartman et Goreau viennent de découvrir plusieurs nouvelles formes dans les récifs de la Jamaïque, possèdent un squelette calcaire aspiculeux supplémentaire offrant de grandes analogies avec celui des Murrayonidae; dans ce cas, il s'agit certainement d'une phénomène de convergence. *Merlia normani* vit à Madère (grotte, et entre 108 et 162 m), en Méditerranée (grottes obscures et semi-obscures) et en Océan Indien (grotte obscure); son squelette principal est en calcite. *Astrosclera willeyana*, dont le squelette principal est en aragonite, a été récoltée dans l'Indopacifique, entre 63 et 180 m et dans des grottes semi-obscures entre 3 et 8 m.

L'apparition d'un squelette supplémentaire rigide chez diverses Eponges fréquemment cavernicoles amène à se demander si les divers modes de renforcement du squelette des Pharétronides ne constitueraient pas une adaptation convergente à la vie dans les grottes. Dans les tunnels des récifs de Madagascar, violemment battus, il pourrait s'agir d'une adaptation à un hydrodynamisme violent; les Lithistides, à squelette siliceux fortement renforcé, sont d'ailleurs abondantes dans ces tunnels; mais ceci n'a guère de raisons d'être dans les grottes méditerranéennes, souvent très calmes, ou en profondeur. Cette coïncidence entre la vie dans les grottes et une hypercalcification demeure difficilement expliquable, mais son importance ne doit pas être exagérée, car, ainsi qu'on l'a vu plus haut, le caractère cavernicole, pour les Silico-calcaires comme pour les Pharétronides, correspond en réalité à une grande sciaphilie allant de pair avec une thermophilie marquée; ces Eponges trouvent actuellement des conditions favorables surtout dans les grottes, mais il en allait sans doute différemment autrefois. L'ancienneté des Silico-calcaires est attestée par leur répartition de type mésogéen, très semblable à celle des Pharétronides actuelles et il est probable que ces Eponges ont dû être beaucoup plus répandues du temps des Pharétronides. Les similitudes d'écologie et de répartition géographique entre les diverses Silico-Calcaires et les Pharétronides correspondent sans doute à leur caractère de reliques tropicales ou subtropicales sciaphiles.

CONCLUSIONS

Les quelques Eponges actuelles que l'on peut considérer comme les survivantes des Pharétronides forment donc un groupe assez hétérogène. Dans cet ensemble, on peut isoler trois familles, Murrayonidae, Paramurrayonidae et Minchinellidae, qui ont de nombreux et importants points communs: spicules en diapason, squelette fortement renforcé, oscule à armature de tétractines, caractère sciaphile-thermophile. *Murrayona* et *Paramurrayona* se distinguent des autres par leurs choanocytes et par leurs larves, mais ceci pose des problèmes uniquement dans la classification des autres Eponges Calcaires. Ce groupe représente certainement les reliques des Pharétronides fossiles et mérite son isolement dans une unité systématique particulière.

Lepidoleucon inflatum, dépourvue de diapasons, possède quelques caractères de Pharétronides (oscule armé, squelette renforcé par un procédé rappelant celui de *Paramurrayona*, vie dans les grottes obscures), mais son appartenance à ce groupe est moins sûre.

Les Lelapiidae n'ont en commun avec les vraies Pharétronides que le spicule en diapason, caractère qui peut fort bien être apparu par convergence. Leur parenté avec elles est donc douteuse. Il en va de même pour *Leuconia nivea* et *L. johnstoni* qui ne possèdent qu'un seul caractère de Pharétronide, l'oscule armé.

RAPPORTS DES PHARÉTRONIDES AVEC LES AUTRES EPONGES CALCAIRES ACTUELLES

Il apparaît chez les Pharétronides, même si on les restreint aux trois familles Murrayonidae, Paramurrayonidae et Minchinellidae, une certaine hétérogénéité vis à vis des caractères sur lesquels est basée la classification actuelle des Eponges Calcaires (Hartman, 1958). En effet, les Minchinellidae et *Petrobiona massiliana* s'apparentent à la sous-classe Calcaronée par leurs triactines sagittaux, par leurs choanocytes à noyau apical et par leurs larves amphiblastula; par contre, *Murrayona* et *Paramurrayona*, tout en ayant des triactines sagittaux de type

Calcaroné se rapprochent davantage de la sous-classe Calcinée par leurs choanocytes à noyau basal et par la larve sans doute blastula de *Paramurrayona. Lepidoleucon*, les Lelapiidae et les deux *Leuconia* ont des caractères de Calcaronées.

Les Pharétronides sont donc impossibles à classer dans le système d'Hartman. Trois solutions sont à envisager: rejeter cette classification, considérer les Pharétronides comme polyphylétiques, isoler les Pharétronides en tant que branche ancestrale ayant évolué de façon originale.

Il serait peu satisfaisant de rejeter, à cause de ce petit groupe relique mal connu, une classification qui reflète de manière particulièrement satisfaisante l'évolution parallèle de deux lignées d'Eponges Calcaires.

Dendy et Row (1913) ont déjà pensé que les Pharétronides étaient diphylétiques, mais ces auteurs, repris par Hartman (1958), réunissaient Minchinellidae et Murrayonidae dans le même groupe, et seules les Lelapiidae en étaient isolées, ce qui ne présentait pas beaucoup d'inconvénients. L'évolution de nos connaissances obligerait maintenant à séparer *Murrayona* et *Paramurrayona* de toutes les autres espèces, ce qui est difficile à défendre étant donné les particularités communes au moins avec les Minchinellidae et *Petrobiona*. Le diapason et l'armature périosculaire de tétractines peuvent être apparus par convergence, le renfort du squelette est obtenu par des moyens différents, les données écologiques n'ont peut-être pas grande signification, mais l'apparition indépendante de l'ensemble de ces caractères dans plusieurs lignées semble peu vraisemblable. De plus, la diphylétie n'explique pas le mélange de caractères de Calcinées et de Calcaronées que l'on observe chez plusieurs Pharétronides.

Ces difficultés me font préférer, peut-être provisoirement, la troisième solution, et j'ai proposé (1967a) de considérer les Pharétronides comme une troisième sous-classe d'Eponges Calcaires. Mais cette solution ne résoud pas tous les problèmes. On peut penser que les Pharétronides se seraient détachées d'une souche commune à toutes les Eponges Calcaires avant les deux lignées Calcaronées et Calcinées, alors que les caractères pleinement réalisés chez les deux sous-classes actuelles étaient alors seulement ébauchés. Malgré leur simplicité anatomique par rapport aux Pharétronides, les Eponges Calcaires actuelles seraient plus récentes que les Pharétronides, ainsi que le montre (Dendy et Row, 1913; Borojević, communication dans cette volume) l'existence de nos jours de tous les stades de complication anatomique dans les deux lignées.

Un autre solution que l'on peut envisager est de considérer les Pharétronides actuelles comme appartenant à la sous-classe Calcaronée,

Murrayona et *Paramurrayona* constituant une branche aberrante de cette sous-classe ayant acquis, par un phénomène de convergence avec la sous-classe Calcaronée, des choanocytes à noyau basal et des larves blastula, mais ayant conservé des triactines sagittaux.

Seule la découverte de formes nouvelles, maintenant que nous savons où trouver ces Eponges, et leur étude détaillée, permettront d'éclaircir ces problèmes.

References

Burton, M. (1963). *A Revision of the Classification of the Calcareous Sponges.* London: British Museum (Natural History).

Dendy, A. and Row, R. W. H. (1913). The classification and phylogeny of the Calcareous Sponges, with a reference list of all the described species, systematically arranged. *Proc. zool. Soc. Lond.* **1913**: 714–813.

Dubosq, O. and Tuzet, O. (1944). L'ovogenèse, la fécondation et les premiers stades du développement du *Sycon elegans Bower*. *Arch. Zool. exp. gén.* **83**: 445–459.

Hartman, W. D. (1958). A re-examination of Bidder's classification of the Calcarea *Syst. Zool* **7**: 97–110.

Hinde, G. J. (1883). *Catalogue of the fossil sponges in the Geological Department of the British Museum. London* **8**: 1–248. London: British Museum.

Kirkpatrick, R. (1908). On two new genera of recent Pharetronid Sponges. *Ann. Mag. nat. Hist.* (8) **2**: 503–514.

Ott, E. (1967). Segmentierte Kalkschwämme (Sphinctozoa) aus der alpinen Mitteltrias und ihre Bedeutung als Riffbildner im Wettersteinkalk. *Abh. bayer. Akad. Wiss. Math.-nat.* Kl. *N.F.* **131**: 1–96.

Reid, R. E. H. (1967). Spicules in the skeleton of *Tremacystia d'Orbignyi* (Hinde). *Nature, Lond.* **215**: 875–876.

Seilacher, A. (1961). Die Sphinctozoa, eine Gruppe fossiler Kalkschwämme. *Abh. math.-naturw. Kl. Akad. Wiss. Mainz* **1961**: 723–790.

Steinmann, G. (1882). Pharetronen-Studien. *Neues. Jb. Miner. Geol. Paläont.* **2**: 139–191.

Vacelet, J. (1960). Interprétation nouvelle de la fibre des Eponges Pharétronides (Calcispongia) fossiles d'après les Pharétronides actuelles. *Neues Jb. Paläont. Mh.* **10**: 433–439.

Vacelet, J. (1964). Etude monographique de l'Eponge Calcaire Pharétronide de Méditerranée, *Petrobiona massiliana* Vacelet et Lévi. Les Pharétronides actuelles et fossiles. *Recl. Trav. Stn. mar. Endoume* **50**: 1–125.

Vacelet, J. (1967a). Description d'Eponges Pharétronides actuelles des tunnels sous-récifaux de Tuléar (Madagascar). *Recl. Trav. Stn. mar. Endoume suppl.* **6**: 37–62.

Vacelet, J. (1967b). Quelques Eponges Pharétronides et "Silico-Calcaires" de grottes sous-marines obscures. *Recl. Trav. Stn. mar. Endoume* **42**: 121–132.

Vacelet, J. and Lévi, C. (1958). Un cas de survivance, en Méditerranée, du groupe d'Eponges fossiles des Pharétronides. *C. r. hebd. Séanc. Acad. Sci., Paris* **246**: 318–320.

Wagner, W. (1964). Zum Skelettbau oberjurassischer Kalkschämme. *Mitt. bayer. Staatssamml. Paläont. hist. Geol.* **4**: 13–21.

Zhuravleva, I. T. and Rezhvoy, P. D. (1956). Contribution à la taxonomie des Eponges fossiles et des Archéocyathes (en Russe). *Dokl. Akad. Nauk SSSR.* **111**: 449–451.

Ziegler, B. (1964). Die Cortex der fossilen Pharetronen (Kalkschwämme). *Eclog. geol. Helv.* **57**: 803–822.

Symp. zool. Soc. Lond. (1970) No. 25, 205–243

JAMAICAN CORALLINE SPONGES: THEIR MORPHOLOGY, ECOLOGY AND FOSSIL RELATIVES

WILLARD D. HARTMAN

Department of Biology and Peabody Museum of Natural History
Yale University, New Haven, Connecticut, U.S.A.

and

THOMAS F. GOREAU

Marine Sciences Research Center, State University of New York
Stony Brook, New York, 11790, U.S.A.

Marine Laboratory, University of the West Indies
Discovery Bay, St. Ann, Jamaica, W.I.

SYNOPSIS

A group of six species of sponges with a compound skeleton of siliceous spicules, organic fibers and aragonite has been found on the fore-reef slope of Jamaican coral reefs. One of these is *Ceratoporella nicholsoni* (Hickson, 1911), originally described as a coenothecalian octocoral; another belongs to the genus *Merlia* Kirkpatrick, 1908; the remaining four represent new genera and species. In life all have a thin veneer of living tissue similar to that of the Demospongiae. The exhalent channels are often raised above the general surface of the sponges and create prominent star-shaped patterns which, in *Ceratoporella*, leave a stellate pattern impressed into the surface of the aragonitic basal mass. During deposition, the aragonite overgrows and entraps the siliceous and organic skeletal elements.

The calcareous skeleton of these coralline sponges bears certain striking resemblances to that of the stromatoporoids as suggested by Kirkpatrick with reference to *Astrosclera* Lister, 1900 and *Merlia*. Support for this hypothesis follows: (1) the presence in some stromatoporoids of astrorhizae comparable to the stellate depressions left in the surface of *Ceratoporella* and *Astrosclera* by the exhalant channels; (2) isolated reports of spicules in stromatoporoids and the likelihood of the occurrence of calcareous pseudomorphs of spicules in others suggest a similarity to the siliceous-aragonitic skeletal complex in coralline sponges; (3) the fasciculate microstructure of the aragonite laid down by recent coralline sponges is identical in pattern to that of certain Paleozoic and most Mesozoic stromatoporoids.

The six coralline sponges are abundant on the most rugged and precipitous parts of the outer reef slope and have been collected by divers at depths ranging from 8 to about 92 m. Although ecologically a part of the ahermatypic cryptofauna of the reef, they are so abundant in places that they must be considered significant contributors of structurally consolidated calcium carbonate to the reef base.

A new class in the Phylum Porifera is proposed, to be called the Sclerospongiae.

PART I. MORPHOLOGY AND SYSTEMATIC AFFINITIES OF THE CORALLINE SPONGES

Introduction

Sponges with a compound skeleton consisting of siliceous spicules, organic fibers and crystalline aragonite have been known since Lister

(1900) described *Astrosclera willeyana* from the Pacific islands of Lifu and Funafuti. Lister believed that the spicules were foreign to the sponge but Kirkpatrick (1910a, 1910c), on the basis of his studies of additional specimens from Christmas Island (Indian Ocean), concluded correctly that the siliceous spicules are secreted by the sponge. At about the same time Kirkpatrick (1910b, 1911) described from Madeira a second sponge, *Merlia normani*, with a similar array of skeletal elements but bearing a tabulate calcareous skeleton, and Hickson (1911, 1912) named a peculiar calcareous organism, *Ceratoporella nicholsoni*, which he classified as a coenothecalian octocoral. This last mentioned organism, dredged originally off Cuba at a depth of 200 m, has been rediscovered recently off Jamaica during the course of investigations of the fore-reef slope environment (Goreau and Hartman, 1963) on the north coast of that island. It has now been established that *Ceratoporella nicholsoni* is a sponge and not a cnidarian as thought by Hickson; it possesses the same three skeletal components as do *Astrosclera* and *Merlia*. Living with it in the Jamaican fore-reef slope environment are six additional species of similar structure, including a new species of *Merlia*. It is the purpose of this paper to describe the morphology of these remarkable coralline sponges, to discuss their environmental relationships and to present evidence that they are relicts of an animal group that once flourished in Paleozoic and Mesozoic seas.

The morphology of Ceratoporella nicholsoni

Ceratoporella nicholsoni (Hickson, 1911) is the largest and most distinctive of the Jamaican coralline sponges. The massive, dome-shaped aragonitic skeletal masses (Fig. 1) range up to one meter in diameter and up to 50 cm in thickness. Young individuals are occasionally pedunculate; the larger ones often resemble botryoidal malachite with the surface covered with rounded bosses. Frequently a series of two or three dome-shaped masses are found growing one above the other with only the topmost one bearing living tissue. The individuals are cemented securely to the substrate and are extraordinarily tough and heavy. Sometimes serpulid worm tubes overgrow the masses and are in turn overgrown by the sponge.

The surface of the calcareous mass is marked by closely spaced pits (Fig. 2) varying from 0·2 to 0·5 mm across and 1·0 to 1·2 mm deep. The pits taper proximally to a point. In favorable thin sections, tangential to the surface, it can be seen that channels, about 135 μ in diameter, continue from the inner pointed ends of the pits deep into the interior of the calcareous skeleton. Distally, irregular spinelike processes extend to a height of 150 to 200 μ above the edge of each pit. The size

Fig. 1. *Ceratoporella nicholsoni* with prominent mamelons and astrorhizae. The colony is about 15 cm long and is shown *in situ* in a sub-reef tunnel at a depth of 30 m off Runaway Bay, Jamaica.

and shape of the pits in surface view vary considerably. Some are circular in outline, others are elliptical, and still others are irregularly polygonal. The surface diameter of circular pits varies from 170 to 280 μ; elliptical and irregular pits have diameters of 390 to 400 μ by 220 to 250 μ. There is no evidence of a definite dimorphism in pit structure, the range of variation in size being continuous. The largest pits are sometimes partially subdivided by incomplete walls growing from the outer boundaries of the pits.

Transverse thin sections of the skeleton of *Ceratoporella nicholsoni* (Figs 3 and 4) reveal that the walls of the pits continue deep into the calcareous skeletal mass, some reaching the base. The lumen of the pits is filled in secondarily with aragonite as the organism grows upward so that the gross appearance of the skeleton is that of a solid mass of calcareous material bearing at its surface numerous shallow pits bordered by spinelike processes.

In fine structure the aragonitic skeleton of *C. nicholsoni* is similar to that of scleractinian corals. From centers of calcification that usually

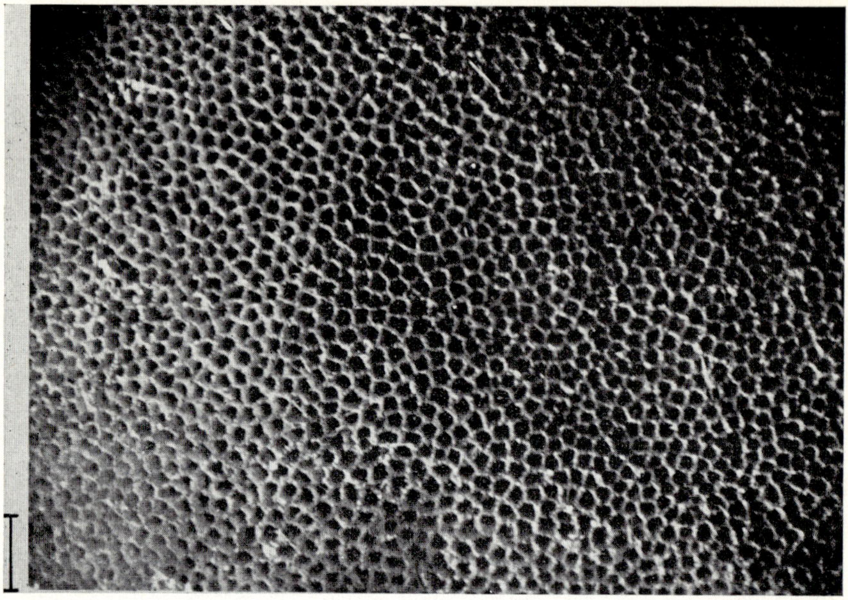

FIG. 2. Enlarged view of surface of *Ceratoporella nicholsoni* (Hickson), Scale: 1 mm.

appear to be organic fibers, crystals of aragonite radiate towards those of adjacent centers (Fig. 17). Each center of calcification with its radiating crystals may be called a sclerodermite, to borrow the terminology of coral specialists. The sclerodermites constitute the primary units of the calcareous skeletal mass.

The living tissue of the sponge, orange, peach or rufous in color, forms a thin veneer over the surface of the extensive calcareous skeleton and is organized into functional units, each of which fills one of the pits in the surface of the skeleton. One or two incurrent pores or ostia, 100 to 150 μ in greater diameter, open into subdermal cavities above each tissue-filled pit. One or, more frequently, two incurrent canals, 60 to 70 μ across, course down through the tissue and give off branches that communicate with flagellated chambers, 15 to 23 μ in greater diameter. These in turn open into fine excurrent canals that become larger gradually and run to the surface where they join other such canals and, increasing in size, converge upon oscular openings. In living specimens observed *in situ* the surface of the sponge is covered with star-shaped patterns of dilated excurrent canals (Fig. 5), each converging upon an osculum.

The continual flow of water through these channels often inhibits the upward growth of the calcareous material beneath them. Stellate

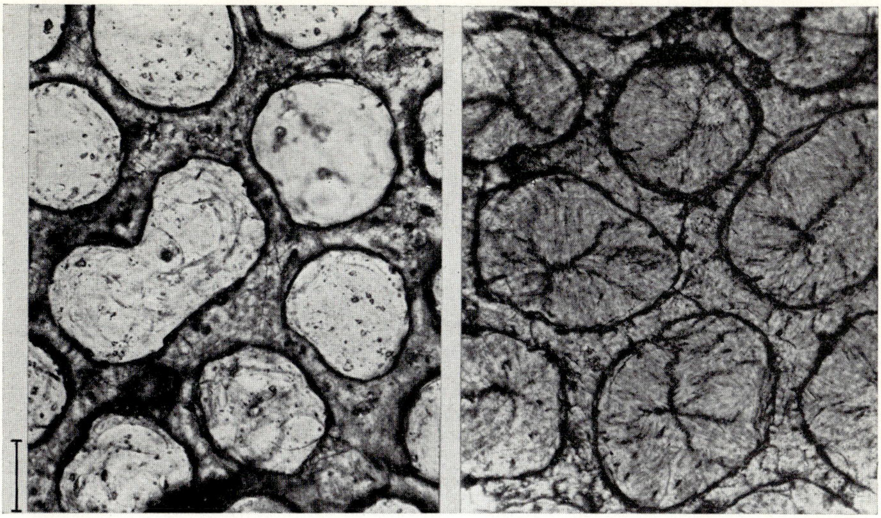

FIG. 3 FIG. 4

FIG. 3. Section of calcareous skeleton of *Ceratoporella nicholsoni* (Hickson), parallel to and just below surface. One pit is dividing. Scale: 100 μ.

FIG. 4. Section of calcareous skeleton of *Ceratoporella nicholsoni* (Hickson), parallel to and 5 cm below surface. Scale: 100 μ.

patterns of depressions on the surface of the skeletal mass give evidence in many but not all individuals of the site of excurrent channels in the living sponge. In some specimens, mamelons rise to heights of several millimeters to one centimeter above the surface of the calcareous skeleton beneath the site of each osculum (Fig. 6). Commensal zoanthideans may grow on the surface of the sponge, leaving circular depressions in the skeleton, about 1 mm in depth and 3 mm in diameter.

Within the tissue of the sponge is secreted a peculiar organic skeletal component, made up of diaphanous anastomosing sheets of an as yet unidentified substance. Very fine fibrils are associated with the material (Fig. 7). This organic skeletal component serves as a matrix for the aragonitic basal mass of the sponge. A particularly prominent strand of organic skeletal material continues from the base of each pit into the interior of the calcareous skeleton.

Also secreted in the sponge tissue are siliceous spicules in the form of long, thin styles (Fig. 8) bearing whorls of spines, three to five per whorl. Spicule dimensions are given in Table I. The head of each spicule is embedded in the organic matrix and the pointed end is in each case directed toward the surface of the sponge. The siliceous spicules and the organic matrix with which they are intimately associated become

FIG. 5. Underwater photograph of surface of *Ceratoporella nicholsoni* (Hickson), showing dilated excurrent channels converging upon oscules. Scale: 1 cm.

entrapped in aragonite (Fig. 17) as the sponge grows upward and outward, leaving behind an increasingly thick mass of calcium carbonate.

Many of the siliceous spicules begin to erode, or dissolve, after entrapment in the calcareous basal mass. Frequently it is the head of

FIG. 6. Surface view of calcareous skeleton of *Ceratoporella nicholsoni* (Hickson). Mamelons at sites of oscules and radiating impressions of excurrent channels are shown. Scale: small square = 1 mm side.

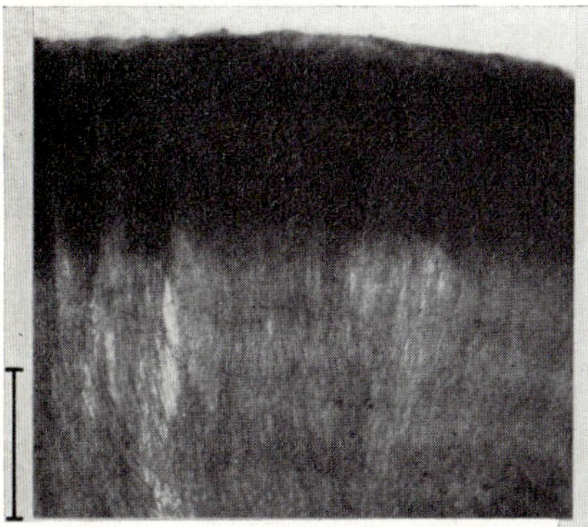

FIG. 7

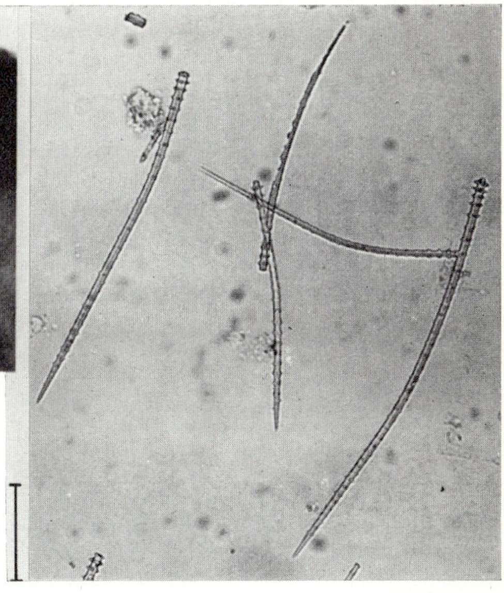

FIG. 8

FIG. 7. Decalcified, cleared vertical section of *Ceratoporella nicholsoni* (Hickson). Sponge tissue above; organic matrix and siliceous spicules below. Note concentric dark bands in matrix. Scale: 1 mm.

FIG. 8. Siliceous spicules of *Ceratoporella nicholsoni* (Hickson). Spicule in top center is eroded. Scale: 50 μ.

TABLE I

Spicule dimensions of Jamaican coralline sponges

Species	Length range of means with standard errors (microns)	Length overall range (microns)	Width range of means with standard errors (microns)	Width overall range (microns)	No. of specimens	No. of spicules measured per specimen
Ceratoporella nicholsoni (Acanthostyles)	206(±7·4)–298(±5·8)	64–394	3·1(±0·07)–4·0(±0·1)	1·7–5·2	3	50
Stromatospongia vermicola (Acanthostyles)	165(±7·0)–187(±7·7)	75–519	6·2(±0·15)–8·0(±0·18)	3·3–13·0	3	100
Stromatospongia norae (Acanthostyles)	195(±10·6)–215(±9·5)	90–589	5·5(±0·11)–6·1(±0·16)	2·7–9·1	3	50
Hispidopetra miniana (Styles)	269(±6·9)–301(±10·8)	125–818	5·4(±0·11)–7·4(±0·16)	1·3–10·4	3	100
Goreauiella auriculata (Acanthostrongyles)	60(±1·5)–68(±2·3)	35–124	2·3(±0·06)–2·7(±0·08)	1·3–3·9	3	50
Merlia sp.—Tylostyles	160(±3·3)	142–171	1·8(±0·32)	1·3–2·0	1	10
Clavidiscs	35(±1·5)	28–44	26(±0·32)	24–28	1	10

the spicule that dissolves first; in other cases erosion occurs on one side of the spicule (Fig. 8). In still others, the central canal of the spicule becomes enlarged before any external evidence of erosion is apparent. Hickson (1911) was undoubtedly referring to eroded spicules of this sort when he remarked that dilute nitric acid causes the spicules of *Ceratoporella nicholsoni* to dissolve away. He concluded incorrectly from his observations that the spicules are calcareous. Eroded spicules have not been observed among those that are still enclosed within the tissue of the sponge. In some specimens examined, certain regions of the calcareous skeleton of the sponge are devoid of siliceous spicules.

Examination of stained sections of the sponge tissue reveal no especially differentiated cells that may be responsible for the deposition of the aragonitic mass. The possibility has been considered that the calcareous component of the sponge skeleton is laid down by or in association with a symbiotic organism. Kirkpatrick (1912a, 1912b) concluded that red algal tetraspores and carpospores serve as centers of calcification in *Astrosclera* and that the calcareous skeleton of *Merlia* consists of the "calcified corpses" of "zooxanthella monads". We can find no evidence of algal symbionts serving as centers of calcification in *Ceratoporella nicholsoni*. Instead, the sclerodermites tend to form around the blobs of organic matrix that surround the heads of the siliceous spicules. However, a filamentous organism, possibly a species of boring Cyanophyceae, with a width of about 1·5 μ and bearing infrequent septa is a common occurrence within the calcareous skeleton. Since the filaments have no regular arrangement with respect to the sclerodermites, we do not believe that they are involved in the process of calcification, which we conclude is intrinsic to the sponge.

The morphology of other species of Jamaican coralline sponges

Six additional species of coralline sponges occur in the fore-reef slope environment off the north coast of Jamaica (Hartman, 1969). Of these, four are discussed briefly here, while the remaining two, collected alive only recently, will be described in a subsequent contribution. The four species under consideration share the presence of a basal skeletal mass of aragonite that entraps siliceous spicules and an organic matrix secreted by the living cells of the sponge. Sclerodermites, comparable to those described for *Ceratoporella nicholsoni*, are the basic units of the calcareous skeleton in each case. In each species some of the siliceous spicules, after being embedded in the aragonitic basal mass, show partial erosion or dissolution.

Stromatospongia vermicola Hartman, 1969 is an encrusting species that spreads out over the substrate and always grows in association with

FIG. 9. *Stromatospongia vermicola in* −45 m off Runaway Bay, Jamaica. The numerous spiky projections are serpulid worm tubes encrusted and consolidated into a single mass by the sponge. This colony was found under a large coral overhang and is about 12 cm high. In the left foreground is *Halimeda cryptica* Colinvaux and Graham, 1964, a deep water alga which occurs only in very shaded habitats.

masses of tubes of a serpulid worm (Fig. 9). Individuals reach a diameter of at least 40 cm. The aragonitic skeleton is laid down around serpulid tubes and there forms a basal mass, 3 mm in thickness, from which extend numerous multibranching processes, 1·5 to 2 mm high. The living tissue extends down into irregular somewhat meandroid spaces left between these processes. Not infrequently the calcareous skeleton extends into the interior of the serpulid tubes. The siliceous acanthostyles vary greatly in length and are thicker than those of *C. nicholsoni*. They often echinate the calcareous processes at the surface of the basal mass. Dimensions are given in Table I.

The living tissue is apricot to light salmon pink in color. Closely spaced ostia, varying in diameter from 32 to 68 μ, perforate the entire epidermis which is supported above subdermal cavities by upright tracts of spicules. The incurrent subdermal cavities lead into vertical channels that subdivide and open into spheroidal, eurypylous flagellated chambers, 16 to 20 μ in diameter. The excurrent channels course to the surface and often form stellate patterns as they converge upon oscular areas where one to three oscules open to the exterior. Oscules, up to 500 $\times$ 750 μ in major axes, may open out anywhere along the length of the excurrent channels as well as at points of convergence of many channels. Egg cells, 25 $\times$ 40 μ in chief diameters with nuclei 12 μ in diameter and nucleoli 4 μ in diameter, occur rarely in sections. Several ellipsoidal cleaving embryos, probably at the 32-cell stage, and measuring 44 $\times$ 32 μ in major axes, have also been observed in sections, suggesting that this species incubates its larvae. Fully developed larvae have not been found as yet.

Stromatospongia norae Hartman, 1969 takes the form of nodular or mammillate masses (Fig. 10) up to 5 cm in diameter and 4 cm in height. In many cases the sponge grows out from a rounded mass along serpulid worm tubes that protrude from the base like horns. The surface of the calcareous skeleton is marked by upright lamellae ranging up to 0·8 mm high and up to 1 mm or more in length and terminating in low, rounded processes. In some instances several such lamellae run parallel to one another and thus mark off meandroid furrows into which the living tissue extends. In other cases the lamellae are shorter and are arranged irregularly, leaving depressions of varying shapes. The basal mass of aragonite is solid and may reach a thickness of 4 cm or more. The siliceous spicules are acanthostyles, very similar in form to those of *S. vermicola* but somewhat larger and thinner. Dimensions are given in Table I.

The living tissue of *S. norae* varies in color from cream to ecru beige. Ostia, 62 to 186 μ in diameter and circular to elliptical in outline, are

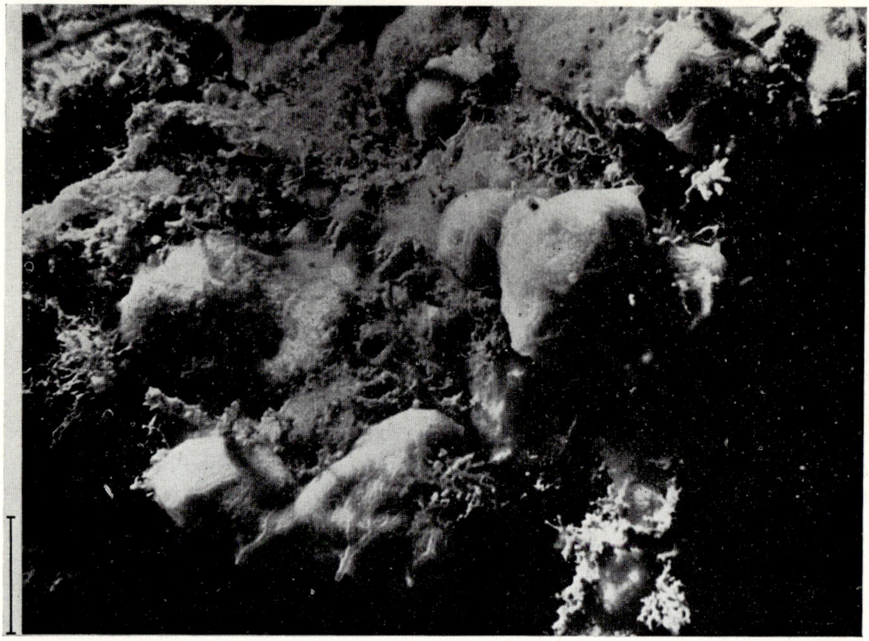

FIG. 10. Underwater photograph of *Stromatospongia norae* Hartman showing excurrent channels and oscules. Scale: 1 cm.

distributed evenly over the surface. The ostia lead into subdermal cavities from which vertical inhalant canals lead down into the tissue masses that fill the spaces between the superficial lamellae of aragonite. The spheroidal flagellated chambers are 15 to 17 μ in diameter. The excurrent channels form radiating patterns on the surface (Fig. 10) and converge upon elliptical oscules, 350 $\times$ 500 μ in major axes. The excurrent channels leave faint stellate depressions on the surface of the calcareous skeleton in some specimens.

Hispidopetra miniana Hartman, 1969 is encrusting in early life and grows into irregular masses up to 15 cm in diameter and 5 cm high. The surface of the aragonitic skeleton is provided with numerous processes, lamellate, knob-like or arborescent in form and up to 7 mm high. The processes are covered with rounded spines, 25 to 50 μ high, and the pointed ends of numerous long stylote siliceous spicules protrude from the processes (Fig. 11). Serpulid worm tubes, bearing a keel, 3 to 4 mm high, invariably overgrow older specimens, and the sponge overgrows them in turn, often causing the keel to look like a cockscomb.

The siliceous spicules are smooth, slightly curved styles varying greatly in length. Dimensions are given in Table I. The heads of the

FIG. 11. *Hispidopetra miniana* Hartman. Close-up of calcareous skeleton with protruding siliceous spicules. Scale: 1 mm.

spicules are enclosed in an organic material. A fibril-bearing, flocculent substance serves as a matrix for the aragonitic basal mass.

The living tissue of *H. miniana* varies from vermilion or cherry red to carmine in color. The epidermis is thin, diaphanous and unpigmented and is raised above the tissue, stretching between the tips of the vertical calcareous processes of the basal mass, and is pierced by spicules. Ostia, 50 to 100 μ in diameter in living specimens, are rather sparsely distributed but occur occasionally in groups of 50 or more. The extensive subdermal cavities lead into wide incurrent canals that subdivide repeatedly and eventually communicate with eurypylous flagellated chambers, 20 to 22 μ in diameter. A complicated network of excurrent canals, 0·5 to 3·0 mm in width, runs along the epidermis. Oscules, varying in outline from circular to elliptical and 62 μ to 2·0 × 3·0 mm in diameter, open out from these channels at any point and not necessarily where anastomoses occur.

Goreauiella auriculata Hartman, 1969 is saucer-shaped to auriculate in form (Fig. 22) with edges upturned or curled downward and is attached to the substratum by a broad peduncle. Individuals range up to 16 cm in diameter but the aragonitic skeletal base is very thin, seldom exceeding 3 mm. From the base arise numerous closely set, delicate, arborescent, calcareous processes varying in height from 0·5 to 1·1 mm. The processes are covered with rounded spines, 12 to 40 μ high. The surface of the calcareous skeleton of most specimens shows raised dendritic patterns each converging at the edge of the skeleton (Fig. 12). These raised areas result from an increased height of the arborescent processes.

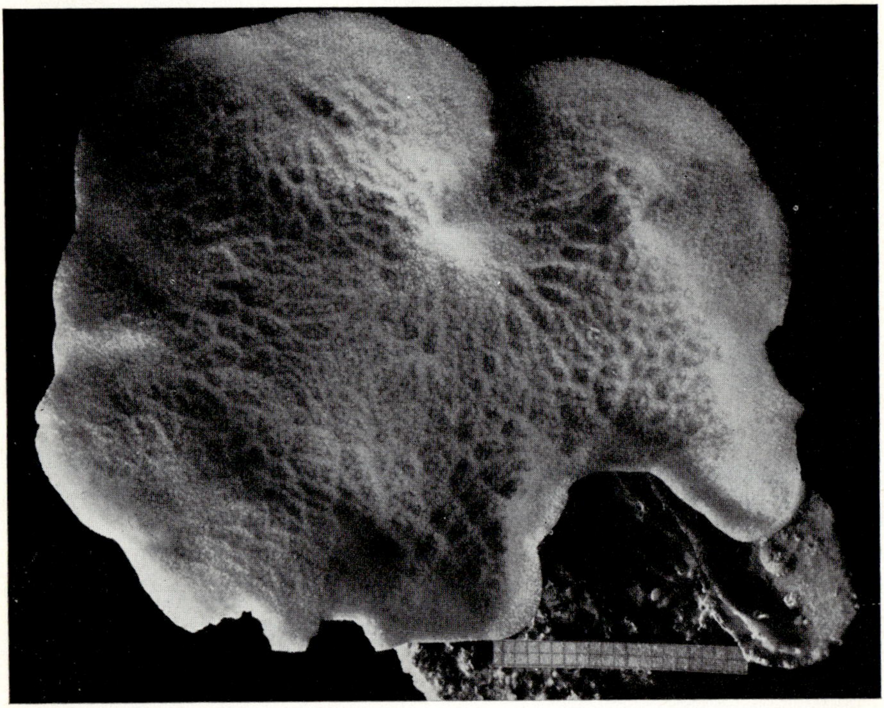

FIG. 12. Calcareous skeleton of *Goreauiella auriculata* Hartman showing raised reticulate patterns that follow beneath excurrent channels. Scale: small square = 1 mm side.

The basic units of the calcareous skeleton, as in the other coralline sponges described, are sclerodermites made up of crystals of aragonite radiating out from centers of calcification. The aragonite is colorless in ground thin sections.

The siliceous spicules are relatively short rods of equal diameter throughout, straight, slightly curved or with a pronounced curve. They are ornamented with whorls of spines that vary greatly in size and orientation. The details of the spination are similar to those described for *S. vermicola* but there is a terminal whorl of spines present. The ends with blunt knobs are enclosed in organic material. Spicule dimensions are given in Table I.

The living tissue of *G. auriculata* varies in color from straw yellow to chrome lemon or yellowish-grey. Ostia, 50–150 μ in diameter in the living sponge, are evenly distributed over the epidermis between the branching excurrent channels. The ostia open into subdermal cavities which lead into wide incurrent canals. The latter, after branching, open into eurypylous flagellated chambers, 18–20 μ in diameter. The oscules, about 300 μ in diameter, open out along the edge of the sponge; their positions are usually indicated by indentations in the periphery of the calcareous skeleton. The exhalant channels are transparent and dotted with fine white spots; they receive short side branches from the adjacent tissues as they run along the surface of the sponge. The systems of branching exhalant channels lie directly above the raised dendritic patterns of the calcareous skeleton mentioned above. A fibril-bearing flocculent material remains as a matrix after decalcification.

Relationship of coralline sponges to other Porifera

The genera *Astrosclera*, *Ceratoporella* Hickson, 1912, *Stromatospongia*, *Hispidopetra*, *Goreauiella* and *Merlia* share a compound skeleton of siliceous spicules, organic fibrous material and aragonite. Sponges of the Class Demospongiae possess the first two skeletal elements. The presence of aragonite, however, makes the coralline sponges unlike any other Porifera known. The Calcarea, including the pharetronids (of which the skeleton of *Petrobiona massiliana* Vacelet and Lévi, 1958 has been studied by X-ray diffraction analysis) all lay down a skeleton of calcium carbonate in the form of magnesium calcite (Vinogradov, 1953; Chave, 1954).

The general morphology of the water current system and the presence of siliceous spicules associated with an organic fibrous material appear to place the coralline sponges in the Class Demospongiae. Their relationships to specific members of the class are difficult to ascertain, however. The spicules of *Astrosclera*, *Ceratoporella*, *Stromatospongia* and *Goreauiella* are all verticillately spined styles, albeit that the spines on the spicules of *Astrosclera* are somewhat irregular at times (cf. Vacelet, 1967a) and the spicules of *Goreauiella* are truncate with a terminal whorl of spines. All share the characteristic of having the head ornamented with rounded or knob-like processes while the more distal whorls of spines are recurved toward the head. Demospongiae with spicules of this general sort occur within the families Hymedesmiidae Topsent, 1928 and Euryponidae Topsent, 1928. [Kirkpatrick (1910c) suggested a relationship of *Astrosclera* to *Hymeraphia* Bowerbank, 1862.] However, in both of these families the spines of the styles are arranged at random, not in whorls, and additional categories of megascleres and often microscleres as well are present. Among Demospongiae, only the genus *Agelas* Duchassaing and Michelotti, 1864 is characterized by a spiculation made up exclusively of verticillately spined styles. Here, however, the spines are straight, not recurved, and there is a strongly developed network of spongin. Thus no precise homology can be found between the spicules of the coralline sponges mentioned above and those of any known member of the Demospongiae. The presence of incubated parenchymellae in *Astrosclera* (Lister, 1900) and presumably also in *Stromatospongia vermicola* provides a character that is not inconsistent with the placement of these genera in the subclass Ceractinomorpha and within this group their spicule complement is closest to sponges of the order Poecilosclerida.

Merlia normani, with siliceous spicules in the form of tylostyles, raphides (some are toxiform) and clavidiscs, seems far removed from the

FIG. 13. Astrorhizae on the surface of an unidentified Silurian stromatoporoid from Sweden. Scale: 5 mm.

other genera of coralline sponges; its tabulate calcareous skeleton and hymenopylous flagellated chambers (Kirkpatrick, 1911) also serve to set it apart. Kirkpatrick (1911) pointed out the similarity of the clavidiscs of *Merlia* to the diancistras of *Hamacantha* Gray, 1867 and recognized a general similarity between the two genera. His views were supported by Topsent (1928) who placed the genera in a subgroup of his family Desmacidonidae Schmidt, 1870, order Poecilosclerida. Vacelet (1964), however, related *Merlia normani* to the Sigmaxinellidae Lévi, 1955, order Clavaxinellida Lévi, 1955, on the basis of spicule similarities. Eggs and larvae of *Merlia* are unknown at this time and hence the character of oviparity *versus* viviparity used by Lévi (1956) to divide the Demospongiae into two subclasses can be of no assistance in choosing between these two suggestions. On the basis of spiculation alone the hypothesis of Kirkpatrick and Topsent seems more likely since the diancistras of *Hamacantha* are intermediate between sigmas and clavidiscs, whereas sigmas modified in the direction of clavidiscs are unknown among the sigmaxinellids.

If the relationships of the coralline sponges outlined above are accepted, we must assume that the phenomenon of laying down a basal skeletal mass of aragonite has arisen independently within several different phylogenetic lines of the Demospongiae. An alternative hypothesis is that sponges with this modification separated from the Demospongiae at some remote time in the past and that the existing species represent relicts of a group that was once more abundant than it is in modern seas. It is of interest at this time to explore the possible relationship of the coralline sponges to certain enigmatic calcareous fossil organisms.

Possible fossil relatives of coralline sponges

When *Ceratoporella nicholsoni* was first dredged off the coast of Cuba by the U.S. Coast Survey Steamer "*Blake*", Agassiz (1878) thought that it was a tabulate coral of the genus *Favosites* Lamarck, 1816. Later he (1888) changed his mind and assigned it to a bryozoan genus allied to *Heteropora* Blainville, 1830. Thus the resemblance of these organisms to various fossil groups was recognized long before they were described as sponges.

In a series of papers Kirkpatrick (1909, 1910b) pointed out the resemblances between the calcareous skeleton of *Merlia normani* and certain tabulate corals of the family Chaetetidae Milne-Edwards and Haime, 1850 [specifically *Rhaphidopora stromatoporoides* (Roemer, 1883) belonging to a genus not listed by Hill and Strumm, 1956] and trepostomatous bryozoans of the genus *Monticulipora*. He regarded these resemblances as superficial in his detailed account of *Merlia* (Kirkpatrick, 1911). However, a year later (1912c,d,e) he seemed convinced that trepostomatous bryozoans of the *Monticulipora* d'Orbigny, 1850 type, tabulate corals of the genera *Chaetetes* Fischer in Eichwald, 1829, and *Rhaphidopora* Nicholson and Foord, 1886, and stromatoporoids as well, are all siliceous sponges with a supplementary calcareous skeleton. He stated that he had found siliceous ***C***-shaped spicules in ground thin sections of specimens of each of these diverse groups. Regrettably, he never published figures and a detailed documentation of these tantalizing statements.

Kirkpatrick's comments have led us to compare the coralline sponges described above with representatives of several groups of organisms known from the fossil record. It is well to dismiss the trepostomatous bryozoans from the discussion at once. Kirkpatrick (1911) was certainly correct when he doubted that *Merlia* would ". . . throw light on the problems concerning the nature of these organisms". Among other details of structure, the presence of heterozooids in the colonies finds no parallel among the coralline sponges. However, the possible relationship of stromatoporoids and certain chaetetids deserves further examination.

It has been supposed by most students of stromatoporoids in recent years that these enigmatic calcareous organisms are somehow related to the Hydrozoa (see, e.g., Lecompte, 1956). A minority of workers, however, have seen resemblances to the sponges. Stromatoporoids are the calcareous skeletal remains of organisms that lived from Cambrian to Cretaceous times. One species has been reported from Middle Oligocene strata in Germany (Birenheide, 1967). The organisms vary in form from laminar to massive and dendroid and range in size from small ones, 1 cm across, to huge masses, 2 m across and 1 m thick. Most forms show a marked layered structure. The surface of some species is provided with mound-like elevations, called mamelons. Some species exhibit stellate depressions, called astrorhizae, on the surface (Fig. 13). Stearn (1966), in an excellent study of the microstructure of stromatoporoids, recognized 14 different types, some of which he regarded as primary (secreted by the organism), and others as secondary (produced by post mortem processes).

Rosen (1869) was the first to develop in great detail evidence for relating stromatoporoids to sponges. He believed that he had found the calcified remains of spongin fiber networks in sections and he also reported the presence of both ostia and oscules on the surface of stromatoporoids. He asserted that the astrorhizae found on certain forms were in fact the exhalant channels of a sponge. Rosen was also the first to imply that the stromatoporoids were Demospongiae. Sollas (1877a,b) was inclined to the view that some but not all species of *Stromatopora* Goldfuss, 1826 represent hexactinellid sponges, the siliceous skeleton of which has been replaced by calcium carbonate.

Nicholson and Murie (1878), who reviewed the early literature on stromatoporoids and their relationships, were inclined to regard them as an extinct group of calcareous sponges with a continuous skeleton composed of non-spicular calcareous matter. They proposed the category Stromatoporoidea as a "new section" of the calcareous sponges. In his later work, however, Nicholson (1886) transferred the Stromatoporoidea to the Hydrozoa.

Heinrich (1914) also supported a relationship between stromatoporoids and sponges, basing his arguments largely on the resemblance between the astrorhizae of the former and the exhalant channels of the latter. He was disturbed by the absence of evidence of spicules in stromatoporoids and concluded that these animals were *sponge-like* but that they cannot be placed in the group of calcareous sponges as we know them today. He predicted that living descendants of stromatoporoids would some day be found and that these would probably belong to the horny sponges.

Twitchell (1929) compared the astrorhizae of stromatoporoids to the stellate patterns of excurrent channels found on certain fresh-water sponges. He was impressed by the observations of Bowerbank (1863) and Potts (1888) that *Trochospongilla leidyi* Bowerbank, 1863 is perennial and each year adds another distinct layer of growth; this, he felt, provides a clue to the mode of formation of the lamellae (or latilaminae) of stromatoporoids. He further described the occurrence of hollow, spined rod-shaped spicules in *Syringostroma centrota* Girty, 1895. For all these reasons, as well as a fancied resemblance between the rhagon of spongillids and certain stromatoporoids, Twitchell concluded that these fossils were the ancestors of the Demospongiae. Newell (1935) found calcareous spicules in a Pennsylvanian stromatoporoid, *Parallelopora mira* Newell, 1935, and concluded that this form "incorporates features of both" sponges and stromatoporoids. More recent authors, up until the present time, have tended to accept with little question the theory first proposed by Carter (1877) that stromatoporoids represent the calcareous skeletons of hydrozoans.

Heinrich's prediction that modern descendants of stromatoporoids would be found seems to have come true in the form of the coralline sponge genera described above. Our arguments in support of this hypothesis are (1) the presence of astrorhizae on the surface of the calcareous skeleton of both; (2) the presence of spicules or pseudomorphs

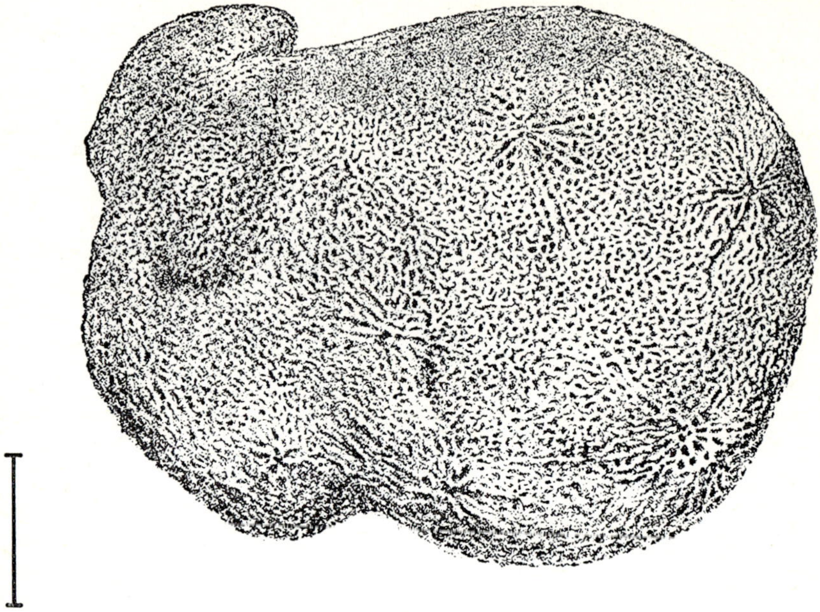

FIG. 14. Surface of *Astrosclera willeyana* Lister showing astrorhizae. From Lister (1900). Scale: ca. 5 mm.

of spicules in occasional specimens of stromatoporoids; (3) the similarity of the microstructure of certain stromatoporoids to that of the coralline sponges.

The astrorhizae of stromatoporoids (Fig. 13) bear a striking resemblance to the stellate patterns impressed into the surface of such coralline sponges as *Ceratoporella nicholsoni*, *Stromatospongia norae* and *Astrosclera willeyana* (Fig. 14). It is the similarity of astrorhizae to the system of excurrent channels of sponges that has been the most frequently cited homology by workers advancing an hypothesis of stromatoporoid-sponge relationships. The astrorhizae of *Ceratoporella*, *Stromatospongia* and *Astrosclera* do not represent the excurrent channels themselves (as many paleontologists have implied in their speculations about stromatoporoids) but are instead depressions in the calcareous skeleton lying beneath the water channels, the presence of which inhibits the upward growth of the skeleton. The channels themselves are cellular and are not fossilized. Those who favor hydrozoan affinities for the stromatoporoids compare astrorhizae to the basal stolonic network of a hydroid like *Hydractinia* Beneden, 1841 or to the coenosarcal tubes of *Millepora* Linné, 1758. However, as Twitchell (1929) has pointed out, these structures are of relatively uniform diameter throughout, whereas the ramifying patterns of astrorhizae gradually increase in size as they approach the center of the system just as do the excurrent channels of sponges as they converge upon the central osculum. It is of interest to note that in some specimens of *Ceratoporella nicholsoni* the oscula open out at the summit of bosses or mamelons from the edge of which extend the impressions of astrorhizal systems. Identical structural features exist in some stromatoporoids.

A common feature among stromatoporoids is the superposition of astrorhizae in each layer of the individual. A vertical canal, representing the center of each astrorhiza, may run from the base to the summit of the specimen. A similar structure is seen only in *Astrosclera* among the astrorhiza-bearing coralline sponges. Lister (1900) noted that the stellate patterns characteristic of the surface extend deep into the interior. He figured the cut surface of a specimen showing an internal astrorhiza on p. 461, Fig. A5. The secondary deposition of aragonite as the skeleton grows up obscures the astrorhizae beneath the surface in *Ceratoporella.*

The lower surface of specimens of *Ceratoporella nicholsoni* is marked by concentric growth lines as in many stromatoporoids. In vertical sections faint concentric lines, parallel to the surface, are also apparent. Curiously, these lines may be seen in the organic matrix left after decalcification (Fig. 7). Discrete layering (latilaminae), so commonly exhibited by Paleozoic stromatoporoids and probably resulting from periodic, perhaps seasonal, cessation of growth, is not evident in coralline sponges, all of which (except *Merlia*) are confined to tropical seas.

Evidence for the occurence of spicules in stromatoporoids is scanty. As mentioned above, Twitchell (1929) reported and figured calcareous, hollow rod-shaped spicules in the Devonian *Stromatopora centrota* Girty, 1895; these could represent partially eroded calcareous pseudomorphs of siliceous spicules. The random orientation of the spicules figured by Twitchell lends credence to the view of Lecompte (1956) that they may be foreign, perhaps belonging to a boring or cavity-dwelling sponge both of which are frequent inhabitants of the calcareous skeleton of the living coralline sponges. Newell (1935) figured vertically oriented calcareous spicules (possibly also pseudomorphs) in *Parallelopora mira* Newell, 1935 of Pennsylvanian age (Fig. 15). In figures of transverse and oblique sections of *Astroporina valangiensis*, a Jurassic stromatoporoid described by Schnorf (1960), pseudomorphs of spicules are clearly apparent, as they are also in sections (Fig. 16) of the Jurassic *Stromatopora japonica* Yabe, figured by Yabe and Sugiyama (1935). In both these cases the spicules are oriented obliquely in the calcareous

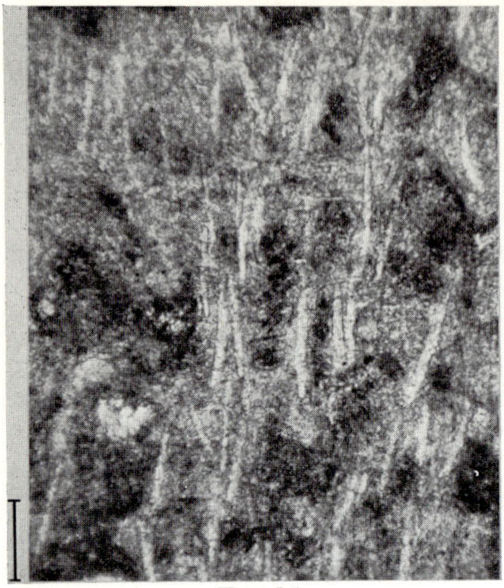

FIG. 15

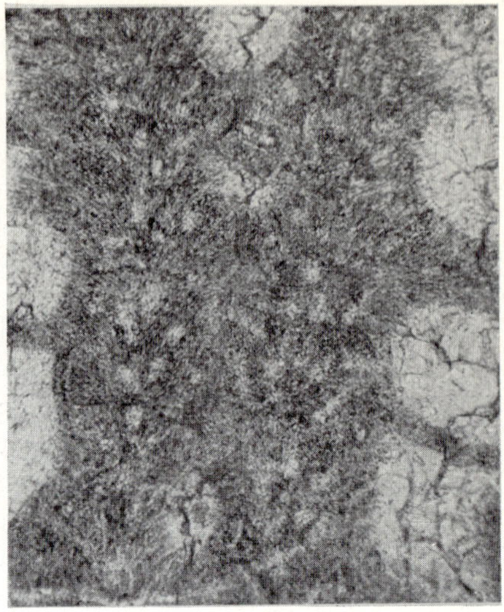

FIG. 16

FIG. 15. Vertical section of *Parallelopora mira* Newell showing spicules. Scale: 100 μ.

FIG. 16. Transverse section of *Stromatopora japonica* Yabe showing supposed pseudomorphs of spicules in section. Scale: 100 μ.

skeleton. That spicules are so rarely apparent in stromatoporoids may find an explanation in two observations on the existing corralline sponges. First, there is a tendency in all species in which the siliceous spicules are embedded in the calcareous matrix (i.e., all except *Merlia*) for the spicules to undergo slow chemical erosion. If many of the spicules in living specimens are already dissolving, it is not surprising to find that complete disappearance of the spicules occurs in fossil remains of Palaeozoic and Mesozoic age. The replacement of siliceous sponge spicules by calcite has been reported by Newell *et al.* (1953) in the Permian limestone reefs of Texas and New Mexico. Second, *Merlia* provides us with an example of a coralline sponge in which the siliceous spicules are not entrapped in the calcareous skeleton but are found only in the living sponge tissue. Such spicules would almost certainly be lost prior to fossilization.

The microstructure of the calcareous skeleton of coralline sponges is similar to that of certain Mesozoic stromatoporoids. Steiner (1932) and Yabe and Sugiyama (1935) have noted two chief types of microstructure in the latter: (1), radial, in which fine needles of calcite radiate perfectly perpendicularly from a dark axis and (2) fasciculate or "water jet", in which the calcite needles arise from a granular axial zone and extend outward at an acute angle to the axis. The second structure is identical to that of the aragonitic skeleton of *Ceratoporella* (Fig. 17), *Stromatospongia*,

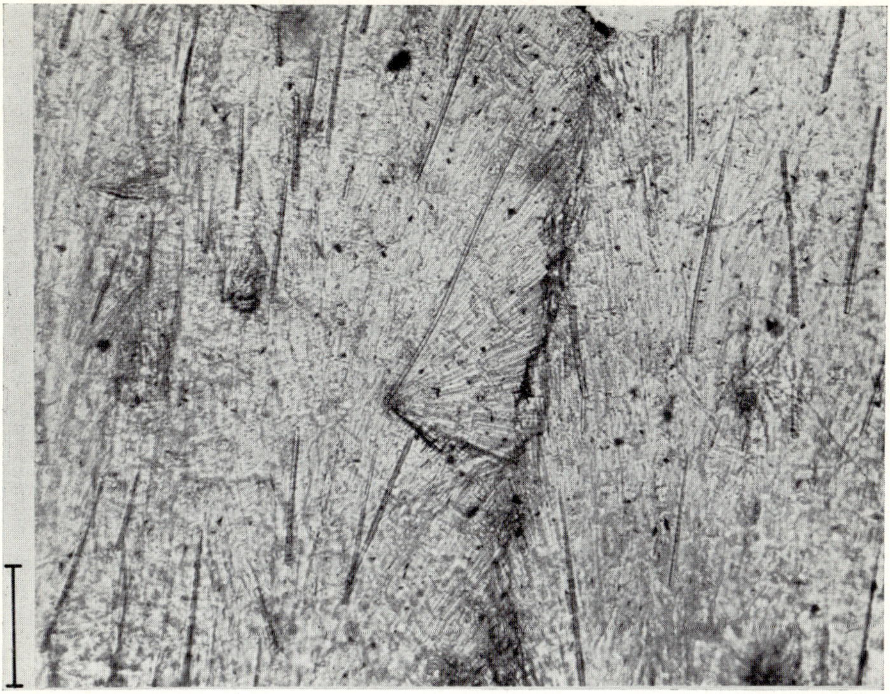

FIG. 17. Vertical section of calcareous skeleton of *Ceratoporella nicholsoni* (Hickson) showing embedded siliceous spicules and center of calcification at head of one of them. Scale: 100 μ.

Hispidopetra, *Goreauiella* and *Merlia*. It is of interest to note that Nicholson (1886) suggested that the stromatoporoid skeleton was originally composed of aragonite which has altered to calcite with time, although Galloway (1957) believes that calcite was the original form of calcium carbonate laid down.

In *Astrosclera* the calcareous skeleton is made up of polyhedrons of aragonite, about 60 μ across, in which crystals radiate out from a central point. The aragonitic spherules were reported by Lister (1900) and confirmed by Kirkpatrick (1910a) as being intracellular in origin. We have found no comparable microstructure mentioned in the literature on stromatoporoids.

Paleozoic stromatoporoids show a diversity of microstructures and Stearn (1966) has described 14 different types, some of which are secondary recrystallization phenomena. Radial and fasciculate types are widespread and may represent the basic microstructures of stromatoporoids since they are of almost universal occurrence among Mesozoic forms, in which recrystallization is rare. There is considerable difference of opinion among specialists, however, about which microstructures are primary and secondary.

Some remarkable structural parallels occur between existing coralline sponges and stromatoporoids. Vertical sections of *Stachyodes* Bargatzky, 1881, a Devonian genus, shows a relationship of skeleton to tissue spaces very like similar sections of *Astrosclera* (compare Lecompte, 1956, Fig. 109, 4b with Lister, 1900, Fig. C).* The skeleton of the Devonian species *Taleastroma*? *confertum* Stearn with contiguous "pillars" of a fasciculate microstructure (Stearn, 1963) is reminiscent of the solid calcareous skeleton of *Stromatospongia norae*.

Interestingly, the skeleton of *C. nicholsoni* finds its closest parallel among certain chaetetid corals. Surface features and transverse sections of the genus *Varioparietes* as figured by Schnorf-Steiner (1963) are virtually identical to similar representations of *C. nicholsoni* (Figs 3, 4). Tabulae are present in the Jurassic genus, however, giving it, along with the Devonian *Rhaphidopora*, as mentioned above, a strong resemblance to *Merlia*. The possibility that many genera now assigned to the family Chaetetidae, order Tabulata Milne-Edwards and Haime, 1850, phylum Cnidaria, are in fact sponges—and hence related to stromatoporoids—should be given serious consideration. *Ceratoporella nicholsoni*, related by way of its astrorhizae to stromatoporoids and by way of the organization of its calcareous skeleton to *Merlia* and certain chaetetids, constitutes a remarkable link between those fossils now called stromatoporoids and those known as chaetetids. There is little doubt in our minds that both groups belong with the sponges.

We suggest that the modern coralline sponges represent surviving remnants of what was once a diverse fauna of calcareous organisms that have been mistakenly included within the phylum Cnidaria. Astrorhiza-bearing genera of stromatoporoids at least, and certain genera of tabulate corals of the family Chaetetidae (and perhaps other tabulate corals with a single skeletal wall between adjacent units of the organism), must be placed within the phylum Porifera. Although these organisms bear resemblances to the Demospongiae, they show an emphasis in their skeletal morphology on the secretion of a calcareous basal mass, a fact which has obscured their poriferan affinities. Those that have been known as stromatoporoids have evolved independently of other sponge classes since Cambrian times; the chaetetids extend back to the Ordovician. To receive this array of fossil organisms along with their living descendants, the coralline sponges, we propose a new class in the phylum Porifera to be called the Sclerospongiae.

* It should be noted that paleontologists have the unfortunate habit of referring to the finer structures of the skeleton of stromatoporoids as "tissue". The word is used here to denote living cellular material.

PART II. DISTRIBUTION AND ECOLOGY OF THE CORALLINE SPONGES

Introduction

This study on coralline sponges is the result of an accidental discovery made during underwater explorations of the coral reef habitat in the vicinity of Discovery Bay, Jamaica. Whilst new species are not an unusual occurrence in such an unsampled environment, the sclerosponges were surprising firstly because of their diversity, abundance and large individual sizes, and secondly because they constitute a new and unexpected element in the formation and consolidation of the basal reef framework.

A brief introduction to the recent geological history of the north coast reefs of Jamaica is desirable as background to the ecology of the sclerosponges since their distribution is so intimately bound up with the structure of their environment. The reader is referred elsewhere for more detailed descriptions of Jamaican reefs (Goreau, 1959, 1961; Goreau and Hartman, 1963; Goreau and Burke, 1966; Goreau and Wells, 1967; Zans, 1959). Only some basic concepts will be defined herein.

The great north coast reefs of Jamaica are perched on the edge of a narrow island shelf, the morphology of which is controlled in its major aspects by tectonic forces such as block faulting, uplift and displacement along an east-west axis. The profile of the upper 200 m of the coastal scarp has, in addition, been remodelled by a succession of eustatically controlled sea level oscillations linked to the growth and decay of the great Pleistocene ice sheets. Each glacial maximum and minimum had its associated sea levels and these left characteristic world wide traces of their presence, both above and below modern sea level. According to Emiliani (1966), the sea surface temperature in the Caribbean during the Pleistocene oscillated from about 26–27°C during glacial minima to 21–22°C during glacial maxima. Hermatypic corals must therefore have grown in the Caribbean throughout the Pleistocene, even during the greatest ice ages, and reefs became established whenever the sea level held steady for any length of time. These reefs were eventually "drowned" or "stranded" by later advances or retreats of the sea. Associated with ancient sea shores in the Caribbean were wave-cut cliffs, tidal notches and offshore reefs. These left a series of characteristic steps and platforms that are the foundation of the modern reef and which control to a large extent its form and zonation.

The most recent major eustatic event was the post-Wisconsin marine transgression which started about 17 000 years ago with the melting of the great continental ice sheets of North America and Europe. According to Curray (1965) this marked the end of the Pleistocene.

During the ensuing 12 000 year period, the Holocene interval, the sea level rose about 130 meters at the average rate of not less than 1·1 cm per year. The rise of the sea was interrupted by several pauses about 14 000, 11 000 and 8 500 years ago; each interval, from 800 to 2000 years long, left well defined reefs or wave-cut notches at −60 m, −40 m and −25 m. The stabilization of the sea at its modern level marks the end of the Holocene and the beginning of the Recent period and is a point in time before which modern reefs could not have become established. Some modern reefs are undoubtedly much younger than 5000 years old because the date of their re-establishment varied according to local conditions the world over; the point is that the living reefs cannot be older than that.

Typical of Jamaican north coast reefs is their steep and rugged outer slope. As shown above, this is due partly to the presence in the antecedent platform of ancient wave-eroded cliffs only thinly veneered with modern reef deposits, and partly to the growth of the Recent reef which tends to exaggerate pre-existing topographic irregularities. Because of the steep offshore gradients, there is no extensive accumulation of unconsolidated reef sediments and the basal framework, with its overhangs, tunnels and caves, remains exposed and available for colonization by a shade-loving fauna of which the most prominent members are the sclerosponges.

Distribution and ecology of the sclerosponges

Most of the sclerosponges so far collected come from a restricted area on the north coast of Jamaica between Runaway Bay and Rio Bueno. These sites, shown on the map in Fig. 18, have been systematically investigated to depths between −70 and −95 m by divers. Other localities from which sclerosponges have been collected are also shown on the map but it has not yet been possible to carry out an island-wide survey of these forms.

Data are lacking from most of the Jamaican south coast where the only sclerosponges so far collected consist of one *Merlia* from −15 m off South East Cay, near Kingston, and two *Hispidopetra miniana* Hartman dredged from −40 m by R. V. "*Gosnold*" south of Great Pedro Bay (Fig. 18).

The sclerosponges are part of the cryptofauna of the reef and first-hand observations on their ecology are considerably hampered by a combination of depth, darkness and inaccessibility. Fortunately, in a few localities it is possible to gain entry into this environment by way of caves and crevices. In other places the living overburden had to be removed, a process made more difficult by the unexpected degree of

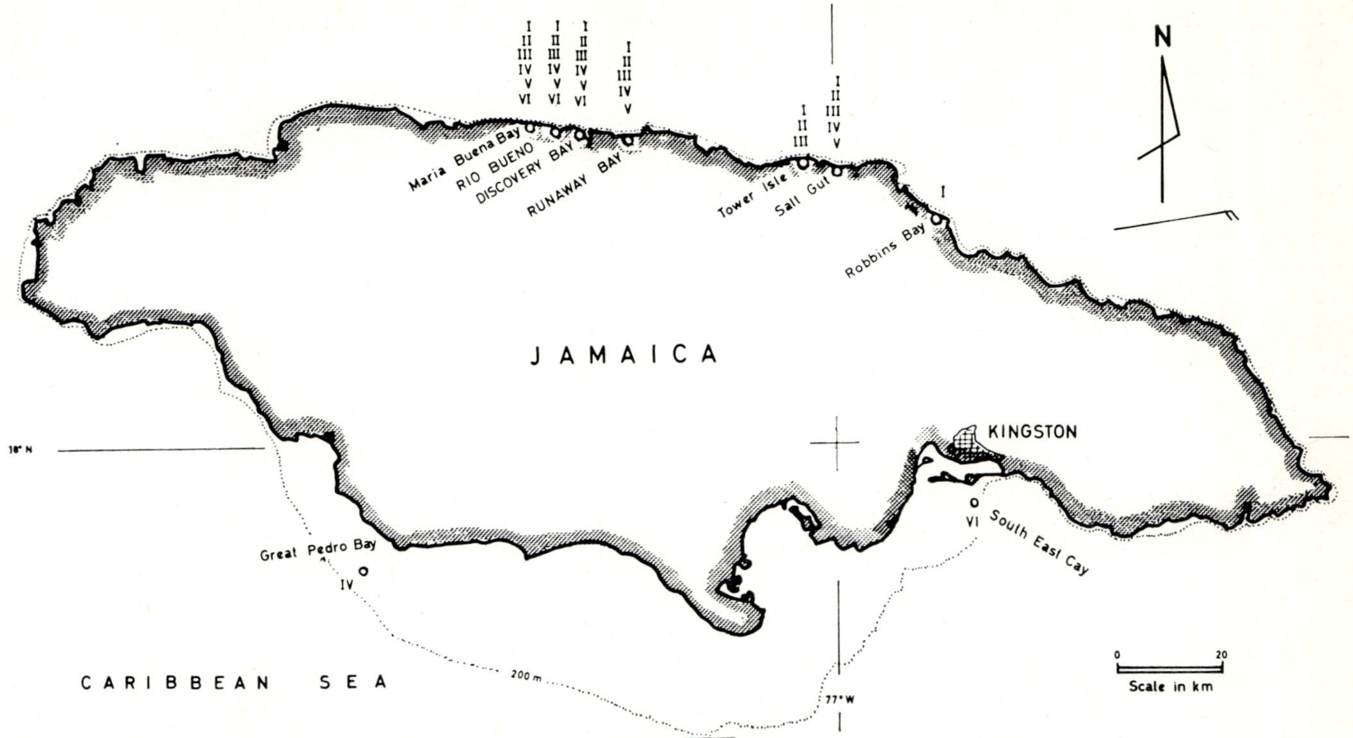

FIG. 18. Map of Jamaica showing localities where sclerosponges have been collected. The roman numerals refer to the species found at each site: I, *Ceratoporella nicholsoni*; II, *Stromatospongi avermicola*; III, *S. norae*; IV, *Hispidopetra miniana*; V, *Goreauiella auriculata*; VI, *Merlia* sp. The dashed line gives the position of the 200 m contour and the flagged arrow shows the direction of the prevailing trade winds.

framework lithification which was encountered just below the surface of the reef at all depths (Land and Goreau, 1969).

In this section we will consider in detail the ecological characteristics of each of the six known Jamaican sclerosponges. Most of the field observations and collections were made by diving with standard compressed air breathing equipment since the extreme steepness and ruggedness of the Jamaican north coast reefs make every other sampling technique impractical. The divers were equipped with collecting tools, geological picks, cold chisels, plastic bags and powerful hand lights for work in caves.

Underwater photography was used extensively for panoramic as well as close-up studies, with synchronized flash as a light source. Close-up photographs sometimes revealed important and unexpected features. For example, the natural appearance of the excurrent canals in *Ceratoporella nicholsoni* was not at first appreciated from observations made on specimens in tanks ashore. Close-up photographs of this species under natural conditions revealed a far greater state of expansion of the canals than was ever seen in the laboratory. This finding led to further investigation which showed that *Ceratoporella* and *Stromatospongia* did not do well under laboratory conditions in running sea water, whereas *Hispidopetra*, *Goreauiella* and *Merlia* survived and preserved their natural appearance in the tanks for an indefinite period.

Ceratoporella nicholsoni Hickson is the most abundant and largest of the Jamaican sclerosponges, ranging in size from 1 mm to about 1 m across and 50 cm in thickness (Fig. 1). *Ceratoporella* is over twice as heavy as reef coral, its bulk density being close to that of solid limestone, and large individuals are almost impossible for divers to collect because of their great weight.

The preferred habitat of *C. nicholsoni* is the twilight zone. Because of its comparatively high tolerance to light, *C. nicholsoni* has a wider optimum depth range and occurs in more exposed localities than other sclerosponges. Below −70 m, *C. nicholsoni* sometimes competes for space with massive hermatypic corals, such as *Madracis pharensis* (Heller, 1868) and *Montastrea cavernosa* (Linnaeus, 1776) whose growth rate the sponge appears to match at those depths.

Wherever the substrate conditions permit, i.e. steep, well drained rock surfaces and low light intensity, *C. nicholsoni* is common to abundant and is occasionally the dominant form. Photographic area surveys (Fig. 19) carried out on the deep fore-reef slope between −70 and −95m show between five and twelve individuals (larger than 10 cm across) per square meter. At that depth *Ceratoporella* grows out in the open but above −60 meters *C. nicholsoni* gradually becomes more cryptic as the

FIG. 19. Group of *C. nicholsoni* growing in the open on a vertical submarine precipice off Discovery Bay, Jamaica, at a depth of 72 m. Eleven individuals between 6 and 15 cm wide are shown scattered over an area of approximately 0·75 m^2. Many smaller colonies are present in the crevices but do not appear on this print. These *Ceratoporella* are a representative sample of the sclerosponge population on the deep fore-reef slope and give a good idea of the abundance of this species.

twilight zone recedes further into the reef and above -40 m it is almost exclusively cavernicolous. Nevertheless, the densest populations and largest individuals of *C. nicholsoni* we have so far studied occur in depths of less than -30 m, in submarine tunnels and caves off Runaway Bay where conditions for its growth appear to be optimal. The walls of these sub-reef tunnels (see Fig. 20) are covered with *Ceratoporella*, many of them very large. Most rock samples from such caves show the common bulk constituent to be dead *C. nicholsoni*, buried and lithified in place. This suggests that *Ceratoporella* grows relatively fast and can contribute significantly to the sub-reef framework in such restricted localities. However, remains of *Ceratoporella* have never been observed in the unconsolidated reef sediments, or in fossil reefs of Pleistocene age, and it is unlikely that the sclerosponges account for more than a very small proportion of the total calcium carbonate production of the reef.

Stromatospongia vermicola Hartman is the second most common of the Jamaican sclerosponges. It forms very irregular encrusting

FIG. 20. Sub-reef tunnel at a depth of 30 m off Runaway Bay, Jamaica, populated by large numbers of *C. nicholsoni*, *S. norae* and *G. auriculata* which live in deeply shaded recesses and crevices on the sides and ceiling. The tunnel was formed by overgrowth of a gully in the old reef framework. The walls are mostly made up of large dead *C. nicholsoni* buried in place and lithified. The passage is from 5 to 6 m high and is floored with sand. Numerous pennate hydroids, antipatharians and sponges also occur here.

masses of varying thickness on rock walls or the reverse sides of large reef corals or coral shingle, always in deeply shaded recesses, at depths from about −10 to more than −95 m. Its optimum range lies deeper than that of *C. nicholsoni*: above −30 m the colonies of *S. vermicola* tend to be relatively small and infrequent but below −60 to −70 m it reaches sizes up to 40 cm and becomes very abundant.

S. vermicola exists in intimate relationship with an unidentified serpulid worm whose tubes are cemented into a solid mass by the sponge and give the colonies their typically twisted spiky appearance (Fig. 9). The relationship of the sponge to the worm tubes suggests that this species can grow rapidly.

Stromatospongia norae Hartman is smaller and more cryptic than *C. nicholsoni* and *S. vermicola*; individuals are generally 1 mm to 7cm across, round to nodular (Fig. 11), or cornuate when growing around serpulid worm tubes. As in *C. nicholsoni*, the relationship to serpulids seems to be accidental and most colonies lack the worms. Although *S. norae* is small, the colonies are well attached and surprisingly massive, with a bulk density somewhat greater than that of *C. nicholsoni*.

The vertical range of *S. norae* is from less than −8 m to about −85 m but the species is most abundant above −35 m. Although it has an unusually shallow optimum depth, *S. norae* is nevertheless one of the least light tolerant of all sclerosponges. It is a true cavernicole, living on the sides and ceilings of deep caverns and sub-reef tunnels. In this zone, the substrate consists of partly lithified old coral framework coated with a dense concretionary mineral crust that has a high iron content and is probably of inorganic origin (Land and Goreau, 1969). It is not known whether this crust is a factor in habitat selection by *S. norae* but the multitude of narrow winding passages that riddle the framework are literally speckled with small *S. norae* (Fig. 21). At least 400 individuals per square meter have been noted. In such environments the associated fauna is reduced in diversity, consisting mostly of another sclerosponge, *Goreauiella auriculata* Hartman and the thecidioid brachiopod *Thecidellina barretti* Davidson, 1864.

Hispidopetra miniana Hartman has a vertical range of about −10 to more than −95 m, the optimum depth being below −60 m. *H. miniana* is restricted almost entirely to overhanging surfaces, from which it grows in a downward direction in caves, crevices and beneath large coral ledges. This habit may be due to its low tolerance for light and sedimentation. The abundance and colony size of *H. miniana* varies strongly with depth; the species is uncommon, small and cryptic above −25 m, becomes somewhat larger and more abundant between −30 and −55 m and reaches a maximum population density and individual size below

FIG. 21. Wall of sub-reef cave in −28 m off Runaway Bay, Jamaica, showing over one hundred *Stromatospongia norae*. The picture covers an area of approximately 0·25 m^2 and the large nodular specimen in the upper center is about 7 cm long. Such dense populations occur only in total darkness in narrow passageways deep inside the reef framework. The colonies are clustered on projections and overhanging surfaces.

– 60 m where large colonies are often in the open, side by side with *C. nicholsoni*.

Goreauiella auriculata Hartman has the most restricted distribution of all the Jamaican sclerosponges. Like *S. norae*, with which it often occurs, *G. auriculata* is a true cavernicole living in extremely reduced ambient light or total darkness. The pedunculate mushroom-like colonies are invariably suspended from overhangs or cave ceilings (Fig. 22) with the concave living surface facing down. The colonies are cup- or dish-shaped, circular when small, usually irregular in outline when larger, and range in size from 3 mm to about 16 cm.

The known vertical distribution of *G. auriculata* is from – 8 to more than – 70 m, with an optimum range between – 15 and – 30 m. The largest populations occur in narrow passages inside the reef framework where *S. norae* is the only other common species. In larger caves, however, *G. auriculata* is associated with a diverse fauna of ahermatypic corals including *Desmophyllum riisei* Duchassaing and Michelotti, 1860, *Caryophyllia flos* Pourtalès, 1878, *C. antillarum* Pourtalès, 1874, *C.*

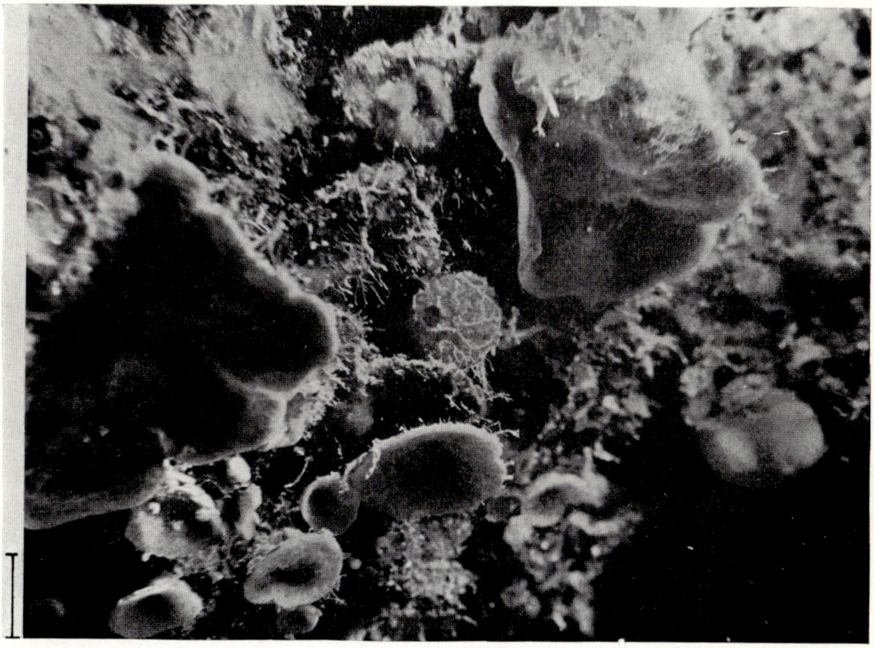

FIG. 22. Group of *Goreauiella auriculata* on ceiling of sub-reef cave, – 28 m, off Runaway Bay, Jamaica. Individuals are invariably oriented with living surface down. Excurrent channels are visible. Scale: 2 cm.

maculata Pourtalès, 1874, *Paracyathus defilippi* Duchassaing and Michelotti, 1860, *Colangia immersa* Pourtalès, 1871, all of which grow in the characteristically dependent position. Lithistid sponges also occur in this habitat.

Merlia sp. is the smallest of the Jamaican sclerosponges. The only living specimens so far collected came from a depth of 25 m on a windward reef just west of Rio Bueno harbour. About 50 cm of coral overburden had to be removed to expose the reef framework which, in this site, consisted of a mixture of unconsolidated and well cemented lithified reef talus. *Merlia* occurs in small fluorescent orange patches in cavities and passages within the framework, often encrusting loose sticks of dead *Madracis mirabilis* (Duchassaing and Michelotti, 1860), and on the walls of crevices together with numerous small *C. nicholsoni* and *S. norae*. Ahermatypic corals such as *Astrangia* sp., *Caryophyllia* sp. and *Colangia immersa* are also numerous here. Judging by the complete absence of any twilight algae such as *Peyssonnelia* sp., these *Merlia* must live in an environment practically devoid of light. The finding of fourteen specimens of *Merlia* from this single site suggests that it is probably common but easily overlooked because of its small size.

Two other specimens collected in 1964 and 1967 from about −75 m off Discovery Bay were much larger and thicker than any of the shallow water *Merlia* described above. Both were encrusting on pieces of coral talus lying on ledges on a very steep submarine precipice. Neither of these specimens was seen alive, one having been cloroxed and the other dried before their identity was realized. If these *Merlia* are the same as the Rio Bueno specimens, it suggests that the optimum habitat of the species is the deep fore-reef slope below −70 m. Unfortunately, repeated dives to this site have failed to turn up any more specimens, and it is still not known whether the deep water *Merlia* is perhaps another species. Thus far, the only indication of a possible difference is that the dried tissues of the deep water form are light greyish green instead of orange in color.

DISCUSSION AND CONCLUDING REMARKS

In pioneering underwater studies of submarine cave habitats of the eastern Indian Ocean, northern Mediterranean and Madeira, Vacelet (1967a) and Vacelet and Vasseur (1966) noted the presence of "silico-calcareous" sponges of the genera *Merlia* and *Astrosclera*. The ecology of *Astrosclera willeyana* Lister appears to be very similar to that of *Stromatospongia norae* in Jamaica; the two species occupy essentially the same niche in different parts of the world. On the other hand, *Merlia normani* and the Jamaican *Merlia*, while obviously very close relatives, differ in their ecological requirements: the West Indian form occurs in shallower more turbulent and exposed habitats than *M. normani* which appears to be restricted to deep, calm environments (Kirkpatrick, 1911; Vacelet, 1967a).

Prompted by his discovery of a remarkable cavernicolous fauna of pharetronid sponges off southern France and northern Madagascar, Vacelet (1967b) speculatively suggested that these forms, along with *Merlia* and *Astrosclera*, may constitute a relict fauna of ancient origin which has taken refuge in dark caves and crevices. If, as we have suggested above, the Jamaican sclerosponges are indeed living representatives of the ancient and supposedly extinct order Stromatoporoidea, it may be relevant to examine the possibility of these forms likewise being relicts, in an ecological sense. That is, they barely hang on in the face of competition from a better adapted and more modern reef fauna and survive only because they have found refuge in dark coral caves where little else will live. Our field observations do not support such a diagnosis. On the contrary, in terms of species diversity, population density and individual size, the sclerosponges are more than holding their own in Jamaica. Indeed, there is evidence that they are still in process of colonizing an environment that has become available to them only in comparatively recent times, i.e., since the end of the Holocene marine transgression, at most 5000 years ago, probably much less. Additional testimony to their vigour of growth is the thickness of the *Ceratoporella* deposits in some of the submarine caves at Runaway Bay. These are hardly the characteristics that one would expect of a "relict" fauna.

The Jamaican sclerosponges are regular members of a widespread and diverse benthonic assemblage which also includes encrusting Foraminifera, lithistid sponges, ellisellid Gorgonacea, caryophylliid, rhizangiid and flabellid corals, ectoprocts, thecidioid brachiopods, serpulid worms, etc. The fauna is normally confined to the benthic habitats of the lower photic zone, below the reef biotope proper, where the ambient light flux is less than 1% of the surface, and competition from phototropic hermatypes is almost absent. However, the entire community will extend its range into very shallow water if habitats of the requisite physical characteristics are available. Such a place is the coral reef framework which, permeated by a complex of caves, tunnels and crevices, provides for colonization an enormous surface area sheltered from light, excessive turbulence and competition. This, then, is the pathway by which this association, migrating as a community, can penetrate into shallow water to gain access to the immensely productive surface ecosystem. Although the precise factors remain to be identified, the move into shallow water may have important advantages for sclerosponges. For example, *Ceratoporella* achieves its greatest population density and individual size not in its normal habitat below -70 m but in less than -30 m if the appropriate substrate conditions are available. It seems no accident that this depth also lies within the optimum range for most of the dominant reef associated Demospongiae, hermatypic corals and gorgonian octocorals; the common factor appears to be the greater availability of food in this part of the reef environment.

Acknowledgements

We gratefully acknowledge the help and advice of F. M. Bayer, K. E. Chave, W. G. Fry, N. I. Goreau, F. W. Goro, K. Hatai, L. S. Land, the late W. J. Rees, S. M. Stone, M. Weidmann, J. W. Wells and H. Winchell at various crucial stages of this work. We also thank R. and P. Chapman, E. A. Graham, the late R. S. Jackson, L. S. Land, J. C. Lang and Y. Neumann whose steady assistance underwater made it possible to carry out the field observations on the Jamaican Sclerospongiae. We are grateful to the Kaiser Bauxite Company for their support in providing facilities at Discovery Bay, Jamaica. This work was supported from the following sources: Contract Nonr 4816(00) between the Office of Naval Research and the University of the West Indies and National Science Foundation grants GB-3542 to the New York Zoological Society, and GB-516 and GB-7343 to Yale University.

References

Agassiz, A. (1878). (Letter No. 1). To C. P. Patterson, Superintendent Coast Survey, Washington, D.C., from Alexander Agassiz, on the dredging operations of the United States Coast Survey, Steamer "Blake", during parts of January and February, 1878. *Bull. Mus. comp. Zool. Harv.* **5**: 1–9.

Agassiz, A. (1888). *Three cruises of the United States Coast and Geodetic Survey Steamer "Blake"*. **2**: 1–220. Boston: Houghton, Mifflin & Co.

Birenheide, R. (1967). Mikrostrukturen zweier Stromatoporen aus dem Mitteloligozän. *Natur Mus., Frankf.* **97**: 313–328.

Bowerbank, J. S. (1863). A monograph of the Spongillidae. *Proc. zool. Soc. Lond.* **1863**: 440–472.

Carter, H. J. (1877). On the close relationship of *Hydractinia, Parkeria,* and *Stromatopora*; with descriptions of new species of the former, both recent and fossil. *Ann. Mag. nat. Hist.* (4) **19**: 44–75.

Chave, K. E. (1954). Aspects of the biogeochemistry of magnesium. 1. Calcareous marine organisms. *J. Geol.* **62**: 266–283.

Curray, J. R. (1965). Late Quaternary continental shelves of the United States. In *The Quaternary of the United States*: 723–735. Wright, H. E. and D. G. Frey (Eds). Princeton: University Press.

Emiliani, C. (1966). Paleotemperature analysis of Caribbean cores P-6304-8 and P-6304-9, and a generalized temperature curve for the past 425 000 years. *J. Geol.* **74**: 109–124.

Galloway, J. J. (1957). Structure and classification of the Stromatoporoidea. *Bull. Am. Paleont.* **37**: 345–470.

Goreau, T. F. (1959). The ecology of Jamaican coral reefs. I. Species composition and zonation. *Ecology* **40**: 67–90.

Goreau, T. F. (1961). The structure of Jamaican reef communities: geological aspects. Final Progress Report to Biology Branch, Office of Naval Research under Contract Nonr (G)-0003-60.

Goreau, T. F. and Burke, K. (1966). Pleistocene and Holocene Geology of the Island shelf near Kingston, Jamaica. *Mar. Geol.* **4**: 207–225.

Goreau, T. F. and Hartman, W. D. (1963). Boring sponges as controlling factors in the formation and maintenance of coral reefs. In *Mechanisms of hard tissue destruction*: 25–54. Sognnaes, R. F. (Ed.). Washington D.C.: AAAS Publ. No. 75.

Goreau, T. F. and Wells, J. W. (1967). The shallow water Scleractinia of Jamaica: revised list of species and their vertical range. *Bull. mar. Sci. Gulf Caribb.* **17**: 442–453.

Hartman, W. D. (1969, in press). New genera and species of coralline sponges from Jamaica. *Postilla*.

Heinrich, M. (1914). Studien in den Riffkalken des rheinischen Mitteldevons. *Inaug.-Dissert. rhein. Friedr.-Wilh.-Univ. Bonn*: 1–58. Freiburg i. Br.: Buchdruckerei Hch. Wuhrmann.

Hickson, S. J. (1911). On *Ceratopora*, the type of a new family of Alcyonaria. *Proc. R. Soc. Lond.* (B) **84**: 195–200.

Hickson, S. J. (1912). Change in the name of a genus of Alcyonaria. *Zool. Anz.* **40**: 351.

Hill, D. and Strumm, E. C. (1956). Tabulata. In *Treatise on invertebrate paleontology*, Pt. F, Coelenterata: F444–F477. Moore, R. C. (Ed.). Lawrence: University of Kansas Press.

Kirkpatrick, R. (1909). Notes on *Merlia normani* Kirkp. *Ann. Mag. nat. Hist.* (8) **4**: 42–48.

Kirkpatrick, R. (1910a). A sponge with siliceous and calcareous skeleton. *Nature, Lond.* **83**: 338.

Kirkpatrick, R. (1910b). Further notes on *Merlia normani* Kirkp. *Ann. Mag. nat. Hist.* (8) **5**: 288–291.

Kirkpatrick, R. (1910c). On the affinities of *Astrosclera willeyana* Lister. *Ann. Mag. nat. Hist.* (8) **5**: 380–383.

Kirkpatrick, R. (1911). On *Merlia normani*, a sponge with a siliceous and calcareous skeleton. *Q. Jl miscrosc. Sci.* **56**: 657–702.

Kirkpatrick, R. (1912a). Note on *Astrosclera willeyana* Lister. *Proc. R. Soc. Lond.* (B) **84**: 579–580.

Kirkpatrick, R. (1912b). [Note on *Merlia*.] *Nature, Lond.* **89**: 353.

Kirkpatrick, R. (1912c). *Merlia normani* and its relation to certain Palaeozoic fossils. *Nature, Lond.* **89**: 502–503.

Kirkpatrick, R. (1912d). On the nature of stromatoporoids. *Nature, Lond.* **89**: 607.

Kirkpatrick, R. (1912e). Note on *Merlia normani* and the "Monticuloporas". *Proc. R. Soc. Lond.* (B) **85**: 562–563.

Land, L. S. and Goreau, T. F. (1969, in press). Submarine lithification in Jamaican reefs. Abstract. *Symp. Soc. Econ. Paleont. Min.*

Lecompte, M. (1956). Stromatoporoidea. In *Treatise on invertebrate paleontology*, Pt. F, Coelenterata: F107–F144. Moore, R. C. (Ed.). Lawrence: University of Kansas Press.

Lévi, C. (1956). Etude des *Halisarca* de Roscoff. Embryologie et systématique des Démosponges. *Archs Zool. exp. gén.* **93**: 1–181.

Lister, J. J. (1900). *Astrosclera willeyana*, the type of a new family of sponges. In *Zoological results* . . . by Arthur Willey, **4**: 459–482.

Newell, N. D. (1935). Some mid-Pennsylvanian invertebrates from Kansas and Oklahoma: II. Stromatoporoidea, Anthozoa, and Gastropoda. *J. Paleont.* **9**: 341–355.

Newell, N. D., Rigby, J. K., Fischer, A. G., Whiteman, A. J., Hickox, J. E. and Bradley, J. S. (1953). *The Permian reef complex of the Guadalupe Mountains region, Texas and New Mexico: A study in paleoecology*. San Francisco: W. H. Freeman and Co.

Nicholson, H. A. (1886–1892). *A monograph of the British stromatoporoids.* London: Alard and Son.

Nicholson, H. A. and Murie, J. (1878). On the minute structure of *Stromatopora* and its allies. *J. Linn. Soc. Zool.* **14**: 187–246.

Potts, E. (1888). Contributions towards a synopsis of the American forms of fresh water sponges with descriptions of those named by other authors and from all parts of the world. *Proc. Acad. nat. Sci. Philad.* **1887**: 158–279.

Rosen, F. B. (1869). Über die Natur der Stromatoporen und über die Erhaltung der Hornfaser der Spongien im fossilen Zustande. *Verh. russ.-kaiserl. mineral. Ges. St.-Petersb.* (2) **4**: 1–98.

Schnorf, A. (1960). Parastromatoporidae nouveau du Jurassique supérieur et du Valanginien inférieur du Jura. *Eclog. geol. Helv.* **53**: 729–732.

Schnorf-Steiner, A. (1963). Sur quelques "Chaetetidae" du Valanginien du Jura. *Eclog. geol. Helv.* **56**: 1117–1129.

Sollas, W. J. (1877a). On *Stauronema*, a new genus of fossil hexactinellid sponges with a description of its two species, *S. carteri* and *S. lobata*. *Ann. Mag. nat. Hist.* (4) **19**: 1–25.

Sollas, W. J. (1877b). On the structure and affinities of the genus *Siphonia*. *Q. Jl geol. Soc. Lond.* **33**: 790–834.

Stearn, C. W. (1963). Some stromatoporoids from the Beaverhill Lake Formation (Devonian) of the Swan Hills Area, Alberta. *J. Paleont.* **37**: 651–668.

Stearn, C. W. (1966). The microstructure of stromatoporoids. *Paleontology* **9**: 74–124.

Steiner, A. (1932). Contribution à l'étude des Stromatopores secondaires. *Bull Lab. Géol. Géogr. phys. Min. Paléont. Univ. Lausanne* **50**: 1–117.

Topsent, E. (1928). Spongiaires de l'Atlantique et de la Méditerranée, provenant des croisières du Prince Albert Ier de Monaco. *Résult. Camp. scient. Prince Albert I* **74**: 1–376.

Twitchell, G. B. (1929). The structure and relationships of the true stromatoporoids. *Am. Midl. Nat.* **11**: 270–306.

Vacelet, J. (1964). Étude monographique de l'éponge calcaire pharétronide de Méditerranée, *Petrobiona massiliana* Vacelet et Lévi. Les pharétronides actuelles et fossiles. *Recl Trav. Stn mar. Endoume* **50** (Bull. 34): 1–125.

Vacelet, J. (1967a). Quelques éponges pharétronides et "silico-calcaires" de grottes sousmarins obscures. *Recl Trav. Stn mar. Endoume* **58** (Bull. 42): 121–132.

Vacelet, J. (1967b). Descriptions d'Eponges Pharétronides actuelles des tunnels obscurs sousrécifaux de Tuléar (Madagascar). *Recl Trav. Stn. mar Endoume*, Fasc. hors série suppl. No. 6: 36–62.

Vacelet, J. and Vasseur, P. (1966). Les tunnels obscurs sou-récifaux de Tuléar (Madagascar) et leur faune de spongiaires. *Proceedings Second International Oceanographic Congress. Moscow*. No. 443 SIIc p. 378.

Vinogradov, A. P. (1953). The elementary chemical composition of marine organisms. *Mem. Sears Fdn mar. Res.* **2**: 1–647.

Yabe, H. and Sugiyama, T. (1935). Jurassic stromatoporoids from Japan. *Sci. Rep. Tôhoku Univ.* 2nd ser. (Geol.) **14**: 135–192.

Zans, V. A. (1959). Recent stony corals of Jamaica. *Geonotes* **2**: 43–48.

ECOLOGY AND BIOGEOGRAPHY

Symp. zool. Soc. Lond. (1970) No. 25, 247–271.

ADAPTATION TO INTERTIDAL EXISTENCE: REPRODUCTIVE CYCLES AND LARVAL BEHAVIOUR IN DEMOSPONGIAE

PATRICIA R. BERGQUIST, MARY E. SINCLAIR
and
JANFRIE J. HOGG

Zoology Department, University of Auckland, Auckland, New Zealand

SYNOPSIS

A review of published data on sponge larval behaviour, together with new data, confirms that the pattern of diversity of such behaviour reflects more closely the micro-environmental requirements of adult sponges than their taxonomic placings. The intertidal distribution of sponges is not merely the result of selective mortality following unselective settlement.

Attention is drawn to earlier reports of asexual development of free-swimming larvae. The existence of this phenomenon has not been generally accepted. However, in all species of intertidal Ceractinomorpha that we have studied, asexual development is the predominant pathway of larval production. Thus, in these species, dependence upon fertilization is minimized without forfeiture of the larval stage. Several mechanisms of asexual development, including gemmule production, are described for three common intertidal tetractinomorph sponges.

INTRODUCTION

Knowledge of the morphology and morphogenesis of sponge larvae is now sufficiently advanced to be used to advantage in discussion of systematics and evolution. By comparison, the information available on the behaviour of larvae, e.g. longevity, type of movement, response to light and substrate, is fragmentary. While such attributes are unlikely to be relevant to discussions of higher category systematics, they are of vital importance in ecological investigations, in detecting differences between sibling species, and in contributing to the factual basis for zoogeographic assumptions.

The work discussed in this paper was initiated in an attempt to establish whether patterns of behaviour were to be found among sponge larvae. The observations presented relate to five species of intertidal Ceractinomorpha Lévi, 1953.

Monthly samples were taken of fifteen intertidal and shallow subtidal species, chosen to represent most orders of the Demospongiae in order to discover reproductive seasons and to obtain live larvae. From the study of these samples has come evidence of widespread asexual reproduction occurring in some species to the exclusion of sexual processes; in other species asexual reproduction takes place at the same time, and in the same colony, as sexual reproduction. In the eight ceractinomorph species studied, asexual development was directed toward the production of free swimming larvae. This process has been reported previously by Wilson (1891, 1894) and by Sivaramakrishnan (1951). The existence of this type of asexual reproduction has been considered to be an aberration. Wilson's observations, in particular, were received with some scepticism by his contemporaries. However, present evidence indicates that asexual development of larvae is not unusual and is a common occurrence in intertidal Ceractinomorpha.

It is logical to expect that in the Demospongiae, which is a predominantly benthic group, intertidal species will exhibit modification of the normal life cycle. Such modification could be expected to be directed toward conservation of resources, possibly by reducing dependence upon fertilization or, by an abbreviation of the developmental process, particularly the length of the vulnerable free-swimming phase. Asexual development of larvae is known to occur only in those forms with incubated parenchymella larvae (i.e. Ceractinomorpha). However, reliance upon asexual reproduction by several methods has also been observed in intertidal Tetractinomorpha Lévi, 1953, particularly in species of Hadromerida Topsent, 1928 and Epipolasida Sollas, 1888.

LARVAL BEHAVIOUR

Introduction

The five species upon which this discussion is based are *Haliclona* sp*, *Microciona coccinea* Bergquist, 1961, *Ophlitaspongia seriata* Grant, 1826, *Mycale macilenta* Bwk, 1866 and *Halichondria moorei* Bergquist, 1961. These species all belong to genera common in intertidal faunas of temperate and tropical regions throughout the world. Reasons of availability have dictated the choice of intertidal sponges for work requiring living larvae both in the present and in all earlier studies. There is, however, a very obvious limit to the value of information derived from these studies for it is relevant only in a broad sense to the offshore benthic sponges which constitute the majority of the group. This has important implications for the zoogeography of sponges for nothing is known of the longevity, swimming behaviour or frequency of larval production in deep water sponges. Such indications as can be gained from our own incomplete observations and from the work of Borojević (1967) on species of *Polymastia* Bwk, 1866 are that larvae of some deeper forms are relatively long lived.

* This is an undescribed species, thus the name, which will be published shortly, will not be cited here for reasons of priority.

Literature references to various aspects of sponge larval behaviour are numerous, apparently conflicting and, in almost all cases, incomplete in some important respect. We expected initially that a study having a broad systematic base would resolve this confusion. However, analysis of our results at present reveals the same diversity in respect to larval movement, light response, settlement behaviour and longevity as has figured in earlier reports.

Movement

There are a few precise references in the literature to the type of movement effected by sponge larvae. Topsent (1911) noted that *Myxilla* O. Schmidt, 1862 larvae are lethargic, simply falling to the bottom and rotating for several hours after release. Lévi (1956) remarked that larvae of *Haliclona indistincta* Bwk,1866 swim "rapidly but irregularly", *Halisarca dujardini* Johnston,1842 larvae swim rapidly and have a tendency to swim toward the substrate and *Halisarca metchnikowi* Lévi,1953 larvae move slowly, rotating about their long axis, and tending to swim near the surface. Bakus (1964) noted that *Tedania gurjanovae* Koltun,1958 larvae swim either anterior or posterior pole forward while completing spiral gyrations. Warburton (1966) has furnished the most precise account of movement. He described a bouncing type of swimming in *Microciona prolifera* Ellis and Solander, 1786 larvae. This is effected only at the surface where the larva swims in an open spiral and bounces off the interface every two to three centimetres.

Three distinct types of movement are displayed in the larvae we have studied.

1. Directional swimming with constant rotation. This is performed by *Haliclona* larvae which, aided by the posterior ring of long cilia, are capable of fast (1 cm/sec) directional movement.

2. Swimming on a spiral path while rotating constantly. This type of movement occurs in all species studied. *Halichondria* larvae rarely swim but when they do they swim in this "corkscrew" fashion. In *Microciona*, *Ophlitaspongia* and *Mycale* this was the only swimming pattern shown until immediately before settlement. The larvae, over a horizontal distance of ten centimetres, reached the top of the spiral two to three times per centimetre. This movement can involve bouncing off the interface at the top of the spiral but is more usually effected without contact with the surface.

3. Slow creeping along the substrate without rotation. This is

typical of *Halichondria* larvae but was observed in all other species immediately prior to settlement.

There is variation in behaviour of the larvae of any species during the swimming period, even among individuals from the same colony. Some larvae are always lethargic—owing, presumably, to premature release. Others, held at the surface by surface tension or adhesion to floating detritus, can begin to metamorphose but more frequently disintegrate. However, in each species the great majority of larvae exhibited one or more of the types of movement described. This constitutes a definite pattern of activity and is considered to characterize the species. The observations are based on samples of fifty to one hundred larvae in two consecutive year classes. Some of the confusion in the literature on sponge larval movement has arisen from overemphasis of the negative and accidental aspects of larval life (Wilson, 1935; Ali, 1956). Premature release of larvae must occur in many cases as larvae, as can be seen in sections, are never all at precisely the same stage of development. Release, when it occurs, is vigorous and must carry some inactive larvae along with the many ready for liberation. Also, larvae which swim near the surface, which in nature will always be moving, will sometimes break through the interface and not be able to swim down again. Likewise, contact with surface floating objects is a natural hazard. These are some of the chance factors which operate on any free swimming larva; the number produced is sufficient to compensate for such hazards and it is the behaviour of the majority that we have described.

Reaction to light

Siliceous sponge larvae have been reported to "swim away from the light" (Ali, 1956, *Lissodendoryx* Topsent,1888), to be "negatively phototropic" and then "less distinctly phototropic" (Wilson, 1935, *Mycale*), to show no particular taxis (Lévi, 1956, *Haliclona;* Bakus, 1964, *Tedania*), and to "avoid light" (Sivaramakrishnan, 1951, *Callyspongia*).

Definite phototactic reactions were shown by two of the larval species investigated.

Haliclona sp. larvae were positively phototactic. After release they congregated in the lighted area of a tank and remained actively swimming in that area until settlement began nine to ten hours later. These larvae resort to creeping over the bottom before settlement and in this period when the larvae have ceased to swim freely they become indifferent to the light gradient or are negatively phototactic. This is similar to the reversal of phototaxis which Wilson (1935) reported as occurring in the larva of *Mycale syrinx* Schmidt,1862.

Mycale macilenta larvae were negatively phototactic. The reaction of these larvae to a light/shade boundary was spectacular. Whenever the larvae approached the boundary they turned through 180° and swam away from the light. This behaviour was repeated constantly as the larvae swam within the darkened area of the tanks. *M. macilenta* larvae did not settle in the experimental tanks. Thus it is not known whether the photonegative condition persists or is reversed before metamorphosis.

Microciona, *Ophlitaspongia* and *Halichondria* larvae were indifferent to light gradients.

There were early attempts to ascribe light sensitivity of sponge larvae to differential pigmentation. Minchin (1900) suggested that the pigmented posterior pole of the larva was turned toward the light, thereby causing the larva to move toward the dark. As larvae of more species have been described it has become obvious that this simple hypothesis will not suffice. There is no pigment differentiation in *Haliclona* sp. larvae which could account for the positive phototrophism. As in all described *Haliclona* larvae there is a pigment cap at the posterior pole only. In our species this is always oriented away from the light. In *M. macilenta* the posterior pole of the larva is unpigmented, and the remainder of the body uniformly pigmented, except for a pale ring which surrounds the anterior pole. It is possible that this differential pigmentation has some connection with the marked photonegative response of *M. macilenta* larvae.

Settlement and longevity

Most authors who have described settlement in larvae of Demospongiae agree that attachment is by the anterior pole and is preceeded by a period of spinning with the long axis vertical. Wilson (1935) and Ali (1956) have recorded attachment of larvae by the posterior pole but do not mention whether these larvae were able to metamorphose. Lévi (1956) suggested that settlement by the anterior pole is a necessary prerequisite for metamorphosis and all of our observations support this interpretation.

There is, however, no agreement to be found in the literature on the longevity of larvae; every report is different. Wilson (1935) recorded that *Mycale syrinx* larvae swam freely for "some hours" and attached sometimes "within twenty four hours but more commonly, two to three days". The bottom-living phase in this species could be extended for up to 10 days, during which time the unattached larvae began to metamorphose. Sivaramakrishnan (1951) recorded *Callyspongia* Ridley,1884 larvae as swimming for 3 h and then crawling or rotating on the bottom for 3 to 8 h depending upon the nature of the available substrate. Settlement was quicker on algae, slower on glass. Ali (1956) stated that *Lissodendoryx* larvae swim 20 h at the surface, crawl at the bottom for 2 to 3 h, rotate for about 2 h and then fall on one side and lie for about 15 h before attachment. Bakus (1964) recorded the free life of *T. gurjanovae* larvae as 6 to 7 h but this estimate is very dubious since the larvae were liberated artificially and, in our experience, such larvae are often lethargic or totally immobile. Lévi (1956) noted that larvae of *Halisarca metchnikowi* swim from 6 h to 2 days while those of *Halisarca dujardini* rarely swim longer than 24 h. Warburton (1966), dealing with *Cliona celata* Grant, 1826 and *Microciona prolifera,* does not differentiate between the two species, but states that the swimming period is 20 to 30 h and the period of creeping and rotation in contact with the substrate is of similar duration.

Borojević (1967), in an excellent description of the development of *Polymastia robusta* Bwk,1861 (Hadromerida), recorded that larvae begin to settle 18 to 20 days after extrusion of the fertilized eggs and that settlement is complete within 7 days. The definitive larval form is attained 24 h after extrusion and from this time until settlement the larva glides over the bottom. These observations are particularly valuable as they represent the only relatively complete account of larval life in a tetractinomorph sponge. Our own observations on the development of *Polymastia granulosa* Brøndsted,1923 are incomplete, as these are based on a single sponge colony collected from a depth of 13 metres. In our material, settlement occurred 20 to 25 days after extrusion of the eggs. The larvae were inactive except for crawling across the bottom, were indifferent to light gradients and showed no tendency to move upward. It is clear that larval life in *Polymastia* is considerably longer than the 2 to 3 days so frequently quoted as the period of free life for siliceous sponge larvae.

In our investigations all settling larvae passed through a creeping phase of 2 to 3 hours duration which could be interrupted by the larva leaving the bottom and swimming for a short time before resuming creeping. Only *Halichondria* larvae were observed to spin slowly while attached by the anterior pole. In *Microciona, Ophlitaspongia, Haliclona* and many specimens of *Halichondria,* the larvae rested in an inclined position on the anterior pole while the cells in that region flattened against the substrate.

Microciona and *Ophlitaspongia* larvae began to settle three to four hours after release, crawled for 2 to 3 h and then attached. Cellular reorganization in these forms was rapid, the larva became a flat disc within an hour of attachment and a canal system with a single osculum was functional within 2 days.

Haliclona larvae began to settle 9 to 10 h after release. The brief creeping phase lasts only 2 h and cellular rearrangement to the single osculum stage is completed within 24 h.

Some *Halichondria* larvae began to settle after crawling for 20 h but most extended the crawling phase to as long as 60 h. During this phase the larvae moved continually up the sides of the tank and many settled on the sides. This upward crawling is not in response to a light gradient. The canal system in *Halichondria* is functional 72 h after settlement.

Relationships between larval behaviour and adult habitat

The above observations, in conjunction with the cited reports, present a confused picture of larval behaviour and the justified assumption up to the present has been that sponge larvae do not have consistent behaviour patterns. As a result, it has been assumed that observed habitat specialization in Demospongiae is a result of selective mortality following unselective settlement.

It is true that in cases where highly specific associations between sponges and some substrates are known, e.g. *Microciona coccinea* investing the shell of *Herpetopoma bella* Hutton,1873, *Dactylia palmata* Carter,1885 and *Iophon proximum* Ridley,1881 investing *Chlamys zealandiae* Gray,1843, *Dysidea hirciniformis* Carter,1885 on the carapace of *Notomithrax minor* Filhol,1885, the sponge always has a wide distribution and is not limited to the specific substrate. One exception of course is that boring sponges must settle on calcareous substrates. If the habitat is defined in more general terms, such as exposure to light and dessication, grade and constitution of substrate, current, wave action and turbidity, then particular sponge species certainly can be said to occupy particular habitats. When we consider the results of our investigations on larval behaviour in relation to the habitat of the parent sponges, it appears that larval behaviour has a definite role in determining the location of the colony.

At this point all literature reports are inadequate; not one paper combines specific ecological information with full details of larval behaviour. Where, as in the case of *Microciona prolifera* and *Polymastia robusta*, ecological information can be gleaned from general literature, we cannot draw any conclusions because both species have a wide depth and geographic range, and neither Warburton (1966) nor Borojević (1967) specified the source of their specimens. Lévi (1956), in his

TABLE I

Behaviour of larvae and the ecological situation of the adult sponge

	Swimming pattern	Length of free life	Light response	Ecological situation of adult sponge
Haliclona sp. (HAPLOSCLERIDA)	1. Directional swimming with constant rotation 2. Corkscrew	9–10 h	positively phototactic, moderated when creeping	upper and middle mid-littoral, around and under stones
Microciona coccinea (POECILOSCLERIDA)	Corkscrew	3–4 h	no response	middle mid-littoral to ELWS from full light to deep shade never away from water
Ophlitaspongia seriata (POECILOSCLERIDA)	Corkscrew	3–4 h	no response	littoral fringe to ELWS, full light to deep shade never away from water
Mycale macilenta (POECILOSCLERIDA)	Corkscrew	over 48 h	negatively phototactic	littoral fringe to 4 m; cryptic
Halichondria moorei (HALICHONDRIDA)	Creep along substrate, rarely swim, move upward	20–60 h	no response	fringing middle and upper mid-littoral pools and rocks in standing pools

work on two species of *Halisarca*, came very close to demonstrating the relationship between larval behaviour and adult habitat but the published details are not clear unless one is familiar with the intertidal ecology of the area in which he worked.

On northern New Zealand rocky coasts *Mycale macilenta* occurs from just above MLW to a depth of 4 m, *Microciona coccinea* and *Ophlitaspongia seriata* occur from the lower mid-littoral zone to just below ELWS, *Haliclona* sp. is found in very restricted sheltered areas in the mid- to upper mid-littoral around and under boulders, and *Halichondria moorei* occurs fringing pools and rocks in the middle mid-littoral zone associated with the extensive *Corallina/Hormosira* turf characteristic of sheltered shores.

Within these regions, *Mycale* is a cryptic sponge, always occurring under stones or in other deeply shaded locations. The marked negative phototaxis of the larva would ensure that settlement took place in such a location. *Microciona* and *Ophlitaspongia* range from full light conditions to deeply shaded locations, a distribution in line with the absence of any light response in the larvae. The significant environmental requirement for these species is protection from dessication; they are always found in or around pools, in crevices where water stands or waves reach, or in the area still covered by surge at low tide. The fact that larvae of both species have a very short swimming period would ensure that many larvae were maintained in the intertidal region, particularly so if release of larvae coincided with the falling tide. The specimens on which observations were made came from the upper part of the species range in both cases.

Haliclona larvae show strong positive phototaxis for the 9 to 10 h that they swim. During this time they must be retained in the mid-littoral region; swimming towards well lit regions would certainly assist this but could not be said to ensure it. Reversal or moderation of light response does not occur until onset of the creeping phase. This would allow the larvae to settle under and in the shade of mid-littoral boulders. As for the other species, it would be interesting to know whether larval release takes place at any particular phase of the tidal cycle. No data are available on this subject. Such data can best be obtained from tide simulation in the laboratory and this has not been attempted.

Halichondria larvae are indifferent to light intensity and, after locating the bottom, spend their free life crawling slowly upwards. The adult sponge occurs in a fringe around pools and rocks, and is tolerant of any light regime. Larvae which creep upward, retaining contact with the substrate, would tend to congregate around the fringes of the pools in which they were released. The *Corallina/Hormosira* community with which *H. moorei* is associated is an indicator of conditions of moderate to extreme shelter on northern New Zealand coasts. In such locations there will rarely be severe wave action and larvae could adhere to the fringing coralline algae without being swept off.

These results are preliminary and our survey is incomplete; there are several additional species for which we have partial data. We know the seasons of their larval production but have not been able to obtain larval release in the laboratory. It does appear, however, that when sufficiently accurate statements can be made about both larval behaviour and adult ecology, a pattern does emerge and that this pattern can be expected to follow ecological rather than systematic lines. That such relationships exist is not surprising. Indeed, on the basis of theories of niche diversification and speciation they are to be expected.

MODIFICATION OF REPRODUCTIVE PROCESSES IN INTERTIDAL DEMOSPONGIAE

Introduction

The recognition of adaptive behaviour in the larval phase of the reproductive cycle suggests that investigation of other phases of the developmental cycle might reveal further adaptation to the intertidal environment. As mentioned above, in order to establish the period of larval production and egg release, monthly samples were taken of 15 species, 8 being Ceractinomorpha and 7 Tetractinomorpha. Examination of histological preparations of these samples revealed that in all the Ceractinomorpha ciliated larvae were being produced by sexual and asexual processes, and that the latter (in which ovocytes were never involved) predominated. This phenomenon is, of course, not exhibited in tetractinomorph sponges, which are characterized for the most part by the extrusion of eggs and external development of larvae. However, of the tetractinomorph species studied, the three which are most successful in the intertidal region showed great elaboration of asexual development by either budding or gemmule production.

Larvae of asexual origin were first described by Wilson (1891, 1894). The accuracy of Wilson's observations and his interpretation of what he had observed were challenged by Maas (1893) and Minchin (1897). The topic remained the subject of acrimonious debate between these three students of sponge development for several years without any satisfactory conclusion being reached. Support for Wilson's view came from the work of Ijima (1901) on the development of euplectellids, while the argument against his view was solely that sexual development of larvae had been observed and thus *must* occur in all cases. At the time that Wilson's large monograph detailing his observations on both sexual and asexual development appeared, his viewpoint on the fate of the cell layers during larval metamorphosis was clearly out of date. He maintained that the ciliated larval cells became the pinacoderm of the young sponge and he was obviously unaware of the work of Delage (1892) and Maas (1893). It was unfortunate that instead of restricting their criticisms to this aspect of Wilson's work, his contemporaries rejected his contribution entirely. This attitude ensured that the report of asexual origin of larvae in *Esperella* (*Mycale*) *fibrexilis* Wilson,1894 and *Tedania brucei* Wilson,1894 was discounted completely. Wilson (1902, 1935), however, never found any reason to doubt that asexual development of larvae occurred.

Sivaramakrisknan (1951) described asexual development of larvae in *Callyspongia diffusa*, *Ircinia* sp. and *Tedania nigrescens* O. Schmidt, 1862. His account of the process is clear, and well illustrated, and differs from Wilson's observations only in details. For instance, Wilson (1894) described the initial stage in formation of asexual bodies, which both authors termed "gemmules", as a group of amoebocytes, each laden with yolk material, and which secreted a gemmular sheath around themselves. Subsequent increase in size was achieved by cell division or by fusion with adjacent, smaller gemmules. Sivaramakrishnan, on the other hand, described the gemmule as being formed from the merging of several multinucleate masses, each of which originated from an amoebocyte surrounded by many choanocytes. Fusion with adjacent young gemmules could take place and the spongin envelope which separated gemmules from the mesenchyme developed from mesenchyme cells.

A further point of disagreement between Wilson and Sivaramakrishnan was over the rearrangement of cells during the metamorphosis of gemmule to larva. In *Mycale fibrexilis* the gemmule fragmented into irregular cell masses which moved to their respective positions in the body of the larva (Wilson, 1894). In *Callyspongia diffusa* the nuclei, which were identified by their staining reaction to be of choanocytic derivation, migrated to the periphery of the gemmule and reconstituted flagellated cells (Sivaramakrishnan, 1951).

Sivaramakrishnan did not mention the presence of ovocytes in his material. Apparently, asexual development was the only process leading to production of larvae. In Wilson's study ovocytes were reported to be present in small numbers, suggesting that both sexual and asexual development occurred. The latter situation is found in our material as well, but ovocyte production is always limited seasonally and the number produced is too small to account for the number of developing gemmules and larvae.

Sexual and asexual development of larvae in Ceractinomorpha

Sexual and asexual development to the larval stage has been followed in the following sponges, *Spongia reticulata* Lendenfeld,1886, *Ircinia novaezealandiae* Bergquist,1961, *Haliclona heterofibrosa* Lundbeck,1902, *Callyspongia diffusa*, *Mycale macilenta*, *Halichondria panicea* Pallas,1766, *Halichondria moorei* and *Hymeniacidon perleve* Montagu,1818. In all species asexual processes predominate—in *Halichondria moorei* to the almost total exclusion of sexual development. Ovocytes were found in this species only once during a 4-year sampling period. In the other species, sampled during 2 years, ovocytes were found at definite periods of the year but gemmular stages were present almost continuously.

The details of seasonal occurrence of the various reproductive products are not relevant to this discussion and need not be considered further. Nor is it necessary to digress into the histological details leading to larval development. These have been described in painstaking detail by Wilson (1894) and Sivaramakrishnan (1951) for the asexual process and many workers (e.g. Maas, 1893; Meewis, 1938, 1939, 1940; Lévi, 1956; Tuzet, 1947; Tuzet and Pavans de Ceccatty, 1958) have considered the normal development of Demospongiae. The developmental sequences we have observed will be outlined, with reference to *Spongia reticulata* and *Hymeniacidon perleve*, and related to previous accounts. In the interpretation of events in early ovocyte development up to and including fertilization, the conclusions of Tuzet and Pavans de Ceccatty (1958) have been adopted. These authors were able to observe almost all events except fertilization and early segmentation in *Hippospongia communis* Lamarck, 1814. The early stages of development in *Spongia reticulata* correspond closely with those described for *H. communis*.

In *Spongia reticulata* the ovocytes at first appearance are 9 to 12 microns in diameter, begin slightly larger than the amoebocytes from which they are derived. The ovocytes lie in cavities in the mesenchyme and are distinguished by their prominent nuclei. Larger ovocytes, 45 to 50 μ in diameter and with prominent nuclei and granular cytoplasm, are interpreted as ovocytes which have completed the "petit accroissement" which ensues after sperm have reached the female cells. These larger cells are surrounded by a delicate spongin envelope and some nutritive amoebocytes. No carrier cells or sperm nuclei have been recognized in *S. reticulata*. At this stage the sequence differs from that in *Hippospongia*. Instead of the ovocyte enlarging enormously (that is, undergoing the "grand accroissement", to 300 μ diameter, after meiotic division of the female nucleus has occurred and fertilization is completed) there is little increase in size after the "petit accroissement". Fertilization must follow rapidly after this, for segmentation begins when the egg is 60 to 80 μ in diameter.

Segmentation in *S. reticulata* is more a disorderly fragmentation of the egg than regular cleavage. The masses of fragments increase in size until they form compact bodies, 300 to 400 μ in diameter, composed of uniformly small cells packed with food reserves. These bodies are identical structurally with asexually produced gemmules and for reference we have termed this the gemmular stage. From the gemmular stage the free swimming larva is produced by reorganization and rapid division of the constituent cells. Ultimately, the smaller cells with densely staining nuclei form a peripheral layer, the larger cells being central. Larval histology is typical (cf. Sivaramakrishnan, 1951) and the fibrils of the developing spongin skeleton can be seen in the central region of the larva before its release.

As a result of a glistening white pigmentation, the gemmular stage is easily visible to the eye if the sponge is sectioned and these bodies were observed in *S. reticulata* throughout the sampling period. However, ovocytes were found in abundance only in the two months prior to larval production and release. At all other periods the gemmular stage is reached by a process of aggregation and cell division. The earliest stage is the aggregation of archaeocytes into small cavities in the mesenchyme. These groups increase in size by cell division and fusion with adjacent cell groups until they form compact bodies composed of cells with granular cytoplasm and indistinct cell boundaries. Each of these cell masses is surrounded by a delicate spongin A membrane and by nutritive cells. Except for the presence of nutritive cells, these structures are indistinguishable from the gemmular stage resulting from sexual reproduction.

Reorganization to form the larva is effected by the separation of the cells making up the gemmular stage, so that the cellular structure is again easily discernible. Small cells with deeply staining nuclei move to the periphery, larger amoebocytes and archaeocyte-like cells remaining in the centre. Once the layered structure is established, development to the final ciliated stage is rapid and is identical to that of any parenchymella.

It is interesting to note that in *S. reticulata*, reorganization of the asexual gemmular stage to form the larva is delayed until gemmular

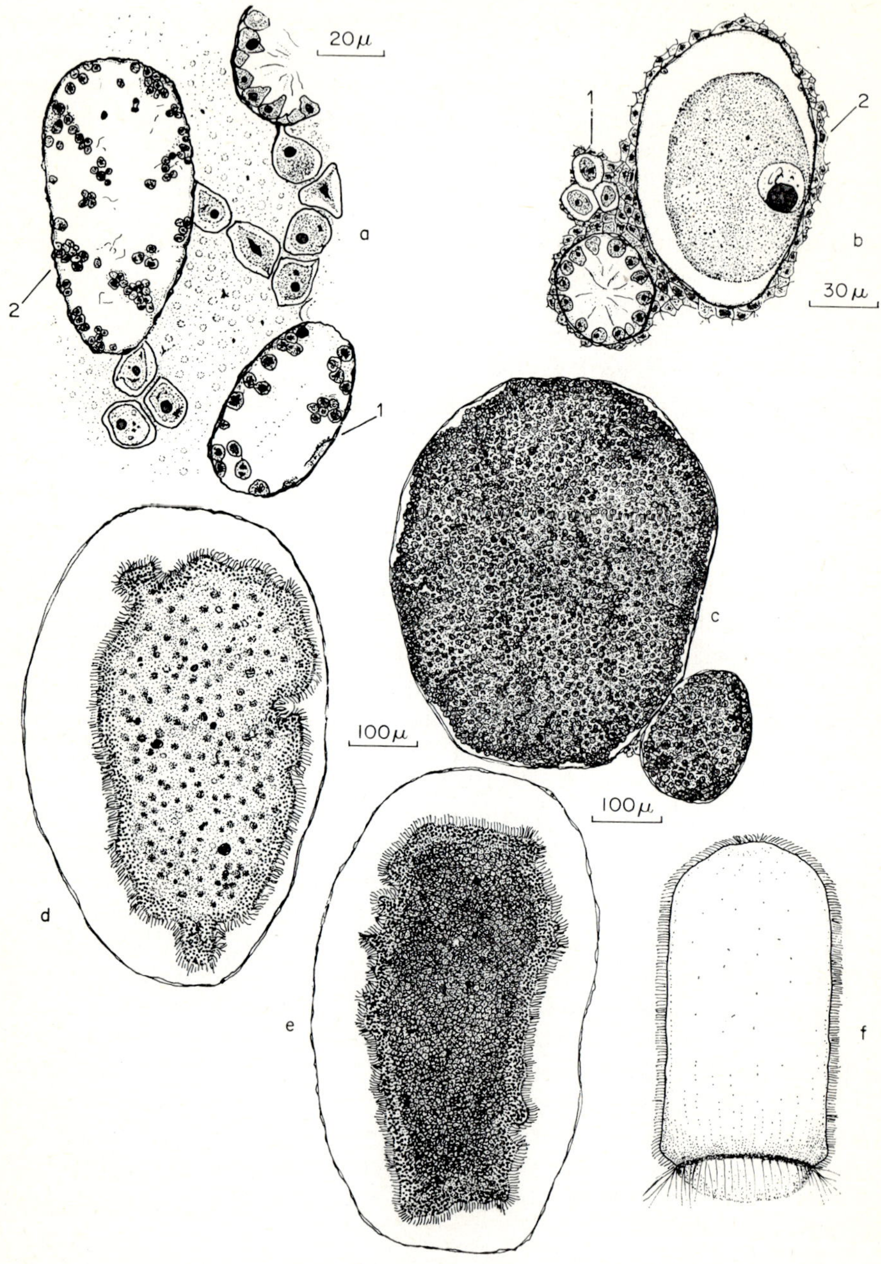
20μ
1
2
a
b
30μ
c
100μ
100μ
d
e
f

stages of sexual origin are developing, for all larvae are produced at the same time, in one month of the year, immediately after the appearance of ovocytes. Asexual stages, as indicated earlier, are present throughout the year.

In *Hymeniacidon perleve* the smallest recognizable ovocytes are 20 to 25 μ in diameter, with large nuclei, prominent nucleolus and granular cytoplasm. These cells are retained within cavities in the mesenchyme. It is at this early stage, before nutritive cells surround the ovocyte, that the transport of sperm by carrier cell to the ovocyte occurs. Amoebocytes then congregate about the ovocyte and supply nutriment, which the enlarging cell incorporates as cytoplasmic spherules. Increase in size continues until the ovocyte is 110 to 120 μ in diameter, at which time fusion of male and female nuclei presumably occurs, since segmentation begins. By this time the layers of nutritive cells have been depleted, only a few cells remaining. Segmentation in *Hymeniacidon* is regular and division continues until a compact mass of small cells is formed. These eventually become tightly packed to form the gemmular stage, which in this species, is 160 to 180 μ in diameter. Differentiation to form the larva is typical. The spicules appear late in development, aligned in a band inside the zone marked by nuclei of the epithelial layer.

The gemmular stage can also be reached in the following manner. Archaeocytes packed with yolk reserves congregate in the basal region of the sponge and become organized into groups. Adjacent groups can fuse to give at first loose, and later compact, aggregates, in which cell boundaries are difficult to distinguish. This gemmular stage is 150 to 200 μ in diameter, and is surrounded by a spongin sheath secreted by mesenchymal elements. Again, reorganization to form the larva is preceded by a slight loosening of the cell mass, rendering the nuclei and cell boundaries more easily visible. The smaller cells with densely staining, small nuclei migrate to the periphery, leaving cells with larger nuclei and granular cytoplasm in the central region.

There is no pause in asexual development at the gemmular stage in *H. perleve* and development to the ciliated larva is uninterrupted.

This sequence of asexual development has been followed in the eight species mentioned earlier. Parts of the sexual sequence involving

---

FIG. 1 (facing page). Developmental stages in *Spongia reticulata*. a, Stages in spermatogenesis, 1 early sperm formation, 2 mature sperm; b, stages in ovogenesis, 1 ovocyte before "petit accroissement", 2 ovocyte before division; c, gemmular stages (reached either by way of sexual fusion and development or from aggregations of archaeocytes); d, e, differentiation of the gemmular stage to form a larva; f, free-swimming larva. (After Hogg, 1967.)

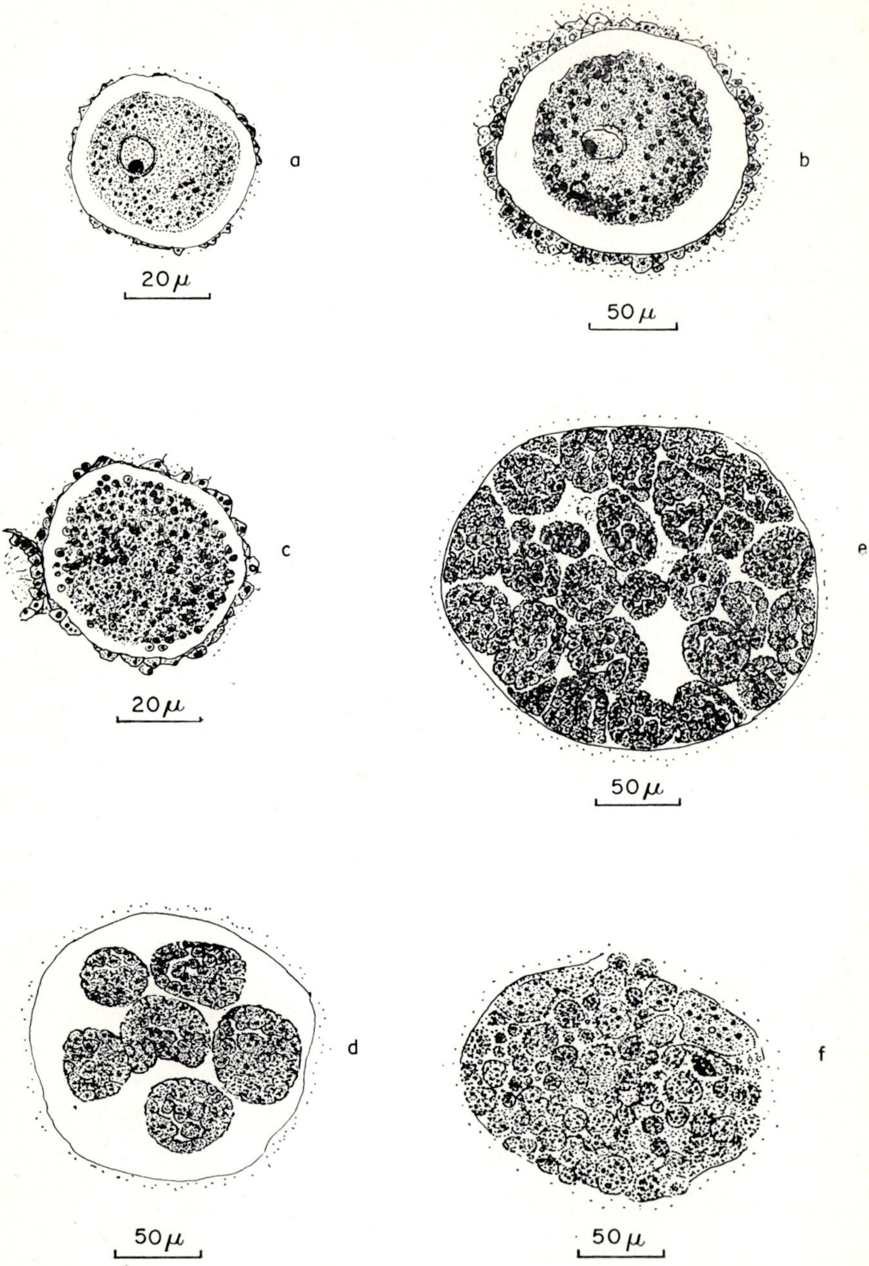
a
20 μ
b
50 μ
c
20 μ
e
50 μ
d
50 μ
f
50 μ

indisputable ovocytes and fragmenting or segmenting eggs have also been followed, although there are certain points which remain obscure. In particular, it has proved difficult to follow the process of fertilization and egg maturation in Demospongiae.

The only possible conclusion from our results is that in intertidal Ceractinomorpha the occurrence of asexual development of larvae is not the exception but the rule.

There is a third developmental alternative which should be mentioned. Unfertilized ovocytes could simply increase in size, drawing upon mesenchymal food reserves until, by fragmentation and reorganization, they reach a gemmular stage. No direct evidence for this developmental pathway can be offered. In the species studied here, developmental sequences involving ovocytes are comparable with the few published accounts in which fertilization is known to have occurred. However, in the absence of cytological evidence of fertilization, the former sequence must remain a possibility.

The development of asexual larvae as we have observed it is identical in most respects with the sequence outlined by Wilson (1894) for *Mycale fibrexilis*, with the exception that the spongin A sheath surrounding the gemmule is secreted by mesenchymal components, and not by the cells making up the aggregate. In this respect we agree with Sivaramakrishnan. We cannot, however, confirm his observation that choanocytes are the major participants in development of gemmular stages. This could presumably be the case where gemmule production is heavy and leads to a breakdown of the flagellated chambers in certain areas of the sponge. Under these conditions choanocytes might become amoebocytic and could be incorporated into developing gemmules. One aspect of development from ovocyte to gemmular stage in *S. reticulata* merits particular comment. This is the irregular cleavage of the egg. Of the eight species we have studied, only *Hymeniacidon* shows regular isomeric cleavage. This could mean that only in *Hymeniacidon* were we observing development after fertilization. If this should prove to be the case, one must assume that asexual production of larvae predominates in intertidal sponges to an even greater extent than we have already suggested. In this connection, Burton (1949) commented that in the course of his extensive systematic work, he had noted few instances of regular cleavage of early embryos. Possibly, regular cleavage is not the normal condition in Demospongiae

FIG. 2 (facing page). Developmental stages in *Hymeniacidon perleve*. a, young ovocyte; b, mature ovocyte; c, sperm ball; d, e, stages in segmentation; f, development of the asexual gemmular stage, the archaeocyte mass becoming surrounded by a sheath. (After Hogg, 1967.)

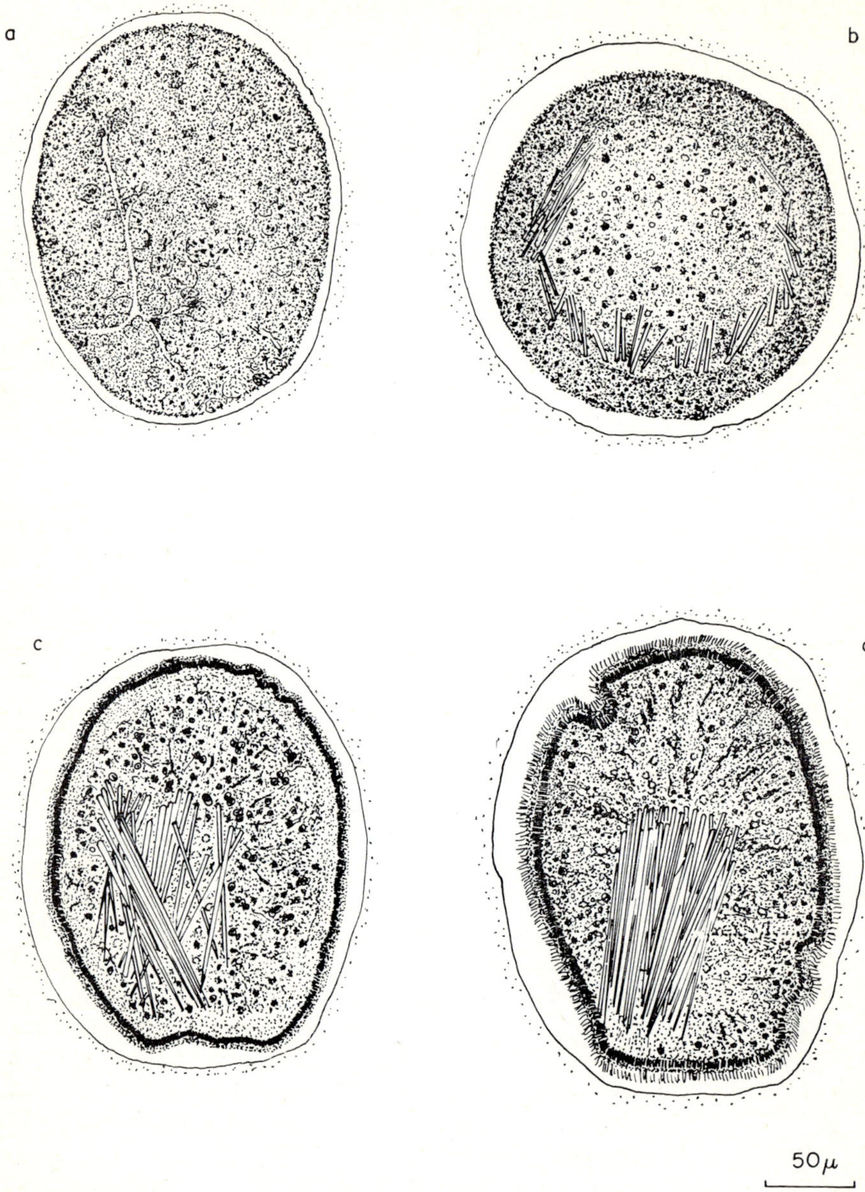

FIG. 3. Developmental stages in *Hymeniacidon perleve*. a–d, successive stages of differentiation of the compact gemmular stage to form a larva. This process is common to both sexual and asexual sequences. (After Hogg, 1967.)

TABLE II

Stages in the development of ciliated larvae

ASEXUAL SEQUENCE	SEXUAL SEQUENCE*	
	SPERM	OVOCYTE
Aggregation of archaeocytes		(from Amoebocyte)
Incorporation of additional archaeocytes and cell division	Enters choanocyte which rounds off and enters carrier cell	increase in size
Separation of cell groups within a fine spongin sheath	Juxtaposition of carrier cell and ovocyte with fusion of cells	
	Increase in size from surrounding nutritive cells	
	Fertilization	
	Segmentation (regular or disorderly)	
Compaction (Gemmular stage)	Compaction (Gemmular stage)	
Reorganization		
Ciliated larva		

* Information derived in part from Tuzet and Pavans de Ceccatty (1958).

or, alternatively, the early stages could be traversed very rapidly. In *Polymastia robusta* a morula is formed within 12 h after fertilized eggs are liberated but there is no information available as to the rate of early development in any ceractinomorph sponge. Thus the question of whether or not irregular fragmentation of an egg or ovocyte is an indication of parthenogenetic development cannot be answered yet.

Asexual reproduction in intertidal tetractinomorpha

Asexual production of stalked buds in species of *Tethya* Lamarck, 1815 is well known and has been the subject of many papers (see Connes, 1967, for review). Carter (1883) first recorded and figured gemmules in the tetractinomorph sponge *Suberites domuncula* Olivi, 1792. Since that time gemmules have been observed frequently in *S. domuncula* and occasionally in other hadromerid sponges, e.g. *Cliona vastifica* Hancock,1849, *Prosuberites epiphytum* Lamarck,1816 and *Terpios fugax* Topsent,1900, *Prosuberites microsclerus* Wells, Wells and Gray,1946. The structure and occurrence of these gemmules was discussed in detail by Muller (1913). Gemmules of hadromerid sponges are concentrated in a basal layer and fused to the substrate by a strong spongin B envelope. They lack the elaborate hilum and spicule armour characteristic of the overwintering gemmules found in fresh water sponges.

The term "gemmule" has been used in sponge literature to describe several different structures. There is, first, the typical gemmule—the complex, spongin B-encapsulated structure found in the Spongillidae. Similar, but less complex, are the gemmules found in intertidal and shallow water sponges chiefly belonging to the Haplosclerida Topsent, 1928. These gemmules, described by Topsent (1888), Lundbeck (1905), Hartman (1958), have a spongin B sheath over which spicules are arranged tangentially. They are obviously overwintering devices. Where gemmules occur on the Hadromerida they are cemented to the substrate and have a spongin coat but lack spicules. These structures occur in both shallow and deep water forms.

In addition to these three types of gemmule (all to some extent comparable in structure and function) there are two other types of structure which have been termed gemmules. The first is the incubated larva of tetillid sponges in which there is no free developmental stage at all and the product of sexual reproduction liberated from the parent is a miniature sponge. The sexual origin of these bodies is clear, however. Sperm, egg and segmentation stages have been observed and thus they cannot be considered as gemmules. The report (Topsent,1909) of gemmules in species of the Cladorhizidae de Laubenfels,1936 is inconclusive; Lundbeck (1905) considered the same structures to be embryos. There is insufficient information available on the development of these bodies to resolve the question at the present time.

The second includes the distinctive structures referred to as gemmules, which we have named "gemmular stages" (see above). They are known only to occur in intertidal ceractinomorph sponges. They are bounded by a thin spongin sheath of mesenchymal origin constituted chiefly of spongin A and are not terminal developmental stages but undergo reorganization to form free-swimming larvae. They are comparable to true gemmules only in their asexual origin.

Of the seven species of Tetractinomorpha we have studied, only four, *Tethya aurantium* Pallas,1766, *Tethya ingalli* Bowerbank,1858, *Aaptos aaptos* O. Schmidt,1864 and *Plakina trilopha* Schulze,1880 are permanent and abundant components of the New Zealand intertidal fauna. The remainder, *Raspailia agminata* Hallmann,1914, *Polymastia granulosa* and *Ancorina alata* Dendy,1924 extend into the sublittoral fringe but have their greatest distribution in deep water.

Tethya aurantium is a common sponge intertidally and forms extensive mats under rocky overhangs from the midlittoral region down. *Aaptos aaptos* has a similar range and habit but is not confined to shaded locations. *Tethya ingalli* is never as abundant as the preceding species and occurs only on the lower part of the shore in the littoral fringe, often in association with the holdfasts of *Carpophyllum maschalocarpum* Turner,1808–1819. *Plakina trilopha* is a common encrusting sponge under rocks in the lower midlittoral zone of moderately sheltered shores.

Tethya species and *Aaptos* show great emphasis on asexual reproductive mechanisms in the intertidal and shallow water part of their range. *Plakina,* however, produces the amphiblastula larvae typical of the genus by normal sexual processes.

Both *Tethya* species and *Aaptos aaptos* produce stalked surface buds. In *Tethya* this occurs regardless of depth but in *Aaptos* only in intertidal specimens. The buds of *Aaptos* differ from those of *Tethya* in becoming much larger before they detach, often over 2 cm in diameter. Some buds never detach fron the parent.

Tethya aurantium and *A. aaptos* also multiply extensively in intertidal locations by the formation of stolon buds. The attachment stolons spread out from the original sponge and expand into new individuals, which then repeat the process. The whole complex remains united by the network of stolons. In this way the sponges increase surface area and thus their feeding capacity and extend over large areas of substrate. Aggregates of up to eighty components have been observed to form from an initial group of three individuals.

Tethya aurantium also produces gemmules in great numbers but at infrequent intervals. On the three separate occasions on which these have been found the sponge endosome was packed with gemmules all at the same stage of development. In two cases the gemmules were 800 to 2,000 μ in diameter; in the other only 100 μ in diameter. The structure was always the same, consisting of a dense, compact, completely homogeneous mass of cells loaded with granular nutritive materials and enclosed within a tough spongin A capsule. *Tethya* gemmules differ from those described in hadromerid sponges in being endosomal rather than basal and in having a spongin A rather than a spongin B capsule.

It is interesting to note at this juncture references by Hallmann (1914) and Burton (1926) to the occurrence in *Tethya fissurata* Lendenfeld,1888 and *Tethya magna* Kirkpatrick,1903 of structures of similar size to the gemmules described above. Both authors considered the structures to be embryos, since in those of *T. magna* a cortical/endosomal differentiation was discernible with scattered megascleres and microscleres present, while in *T. fissurata* in a body 2500 μ in diameter, the stylote spicules were arranged radially, the tylasters being superficial as in the adult while minute developing spherasters were present in the cortex. It is probable that in our material of *Tethya aurantium* we have found earlier stages of structures which would ultimately become miniature adult sponges comparable to those described by Hallmann (1914). The strong indication from our work is, however, that these are of asexual origin. We know (Lévi, 1956) that the sexually produced larva of *Tethya aurantium* is produced in oviparous fashion following extrusion of fertilized eggs (a type of development which characterizes all clavaxinellid sponges for which developmental details are known). There is no possibility that the structures we have described are of sexual origin, as undoubted ovocytes have been found in offshore specimens of *Tethya aurantium* on many occasions and these are comparable in size and structure with sexual products described by other authors. The gemmules of *Tethya* show an interesting parallel to the viviparous development of young sponges in tetillids, the difference being that the former are of asexual and the latter of sexual origin.

In the course of two years sampling of intertidal specimens, no sexual products were found in *Tethya ingalli* and ovocytes were identified once only in *Tethya aurantium* and in *Aaptos aaptos*. In *T. aurantium* and *Aaptos aaptos*, however, specimens from deeper water (3 m or more) produce ovocytes at regular intervals. It is obvious that recruitment to the intertidal populations of all three species is maintained by larval production in benthic populations. Efficient methods of asexual reproduction ensure the numerical increase of each species following establishment in suitable intertidal niches.

DISCUSSIONS AND CONCLUSIONS

On the basis of our own and published observations we have argued in the preceding pages for the existence in Demospongiae of:

1. patterns of larval behaviour which are relevant to the situation of the sessile adult sponge.

2. a high incidence of asexual reproduction in intertidal forms.

We interpret these phenomena as being examples of adaptation to intertidal existence in a group of animals which are predominantly members of the offshore benthos.

Larval behaviour specialized toward ensuring settlement and adult growth in a suitable location has been demonstrated for many groups of sessile or sedentary marine organisms (e.g. Knight-Jones and Moyse, 1961).

That such behaviour, at a relatively unsophisticated level, should exist in sponge larvae is not at all surprising. What the determinants of such behaviour are remain to be discovered. It has been demonstrated several times that competition for space among sessile animals on rocky shores is of great importance in establishing and maintaining zonation patterns. Thus, when one observes sponges recurring from year to year in identical habitats on a rocky shore, it is logical to suppose that either they possess some adaptive mechanism which enables them to locate and compete effectively for this space or larval production is so excessive that after all chance factors have operated on the larva there is still a sufficient number to allow indiscriminate settlement to produce a significant number of colonies located in niches suitable for adult sponge growth. We believe that the first alternative is true. There is need for much more study in order to prove or disprove our interpretation. Until recently systematic problems have hindered ecologists who attempted to incorporate sponges in their studies but this need no longer deter workers around European coasts for which adequate systematic data are now available.

The reliance among intertidal sponges upon asexual reproduction is probably related to the difficulties of fertilization in intertidal locations. Production of larvae by asexual means reduces the dependence of the sponge upon fertilization and at the same time conserves the active dispersal phase, the larva. Other asexual methods seem directed towards passive dispersal, e.g. gemmules and detaching buds, or towards long term occupancy of a niche by vegetative spread.

We are now extending our studies to include pairs of closely related species and ecologically distinct forms of the same species. There is evidence in the Tetractinomorpha at least that deeper water and intertidal populations of the same species follow a very different reproductive cycle. This may also be true of ceractinomorph sponges but those on which we have worked do not extend much beyond the sublittoral fringe. Other species which cover a wider depth range will have to be studied.

It should be noted that the literature reviewed in this paper and the observations made refer only to Demospongiae and that the paper is not intended to be a complete review of the literature on this group.

Acknowledgements

The senior author wishes to acknowledge the financial support provided by the University of New Zealand Research Grants Committee.

References

Ali, M. A. (1956). Development of the monaxonid sponge *Lissodendoryx similis* Thiele. *J. Madras Univ.* **B. 26**: 553–581.

Bakus, G. J. (1964). Morphogenesis of *Tedania gurjanovae* Koltun (Porifera). *Pacif. Sci.* **18**: 58–63.

Bergquist, P. R. (1961). *Demospongiae of New Zealand. Systematics, distribution and relationships*. Ph.D. Thesis University of Auckland.

Bergquist, P. R. and Sinclair, M. E. (1968). The morphology and behaviour of larvae of some intertidal sponges. *N.Z. Jl Mar. Freshw. Res.* **2**.

Borojević, R. (1967). La ponte et le developpement de *Polymastia robusta* (Demosponges). *Cah. Biol. Mar.* **8**: 1–6.

Burton, M. (1926). Descriptions of South African sponges collected in the South African Marine Survey. I. Myxospongida and Astrotetraxonida. *Rep. S. Afr. Fish. Mar. Biol. Surv.* **4**: 1–29.

Burton, M. (1949). Non Sexual reproduction in sponges with special reference to a collection of young *Geodia*. *Proc. Linn. Soc. Lond.* **160**: 163–178.

Carter, H. J. (1883). On the presence of starch granules in the ovum of marine sponges and on the ovigerous layer of *Suberites domuncula*. *Ann. Mag. nat. Hist.* (5) **12**: 30–36.

Connes, R. (1967). Structure et développement des bourgeons chez l'eponge siliceuese *Tethya lyncurium* L. Recherches expérimentales et cytologiques. *Archs Zool. exp. gén.* **108**: 157–195.

Delage, Y. (1892). Embryogénie des Éponges. Developpement post-larvaire des Éponges siliceuses et fibreuses marines et d'eau douce. *Archs Zool. exp. gén.* **10**: 345–498.

Hallmann, E. F. (1914). A revision of the monaxonid species described as new in Lendenfelds "Catalogue of the sponges in the Australian Museum". Pt. 1. *Proc. Linn. Soc. N.S.W.* **39**: 263–315.

Hartman, W. D. (1958). Natural history of the Marine sponges of Southern New England. *Bull. Peabody Mus. nat. Hist.* **12**: 1–155.

Hogg, J. J. (1967). *Approaches to the systematics of the Demospongiae*. Thesis M.Sc. University of Auckland.

Ijima, I. (1901). Studies on the hexactinellida I. *J. Coll. Sci. Tokyo.* **15**: 1–299.

Knight-Jones, E. W. and Moyse, J. (1961). Intraspecific competition in sedentary marine animals. *Symp. Soc. exp. Biol.* **15**: 72–95.

Lévi, C. (1956). Étude des *Halisarca* de Roscoff. Embryologie et systématique des Demosponges. *Archs Zool. exp. gén.* **93**: 1–181.

Lundbeck, W. (1905). Porifera II. Desmacidonidae (pars). *Dan. Ingolf-Exped.* **6**: 1–219.

Maas, O. (1893). Die Embryonal–Entwicklung und metamorphose der Cornacuspongien. *Zool. Jb.* (Abt. 2) **7**: 331–448.

Meewis, H. (1938). Contribution à l'étude de l'embryogenèse des Myxospongidae: *Halisarca lobularis*. *Archs Biol., Paris* **50**: 1–66.

Meewis, H. (1939). Etude comparative des larves d'Eponges siliceuses. *C.r. Ass. fr. Avanc. Sci.* **1939**: 660–669.

Meewis, H. (1940). Contribution à l'étude de l'embryogenèse des Chalinidae: *Haliclona limbata*. *Annls Soc. r. Zool. Belg.* **70**: 201–243.

Minchin, E. A. (1897). The position of sponges in the Animal Kingdom. *Sci. Progr. Twent. Cent.* **1**: 451–459.

Minchin, E. A. (1900). Sponges. In *Treatise on zoology* **11**: 1–178. Lankester, R. (Ed.) London: A. and C. Black.

Muller, K. (1913). Gemmula–Studien und allgemein–biologische Untersuchungen an *Ficulina ficus*. *Wiss. Meeresuntersuch.* **16**: 287–313.

Sivaramakrishnan, V. R. (1951). Early development and regeneration in Indian Marine Sponges. *Proc. Indian Acad. Sci.* **34**: 273–310.

Topsent, E. (1888). Sur les gemmules de quelques Silicisponges marines. *C.r. hebd. Séanc. Acad, Sci., Paris* **106**: 1298–1300.

Topsent, E. (1909). Etude sur quelques *Cladorhiza* et sur *Euchelipluma pristina* n.g. et n. sp. *Bull. Inst. oceanogr. Monaco* No. **151**: 1–23.

Topsent, E. (1911). Sur les affinités des *Halichondria* et la classification des Halichondrines d'après leurs formes larvaires. *Archs Zool. exp. gén.*, (7) **5**: 1–15.

Tuzet, O. (1930). Sur la fecondation de l'éponge siliceuse *Cliona viridis C.r. hebd. Séanc. Acad. Sci., Paris* **191**: 1–3.

Tuzet, O. (1947). L'ovogenèse et la fecondation de l'Éponge calcaire *Leucosolenia coriacea* et de l'Éponge siliceuse *Reniera elegans*. *Archs Zool. exp. gén.* **85**: 127–148.

Tuzet, O. and Pavans de Ceccatty, M. (1958). La Spermatogenèse, l'ovogenèse, la fecondation et les premiers stades du developpement d'*Hippospongia communis*. *Bull. biol. Fr. Belge.* **92**: 333–348.

Warburton, F. E. (1966). The behaviour of sponge larvae. *Ecology* **47** (4): 672–674.

Wells, H. W., Wells, M. J. and Gray, I. E. (1964). Ecology of sponges in Hatteras Harbor, North Carolina. *Ecology* **45**: 752–767.

Wilson, H. V. (1891). Notes on the development of some sponges. *J. Morph.* **5**: 511–519.

Wilson, H. V. (1894). Observations on the gemmule and egg development of Marine Sponges. *J. Morph.* **9**: 277–406.

Wilson, H. V. (1902). On the asexual origin of the ciliated sponge larva. *Am. Nat.* **36**: 451–459.

Wilson, H. V. (1935). Some critical points in the metamorphosis of the halichondrine sponge larva. *J. Morph.* **58**: 285–353.

Symp. zool. Soc. Lond. (1970) No. 25, 273–284

COMPETITION AND COOPERATION IN SPONGE POPULATIONS

MICHELE SARÀ

Istituto di Zoologia ed Anatomia Comparata, Università di Bari, Bari, Italy

SYNOPSIS

In caves and other shallow-water biotopes sponges may constitute continuous populations from which other elements of the sessile macrofauna and macroflora are virtually completely excluded. In this type of population the coaction effects among species of sponges reach their maximal intensity.

The competition phenomena, the only ones hitherto studied, occur in respect to the substratum and also to food supply. However, their importance is probably less than generally thought and there is no clear evidence that some species of sponges may destroy other species by overwhelming them.

On the other hand, the complexity of the continuous sponge populations and the aspects of their species diversity, ecological stratification and temporal stability involve a very important role of cooperation phenomena. In this regard, some significant facts may be indicated.

A positive correlation between density and species diversity is shown by a Ligurian sponge population. A vertical stratification of sponges may be formed on the floor of caves with four ecologically characterized strata of Porifera: boring, hypobiotic, basal and epibiotic.

A considerable temporal stability, both from a quantitative and a qualitative point of view, without sharp seasonal modifications, occurs in the continuous sponge populations of shallow waters (0–1 m), except where unusual ecological conditions exist.

Furthermore, different species of sponges may join together to form single rounded masses which remain stable in shape and size over long periods.

INTRODUCTION

No general survey of sponge population ecology exists and even special works devoted to single aspects of this subject are very scarce. Nevertheless, sponge populations are a very promising material for the investigation of important aspects of benthic synecology, such as substrate colonization, competition for space, environmental conditioning, coaction and integration between species. In this article I shall attempt to elucidate some synecological aspects of shallow-water sponge populations.

SPONGE POPULATIONS

With regard to the relations between sponge species, three types of sponge populations may be considered schematically.

The first type is represented by discontinuous populations of scattered sponge individuals on the bare rock. As Mediterranean examples we may indicate such species as *Crambe crambe* Schmidt, 1862, *Spongia officinalis* Linné, 1759, *Ircinia fasciculata* Pallas, 1776, *Chondrilla nucula* Schmidt, 1862, *Cliona viridis* Schmidt, 1862, etc. Only a limited number of species live in this condition, however. In general, the shallow water sponges inhabit dense organismic associations.

In the second type of population the scattered sponge specimens grow within mixed communities with other elements of the sessile macrofauna or macroflora. Examples are given by the coralligenous banks, *Posidonia oceanica* Delile, 1813 beds, some phytal communities, etc.

The third type is represented by a continuous sponge population, with practically negligible intervention of other elements of the sessile macrofauna and macroflora. This type is well developed on the vertical walls, floors and roofs of submerged caves.

The ecological differences between these three types of sponge populations are manifold. The coactions among sponge species are obviously stronger in the third type of population, which is also formed by the largest number of species.

INTERSPECIFIC COMPETITION

The problem of interspecific competition in sponge populations has been approached from both the points of view of the common niche and of the common food supply. Hartman (1957), considering nine sympatric species of Clionids present in the region of Rovigno, observed that the optimum habitats of the different species show some differentiation and that the distribution patterns suggest the importance of competition for substratum among the species considered. The most striking evidence for such competition was, for Hartman, the occurrence of species with highly specific habitats as *Thoosa mollis* Volz, 1939, which lives almost exclusively in the walls of abandoned burrows of the lamellibranch *Lithodomus* Cuvier, 1817, and *Cliona rovignensis* Volz, 1939, which lives only in the thin plates of limestone separating two closely adjacent *Lithodomus* burrows.

Moreover, Hartman (1957) has suggested that the gradation in the size of the ostia and of the oscular papillae, which seem to bear a direct relationship to the diameters of the ostia, indicates that different foods may be taken. This would indicate alimentary specializations and therefore competition for food. Unfortunately, this interesting hypothesis is based on a very limited set of data. On the other hand, the substratum specialization of clionids is a well known fact which does not necessarily derive from competition phenomena and, anyway, it may be expected that competition for substratum, owing to the clionids' behaviour as burrowers, should play a more significant role for these than for other sponges.

The interspecific competition for space has been studied by Rützler (1965) in Adriatic populations of sponges. Analysing the numbers of species on known surface areas of stones and boulders of different weights, Rützler observed that the species number decreases with the increase of rock weight from 3 to 30 kg, whereas from 30 kg boulder weight to stable rock it remains constant. This fact indicates, according to Rützler, that interspecific competition for space starts between 3 and 30 kg rock weight, when the faster growing or more resistant species begin to overwhelm their neighbours. Nevertheless, it should be noted that the special habitat of stones and small boulders, formed by mobile substrates, does not allow anything other than precarious sponge colonization. Furthermore, the decrease of the number of species with the increase of the weight of the stones may also be due to causes other than direct interspecific competition. For example, it may be due to the fact that on more stable substrate the dominant species eventually exploit the longer time at their disposal for their growth and surface covering. In this manner the larvae of other species may be indirectly excluded from settling on the formerly colonized surfaces. In addition to this, the smaller stones represent, by their mobility, a rapidly changing habitat and may be considered as a more heterogeneous niche than the larger boulders.

In conclusion it may be said that clear evidence is still lacking about the fact that some species may destroy other species by overwhelming them. The observations of Annandale (1915) on the destruction of clionids in their galleries by other infiltrating sponges need confirmation.

On the other hand, much published information indicates that sponges can survive even though covered by other species of sponges. The conditions and the mechanisms that regulate sponge epibiosis on other sponges are practically unknown. Some species, for example those of *Paramyxilla* Dendy,1905, grow constantly on the surface of other sponges. In other cases the epibiosis is species-specific as in the case of *Haliclona cratera* Schmidt, 1862 growing on *Ircinia oros* Schmidt, 1864 and of *Coelectys insinuans* Topsent, 1936 on *Oligoceras collectrix* Schulze, 1879. More often, however, the epibiosis is accidental and depends on the greater growth speed of some species which cover the slowest ones when space is limited. In this regard, at Cape Hatteras, Wells and Gray (1964) have observed that in midsummer *Microciona prolifera* Ellis and Solander, 1786 is overgrown by *Lissodendoryx isodictyalis* Carter, 1882 and *Mycale cecilia* de Laubenfels, 1936. Nothing is said, however, about the subsequent fate of the *Microciona*. In general, the covered sponge continues to live by receiving water with nutrients and oxygen, and by eliminating waste products, through the layer of the covering sponge. Sometimes the condition of the overgrown sponge may be even advantageous as a defense against predators or a protection from light. According to de Laubenfels (1947), in Bermuda *Epallax ajax* de Laubenfels, 1950 lives constantly covered by two other species of sponge and in Beaufort *Calyxabra poa* de Laubenfels, 1947 lives only under *Hymeniacidon sanguinea* (Grant, 1826). Special devices are realized by some sponges in order to maintain a free space between the two superimposed sponge layers (Riedl, 1966). More important than the direct elimination of sponges through overgrowth by other species is probably competition for space for larval settlement.

However, the complexity of continuous sponge populations, with the coexistence of a large number of species, examined from the three aspects of species diversity, community structure and stability in time, suggests that some cooperation phenomena may play a very important role. It must be remembered that a dense population of sponges produces a considerable biotic modification of the microenvironment. This environmental conditioning, which consists chiefly in the physical transformation of the substratum, in the modification of the physicochemical characters of the water and in the reduction of the microclimate fluctuations, especially of the hydrodynamic effects, may produce advantages for the associated sponges. Such advantages may balance and probably outweigh competitive phenomena.

SPECIES DIVERSITY

The number of species living in a continuous sponge population is very high. For example, within a surface of 50 m^2, more than 60 species inhabit the floor of a shallow water cave at Santa Maria di Leuca. In a quantitative survey of the sponges at Cala dell'Olivetta (Riviera Ligure di Levante) (Sarà, 1966), 25 different species have been discovered on a surface of only 2 m^2. In this locality it is interesting to compare the sponge population density and species diversity of five different niches (Fig. 1, Es, Ep, Os, Opc, Ops) distinguished essentially by a light intensity gradient which decreases from Es to Ops. Density is calculated as number of presences of all species in the 150 meshes of 6 quadrats in each niche (Sarà, 1966). Species diversity is evaluated according to the following index: $d = (S-1)/ln\ N$ (see Margalef, 1957).

Along the gradient Ep–Ops the increase in density of the sponge population is clearly accompanied by an increase in the species diversity (Fig. 1). The value of the Es diversity index which is higher than the Ep index is probably due to the fortuitous and scattered type of sponge population growing in a niche directly exposed to the sunlight

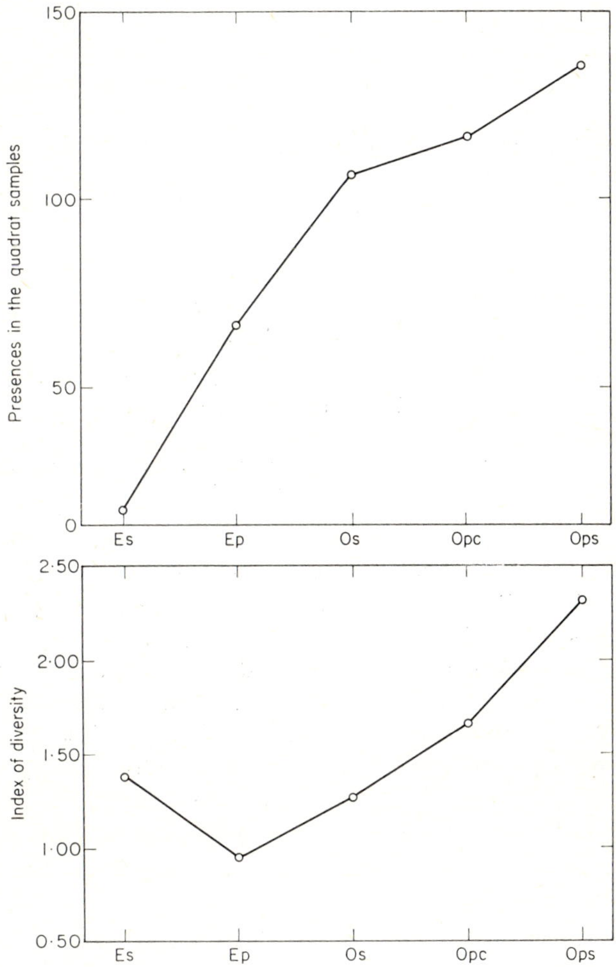

FIG. 1. Sponge population density (above) and species diversity (below) in five niches at Cala dell'Olivetta.

and therefore inhabited also by sea-weeds. Its sponge population is perhaps comparable, for its precariousness, with that of the small stones of Rovigno studied by Rützler, which also showed high species diversity.

The gradient Ep–Ops with its positive correlation between density and species diversity demonstrates that the competition effects, if existing, are overcome by the cooperation effects. This is related to favorable environmental conditioning and also is probably related to greater niche complexity which, in turn, is produced by the denser population itself.

STRATIFICATION

The large number of species in the continuous sponge populations is related also to the different roles which sponges may assume in the community. These differences lead to vertical stratification in the community.

A dense sponge population developed on the floor of a calcareous cave near Leuca, at 0–1 m depth, shows the existence of four strata, of which two are everywhere present (Sará, 1968):

1. A stratum of basal sponges, formed by the substrate-covering species, which constitutes the bulk of the population in surface area and biomass. In the Leuca cave the basal species are encrusting or cushion-shaped but in other habitats they may also be massive or erect.

2. A stratum of boring sponges; to this stratum belong the species of clionids excavating their burrows in the calcareous substrate. In the Leuca cave eight different species have been recognized. The role of boring sponges is very important for the subsequent colonization of the substratum by sponges of the basal stratum. For example, the establishment of *Erylus discophorus* Schmidt, 1862 in the Leuca cave is enhanced by the burrows and holes left by *Alectona millari* Carter, 1879.

The other strata, which are not everywhere developed, are dependent upon the existence of the above mentioned strata.

3. A stratum of epibiotic species, which are small, usually encrusting forms which grow preferably or exclusively on the basal sponges; some Calcarea live in this situation.

4. A stratum of hypobiotic species, which are also generally small, encrusting and live under the basal sponges, not boring but inhabiting the empty burrows of clionids. These sciaphilous species are numerous and may contribute significantly to the species diversity of the sponge population. *Rhabderemia indica* Dendy, 1905, *Dercitus plicatus* Schmidt, 1868, *Myxilla iotrochotina* Topsent, 1892 and many other species belong to this stratum.

The four strata above enumerated have an ecological more than a topographical characterization. For instance, two species of the basal stratum, such as *Crambe crambe* and *Anchinoë tenacior* Topsent, 1925, may cover one another yet always belong to the basal stratum. On the other hand, the hypobiotic species and the boring ones may reach the substratum surface in the interstices between the basal species. Ecologically, the four strata are well characterized and interdependent. I think, on the basis of my field experience, that they are a constant feature of the continuous sponge populations growing on calcareous substrata in shallow waters. Their establishment in the community occurs by steps. The pattern of colonization probably starts with the Clionids, proceeds with the basal sponges and ends with the epibiotic and hypobiotic species. Certainly the cooperation phenomena between the sponges belonging to the different strata are very important. On the other hand, it may be supposed that competition for space plays a significant role only among species of the same stratum, with the exception, of course, of the clionids acting as substratum destroyers.

TEMPORAL STABILITY

Because populations are dynamic phenomena, competition and cooperation phenomena may be more exactly analysed by studying temporal variations in species microdistributions (Labate, 1966). Unfortunately, quantitative data on this important matter are still very scarce. Recently we have compared at intervals the outlines of sponge specimens found within sample quadrats. Different types of sponge populations occurring in shallow waters of 0–1 m depth have been surveyed with this method over a period of more than one year. Preliminary data indicate marked population stability. This is remarkable, considering the wide range of seasonal variations of temperature, hydrodynamics and other factors occurring at 1 m or less of depth. However, in certain cases as, in the "Grotta dei Misteri" in the Bay of Naples and, to a lesser degree, in the "Grotta della Regina" near Monopoli, peculiar ecological shifts give rise to sharp seasonal modifications of the sponge population. In contrast, the caves near Leuca, in a more exposed situation, show a high degree of stability.

The general pattern of the population as determined by the dominant species of the basal stratum is, on the whole, stable both from a qualitative and a quantitative point of view (Fig. 2). Changes are often

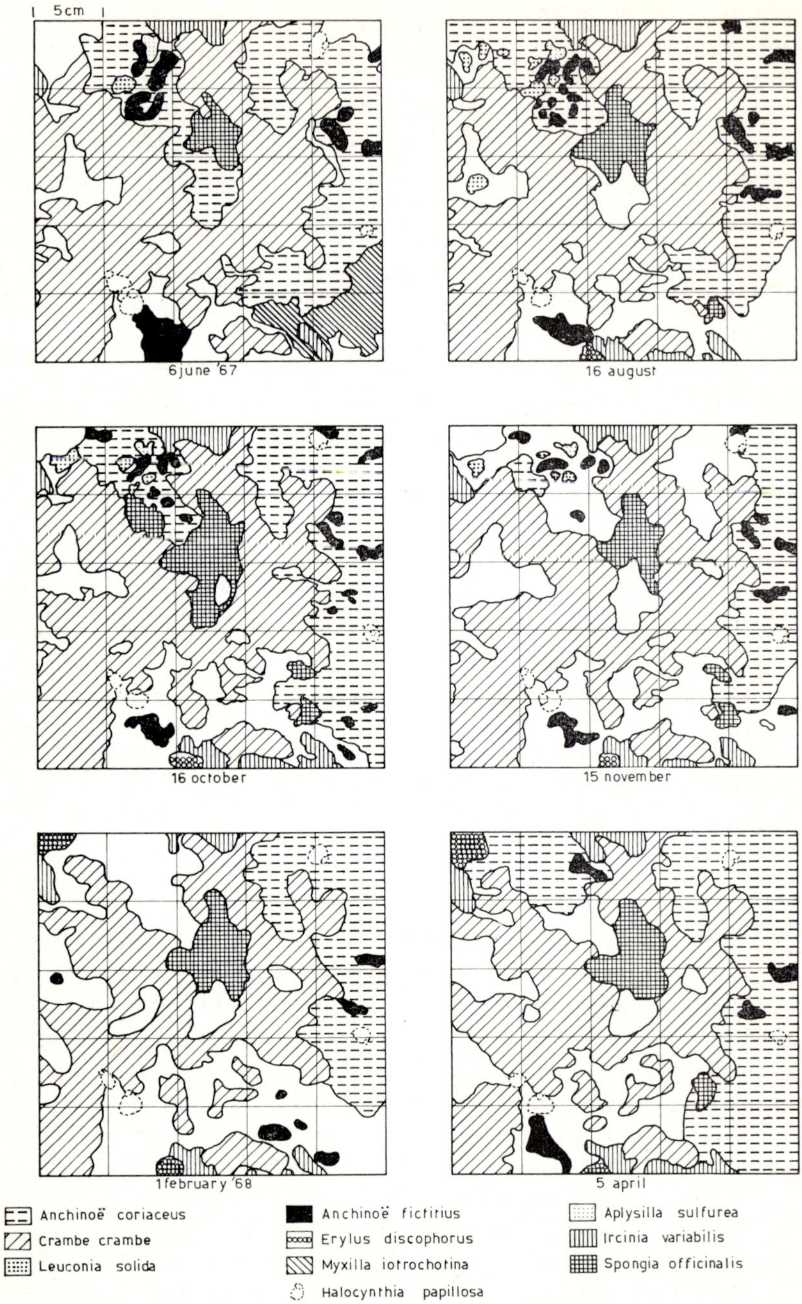

FIG. 2. Variations with time in part of a sponge population in a cave near Leuca.

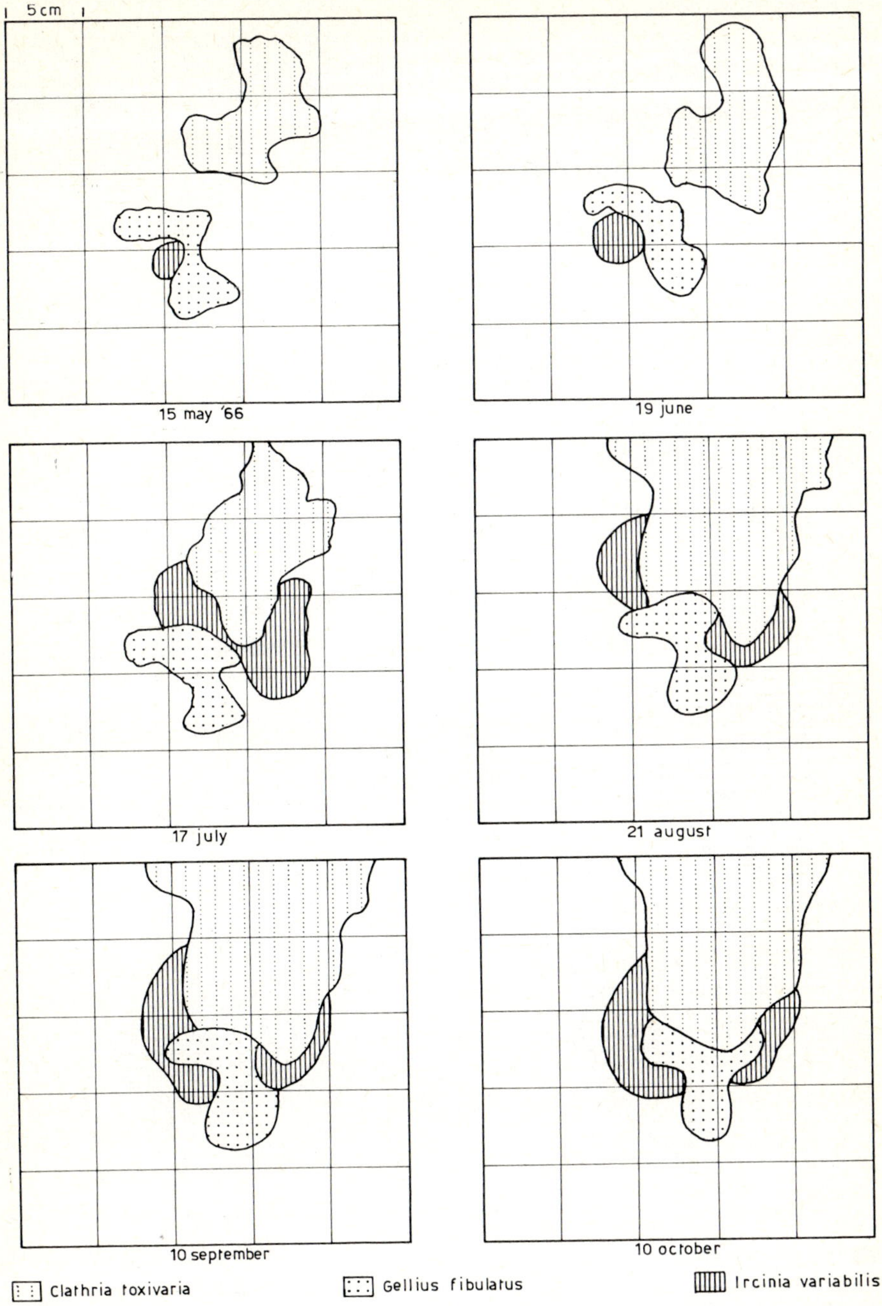

FIG. 3. Variations with time in part of a sponge population in the cave "Grotta della Regina" (May–October, 1966).

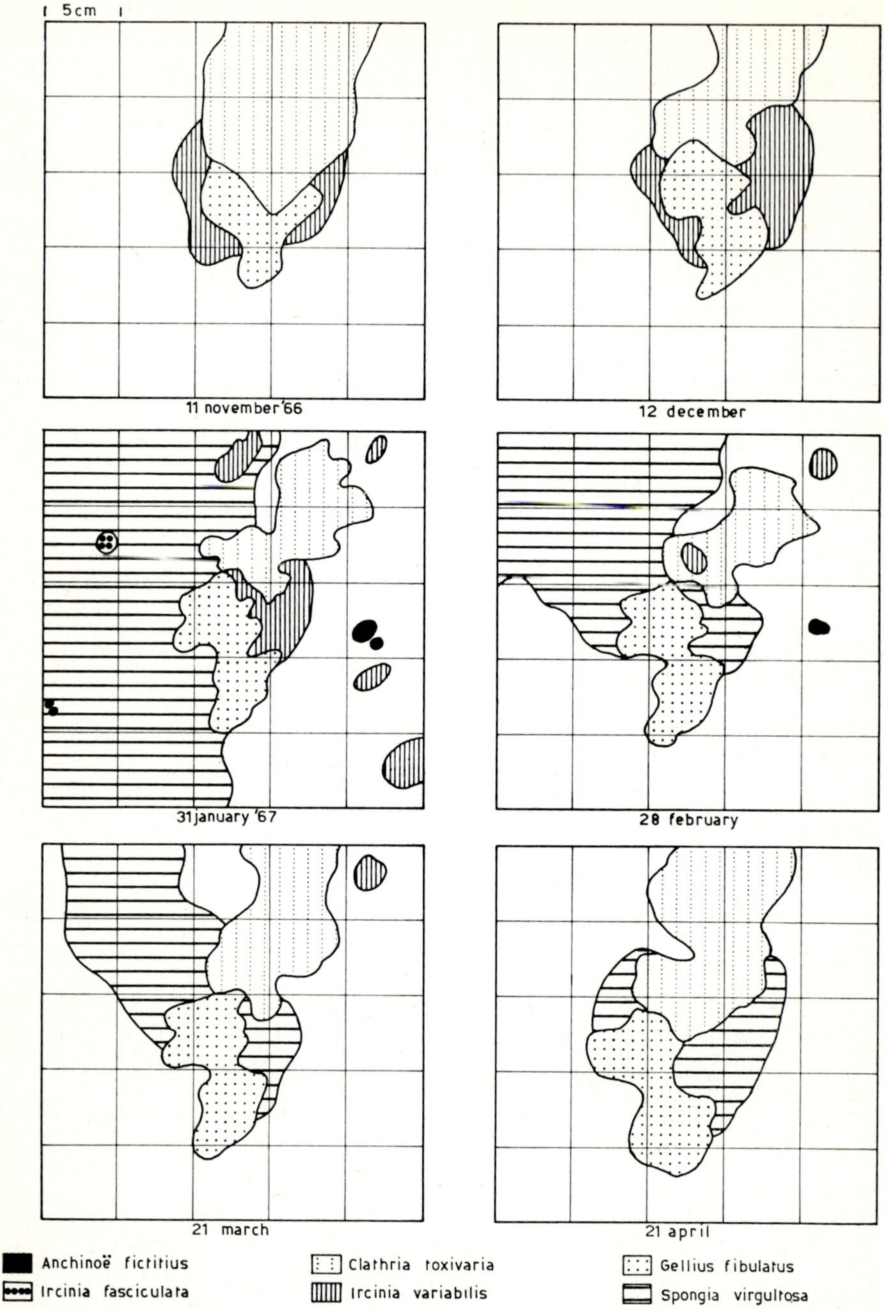

FIG. 4. Variations with time in part of a sponge population in the cave "Grotta della Regina" (November, 1966–April, 1967).

limited to slight variations in the specimens' outlines. On the other hand, some species, e.g. *Clathrina coriacea* Montagu, 1812, *Aplysilla sulfurea* Schulze, 1878 show a large temporal variability in the abundance, extension and distribution of specimens. Significant changes in the population may be also caused by the clionids' development. The considerable and continuous modifications which occur in some places suggest that important competition phenomena may take place until a new equilibrium has been reached (Fig. 2). It is probable, however, that in most cases the changes in size of single specimens and the appearance and disappearance of species are not an effect of competition but of other aspects of the environmental conditioning. In one case (Figs 3, 4) three species (*Clathria toxivaria* Sarà, 1959, *Gellius fibulatus* Schmidt, 1862 and *Ircinia variabilis* Schmidt, 1862) join together to form a single rounded mass remaining stable in shape and size for many months, despite the rapid substitution, occurring during the winter, of another Keratosa, *Spongia virgultosa* Schmidt, 1868, for *Ircinia variabilis* (Schmidt, 1862). It is interesting to note that these two horny sponges appear to act in such an equivalent way in the joined mass that the substitution does not cause any significant modification. A binding action of Keratosa among sponges on other substrata is well known, especially for *Fasciospongia cavernosa* Schmidt, 1862 and *Ircinia* species Nardo, 1833 (Rützler, 1965). The factors controlling growth which leads to the formation of such joined masses require deeper analysis. However, the formation of a single and large mass in the place of three small specimens is evidently an adaptation to ecological conditions which are probably severe for sponge survival, as indicated by the relative scarcity and instability of the sponge population of the "Grotta della Regina", and may be interpreted as a peculiar aspect of cooperation phenomena among sponges. Further examples may be produced in this regard (Sarà, 1968).

References

Annandale, N. (1915). Some sponges parasitic on Clionidae, with further notes on that family. *Rec. Indian Mus.* **11**: 457–478.

Hartman, W. D. (1957). Ecological niche differentiation in the boring sponges (Clionidae). *Evolution* **11**: 291–297.

Labate, M. (1966). Variazioni nel tempo del popolamento di Poriferi di una grotta superficiale. *Boll. Zool.* **33** (1): 181.

de Laubenfels, M. W. (1947). Ecology of the sponges of a brackish water environment, at Beaufort. N.C. *Ecol. Monogr.* **17**: 31–46.

Margalef, R. (1957). Information theory in ecology. *Gen. Syst.* **3**: 36–71.

Riedl, R. (1966). *Biologie der Meereshohlen.* pp. 1–636. Hamburg und Berlin: Paul Parey.

Rützler, K. (1965). Systematik und ökologie der Poriferen aus litoral–Schattengebieten der Nordadria. *Z. Morph. Ökol. Tiere* **55**: 1–82.

Sarà, M. (1966). Studio quantitativo della distribuzione dei Poriferi in ambienti superficiali della Riviera Ligure di Levante. *Arch. Oceanogr. Limnol.* **14**: 365–386.

Sarà, M. (1968). Stratification des peuplements d'Eponges à recouvrement total dans certaines grottes du niveau superficiel. *Rapp. P.v. Réun. Commn int. Explor. Scient. Mer Méditerr.* No. 19: 83–85.

Wells, H. W., Wells, M. J. and Gray, J. E. (1964). Ecology of sponges in Hatteras Harbor, North Carolina. *Ecology*, **45** (4): 752–767.

Symp. zool. Soc. Lond. (1970) No. 25, 285–297.

SPONGES OF THE ARCTIC AND ANTARCTIC; A FAUNISTIC REVIEW

V. M. KOLTUN

Zoological Institute of the Academy of Sciences of the U.S.S.R. Leningrad, U.S.S.R.

SYNOPSIS

The sponge faunas of the Arctic and Antarctic are surprisingly well known. In the Arctic, both specific and generic endemism are very low. The history of the Arctic region has allowed repeated extinctions of the local fauna followed by immigrations of sponges from neighbouring boreal regions. The continuity of the Arctic continental shelf with neighbouring shelves assists these immigrations. The isolation of the Antarctic continental shelf has restricted immigration in this region, where specific endemicity is high but generic endemism relatively low.

Eurybathy is highly developed in both polar regions. It is encouraged by widespread glacial deposition of suitable substrates, and by upwelling of deep ocean water of constant low temperature, but is discouraged by fine-particle continental discharge. Local variations in these phenomena produce a more complex pattern of abundance and diversity in the Arctic than in the Antarctic, where virtual absence of continental discharge other than glacial results in a uniformly greater taxonomic diversity and biomass than in the Arctic and in more frequent circumpolar distributions.

The development of eurybathy on polar continental slopes is probably the reason for the existence of bipolar species, some of which are shown to exist at very great depths in more temperate regions.

INTRODUCTION

The sponge faunas of the Arctic and Antarctic are surprisingly well known. Certainly, the faunas of these remote and inaccessible areas have been investigated in no less detail than those of any other large area of the World Ocean. The data for preparing an inventory of sponges are more adequate for the Arctic, where new species are now detected only rarely in the central bathyal zone. In the Antarctic more unknown sponges will undoubtedly be discovered by detailed study of shallow coastal parts of the shelf and of the abyssal depths. Nonetheless, it seems unlikely that the general picture of the composition of the fauna which we now possess will undergo radical changes in the future. More interesting facts, however, will emerge from thorough revision of the faunas, and researches undertaken in this direction (Burton, 1929, 1930, 1932, 1963; Koltun, 1959, 1964a, 1966, 1967) should be continued with urgency.

Arctic and Antarctic waters have in common many attributes of importance to living organisms which is why comparison of the faunas of these cold water regions commands such interest. In particular, such comparisons allow us to detect those ancient and continuing connections which determine the similarities and differences between the faunas.

THE ARCTIC SPONGE FAUNA

The fauna of the Arctic Ocean contains about 200 species of sponges belonging to 85 genera. Among these, more than 12 species are Calcispongiae, while the Hyalospongiae contribute 7 species and the Demospongiae about 180 species. Regional specificity is relatively insignificant; endemics compose no more than one quarter of all the species and endemic genera are probably totally absent. Given a thorough revision of the Hyalospongiae, the two genera *Schaudinnia* Schulze, 1900 and *Trichasterina* Schulze,1900 which are apparently endemic may prove not to be so.

The Arctic fauna is characterized by a complete absence of Amphidiscophora and is very poor in horny sponges. Sponges are most abundant in regions affected by warm waters of the Gulf Stream and in the bathyal zone of the Arctic Ocean. Good examples are provided by the populations of the Kola Peninsula and West Spitzbergen which contain large spherical *Geodia* attaining diameters of 50 cm. In the Arctic the most abundant and diversified genera are *Tetilla* O. Schmidt, 1868 (4 species), *Polymastia* Bowerbank,1866 (7 species), *Mycale* Gray, 1867 (7 species), *Asbestopluma* Norman,1882 (6 species), *Myxilla* O. Schmidt,1862 (5 species), *Lissodendoryx* Topsent,1892 (7 species), *Hymedesmia* Bowerbank,1864 (18 species), *Ectyodoryx* Lundbeck,1909 (6 species) and *Haliclona* Grant,1841 (19 species). Abundant also are species of *Thenea* Gray,1867 (2 species), *Geodia Lamarck*,1815 (4 species), *Suberites* Nardo,1833 (2 species), and *Phakellia* Bowerbank,1864 (3 species).

This fauna contains many species which also inhabit adjacent temperate areas. There are about 60 species common to the North Pacific and more than 80 species common to the Atlantic. Amongst these one can detect a large group of widespread arctic-boreal species which includes *Mycale lingua* Bowerbank,1866, *Myxilla incrustans* Johnston,1842, *Iophon piceus* Vosmaer,1881, *Phakellia cribrosa* Miklucho-Maclay,1870, *Halichondria panicea* Pallas,1766, *Hymeniacidon assimilis* Levinsen,1886, *Haliclona gracilis* Miklucho-Maclay, 1870, *Thenea muricata* Bowerbank,1858, *Tetilla cranium* Müller,1776, *T. polyura* Schmidt,1870, *Polymastia mammillaris* Müller,1806, *Tentorium semisuberites* O. Schmidt,1870 and *Suberites domuncula* Olivi, 1792.

Highly characteristic of the Arctic are the so-called boreal north-Atlantic species, which are closely associated in their distribution to the north with the warm waters of the Gulf Stream. Some of these species do not extend further than the Norwegian Sea but many others reach the southwestern region of the Barents Sea and even the western and northern coasts of Spitzbergen (Fig. 1). This group of boreal immigrants includes *Tethya aurantium* Pallas,1766, *Polymastia bursa* Müller,1806, *P. robusta* Bowerbank,1861, *P. uberrima* O. Schmidt, 1870, *Geodia macandrewii* Bowerbank,1858, *G. barretti* Bowerbank, 1858, *Vosmaeria crustacea* von Fristedt,1885, *Anchinoë dendyi* Topsent, 1892, *Myxilla fimbriata* Bowerbank,1864 and *Clathria dichotoma* Esper, 1794.

A unique faunal element characterizes the White Sea, where distinct boreal species also occur (*Microciona armata* Bowerbank, 1866, *Tethya*

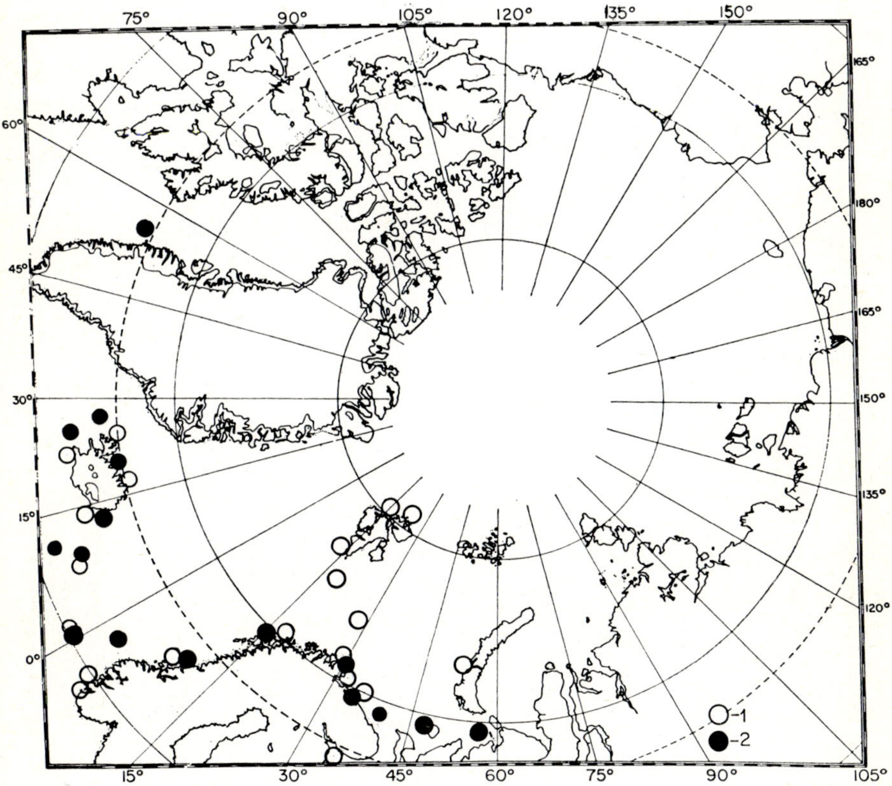

FIG. 1. Penetration of boreal north-atlantic species into the Arctic: 1. *Tethya aurantium*, 2. *Myxilla fimbriata*.

aurantium Pallas,1766, *Vosmaeria crustacea, Polymastia robusta, Hymeraphia stellifera* Bowerbank,1866 and others). These are relicts from former periods of general warming of the Arctic.

Typical Arctic species are for the most part concentrated in the bathyal zone of the Arctic Ocean. They are represented by a small number of ancient autochthonous forms but the majority of them are derivatives of Pacific and Atlantic species which migrated thence in pre-Quaternary times and even later. Amongst the latter a special taxonomic interest attaches to certain species of very recent formation, which have remained distinct from earlier Atlantic immigrants, and themselves probably became distinct during a later boreal transgression and subsequent cooling. This group of species contains *Geodia mesotriaena* Hentschel,1929, *Polymastia thielei* Koltun,1964 and

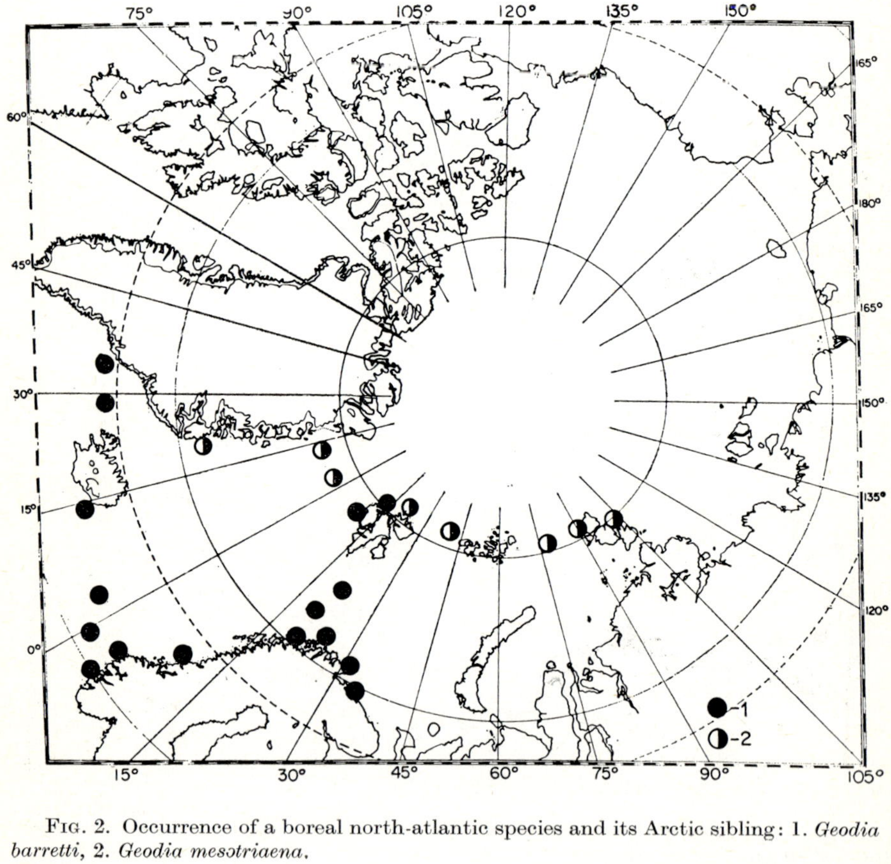

FIG. 2. Occurrence of a boreal north-atlantic species and its Arctic sibling: 1. *Geodia barretti*, 2. *Geodia mesotriaena*.

Rhaphidotheca arctica Hentschel, 1929. Boreal siblings to these species are *G. barretti*, *P. uberrima* and *Rh. marshallhalli* Kent, 1870.

The true abyssal Arctic fauna is very poor and consists only of the five species *Caulophacus arcticus* Hansen,1885, *Thenea abyssorum* Koltun,1964, *Cladorhiza gelida* Lundbeck,1905, *Cl. arctica* Burton, 1946 and *Forcepia topsenti* Lundbeck,1905 (Koltun, 1964b).

THE ANTARCTIC SPONGE FAUNA

About 300 species of sponge are recorded from the Antarctic, including the islands of the Scotia arc and Bouvet, Marion, Crozet, Prince Edward, Kerguelen, Heard and Macquarie Islands. Among the 300 species, more than 8 are Calcispongiae, the Hyalospongiae contribute 26 species and the Demospongiae contribute about 260 species. The majority of these species are cold-loving forms well adapted to extremely low temperatures.

The Antarctic sponge fauna is characterized by a very high level of specific endemism. More than half of the known species are confined within the limits of the Antarctic. On the other hand, the fauna's generic endemism is almost negligible. Of the recorded 116 genera only 9 can be regarded as endemic. These genera include *Scolymastra* Topsent,1916, *Acanthorhabdus* Burton,1929, *Hoplakithara* Kirkpatrick, 1907, *Cercidochela* Kirkpatrick, 1907, *Meliiderma* Ridley and Dendy, 1887, *Crellina* Hentschel,1914, *Bipocillopsis* Koltun,1964, *Acanthoxa* Hentschel,1914 and *Cladothenea* Koltun,1964.

The Antarctic fauna is distinguished by the rarity, or complete absence, of members of such families as the Geodiidae, Theneidae, Tethyidae and, like the Arctic fauna, is extremely poor in horny sponges. On the other hand the abundance of glass sponges belonging to the family Rossellidae is notable. Some of these, such as the barrel-like *Scolymastra joubini* Topsent, 1916, can reach gigantic size, sometimes being much more than 100 cm high.

Most of the species of sponges belong to genera well represented in other parts of the World Ocean, such as *Mycale* (8 species), *Asbestopluma* (5 species), *Myxilla* (9 species), *Tedania* Gray, 1867 (8 species), *Iophon* Gray,1867 (6 species), *Hymedesmia* (6 species), *Isodictya* Bowerbank,1858 (9 species), *Anchinoë* Gray,1867 (5 species), *Haliclona* (21 species), *Adocia* Gray,1867 (6 species) and *Hymeniacidon* Bowerbank,1858 (6 species). Also typical of the fauna are species belonging to less widely dispersed genera with representatives only in one or two neighbouring ocean basins or in quite separate regions of the northern hemisphere. For example, the genus *Monosyringa*

Brøndsted, 1924, as well as occurring in the Antarctic, is found in New Zealand waters, the genus *Cinachyra* Sollas, 1886 occurs also in the West Indies, *Axociella* Hallmann, 1919 around the coasts of South America, Australia and the West Indies and *Rossella* Carter,1872 which is mainly an Antarctic genus, is found in the northern hemisphere only off the Azores.

Many Antarctic sponges have a circumpolar distribution (Fig. 3). Among the most common are *Rossella antarctica* Carter,1872, *R. racovitzae* Topsent,1901, *Scolymastra joubini*, *Cinachyra antarctica* Carter, 1872, *Tetilla leptoderma* Sollas, 1886, *Mycale tridens* Hentschel, 1914, *M. acerata* Kirkpatrick,1907, *Asbestopluma belgicae* Topsent, 1901, *Iophon radiatus* Topsent,1901, *I. spatulatus* Kirkpatrick, 1907, *Acanthorhabdus fragilis* Burton,1929, *Cercidochela lankesteri* Kirkpatrick, 1907, *Tedania massa* Ridley and Dendy, 1886, *T. oxeata* Topsent, 1916, *Axociella nidificata* Kirkpatrick,1907 and *Isodictya erinacea* Topsent, 1916 (Koltun, 1964a).

There are some other species with more restricted distribution. Thus the region of McMurdo Sound (Victoria Land) is to date the only place in the Antarctic where representatives of the following genera have been found—*Stelletta* O. Schmidt, 1868, *Geodinella* von Lendenfeld, 1903, *Jaspis* Gray, 1867, *Penares* Gray, 1867, *Hemiasterella* Carter, 1879, *Rhabderemia* Topsent, 1890, *Chondropsis* Carter, 1886 and *Dysidea* Johnston, 1842 (Burton, 1929). *Hoplakithara dendyi* Kirkpatrick, 1907, *Plocamia gaussiana* Hentschel, 1914, *Clathria pauper* Brøndsted, 1926, *Sphaerotylus schoenus* Sollas, 1882, *Dolichacantha macrodon* Hentschel, 1914, *Desmacella vestibularis* Wilson, 1904, and *Monosyringa longispinum* von Lendenfeld, 1907 have a wider distribution range but are still restricted to the eastern sector of the Antarctic.

Species characteristic of the western sector are far fewer. To this group belong *Mycale macrochela* Burton, 1932, *M. tylotornota* Koltun, 1964, *Isodictya toxiphila* Burton, 1932, *Myxilla pistillaris* Topsent, 1917, and *Hymeniacidon torquata* Topsent, 1917. The differences between the faunas of the eastern and western sectors cannot be accounted for solely by our incomplete knowledge of the fauna.

Analysis of the species composition of the Antarctic fauna indicates its close relationship with that of the Falkland Islands and South America. More than 50 species are considered as common to both regions. Connections with the fauna of New Zealand and Australia appear to be much weaker. Australasia has only 30 species in common with the Antarctic.

The fauna of the abyssal depths of the Southern Ocean contains

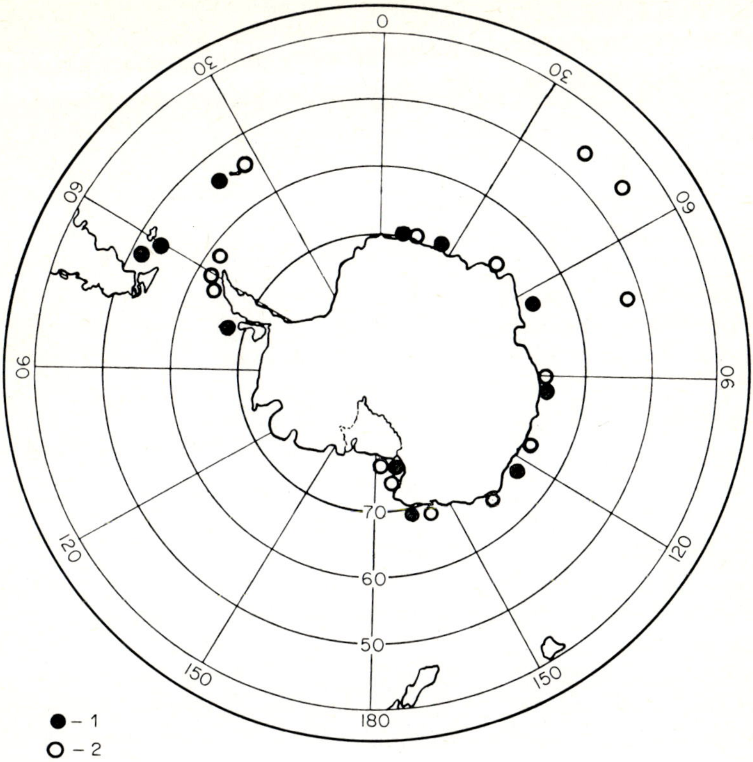

FIG. 3. Circumpolar distribution of sponges in the Antarctic: 1. *Anchinoë glaberrima* (Topsent, 1917), 2. *Rossella antarctica*.

more than 25 deep water species belonging mainly within the genera *Caulophacus* Schulze, 1887, *Malacosaccus* Schulze, 1886, *Holascus* Schulze, 1886, *Aulocalyx* Schulze, 1886, *Hyalonema* Gray, 1832, *Asbestopluma* and *Cladorhiza* O. Schmidt, 1875.

COMPARATIVE ANALYSIS OF THE TWO FAUNAS

The Antarctic sponge fauna, judging by the number of species and genera it contains, is about one and a half times as rich, and much more varied, than is the Arctic sponge fauna. Moreover, sponges, together with Bryozoa and Echinodermata, are the most characteristic components of Antarctic benthos.

Comparatively uniform conditions on the Antarctic shelf favour the mainly circumpolar distribution. In fact, around the whole continent there is a continuous belt of sponge populations extending from shallow waters to the upper parts of the continental slope. It is difficult to find another region of the World Ocean where there is such a continuous abundance of sponges covering such a large area. Undoubtedly, this abundance, and the wide distribution of individual species within the Antarctic, are due both to favourable trophic and physical factors and to the presence of stony and mixed substrates favourable to the settlement and growth of these animals.

In the Arctic, sponges do not show such prevalence in the bottom fauna and their qualitative and quantitative distributions are more uneven. There appear to be two main causes for this pattern of distributions. On the one hand, a warm Atlantic current, penetrating far into the Arctic and transporting thither boreal forms, plays an important role. On the other hand, the presence of intensive continental discharge has a profound effect in restricting the distribution of oceanic forms over a considerable part of the shelf. This is seen clearly in the progressive impoverishment of the fauna from the west to the east. For example, the major contributing group to the fauna, the Demospongiae, number 125 species in the Barents Sea, 61 species in the Kara Sea, 53 species in the Laptev Sea, 14 in the East Siberian Sea and only 12 in the Chukotsk Sea.

Continental discharge is largely absent in the Antarctic and oceanic waters come close to the continental shores, covering the shelf. This results in a general ascent of the fauna from the bathyal zone and lower parts of the sublittoral to shallower waters. For example, members of the Rossellidae occur in other parts of the World Ocean as a rule at depths of more than 100 m. In the Antarctic these organisms are highly abundant around the coasts even at depths of only 10 m. Reflecting the inhibitory effect of continental discharge in the Arctic Ocean, representatives of the Rossellidae and other deep-ocean forms are found mainly in the bathyal region of the central basin and in the Greenland and Norwegian Seas. The occurrence of some bathyal species in shallow parts of the Arctic Ocean can be accounted for by localized ascent of deep oceanic waters to outer regions of the shelf.

Considering the sponge faunas of the Antarctic and Arctic together we find that of 145 genera which contribute to the two regions, only 56 genera have representatives in both. The remaining 89 genera have representatives in only one or other of the regions. These genera define the essential distinctness of the two faunas. Sixty of these genera are represented in the Antarctic, while the Arctic sponge fauna contains only 29 genera which are not represented in the Southern Ocean. Amongst these 29 are *Stryphnus* Sollas, 1886, *Pachymatisma* Johnston,

1842, *Quasillina* Norman, 1868, *Vosmaeria* von Fristedt, 1885, *Rhaphidotheca* Kent, 1870, *Phakellia* Bowerbank, 1864, *Coelosphaera* W. Thomson, 1873, *Cornulum* Carter, 1876, *Hamacantha* Gray, 1867, *Melonanchora* Carter, 1874, *Forcepia* Carter, 1874, *Sycolepis* Haeckel, 1870, *Amphoriscus* Haeckel, 1870, *Sycettusa* Haeckel, 1872, *Asconema* Kent, 1870, *Mellonympha* Schulze, 1897 and *Schaudinnia* Schulze, 1900.

Almost complete absence of endemic and autochthonous genera, and only weakly expressed specific endemism (about 25%), indicate a comparative youthfulness for the Arctic fauna and its close connections with adjacent boreal faunas. At the same time, the peculiar history of the Arctic Ocean, which has involved repeated lowerings of salinity and extinctions of the greater part of the ancient local fauna, together with the half-closed nature of the basin, account for the general poverty of the Arctic fauna. At the present time the isolation of the Arctic Ocean from neighbouring abyssal regions continues to prevent the enrichment of the fauna by certain forms. This is particularly true for the glass sponges, which are not abundant in the Arctic, numbering only seven species. This example is the more striking because it is known, from the distribution of Hyalospongiae in other regions (particularly within the Antarctic where there are 26 species), that low temperature of the bottom water does not hamper the dispersal of these sponges. The extreme poverty of the Arctic abyssal fauna (5 species in all) is due to these ancient and continuing phenomena.

Amongst those Antarctic genera which are absent from the Arctic, the most characteristic are *Plakina* Schulze, 1880, *Monosyringa*, *Cinachyra*, *Desmacidon* Bowerbank, 1862, *Guitarra* Carter, 1874, *Hoplakithara*, *Cercidochela*, *Acanthorhabdus*, *Axociella*, *Ophlitaspongia* Bowerbank,1864, *Rhaphidophlus* Ehlers,1870, *Myxodoryx* Burton, 1929, *Eurypon* Gray, 1867, *Homaxinella* Topsent, 1917, *Raspailia* Nardo, 1833, *Leucettusa* Haeckel, 1872, *Achramorpha* Jenkin, 1908, *Rossella*, *Scolymastra*, *Bathydorus* Schulze, 1888, *Hyalonema* Gray, 1832, *Farrea* Bowerbank, 1862, *Acoelocalyx* Topsent, 1910, *Docosaccus* Topsent, 1910 and *Malacosaccus*.

The distribution of some of these genera (e.g. *Cinachyra*, *Plakina*, *Guitarra*, *Axociella*, *Ophlitaspongia*) which occur in both Northern and Southern Hemispheres is evidence of the wide and ancient connections of the Antarctic sponge fauna. On the other hand, the considerable proportion of endemic species (about 50%) and the presence of some endemic genera point to a continuing isolated development of the fauna. The present isolation of the Antarctic shelf from the neighbouring continents, which has produced such marked endemism, restricts the possibilities for enrichment by immigrants. In contrast, many boreal species penetrate the Arctic because of the continuity of the Arctic shelf with those of the Atlantic. These contemporary immigrant species make up a quarter of the entire sponge fauna of the Arctic.

Besides these differences there are certain similarities between the Arctic and Antarctic sponge faunas which are due to identical environmental factors. Especially noticeable is the pronounced eurybathy of many representatives of both regions. In the Arctic and Antarctic, as nowhere else in the World Ocean, can be seen the ability of sponges to thrive in a large range of depths.

Uniformity of temperature, irrespective of depth, is a major cause of this phenomenon. A second important cause is the role played by icebergs and glaciers in these regions. Because of their activity, boulders, pebbles, and other hard materials are distributed at all depths on the continental shelves. This undoubtedly favours the settlement and growth of sponges.

In addition, the absence of continental discharge from the Antarctic favours eurybathy by allowing the ascent of deep-ocean forms up the shelf. Similar, but more localized, absence of continental discharge favours more restricted ascent in the Arctic.

The great depth of most of the Antarctic shelf, the deep depressions within its topography and deep currents directed from the coast to the open ocean apparently have a selective influence on the development of eurybathy. Some species have acquired wide but precise vertical distributions, e.g. *Myxilla asigmata* Topsent, 1901, 310–2628 m, *Chonelasma lamella* Schulze, 1886, 430–3397 m. We may compare these distributions with the vertical distributions of *Tentorium semisuberites* O. Schmidt, 1870, 36–2800 m, *Polymastia sol* O. Schmidt, 1870, 55–3200 m and *Poecillastra compressa* Bowerbank, 1866, 36–3500 m in the Arctic.

Because of the pronounced eurybathy of many species, we do not observe distinct vertical zonation of sponge species on the greater part of the Antarctic shelf and upper parts of the slope, i.e. between 100 and 900 m. Such zonal transgressions are clearly visible in the Arctic also although to a smaller extent. They are most marked in the central region of the Arctic Ocean and off the coasts of Greenland, where conditions are most similar to those existing in the Antarctic (Koltun, 1964b).

Because of the extreme severity of polar ocean conditions, the horny sponge fauna is very poor. *Spongia carteri* Burton, 1930 occurs in the north and *Aplysilla minima* Hentschel, 1914 around the coasts of the Antarctic. These are both dwarf forms, being seldom more than 3 cm in height. They are very similar and probably should be assigned to a single genus. Poverty of horny sponges in these polar regions is not surprising, as it is well known that the rate of deposition of their organic skeleton is directly dependent upon water temperature. The importance of horny sponges in the bottom fauna, and also the contents of spongin in other forms of Cornacuspongida, increases as one approaches the equator (Burton, 1932).

Typical Tetraxonida are only poorly represented in Antarctic waters. The genera *Tetilla* and *Cinachyra*, and, to a lesser extent *Monosyringa* and *Plakina*, are exceptions to this generalization. In the Arctic this group of sponges occupies a similar position. Many of their species are referable to a number of contemporary boreal immigrants and only the genera *Tetilla*, *Thenea*, and, to some extent, *Geodia*, have moderately wide distribution in Arctic waters proper. This pattern of distribution permits us to distinguish the Tetraxonida as a whole as warmth-loving organisms of oceanic type.

When comparing the Arctic and Antarctic faunas it is of especial interest to examine the bipolar distribution of sponges.

THE BIPOLAR DISTRIBUTION OF SOME SPECIES

The modern sponge fauna of the world contains many species which, while occurring in both northern and southern hemispheres, are absent from equatorial regions. Among these forms the most interesting are those showing extreme bipolarity, which we may define as presence within both Arctic and Antarctic waters, with only rare excursions beyond the limits of these areas (Fig. 4).

In the polar sponge faunas there are at least three such remarkable species: *Sphaerotylus schoenus*, *Sph. borealis* Svarczevsky, 1906 (= *Sph. antarcticus* Kirkpatrick, 1907) and *Asbestopluma infundibulum* Levinsen, 1886 (= *A. calyx* Hentschel,1914). Arctic and Antarctic populations of these species are morphologically almost indistinguishable. On the basis of prolonged spatial isolation and of differing ecological positions we might regard the northern and southern populations as subspecies.

Sph. schoenus occurs in the southwestern part of the Barents Sea, in the Norwegian and North Seas and also off the Atlantic coast of Canada at depths between 50 and 440 m. In the Southern Ocean the species has been recorded off the coasts of Wilhelm Land and Victoria Land at depths between 18 and 385 m. Another species of this genus, *Sph. borealis*, occurs in the White Sea and Barents Sea, in the bathyal zone of the Arctic basin, and in the Norwegian Sea, at depths between 73 and 500 m. In the Antarctic it has a circumpolar distribution, inhabiting depths between 18 and 450 m. The third species, *A.*

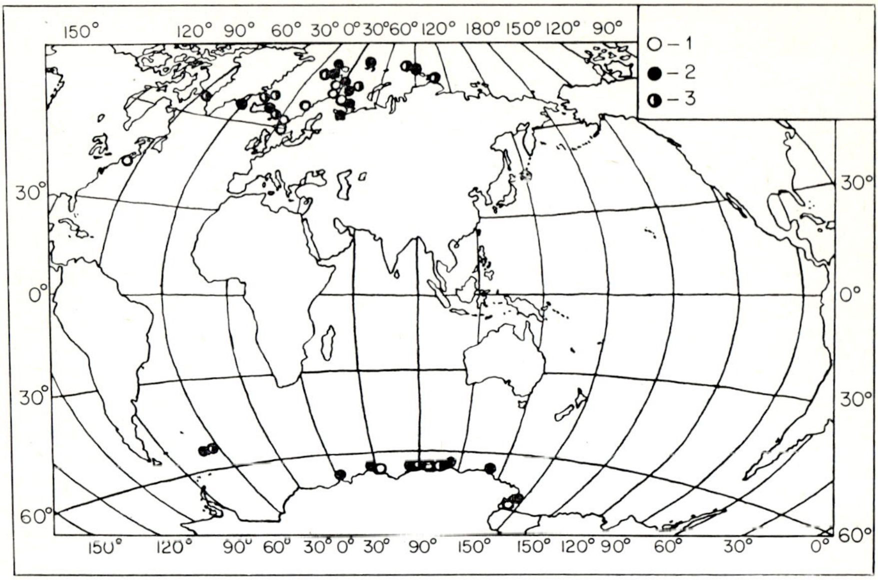

FIG. 4. Bipolar distribution of some species of sponges: 1. *Sphaerotylus schoenus*, 2. *Sphaerotylus borealis*, 3. *Asbestopluma infundibulum*.

infundibulum, is widespread in the Arctic Ocean, occurring at depths of from 91 to 710 m. In the south this species occurs off the coasts of Antarctica and South Georgia at depths between 46 and 450 m.

Thus, both the southern and the northern populations of these species inhabit relatively shallow regions.

The formation of such extreme bipolarity would indeed be a mystery were it not for the fact that among sponges generally there are many species which show a clear tendency to move into deeper water, or at least to develop considerable eurybathy. Thus, the bipolar *Suberites montiniger* Carter, 1880 occurs in the northern hemisphere and off the coasts of Antarctica mainly down to depths of 300 to 400 m but in the northern part of the Pacific it sometimes occurs at depths of more than 1000 m. The possibly bipolar *Poecillastra compressa* Bowerbank, 1866 shows greater development of eurybathy, descending in the western Pacific to 2400 m and in the Norwegian Sea to 3500 m. Finally, the widespread *Stylocordyla borealis* Loven, 1868 occupies a vertical range between 13 and 2880 m, its deepest record being off Crozet Island.

A list of bipolar species could be lengthened considerably but these examples show clearly that bipolar distribution could arise by the development of eurybathy and migration on the ocean floor.

Additional data on the sponge fauna of the bathyal zone near the South Kuril Islands appear to favour such a method of acquiring bipolar distribution in *Sph. schoenus, Sph. borealis* and *A. infundibulum*. Recently I have found in this region two new species of *Sphaerotylus*, similar to *Sph. borealis* and *Sph. schoenus*, at a depth of 1500 m and also, at depths of 2665 and 5450 m, the north Pacific subspecies of *Asbestopluma infundibulum*.

The consideration of such extreme bipolar distributions is yet more evidence for arguing that sponges, despite their great morphological plasticity—or possibly because of it—are evolutionarily extremely conservative organisms. Indeed, a considerable number of modern sponges are justifiably regarded as "living fossils", having survived almost unchanged since very ancient times.

References

Burton, M. (1929). Porifera. Part II. Antarctic Sponges. *Br. Antarct. Terra Nova Exped.* 1910. (Zool.) **6**: 393–458.

Burton, M. (1930). Norwegian sponges from the Norman Collection. *Proc. zool. Soc. Lond.* **1930**: 487–546.

Burton, M. (1932). Sponges. *Disc. Rep.* **6**: 237–398.

Burton, M. (1963). *A revision of the classification of the calcareous sponges.* London: British Museum (Natural History).

Koltun, V. M. (1959). Horny silica sponges of the Northern and Far Eastern seas of the U.S.S.R.

Koltun, V. M. (1964a). Antarctic sponges. I. Tetraxonid and horny silica sponges. Results of a biological investigation by the Soviet Antarctic expedition (1955–1958) *Issled. fauny morei.* **2** (10): 6–116.

Koltun, V. M. (1964b). Contribution to the study of the bottom fauna of the Greenland sea and the central portion of the Arctic basin. *Trudӯ arkt. nauchno-issled Inst* No. 259: 13–78.

Koltun, V. M. (1966). Tetraxonid sponges of the Northern and Far Eastern seas of the U.S.S.R. *Opred. Faune SSSR* **90**: 1–111.

Koltun, V. M. (1967). Vitreous sponges of the Northern and Far Eastern seas of the U.S.S.R. *Opred. Faune SSSR* **94**: 1–124.

BIOCHEMISTRY AND THE CELL SURFACE

Symp. zool. Soc. Lond. (1970) No. 25, 299–324.

POLYSACCHARIDES FROM SPONGES AND THEIR POSSIBLE SIGNIFICANCE IN CELLULAR AGGREGATION

ALASTAIR P. MACLENNAN

Microbiological Research Establishment, Porton, Nr. Salisbury, England

SYNOPSIS

Antigenic substances serologically related to the sponge cell surface were extracted with cold 0·5 N trichloroacetic acid from seventeen species of sponge broadly representative of the phylum. The yields were 0·2–0·8 % of the organic dry weight. The extracts were composed principally of heteropolysaccharide and peptide, possibly partly in combination as glycopeptide. The major sugar constituents of the carbohydrate moiety were galactose, fucose, hexosamine, and hexuronic acid. Other sugars uniformly present were glucose, mannose, rhamnose, xylose, arabinose and two unidentified sugars. The preparations were serologically species specific. They were examined for their possible role in specific cellular reaggregation, in the light of the various mechanisms postulated in explanation of this phenomenon. No positive role for specific polysaccharides was found. These results are discussed.

INTRODUCTION

The observations to be reported here reflect an interest in the mechanism of sponge cell reaggregation and a belief that a knowledge of the chemistry of the sponge cell surface must form part of any final explanation of this phenomenon. Additional to this primary motivation, but related to it, is an interest in the occurrence of polysaccharides in sponges, since it will be shown that it is carbohydrate that dominates some aspects of the behaviour of the sponge cell surface.

The chemistry of mucins (see Gottschalk, 1966) attests to the likely presence of carbohydrates among the numerous slimes and mucins so characteristic of sponges (de Laubenfels, 1934). For example, the mucin of *Myxilla incrustans* Johnston,1842 contains substantial amounts of galactose, glucose and fucose (A. P. MacLennan and D. C. Hawkins, unpublished observations). Invertebrate collagens commonly have a substantial carbohydrate content (Gross and Piez, 1960) and a variety of sugars, including galactose, glucose, mannose, fucose, arabinose, glucosamine, galactosamine and hexuronic acid have been found as components of the tightly bound carbohydrate moiety of collagen from a single species of sponge (Gross, Sokal and Rougvie,

1956). A similar range of sugar units was found by MacLennan (1963) in polysaccharides extracted with trichloroacetic acid from seventeen species of sponge, including species of Calcispongiae and others that were not frankly collagenous. Certain of these sugars occur also in high molecular weight aggregation-promoting materials (Margoliash *et al.*, 1965; Gasic and Galanti, 1966). It appears likely from the presence of aldopentoses in the compounds referred to above that the xylose detected by Tekman and Oner (1963) in hydrolysates of *Euspongia officinalis* Linnaeus,1759 arose from such substances and, more interestingly, that the arabinose nucleosides earlier characterized by Bergmann and Burke (1955) could be intermediates in their biosynthesis (see Leloir, 1964).

The present paper describes the isolation of carbohydrate substances possessing the immunological specificity of the sponge cell surface. It describes the gross chemical properties of these substances and considers their activity in promoting or inhibiting sponge reaggregation, in the light of the several mechanisms proposed in explanation of this phenomenon; mechanisms which commonly make specific assumptions about the chemical nature of the sponge cell surface.

MacLennan (1963, 1969) reported that antisera prepared in rabbits against the cells of each of six species of sponge agglutinated homologous cells and cross-reacted to varying degrees with the cells of the other five species. The agglutinating antibodies were specifically neutralized by the carbohydrate-containing materials that are described in detail in the present communication. These substances were themselves antigenic and induced in rabbits precipitating and agglutinating antisera having the same agglutinating specificity as the homologous cell antisera. When fowl erythrocytes were sensitized with sponge carbohydrates to agglutination by antisera against homologous sponge carbohydrate or cells, the pattern of cross reaction of these antisera with heterologously sensitized erythrocytes was similar to their pattern of cross-agglutination of heterologous sponge cells. This indicates that the carbohydrate substances which selectively sensitize erythrocytes and bring about their agglutination by antisera are similar to those which are normally present on the sponge cell surface and dominate its serological reactions.

It follows that the cell surface is here defined operationally as that region which binds antibody so as to cause agglutination of cells. Antibody agglutination must not be confused with aggregation. Substances with cell surface specificity are defined in terms of their relationship to antibody agglutination as described above.

MATERIALS AND METHODS

Sponges

Seventeen British species were obtained in intertidal areas or by SCUBA diving off the South Devon coast. The specific names of the sponges (Table I) are used in the sense of those listed in "Marine Biological Association, 1957, Plymouth Marine Fauna". They will be referred to hereafter by their generic names alone. The relationships of the higher taxa of Porifera set out in Table I are those proposed by de Laubenfels (1936); but see also Lévi (1956, 1957).

Processing of sponges

Sponges for chemical processing were either frozen at $-20°C$ within 24 h from the time of collection until required or they were freeze-dried after maceration in distilled water. For reaggregation studies, sponges were used within 24 h of collection.

Preparation of materials for analysis

Samples were dried to constant weight *in vacuo* over phosphorus pentoxide at room temperature.

Protein

The method of Lowry, Rosebrough, Farr and Randall (1951) was used, with a standard of bovine plasma albumin (Armour & Co., Fraction V).

Carbohydrates

A rough value for total neutral carbohydrate was obtained from an orcinol method (Synge and Wood, 1958), using the difference in absorption at 420 and 480 mμ as a measure of carbohydrate and subtracting the contribution of independently determined hexuronic acid. Hexuronic acid was determined in terms of glucuronolactone by the 87% sulphuric acid-carbazole reaction (Dische, 1955). Hexosamine was measured in terms of glucosamine free base (Rondle and Morgan, 1955) on samples hydrolysed in 6 N hydrochloric acid for 16 h at 100°C.

Paper chromatography

In preparation for chromatography 10–50 mg samples of material were hydrolysed in N sulphuric acid for 16 h at 100°C and then neutralized with barium hydroxide. After centrifugation and filtration to remove barium sulphate the hydrolysate was evaporated to small volume at 100°C in a stream of nitrogen. The concentrate was percolated through a small column ($5 \times 0{\cdot}5$ cm) of MB-3 resin (The Permutit

Co. Ltd., London), evaporated to dryness *in vacuo*, and the syrup redissolved in a small volume of water (0·05–0·1 ml) for use. Sugars were separated by one-dimensional descending chromatography on Whatman No. 1 or 3 mm paper with a solvent front travel of 50 cm from the base line. The solvent systems used were: solvent A, butan-1-ol-pyridine-water (6 : 4 : 3, by vol.); solvent B, butan-1-ol-ethanol-water (4 : 1 : 5, by vol., upper phase); solvent C, butan-1-ol-acetic acid-water (4 : 1 : 5, by vol., upper phase); solvent D, phenol-rich phase of aq. 90% (w/v) phenol. Developed chromatograms were sprayed with *p*-anisidine hydrochloride (Hough, Jones and Wadman, 1950) which distinguishes by colour the various classes of reducing sugar.

Ultracentrifuging

Preparative ultracentrifuging was carried out in a Spinco model L ultracentrifuge at 2–4°C.

Antisera and serological tests

Antisera against DR carbohydrates (below) of the sponges *Halichondria*, *Hymeniacidon*, *Suberites*, *Cliona*, *Pachymatisma* and *Grantia* were those whose preparation has been described in full by MacLennan (1969). The Ouchterlony method of double diffusion precipitin analysis in agar was used, as described by Crumpton and Davies (1956). The distance between the centres of antigen and antiserum reservoirs was 1·7 cm and the reservoir dimensions were 0·8 cm diameter × 0·4 cm depth.

Reaggregation

Full details of materials and procedures used here to study the promotion of reaggregation of "chemically dissociated cells" by "Aggregation Factors" in a rotation-mediated system are given by MacLennan and Dodd (1967) and, earlier, by Humphreys (1963).

Reaggregation was also studied on simple suspensions of cells expressed from sponges (Wilson, 1907) and serially diluted in sea water. Suspensions were observed in large, covered cavity slides over a period of 16 h.

EXPERIMENTS

Extraction of polysaccharides

Cell preparations were used in preliminary attempts to extract from sponges carbohydrate substances possessing cell surface specificities. These were prepared by filtering macerated whole sponge through nylon net (mesh size 100 μ), dialysing and finally freeze-drying the filtered cell-suspension. Samples of this material from *Suberites* were extracted with trichloroacetic acid (Boivin, Mesrobeanu and Mesrobeanu, 1933), or with formamide (Fuller, 1938) or 45% aqueous phenol (Westphal, Lüderitz and Bister, 1952). The products of each extraction procedure contained qualitatively similar heteropolysaccharide and the trichloroacetic acid product was particularly rich in this material. When it was established by immunochemical methods that the carbohydrate was related to the sponge cell surface (MacLennan, 1963, 1969), seventeen species of sponge were submitted to a standard procedure of trichloroacetic acid extraction, employing freeze-dried macerated whole sponge rather than cellular preparations. Figure 1 outlines the general procedure. A particular example will be described.

Freeze-dried sponge macerate (*Cliona*, 1·5 kg) was suspended in water (6 l) at 2°C. An equal volume of chilled N trichloroacetic acid was added and the mixture was stirred for 3 h at 2°C. The suspension was filtered through nylon mesh and the filtrate was passed through an Alfa-Laval continuous-flow centrifuge so that the dark supernatant fluid was clear or nearly so. The supernatant fluid IA (Fig. 1) was dialysed against flowing distilled water until neutral in pH (Universal Indicator). There was a heavy precipitate but the whole suspension was first concentrated under reduced pressure at 18°C (Reich Still) to less than $1\frac{1}{2}$ l and then centrifuged (Spinco Batch Type Rotor No. 1600; 30 000 g for $\frac{1}{2}$ h). The sediment was discarded and the supernatant fluid IB was poured into 10 vol. acetone at -20°C. After 48 h at -20°C the heavy flocculent precipitate was recovered by centrifuging and washed twice with cold acetone. The sediment was suspended in water and dialysed against water at 2°C for 2 days. The suspension was centrifuged (Spinco Batch Type Rotor 1600, 30 000 g for $\frac{1}{2}$ h) and the sediment washed once by suspending in water and again centrifuging. The dark brown sediment was freeze-dried from aqueous suspension and retained as fraction **Insoluble Glycogen.**

The pooled supernatant fluids IC were freeze-dried (5·5 g) and then dissolved in water (2%, w/v). Centrifuging this opalescent solution (Spinco Rotor 40; 100 000 g for 16 h) yielded a small sediment of stiff translucent jelly. The sediment was again suspended and centrifuged as before, the washings IE being added to the first supernatant fluid. The washed sediment was freeze dried from aqueous solution and retained as fraction **Spinco Sediment** (0·7 g).

The pooled supernatant fluids ID + IE were freeze-dried (4·3 g), dissolved in water as 5% (w/v) solutions and dialysed for three periods

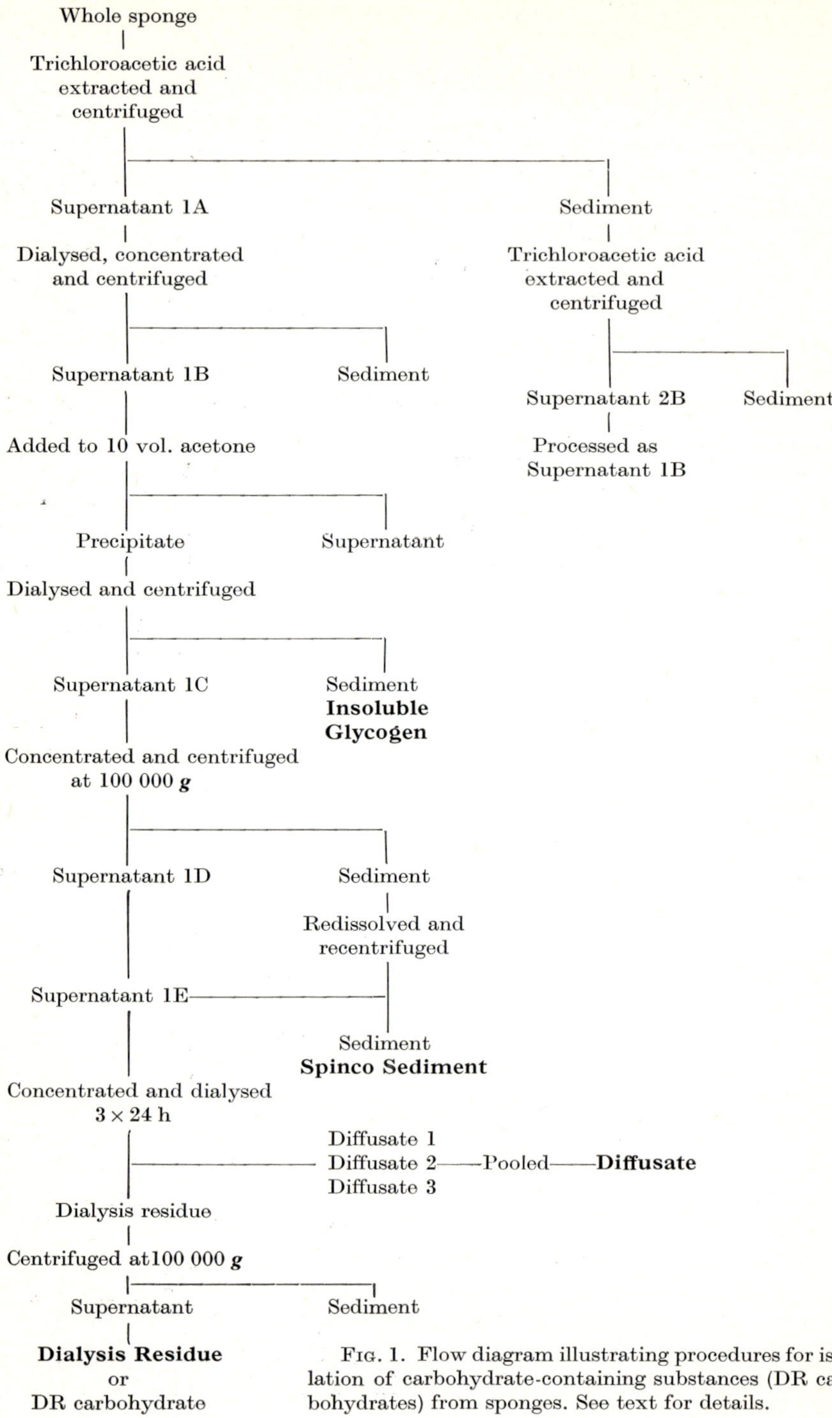

FIG. 1. Flow diagram illustrating procedures for isolation of carbohydrate-containing substances (DR carbohydrates) from sponges. See text for details.

of 24 h at 2°C against 20 vol. water, yielding 0·8, 0·2 and 0·1 g successively of freeze-dried diffusate. The diffusates were pooled in aqueous solution and freeze-dried as fraction **Diffusate**. The dialysis residue was freeze-dried as fraction **Dialysis Residue** (2·4 g)—hereafter DR carbohydrate—following a final centrifuging (100 000 ***g*** for 16 h) which removed a very small amount of sedimentable material.

In the trichloroacetic acid extraction of the calcareous sponges *Grantia* and *Sycon*, concentrated hydrochloric acid was added slowly, with stirring, to the cold aqueous suspension of sponge until further addition did not cause effervescence. Trichloroacetic acid was then added in the usual manner.

It is difficult to generalize from the results of extraction of seventeen species of sponge. There was great variation from species to species, and even from batch to batch of *Cliona*, in the relative amounts of the four named fractions of Fig. 1. The reasons for this variation and the nature of fractions other than DR carbohydrates will be discussed later. The DR carbohydrate fractions, to be considered in detail now, can be defined as being soluble in trichloroacetic acid and water, insoluble in acetone, non-diffusible on dialysis and not sedimenting when centrifuged at 100 000 ***g*** for 16 h. It has already been shown (MacLennan, 1963, 1969) that these materials will induce in rabbits the formation of antibodies that will agglutinate sponge cells and will also agglutinate fowl erythrocytes sensitized with the DR carbohydrates and that they will react with such antibodies to inhibit these reactions.

Yields of DR carbohydrates

For six species of sponge examined in detail (see MacLennan, 1969), the dry weight of the sponge (freeze-drying) represented 10–20% of the wet weight. The contribution made by spicules to the dry weight was variable, being 90% for *Grantia*, and between 35 and 70% for the other species. DR carbohydrates formed 0·2–0·8% of the organic dry weight of the sponge. A second extraction of the sediment left after trichloroacetic acid extraction (Fig. 1) never gave more than an additional 10% yield of DR carbohydrate.

Composition of DR carbohydrates

It follows from their manner of preparation that the DR carbohydrates were all readily soluble in water, forming 1% (w/v) solutions with ease. A few preparations (*Desmacidon*, *Polymastia*, *Tethya*, *Ophlitaspongia* and *Sycon*) gave faintly turbid solutions, possibly due to the insolubilizing of contaminating protein by the final freeze-drying process. The colour of the aqueous solutions varied from colourless (*Cliona*) to almost black (*Axinella, Myxilla, Stylostichon*).

Quantitative determinations of protein and carbohydrate were made on the products of the seventeen species. Hexuronic acid and hexosamine were also determined, since their presence in at least one preparation had been indicated by preliminary analysis.

The DR carbohydrates contained very variable amounts of peptide or protein (Table I), ranging from 21 to 80%. Hexuronic acid and hexosamine were constant and substantial components of the sponge products but there was wide variation in their ratio from species to species (Table I).

With respect to total carbohydrate, glucosamine does not react in the orcinol test and glucuronic acid gives about half the value of an equal weight of galactose. The "total neutral carbohydrate" values of Table I were, therefore, derived directly from orcinol values by subtracting the contribution of the independently determined hexuronic acid.

Chromatography

To analyse the neutral carbohydrate component, hydrolysates of DR carbohydrates were chromatographed in Solvent A. Samples of authentic sugars, from 1 to 130 μg in a two-fold series, were chromatographed concurrently. The intensity of the reactions with anisidine provided a scale from which the relative amounts of the various sugars in a preparation could be judged roughly, arbitrarily taking the intensity of the galactose reaction as the standard.

The DR carbohydrates were qualitatively almost identical in sugar composition. Six sugars, namely galactose, glucose, mannose, fucose, rhamnose and arabinose were present in all fractions. In addition, two unidentified sugars with high Rf values were always found and xylose was present in many, perhaps all, preparations. To confirm identifications a large sample of *Cliona* DR carbohydrate (100 mg) was hydrolysed and chromatographed as bands on washed (MacLennan, Randall and Smith, 1961) Whatman No. 3 mm paper. The sugars, with the exception of xylose, were recovered individually, concentrated and examined chromatographically with appropriate standards in solvent systems A–D (see Methods above). The identities were fully confirmed. Other samples of the *Cliona* preparation were hydrolysed in 1 N sulphuric acid at 100°C for periods ranging from 1 min to 24 h and neutralized with and without the use of MB-3 resin. There was no indication in short-time hydrolysates of additional acid-labile sugars and no suggestion that the "high Rf sugars" were acid-labile.

Although perhaps all the sugars found were present in all preparations, there was large quantitative variation both from sugar to sugar within a species and from species to species with respect to individual sugars (Table I). In general the neutral sugar composition of all species was dominated by galactose and fucose, which were commonly present in similar amounts. Mannose and rhamnose occurred in much smaller quantities and glucose showed great variation from one species to another. Since the trichloroacetic acid extracts of sponges may contain large amounts of glycogen, as will be discussed later, it appears that a high glucose content in DR carbohydrates could be due to the failure of glycogen to sediment completely during high speed centrifuging (Fig. 1). The "high Rf sugars" were certainly minor components of the DR carbohydrates. The aldopentoses were also, in general, minor components but they showed interspecies variation to a high degree. In particular, xylose was strong in *Pachymatisma, Tethya* and *Adocia* products but weak in all others. Sialic acid was absent.

The data in Table I are presented in relationship to the taxonomy of the phylum Porifera (de Laubenfels, 1936). Of the three Classes only the Hexactinellida is not represented. One of the two orders of Calcispongiae and one of the eight orders of Demospongiae have not been included. It is nevertheless clear that polysaccharides of a qualitatively similar nature are of widespread occurrence in the phylum Porifera.

Further characterization of DR carbohydrates

Examination of *Cliona* DR carbohydrate in the analytical ultracentrifuge (Spinco model E) in the synthetic boundary cell showed a single, slowly sedimenting component (S_{20}, W 1·9), the boundary being slightly asymmetrical (Fig. 2). During the run, 93–98% of the sample could be accounted for in terms of the refractive increment across the boundary. *Hymeniacidon* and *Grantia* DR carbohydrates showed a similar major component but with definite indications of a partially resolving second component.

The molecular weight of *Cliona* DR carbohydrate (dissolved in 0·02 M Tris/hydrochloric acid buffer, pH 7·5; potassium chloride, I 0·1) was determined by the Archibald method, using a pycnometrically determined partial specific volume value of 0·635. The weight-average molecular weight of the material at the meniscus decreased during the period of centrifuging from 13 800 to 7500 as heavier molecules resolved from the lighter ones, providing clear evidence of polydispersity within this molecular weight range.

A sample (1·2 g) of *Cliona* DR carbohydrate (21% protein) was

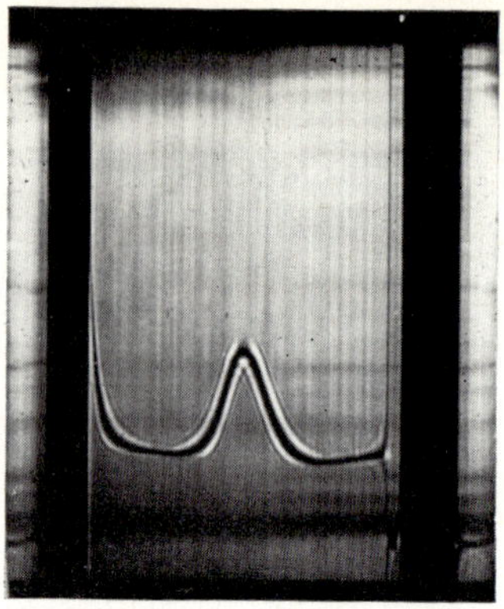

FIG. 2. Sedimentation diagram of *Cliona* DR carbohydrate in synthetic boundary cell. Initial concentration 1·0% (w/v) at pH 8·0 (phosphate, I 0·1; sodium chloride, I 0·1). Centrifugal force 250 000 *g*. Migration from right to left. Exposure 35 min after reaching full speed. Phase plate angle = 60°.

dissolved in 15 ml of 0·05 M Tris/hydrochloric acid buffer, pH 9·2 and chromatographed on a column of DEAE Sephadex A-50 (3 cm × 90 cm) equilibrated in the same buffer. A step-wise elution was made with 0·05, 0·10, 0·15 and 0·30 M sodium chloride in the Tris buffer, using a volume of each eluant greater than the bed volume of the column. A total of 730 mg was recovered. Of this a glycogen-rich fraction (30 mg) was eluted in the starting buffer. A major fraction (335 mg) appeared in the 0·10–0·15 M sodium chloride buffers, having been firmly bound at lower salt concentrations. This fraction contained only 1·2% protein. However, all the sugars listed in Table I, with the exception of xylose, were detected in this fraction by the appropriate tests, including paper chromatography of hydrolysed samples. The material possessed the serological specificity of *Cliona* DR carbohydrate in a haemagglutination-inhibition test (MacLennan, 1969) but was less active in this respect than protein-containing fractions finally eluted from the DEAE-Sephadex column by the application of pH 5·4 acetate buffer.

When *Hymeniacidon* DR carbohydrate (56% protein) was chromatographed on DEAE Sephadex A-50 in 0·0175 M phosphate buffer, pH 6·3, several discrete peaks were obtained on step-wise elution with sodium chloride. Each peak contained both protein (up to 40%) and carbohydrate.

Serological properties of DR carbohydrates

The heterogeneity of DR carbohydrates displayed on anion exchange chromatography was shown also when these substances were examined by immunodiffusion against concentrated antisera to cells and DR carbohydrates. From two to six discrete antigens were found in the DR carbohydrates of the six species examined (see Methods above). The line pattern of the *Pachymatisma* preparation is illustrated in Fig. 3. All reactions were strictly species specific, cross reactivity of antisera with the DR carbohydrates of other species was negligible.

It appears reasonable to assume that strongly thermostable serological specificities displayed by DR carbohydrates would be an expression of their carbohydrate components. Therefore, aqueous solutions of DR carbohydrates (1% w/v) were heated for 2 h at 100°C in

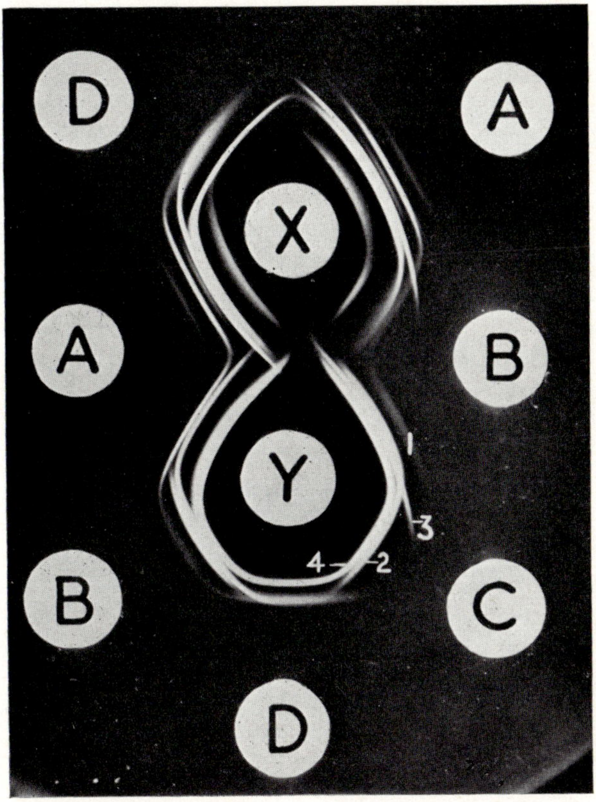

FIG. 3. Agar diffusion precipitin analysis of *Pachymatisma johnstonia*, cells and DR carbohydrate. Peripheral antigen reservoirs contain A, packed cells; B, DR carbohydrate, 1 mg; C, heated (100°C/2 h) DR carbohydrate, 1 mg; D, freeze dried, dialysed supernatant fluid from centrifugation of mechanically disintegrated cells, 1 mg.

Central reservoirs contain four-fold concentrated antisera: X, cell antiserum; Y, DR carbohydrate antiserum (see MacLennan, 1969). The numbers 1 and 3 indicate lines due to thermolabile antigens. Lines 2 and 4 represent thermostable antigens.

sealed ampoules and two-fold dilution sequences were compared on diffusion plates with similar dilutions of unheated preparations. In this manner two or more thermostable antigens were detected in the DR carbohydrates of each of the six species, except *Cliona*, against which antiserum was available (Fig. 3).

In general, the thermostable antigens appeared to be normal constituents of the sponge, rather than artefacts of preparation, since they could be detected in simple extracts of cells or even in packed homologous cells prepared by centrifuging cell suspensions (Fig. 3). It could be argued that the occurrence in individual species of multiple thermostable antigens might represent intra-specific variation, since many individual sponges contributed to the pooled material from which DR carbohydrates were isolated. No support for this idea was provided by the examination on immunodiffusion plates of the cells from fifty individual sponges of *Hymeniacidon*; each sponge gave "thermostable lines" similar to all others.

The **Diffusate** *and* **Spinco Sediment** *fractions*

The chemical and serological properties of DR carbohydrates have been described in some detail. However, the soluble products of trichloroacetic acid extraction also included **Diffusate** and **Spinco Sediment** fractions (Fig. 1). These substances comprised substantial proportions of the total extract in many species.

The **Diffusate** fractions of *Suberites*, *Pachymatisma* and *Grantia* were found to contain all the sugars of the DR carbohydrates but in lesser amount. Although none gave a line of precipitation on immunodiffusion plates, they inhibited specifically the formation of lines against homologous DR carbohydrates when incorporated (at 2% w/v) into the molten agar used to prepare the plates.

The **Spinco Sediments** of *Grantia* and *Pachymatisma* were isolated as clear gels that readily redissolved to form opalescent solutions. They contained all the sugars of the DR carbohydrates. It is of special interest that they both contained about 30% protein since this could not, from its manner of isolation, be contaminating protein of small molecular weight.

Sponge glycogens

The **Spinco Sediments** of *Hymeniacidon*, *Suberites*, *Cliona* and some other species were composed largely of glycogen. However, slight variation in the manner of extraction of the DR carbohydrates (Fig. 1) greatly influenced the amounts of glycogen that appeared in **Spinco Sediments**. Thus, in six separate preparations of **Spinco Sediments** from equal amounts (1 kg) of a single batch of freeze-dried, macerated *Cliona* the yields were 0·1, 0·4, 0·5, 1·0, 4·6 and 5·7 g. Where the yield of **Spinco Sediment** was low, a correspondingly greater amount of glycogen was found in insoluble form at an earlier stage of fractionation (**Insoluble Glycogen** of Fig. 1). Extraction of **Insoluble Glycogen** with trichloroacetic acid, under the conditions described for extraction of whole sponge, gave a water-soluble fraction from which an opalescent sediment was obtained by centrifuging at 100 000 ***g*** for 16 h. A preparation of this sediment was analysed without further purification (found: Carbon, 42·1; hydrogen, 6·76; nitrogen, 0·1. $(C_6H_{10}O_5)_n$ required carbon, 44·4, hydrogen, 6·18%). The preparation had $[\alpha]_D^{20}$ $+173°$ (C, 0·2 in water) and the opalescent solution reacted as glycogen with iodine. Only glucose was detected in hydrolysates chromatographically and the preparation assayed roughly 100% as glucose. Incubation with β-amylase (L. Light & Co. Ltd., Colnbrook) yielded a reducing sugar with an R_f value equal to that of maltose on chromatograms developed with solvent A.

It is probable that the relative proportions of glycogen that appeared in **Spinco Sediment** and **Insoluble Glycogen** fractions were determined by the duration of exposure to acidic conditions during the dialysis and concentration steps (Fig. 1) that preceded precipitation with acetone (see Drochmans, 1963). Glycogen that was not made fully soluble prior to acetone precipitation was presumably rendered completely non-dispersible in water by the acetone treatment.

DR CARBOHYDRATES AND CELLULAR REAGGREGATION

The observations recorded here represent collaborative studies with Mr. R. Y. Dodd.

"Simple" reaggregation

The incorporation of *Hymeniacidon* DR carbohydrate at 1 mg/ml into cell suspensions prepared by squeezing *Hymeniacidon* through nylon net (100 μ mesh) did not influence the process of reaggregation in cavity slides. Neither was there inhibition of *Hymeniacidon* reaggregation by the heterologous DR carbohydrate from *Pachymatisma*. This lack of heterologous inhibition is contrary to what might have been expected if aggregation proceeded by the mechanism suggested by MacLennan (1963). This mechanism had been proposed principally on the evidence that reaggregation was accompanied by mucin formation (de Laubenfels, 1934), together with the evidence that DR carbohydrates exist in a range of molecular sizes and possess chemical resemblances to the carbohydrate moiety of sponge collagen (Gross *et al.*, 1956). On analogy with the extracellular polymerization of collagen monomers to form collagen fibres, it had been suggested that sponge DR carbohydrates might occur as small glycopeptides that polymerize into viscous mucins in the immediate vicinity of the cell surface. Species specificity of reaggregation in binary mixtures could then arise from the mutual inhibition of polymer growth, giving rise to low adhesion, when contiguous cells were secreting different monomers.

Aggregation Factor—promoted reaggregation

Humphreys (1963) and Moscona (1963) found that thermolabile "Aggregation Factors" obtained from sponges can selectively promote the reaggregation of homologous "chemically dissociated cells" in a rotation-mediated aggregation system at low temperature. Since "Aggregation Factors" are specifically adsorbed by homologous cells and since they can selectively promote the reaggregation of one species in a binary mixture (Moscona, 1963; Humphreys, 1963, 1965), there is a clear implication of a specific complementary mechanism, involving factor and a cell surface component with which it combines, echoing the earlier explicit "antigen-antibody hypotheses" of Tyler (1946), Weiss (1947) and Spiegel (1955). We have examined the possibilities that DR carbohydrates form a part of "Aggregation Factor" on the one hand, or a part of its complementary cell surface site on the other.

Hymeniacidon DR carbohydrate, at final concentrations as high as 1 mg/ml, could not replace *Hymeniacidon* "Aggregation Factor" (MacLennan and Dodd, 1967) in promoting the reaggregation of "chemically dissociated cells". It remains possible that polysaccharide is a necessary component of "Aggregation Factors" (Moscona, 1963; Humphreys, 1963, 1965), if not itself sufficient to bring about aggregation. Indeed, when *Hymeniacidon* "Aggregation Factor" was purified ten-fold as the excluded fraction from Sepharose 4B gel (Pharmacia Co. Ltd., London), it contained carbohydrate with the qualitative monosaccharide composition and serological specificity of *Hymeniacidon* DR carbohydrate (R. Y. Dodd, personal communication). However, similarly purified but inactive "Control Factors" (see MacLennan and Dodd, 1967) had a composition similar to the active material, so that the relevance of carbohydrate to biological activity remains undecided.

MacLennan (1969) reported that the antigenic character of the sponge cell surface is dominated by specificities shared with the DR carbohydrate from the same species. Carbohydrate antigens with DR carbohydrate specificity could thus constitute the site "recognized" by "Aggregation Factors". If so, then free DR carbohydrate in solution might compete with the cell surface receptor for combination with added "Aggregation Factor" and thus inhibit the aggregation-promoting activity of the latter substance. This is formally analogous to the inhibition of agglutinating antibodies by isolated products from cell surfaces and the test methods used are familiar to immunologists (see Kwapinski, 1965).

Hymeniacidon "Aggregation Factor" was incorporated into the rotation-mediated test system at a final dilution of one quarter of that which had a detectable aggregation promoting effect. The cells were thus exposed to "four aggregating doses" of "Aggregation Factor", a system more sensitive to inhibitory effects than larger doses. Nevertheless, there was no inhibition of aggregation by DR carbohydrates of *Hymeniacidon*, and other species, incorporated at final concentrations up to 0·5 mg/ml.

Erythrocytes of fowl or sheep exposed to solutions of DR carbohydrates adsorb sufficient antigen to become specifically agglutinable by the appropriate antiserum (MacLennan, 1969). Indeed, mixtures of the morphologically distinct erythrocyte species differentially sensitized with DR carbohydrates from distinct sponge species are selectively agglutinated on the addition of mixed antisera. It was, therefore, of interest to see if *Hymeniacidon* "Aggregation Factor" could simulate antibody and agglutinate erythrocytes sensitized with DR carbohydrate. The results were again clearly negative, both when sensitized erythrocytes were tested in sea water in the rotation-mediated system and when tube agglutination tests were performed in physiological saline. In the latter test calcium chloride was added to the saline to preserve the aggregation-promoting activity of the "Aggregation Factor". Agglutination was not observed with eight-fold concentrated "Aggregation Factor" (i.e. 64–128 "aggregating doses"), whereas antiserum was active at a dilution of 1/320.

DISCUSSION

General nature and cellular location of sponge carbohydrates

It is necessary to consider whether or not substances isolated in small yields from whole sponge are truly sponge products, since the sponge body is a complex system of canals and chambers intimately exposed to an actively maintained current of water which contains small fauna and microorganisms. It must also be considered possible that undigested residues of the supposed microbial diet of sponges could accumulate within the body substance. The sponge DR carbohydrates are, however, qualitatively alike whether isolated from fresh water or marine species, from small, shallow-water calcareous forms or from massive, deep water, siliceous sponges. They possess serological features which may well have a taxon specificity and certain of their broader serological specificities can be related to surface characters of washed sponge cells (MacLennan, 1969). Moreover, the monosaccharide units of the DR carbohydrates are also found in the tightly bound carbohydrate moiety of sponge collagens (Gross *et al.*, 1956), substances which are indisputably part of the sponge substance. It can be concluded that the DR carbohydrates, too, are sponge products.

Ultracentrifugal analysis showed that in some, at least, of the DR carbohydrates the material was principally of a molecular weight less than 14 000 and was polydisperse within the range 7500 to 14 000. The lower limit of this range is probably set, rather arbitrarily, by the dialysis steps employed in the isolation of DR carbohydrates (Fig. 1). There was, in fact, a "spillover" of serologically active carbohydrate into **Diffusate** fractions, so that the true limit of polydispersity at the low molecular weight end of the scale is likely to be lower than that suggested by ultracentrifugal analysis. Beyond the upper molecular weight limit of 14 000 or so there is a large discontinuity before meeting, in the **Spinco Sediments** of some species, soluble materials with high contents of the sugars associated with DR carbohydrates. In addition the firmly bound carbohydrate of sponge cell surfaces (MacLennan, 1969) and of "Aggregation Factors" (Humphreys, 1965; Margoliash *et al.*, 1965; MacLennan and Dodd, 1967) may represent high molecular weight polysaccharide or glycoprotein, although there is the alternative possibility that smaller molecular weight carbohydrate or glycolipid is integrated into membrane structures in these substances.

The isolation of a protein-free, serologically reactive polysaccharide from *Cliona* DR carbohydrate by chromatography on anion-exchange columns is of significance in relation to the molecular form of sponge carbohydrates. Retention of this polysaccharide on the column at low salt concentrations must be attributed to a free carboxyl group on the hexuronic acid component. Since all the other uncharged or basic sugar components are also retained on the column in the same fraction, they can be presumed to be linked in a polysaccharide of which the hexuronic acid is a part.

Although protein-free carbohydrate is present in *Cliona* products, there are several suggestions that carbohydrate may also be linked to protein in this, and other, species. Protein-containing fractions from the chromatography of *Cliona* DR carbohydrate were serologically more active than the pure carbohydrate. Chromatography of *Hymeniacidon* DR carbohydrate yielded several discrete fractions each containing both protein and carbohydrate. The highly immunogenic character of DR carbohydrates is not a property commonly associated with carbohydrates of low molecular weight. Finally, the **Spinco Sediments** of *Pachymatisma* and *Grantia* contained substantial proportions of protein which, from their manner of preparation, could not be uncombined small molecular weight protein. If it is uncombined protein of high molecular weight it is surprising that it was not precipitated by trichloroacetic acid during the isolation procedure and that it was found only in those species yielding **Spinco Sediments** with a high heteropolysaccharide content.

Interest in DR carbohydrates is greatly enhanced by their possession of a high degree of serological specificity detectable by immunodiffusion, as reported above, and by inhibition of antibody agglutination of cells (MacLennan, 1969). The latter reactivities are of broader specificity than the former, in that cross-agglutination occurs but specific agglutination is also always detectable. Inhibition of agglutination by DR carbohydrates is strong evidence that they are structurally related to components of the sponge cell surface. The evidence that it is carbohydrate that is responsible for the serological activity of the crude DR carbohydrate preparations can be summarized as follows:

1. Carbohydrate is a major component of the serologically active DR carbohydrates.

2. Serological activity of DR carbohydrates as expressed in immunodiffusion and in inhibition of antibody agglutination is thermostable.

3. Erythrocytes can be directly sensitized by DR carbohydrates to agglutination by antibodies. This property of direct sensitization is commonly associated with carbohydrates rather than proteins (see Kabat and Mayer, 1961).

4. Protein-free polysaccharide from *Cliona* DR carbohydrate is serologically active.

5. There is good precedent in bacterial systems for a high degree of serological specificity to be founded on structurally different polysaccharides containing a limited number of monosaccharide components (see Lüderitz, Staub and Westphal, 1966). In this connection it should be added that all seventeen of the DR carbohydrates have now been found to be serologically distinct from each other (A. P. MacLennan and D. C. Hawkins, to be published).

A second sort of serological specificity possessed by DR carbohydrates distinguishes thermostable, and therefore presumptively carbohydrate, antigens from each other within the products of a single species. Of six species examined, five had more than one thermostable antigen. These different thermostable antigens clearly could arise in a number of ways. They might represent degradation products of a larger native molecule of unitary biological function, but with multiple antigenic sites, or they could be substances on the path of biosynthesis to such a molecule. Another possibility is that individual sponges which are classed as members of a single species on conventional criteria may differ at the biochemical level and therefore that the pooled specimens used as a source of material could include sponges of different natures, each contributing their own distinctive specific substances to the final product. This possibility is most real where very large numbers of individuals contribute to the pools, as in *Halichondria*, *Hymeniacidon* and *Grantia*, and where the gross individual form is very variable. It is partially discounted by the observation that "individual" specimens of *Hymeniacidon* possessed multiple thermostable antigens in common.

A fourth possibility is that the specificities within a species are cell-type specificities, representing perhaps the choanocytes and archaeocytes, the basic cell types of the sponge (Agrell, 1951). If this interpretation is correct, it could afford a unified chemical basis for the primary species specific reaggregation of sponge cells and for their subsequent sorting out into cell types. However, it must be noted that the narrow serological specificities which distinguish products within a species have been detected only by immunodiffusion analysis and have not been shown directly to be related to cell surface structures.

Concerning cellular location of sponge carbohydrates, it appears that the serologically detected components of DR carbohydrates are present in a similarly freely diffusible form in untreated sponge cells. Some of the thermostable specificities are present also on the cell surface in tightly bound form (MacLennan, 1969). The relationship between the diffusible and the cell-bound specificities may be similar to that between certain of the blood group mucopolysaccharides obtained from cell secretions and the corresponding blood group substance of the cell surface which appears to be glycolipid (Kościelak, 1965).

Relationships of specific carbohydrates to reaggregation

The possible function of extracellular cell-binding substances in the specific interactions of cells has been suggested repeatedly (see Lilien,

1968). The narrow serological specificity of the DR carbohydrates, together with their possible relationships to cell surface structures and to mucins, are properties in keeping with the requirements of specific cell-binding substances. Moreover, the free hexuronic acid function of DR carbohydrates adds a further plausibility in allowing for the involvement of ionic calcium, the importance of which in cell adhesion has been repeatedly demonstrated and made part of hypothetical mechanisms of specific adhesion (see Manery, 1966; Curtis, 1967).

Before considering experimental evidence concerning the role of carbohydrates in reaggregation, it is relevant to consider the actual degree of specificity that has been demonstrated in the reaggregation of binary mixtures of sponge cells. This matter has been called into question in recent years (Sara, 1956; Curtis, 1962; Sara, Liaci and Melone, 1966a, 1966b). Tables II and III bring together the data from eleven investigations, illustrating that reaggregation is not always

TABLE II

Taxonomic relationships of sponges that form bispecific aggregates

Taxon*	Bispecific mixtures
Variety	
Species	*Leucosolenia complicata/L. botryoides* (Sara 1956); pairs from three species of *Pachychalina* (de Laubenfels 1927, 1928).
Genus	
Family	*Halichondria/Hymeniacidon* (MacLennan and Dodd 1967);† *Iotrochota/Haliclona* (de Laubenfels 1934)‡
Order	*Hymeniacidon/Suberites*, *Microciona/Halichondria*,§ *Microciona/Suberites*,§ *Halichondria/Suberites*,§ (Curtis, 1962); *Tethya/Hymeniacidon*, *Tethya/Haliclona*, *Tethya/Petrosia*, *Hymeniacidon/Polymastia* (Sara *et al.*, 1966a),‖ *Hymeniacidon/Suberites* (MacLennan and Dodd 1967),†
Class	*Hymeniacidon/Clathrina*, *Haliclona/Clathrina*, *Tethya/Clathrina* (Sara *et al.*, 1966a).
Phylum	The anthozoan *Anemonia sulcata* with sponges *Clathrina*, *Tethya*, *Haliclona* (Sara *et al.*, 1966b).

* According to the classification of de Laubenfels (1936).

† Under the influence of "Aggregation Factor" in a rotation-mediated system. See text.

‡ Complicated by the action of heteroagglutinins.

§ Fully dissociated cells of one species added to aggregating mixture of another.

‖ Also some monospecific aggregates.

TABLE III

Taxonomic relationships of sponges that form monospecific aggregates from bispecific mixtures

Taxon*	Bispecific mixtures
Variety	Pairs from four coloured varieties of *Clathrina coriacea* (Lévi, 1956).
Species	All pairs from *Haliclona rubens, H. longleyi, H. viridis* (de Laubenfels 1934)†; *Leucosolenia complicata/Clathrina*‡ *coriacea* (Sara, 1956).
Genus	
Family	*Microciona/Lissodendoryx* (Wilson, 1907, 1910); *Microciona/Petrosia, Haliclona/Iotrochota* (Galtsoff 1925); *Halichondria/Hymeniacidon* (MacLennan and Dodd, 1967)§
Order	*Microciona/Stylotella* (Wilson, 1907, 1910); *Microciona/Halichondria, Microciona/Reniera* (Galtsoff, 1925); *Iotrochota/Halichondria* (de Laubenfels, 1934); *Microciona/Cliona* (Spiegel, 1955); *Microciona/Halichondria, Halichondria/Suberites* (Curtis, 1962); *Microciona/Haliclona* (Humphreys, 1963, Moscona, 1963)§; *Halichondria/Reniera* (Yazhikov, 1965); *Tethya/Ircinia* (Sara *et al.*, 1966a); *Hymeniacidon/Suberites* (MacLennan and Dodd, 1967).§
Class	*Hymeniacidon/Clathrina*‡, *Haliclona/Clathrina*‡, *Tethya/Clathrina*‡ (Sara *et al.*, 1966a).‖

* According to the classification of de Laubenfels (1936).
† Monospecific aggregates did not fuse with those of the other species under conditions where monospecific aggregates fused with each other.
‡ De Laubenfels (1936) does not distinguish *Clathrina* from *Leucosolenia*.
§ Rotation-mediated aggregation.
‖ Also a significant intermingling of cells of both species in some aggregates.

specific but may cross broad taxonomic divisions (Table II) and that, contrary to common usage, true species specificity within a genus has been shown in only one or two instances (Table III). The footnotes to both tables illustrate that qualifications must be made in comparing the data. It is not possible then to correlate the specificity of re-aggregation shown in Tables II and III with the finer specificity of DR carbohydrates as recognized by immunodiffusion.

In relation to the particular proposal that aggregation is promoted by cell-bound "Aggregation Factors", it must be remarked that specific activity has been demonstrated so far only in a binary mixture of *Microciona prolifera* and *Haliclona oculata* (Moscona, 1963; Humphreys, 1963). The similarly prepared "Aggregation Factor" of *Hymeniacidon* (MacLennan and Dodd, 1967) was found to be relatively non-specific in that it promoted the aggregation of other species, serologically distinct, and in doing so suppressed the natural capacity of these species to segregate from *Hymeniacidon* in simple, unpromoted reaggregation. It is with *Hymeniacidon* that we have failed to find evidence that implicates carbohydrate either as a recognition site on the cell surface or, and this point remains more open, as a necessary component of "Aggregation Factor". It is, moreover, necessary to stress that the inhibition tests on which our conclusions were partly based are strongly dependent on the molecular size and concentration of inhibitors, as in comparable antibody inhibition systems (Kabat and Mayer, 1961).

The finding of glycogen in **Spinco Sediments** (Fig. 1) of sponge extracts is of unexpected interest in regard to the characterization of *Microciona prolifera* "Aggregation Factor" as a particulate glycoprotein with some degree of ultrastructural organization as revealed by electron microscopy (Margoliash *et al.*, 1965). *Microciona prolifera* contains particulate glycogen and sponge glycogens may form negatively staining rosettes of dimensions 600–1300 Å (Lévi, 1966). The preparations of Margoliash *et al.* (1965) were obtained by centrifugation at 100 000 ***g*** (cf. **Spinco Sediments**, Fig. 1), contained 17–40% of polymeric glucose and showed negatively staining rosettes of dimensions 200–2000 Å, mostly in the range 500–1000 Å. The disaggregation of these particles on dialysis at 0–4°C was attributed by Margoliash *et al.* (1965) to a specific effect of calcium ion depletion. In the light of the lability of mammalian particulate glycogen to low pH and to contaminating enzyme (Drochmans, 1963), the breakdown of the sponge material could be explicable in similar terms. R. Y. Dodd and A. P. MacLennan (unpublished observations) found no evidence for 200–2000 Å particulates in the "Aggregation Factor" of *Hymeniacidon*.

The evidence in favour of specific adhesion as the basis of specific reaggregation is not strong. This directs attention not only towards the acquisition of more data but towards those hypotheses (Curtis, 1962; Steinberg, 1964) which explain specific reaggregation in terms other than those of specific cell-binding substances. Curtis (1962), on the basis of experiments with a number of species, suggested that sponge cells expressed from the parent sponge "tissues" acquire the ability to adhere and form aggregates at a rate that is different from one species to another but that the adhesive property thus acquired is itself without specificity. This hypothesis draws attention to the basically artificial character of sponge cell segregation from binary mixtures. It suggests that segregation will occur only when the two species have different aggregation rates. By suitably delaying the preparation and addition of the cells from a faster aggregating species until the cells of a slower species were already adhesive and aggregating, bispecific conglomerates could be produced (Curtis, 1962). A number of criticisms can be made of the experimental foundation for this hypothesis of temporal specificity. Three investigators (Wilson, 1910; de Laubenfels, 1934; Yazhikov, 1965) have reported that actively aggregating masses of cells from one species, presumably in an adhesive state, will fuse with other masses of the same species but not with those from other species. Curtis (1962) employed ethylenediaminetetraacetate to disaggregate cells, a procedure that might irreversibly destroy "Aggregation Factors" (Humphreys, 1965) and so delay the expression of a specific adhesion mechanism. The addition of cells of one species to aggregating masses of cells of other species raises the possibility that secreted products of the aggregates may influence the aggregative behaviour of the added cells. For example, Galtsoff (1925) and Galtsoff and Pertzoff (1926) refer to the coating of aggregates of one species by the cells of another at alkaline pH, a condition among those described by Curtis (1962). Moreover, sponges produce heteroagglutinins (Galtsoff, 1929; MacLennan and Dodd, 1967; Dodd, MacLennan and Hawkins, 1968) that agglutinate the cells of other species. de Laubenfels (1934) was explicitly of the view that the formation of bispecific conglomerates was promoted by the heteroagglutinin of one partner immobilizing the cells of the other so that segregation could not occur or, significantly, was delayed for a considerable period. Candelas and Candelas (1962), in noting that the culture supernatant of one species inhibited the aggregation of cells of another, further suggest that heteroagglutinins or other secreted products might possibly have influenced Curtis's results.

In conclusion it must be acknowledged that serological analysis may provide a misleading caricature of the cell surface in which minor, but highly immunogenic and serologically reactive, components assume an importance disproportionate to their physiological activity. No final conclusions can be drawn at present about the involvement of carbohydrate in the mediation of cellular reaggregation, beyond noting that carbohydrates of sponges are endowed with more than sufficient structural specificity to underwrite the observed specificities of reaggregation. They might equally serve as the non-specific mediators of adhesion invoked by Curtis (1962) and Steinberg (1964) in which their specific serological character is subordinated to non-specific physical properties. Apart from all considerations of reaggregation mechanisms, we believe it to be of some biochemical and, perhaps, taxonomic interest that the 5000 species of Porifera may prove to produce a diversity of carbohydrate structure to rival that of the bacteria.

Acknowledgements

I thank Mr. D. C. Hawkins and Miss Valerie Hart for valuable technical assistance and the Bath branch of The British Sub-Aqua Club for supplies of sponges. Mr. G. R. Forster of The Marine Biological Association of the United Kingdom, Plymouth, kindly advised on the identification of sponges and was helpful in many other ways. Analytical ultracentrifuge studies were performed and interpreted by Dr. K. A. Cammack. My particular thanks are due to Dr. D. A. L. Davies for introducing me to the problems of cellular aggregation and for his advice and encouragement during the earlier part of this work.

References

Agrell, I. (1951). Observations on cell differentiation in sponges. *Ark. Zool.* **2**: 519–523.

Bergmann, W. and Burke, D. C. (1955). Contributions to the study of marine products. XXXIX. The nucleosides of sponges. III. Spongothymidine and spongouridine. *J. org. Chem.* **20**: 1501–1507.

Boivin, A., Mesrobeanu, I. and Mesrobeanu, L. (1933). Extraction d'un complexe toxique et antigénique à partir du Bacille d'Aertrycke. *C. r. Séanc. Soc. Biol.* **114**: 307–310.

Candelas, G. A. and Candelas, G. C. (1962). Studies on cell reaggregation in marine sponges. *Am. Zool.* **2**: 510.

Crumpton, M. J. and Davies, D. A. L. (1956). An antigenic analysis of *Pasteurella pestis* by diffusion of antigens and antibodies in agar. *Proc. R. Soc. B*, **145**: 109–134.

Curtis, A. S. G. (1962). Pattern and mechanism in the reaggregation of sponges. *Nature, Lond.* **196**: 245–248.

Curtis, A. S. G. (1967). "*The cell surface: its molecular role in morphogenesis*". Academic Press, London and New York.

Dische, Z. (1955). New color reactions for determination of sugars in polysaccharides. *Meth. biochem. Analysis* **2**: 313–358.

Dodd, R. Y., MacLennan, A. P. and Hawkins, D. C. (1968). Haemagglutinins from marine sponges. *Vox Sang.* **15**: 386–391.

Drochmans, P. (1963). Techniques for the isolation of particulate glycogen and its examination in the electron microscope. *Biochem. Soc. Symp.* no. **23**: 127–137.

Fuller, A. T. (1938). The formamide method for the extraction of polysaccharides from haemolytic streptococci. *Br. J. exp. Path.* **19**: 130–139.

Galtsoff, P. S. (1925). Regeneration after dissociation (an experimental study on sponges)—I. Behaviour of dissociated cells of *Microciona prolifera* under normal and altered conditions. *J. exp. Zool.* **42**: 183–221; 1 pl., 10 figs.

Galtsoff, P. S. (1929). Heteroagglutination of dissociated sponge cells. *Biol. Bull. Mar. biol. Lab. Woods Hole* **57**: 250–260.

Galtsoff, P. S. and Pertzoff, V. (1926). Some physico-chemical properties of dissociated sponge cells. *J. gen. Physiol.* **10**: 239–255.

Gasic, G. J. and Galanti, N. L. (1966). Proteins and disulfide groups in the aggregation of dissociated cells of sea sponges. *Science, N.Y.* **151**: 203–205.

Gottschalk, A. (1966). *Glycoproteins, their composition, structure and function* Biochemica biophysica Acta Library No. 5 Elsevier: Amsterdam.

Gross, J. and Piez, K. A. (1960). The nature of collagen. 1. Invertebrate collagens, in *Calcification in biological systems.* American Association for the Advancement of Science, Washington, D.C.

Gross, J., Sokal, Z. and Rougvie, M. (1956). Structural and chemical studies on the connective tissue of marine sponges. *J. Histochem. Cytochem.* **4**: 227–246.

Hough, L., Jones, J. K. N. and Wadman, W. H. (1950). Quantitative analysis of mixtures of sugars by the method of partition chromatography—Part V. Improved methods for the separation and detection of the sugars and their methylated derivatives on the paper chromatogram. *J. chem. Soc.* **1950**: 1702–1706.

Humphreys, T. (1963). Chemical dissolution and *in vitro* reconstruction of sponge cell adhesions. *Devl Biol.* **8**: 27–47.

Humphreys, T. (1965). Cell surface components participating in aggregation: evidence for a new cell particulate. *Expl Cell Res.* **40**: 539–543.

Kabat, E. A. and Mayer, M. M. (1961). *Experimental immunochemistry*, 2nd ed. C. C. Thomas, Springfield, Illinois.

Kościelak, J. (1965). ABO blood group substances from erythrocytes as lipid antigens. *Proc. 10th Congr. Int. Soc. Blood Transf.*, Stockholm: 494–499.

Kwapinski, J. B. (1965). *Methods of serological research.* John Wiley and Sons, Inc., London and New York.

de Laubenfels, M. W. (1927). Bi-specific conglomerations of sponges. *Carnegie Inst. Wash. Yr. Book* **26**: 219–222.

de Laubenfels, M. W. (1928). Interspecific grafting using sponge cells. *J. Elisha Mitchell Sci. Soc.* **44**: 82–86.

de Laubenfels, M. W. (1934). Physiology and morphology of Porifera exemplified by *Iotrochota birotulata* Higgin. *Pap. Tortugas Lab.* **28**: 37–66.

de Laubenfels, M. W. (1936). Sponge fauna of the Dry Tortugas with material for a revision of the families and orders of the Porifera. *Pap. Tortugas Lab.* **30**: 1–225.

Leloir, L. F. (1964). The biosynthesis of polysaccharides. Sixth International Congress of Biochemistry, New York City, *Proceedings of the Plenary Sessions* **33**: 15.

Lévi, C. (1956). Étude des *Halisarca* de Roscoff embryologie et systematique des Démosponges. *Archs. Zool. exp. gén.* **93**: 1–181.

Lévi, C. (1957). Ontogeny and systematics in sponges. *Syst. Zool.* **4** (4): 174–183.

Lévi, C. (1966). Le glycogène chez les spongiaires. *C. r. Séanc. Soc. Biol.* **160** (3): 651–652.

Lilien, J. E. (1968). Specific enhancement of cell aggregation *in vitro*. *Devl Biol.* **17**: 657–678.

Lowry, O. H., Rosebrough, N. J., Farr, A. L. and Randall, R. J. (1951). Protein measurement with the Folin phenol reagent. *J. Biol. Chem.* **193**: 265–275.

Lüderitz, O., Staub, A. M. and Westphal, O. (1966). Immunochemistry of O and R antigens of *Salmonella* and related enterobacteria. *Bact. Rev.* **30**: 192–255.

MacLennan, A. P. (1963). The sponge cell surface in relation to aggregation specificity. *Biochem. J.* **89** (3): 99 pp.

MacLennan, A. P. (1969). An immunochemical study of the surfaces of sponge cells. *J. exp. Zool.* (In press).

MacLennan, A. P. and Dodd, R. Y. (1967). Promoting activity of extracellular materials on sponge cell reaggregation. *J. Embryol. exp. Morph.* **17**: 473–480.

MacLennan, A. P., Randall, H. M. and Smith, D. W. (1961). The occurrence of methyl ethers of rhamnose and fucose in specific glycolipids of certain mycobacteria. *Biochem. J.* **80**: 309–318.

Manery, J. F. (1966). Effects of Ca ions on membranes. *Fedn Proc. Fedn Assn Socs exp. Biol.* **25**: 1804–1810.

Margoliash, E., Schenck, J. R., Hargie, M. P., Barokas, S., Richter, W. R., Barlow, G. H. and Moscona, A. A. (1965). Characterization of specific cell aggregating materials from sponge cells. *Biochem. biophys. Res. Commun.* **20**: 383–388.

Moscona, A. A. (1963). Studies on cell aggregation: demonstration of materials with selective cell-binding activity. *Proc. natn. Acad. Sci. U.S.A.* **49**: 742–747.

Rondle, C. J. M. and Morgan, W. T. J. (1955). The determination of glucosamine and galactosamine. *Biochem. J.* **61**: 586–589.

Sara, M. (1956). Esperienze di aggregazione cellulare mista dopo dissociazione nelle Calcispongie. *Boll. Zool.* **23**: 113–119.

Sara, M., Liaci, L. and Melone, N. (1966a). Bispecific cell aggregation in sponges. *Nature, Lond.* **210** (5041): 1167–1168; 1 fig.

Sara, M., Liaci, L. and Melone, N. (1966b). Mixed cell aggregation between sponges and the anthozoan *Anemonia sulcata*. *Nature, Lond.* **210** (5041): 1168–1169; 2 figs.

Spiegel, M. (1955). The reaggregation of dissociated sponge cells. *Ann. N.Y. Acad. Sci.* **60**: 1056–1076.

Steinberg, M. S. (1964). The problem of adhesive selectivity in cellular interactions, in *Cellular membranes in development*, Ed. M. Locke, pp. 321–366. Academic Press: London and New York.

Synge, R. L. M. and Wood, J. G. (1958). Bound amino acids in protein-free extracts of Italian ryegrass. *Biochem. J.* **70**: 321–339.

Tekman, S. and Oner, N. (1963). A pentose identified in the carbohydrate group of spongin. *Nature, Lond.* **200**: 77.

Tyler, A. (1946). An auto-antibody concept of cell structure in growth. *Growth* (*Suppl.*) **10**: 7–19.

Weiss, P. (1947). The problem of specificity in growth and development. *Yale J. Biol. Med.* **19**: 235–278.

Westphal, O., Lüderitz, O. and Bister, F. (1952). Uber die extraktion von bakterien mit phenol/wasser. *Z. Naturf.* **7B**: 148–155.

Wilson, H. V. (1907). On some phenomena of coalescence and regeneration in sponges. *J. exp. Zool.* **5**: 245–258.

Wilson, H. V. (1910). Development of sponges from dissociated tissue cells. *Bull. Bur. Fish., Wash.* **30**: 1–30.

Yazhikov, A. A. (1965). Surface antigens in cell aggregation of *Halichondria panicea* (Pallas) and *Reniera cinerea* (Grant). *Zh. Obshch. Biol.* **26**: 96–101.

Symp. zool. Soc. Lond. (1970) No. 25, 325–334.

BIOCHEMICAL ANALYSIS OF SPONGE CELL AGGREGATION

TOM HUMPHREYS

Department of Biology, Revelle College, University of California, La Jolla, California, U.S.A.

SYNOPSIS

A factor, called an *Aggregation Factor*, has been isolated from sponge tissue and appears to function in the normal aggregation of sponge cells. It is demonstrated by its ability to make cells aggregate at low temperatures which normally prevent aggregation. Under these conditions the factor can mimic a very striking normal cellular reaction by causing a mixture of cells from two different species to aggregate species specifically. Biochemical experiments indicate that the factor is a large macromolecule which sediments at about 90 S and contains protein and polysaccharide.

The question of the specificity of sponge aggregation is discussed. It is pointed out that, in general, sponge aggregation is not species specific. Only in those cases where normal aggregation is completely species specific does the aggregation factor exhibit specificity.

INTRODUCTION

Sponges are the classical material for the study of cell aggregation. Over 60 years ago Wilson (1907) noted that sponge cells squeezed mechanically out of their matrix and allowed to settle on glass would crawl about, aggregate, and reform a miniature sponge. Analysis of this phenomenon by Galtsoff (1925) showed, as suggested earlier by Huxley (1912), that the dissociated and randomly mixed cells of various types would sort out relative to one another and take up their proper positions in the aggregate such that the new sponge was formed. This principle, that cells whose normal relationships had been experimentally confused would maintain their tissue specific differentiation and reorganize themselves into their typical patterns of association, has been extended to vertebrate cells during the last 30 years (Holtfreter, 1939; Moscona, 1952) and has stimulated considerable interest in the field of developmental biology.

When, as a developmental biologist, I decided to attempt to study the biochemical reactions which were involved in these cell ordering processes, it was natural that I should return to this classical material for the study of cell aggregation. Indeed, sponges offered many advantages. They were the most simple of metazoan organisms. Their cells aggregated rapidly and well in simple, defined, non-sterile salt solutions. In certain cases when cells of two different species were mixed, they sorted completely from one another and thus exhibited the most diagrammatic cases of specific cell association. Moreover, this specific association could be assayed very easily by using cells from species of two different colors and observing the color of the resulting aggregates. These advantages, which simplified the biological parameters of the system, allowed me to concentrate on the biochemical problems of isolating and characterizing the macromolecules which were involved in cell association. I believe that this choice was the best possible system and has contributed much to whatever success I may claim. I hope that, as in the past, what is learned from sponges will point to appropriate approaches for a profitable analysis of cell association in higher organisms (Lilien and Moscona, 1967; Kuroda, 1968).

I began by trying to develop what I called a chemical dissociation procedure for sponge cells (Humphreys, 1963). I wanted a method which would not destroy the macromolecules which were involved in the adhesion of one cell to another but which would break the chemical bonds between the macromolecules that function in cell adhesions. This proved to be relatively easy because sponge cells dissociated upon soaking a piece of sponge tissue in calcium and magnesium-free sea water. Within a few hours the cells would come out of the matrix, producing a suspension of virtually single cells. When these cells were returned to sea water and allowed to settle on glass, they would aggregate and reform small sponges just as described by Wilson (1907) many years ago for cells dissociated mechanically by squeezing the tissue through fine cloth. In order to eliminate the complications of cell migration, which are inherent in studying aggregation on glass, study of the chemically dissociated cells was carried out in suspension in shaker flasks (Moscona, 1961; Humphreys, 1963). Under these conditions the cells were brought together by the random movement and collision caused by the agitation of the medium. They would aggregate if they were adhesive or otherwise would remain single. Using chemically dissociated cells in these shaker flasks, I was able to demonstrate a factor that apparently functions in cell aggregation. This paper describes briefly the demonstration of the factor, discusses some aspects of species specificity of sponge aggregation and its reflection in species specificity of the factor and briefly describes current evidence of the chemical nature of the factor.

THE AGGREGATION FACTOR

The demonstration of the factor is outlined in Fig. 1 and was initially based upon the observation that chemically dissociated cells of *Microciona prolifera* Ellis and Solander,1786 could not aggregate after they

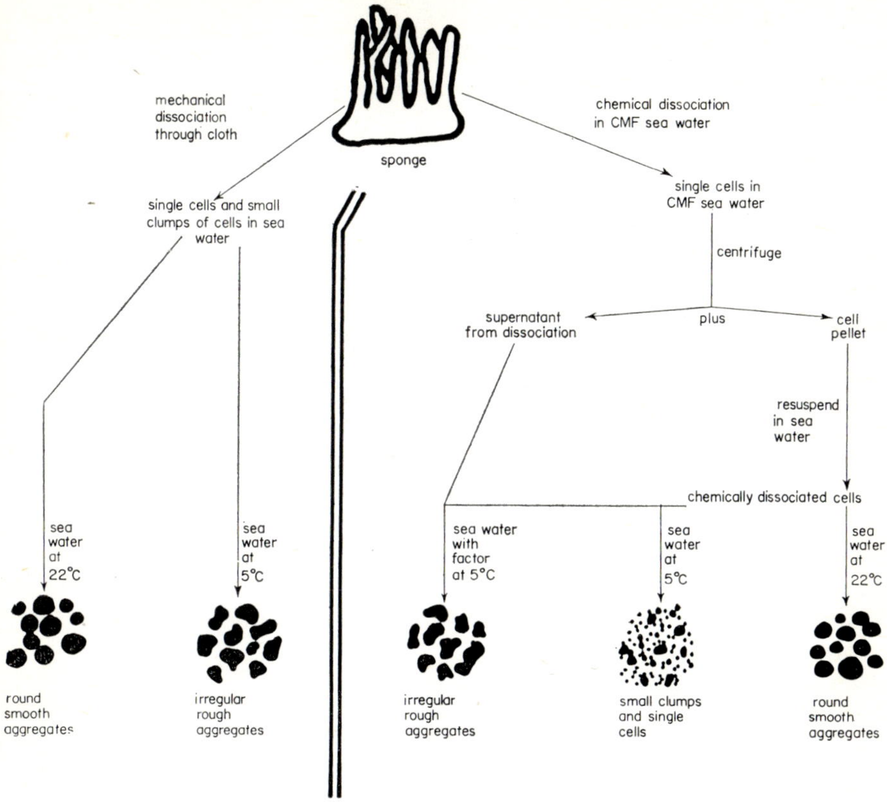

FIG. 1. Scheme for isolation and assay of sponge aggregation factor. As depicted respectively on the right and left of the figure, sponge tissue can be dissociated chemically in calcium- and magnesium-free sea water or mechanically by squeezing it through fine cloth. Cells obtained by either procedure aggregate rapidly in sea water at 22°C and form round, smooth aggregates (extreme right and left). Chemically dissociated cells, however, do not aggregate well in sea water at 5°C (middle right). This failure to aggregate is apparently due to the inability of the cells at 5°C to synthesize a factor which was left behind in the CMF sea water supernatant from dissociation since, if this supernatant is added to the chemically dissociated cells, they can aggregate at 5°C (inside right). This factor is apparently on the cells before washing in CMF sea water since mechanically dissociated cells can aggregate at 5°C (inside left).

were removed from the calcium- and magnesium-free dissociation medium and returned to sea water which was maintained at low temperatures. The inability of the cells to adhere at low temperatures was reversed when some of the supernatant from dissociation was added to the normal sea water in which the cells were suspended. Presumably, this activity of the supernatant indicated the presence of a material which had been removed from the cells during dissociation. This material appeared to be on the cells before chemical dissociation because mechanically dissociated cells could aggregate at low temperatures.

These results, and others I have obtained (Humphreys, 1963, 1965a, 1965b, 1967), fit into an internally consistent interpretation. This interpretation is that the factor is a cell surface material which holds the cells together physically. When calcium and magnesium are removed, this material dissolves from the cell surfaces and leaves the cells non-adhesive. When the cells are kept cold they cannot resynthesize the material and remain non-adhesive, even when calcium and magnesium are returned. However, when this dissolved cell surface material is returned, together with calcium and magnesium, it binds again to the cell membranes and makes the cells mutually adhesive and they aggregate. This interpretation has many assumptions, which I have discussed in previous papers (Humphreys, 1963, 1967). Even though I have made numerous observations which support these assumptions, I do not yet feel that the proof of these assumptions is sound and have thus continued to call the material "aggregation factor", or just "factor" until a more definite demonstration of its nature is available.

SPECIES SPECIFICITY OF THE FACTOR

The most striking evidence that the factor is a normal participant in cell adhesion has been a demonstration for one pair of species that the factor can cause species specific aggregation of cells from the two species. Since I believe my previous failure to outline in detail my experience with species specific aggregation has added to the misunderstanding concerning species specificity of aggregation in sponges, and because some recent results have again demonstrated non-species-specificity of aggregation (Sara, Liaci and Melone, 1966; Curtis, 1962) and of factor (MacLennan and Dodd, 1967), I wish to elaborate on my own experiences concerning species specificity in sponge aggregation.

I became interested in species specific aggregation after the factor had been demonstrated in *M. prolifera* because a test was needed which would rule out the possibility that the factor was a non-physiological product of cell degeneration, such as DNA, which was causing a non-functional agglutination of cells. It was reasoned that, if the cells of two species recognized each other and aggregated completely separately, their factors should be equally species specific if the factors normally functioned during cell aggregation. Therefore, I set about surveying the sponges of Woods Hole for species whose cells exhibited species specificity. I tested seven species which I could personally collect by skin diving so that I could be sure the specimens were healthy. These included *M. prolifera*, *Haliclona oculata* Pallas,1766, *Cliona celata* Grant,1826, *Scypha lingua* Haeckel,1872, *Halichondria panicea* Pallas,1766, *Suberites* sp. and an undetermined species. Of these *Scypha lingua*, *Cliona celata*, and *Suberites* sp., and the unknown species did not form compact, round aggregates or did not even aggregate at all. However, cells of each species were mixed with cells of most of the other species. In every case except two, when the cells of at least one of the species aggregated, the cells of the two species formed mixed clumps. Only in the combinations of *M. prolifera* with *Haliclona oculata* and *Halichondria panicea* with *Haliclona oculata* did the cells of the two species aggregate completely separately. In several combinations the cells of the two species sorted out from one another with time but the initial adhesions of the combinations were non-specific. I have recently been surveying sponges at La Jolla, California, and this rule that sponge aggregation is generally not species specific has been followed. After examining aggregation of cells in shaker flasks from various combinations of 18 different species including *Microciona parthena* de Laubenfels,1930, *Microciona microjoana* de Laubenfels,1930, *Halichondria* sp. and several species of *Haliclona*, I have not found a single pair of sponges in which aggregation is species specific from the beginning of aggregation. As at Woods Hole, there were some combinations in which the cells of the two species separated after initially forming a mixed clump. Thus, sponge aggregation, as I have observed it, is generally not totally species specific but can be so in certain rare exceptions. Some papers in this volume appear to agree (A. P. MacLennan, A. S. G. Curtis).

While I am discussing general aspects of the aggregation of sponges I wish to emphasize two other points which might not be clear in the literature. The cells of very few species of sponges have the capacity to reform a complete functional sponge. Of the species I have examined, only the cells of *M. prolifera* can do this. Also, an aggregation factor cannot be isolated from many species.

I realized all of these generalizations about sponges from the beginning of my experiences with sponges but ignored them because they were not pertinent to my needs. I could use the species specificity exhibited by the exceptional species to test the specificity of the factor. The fact that aggregation and the factor were non-specific in most cases did not matter for my purposes. Therefore, upon finding that mixtures of red cells from *M. prolifera* and lavender-brown cells from *H. oculata* aggregated species specifically, and that I could isolate a factor from *H. oculata,* I proceeded to demonstrate that the factors of these species could cause species specific aggregation (Humphreys, 1963).

First, the specificity of the factor was tested. A large concentration of factor from one species was added to chemically dissociated cells of the opposite species held at low temperatures. In both combinations the factors had no effect, the cells behaving as if they were in plain sea water. Second, the ability of the factors to cause cells of their own species to aggregate within a mixed suspension of cells from both species was examined. Three types of flasks with mixed suspensions containing equal portions of *M. prolifera* and *H. oculata* cells were prepared, one flask each with factor from one species separately, and one with both factors. In flasks to which factor from only one species was added, the cells of that species aggregated separately, while the cells of the other species did not respond and remained unaggregated. The cells in the flasks with both factors all aggregated, but *M. prolifera* cells aggregated only with those of their own species and formed separate red aggregates, while cells of *H. oculata* aggregated only with others of their species and formed lavender aggregates. Thus the factors were able to cause species specific aggregation. This result is reassuring and suggests that the factors can reproduce a very striking normal reaction and so may indeed be a material which normally functions in the association of one sponge cell with another.

Although *M. prolifera* factor is specific in combination with *H. oculata* cells, it seems worth emphasizing that it is not species specific in combination with cells of several other species which do not aggregate species specifically with *M. prolifera*. Its effect on cells of *Halichondria panicea* provide a good example. If cells of *M. prolifera* and *H. panicea* are mixed in a shaker flask at 22°C, they both begin to aggregate together non-specifically, forming finely mixed aggregates. Within 15–30 min, however, the cells of both species always show a separation from one another and are found mainly in separate aggregates. The factor of *M. prolifera* also shows the non-specificity since it causes cells of *H. panicea* to form very tenuous and loosely adhering aggregates. In mixtures of cells of the two species it caused mixed aggregates to form. Unfortunately *H. panicea* does not yield a factor and thus reciprocal experiments were not possible. These results show, however, that a factor which is entirely specific in one combination does not exhibit specificity in other combinations in which specificity would not be expected.

BIOCHEMICAL NATURE OF THE FACTOR

The factor isolated from *M. prolifera* sediments at about 90 S (Humphreys, 1965a, 1967), indicating a molecular weight of at least five million, if the particle is spherical, and more, if it is fibrous. Preparations of the factors, which had been purified 15- to 30-fold over the crude extract, but which were not yet proven to be pure, contain polysaccharide and protein. These results indicated a rather large particle of protein and polysaccharide.

Using a factor isolated from *Microciona parthena*, the nature of the active particle has been investigated more extensively. The factor from this species can be purified such that most of the physical material in the preparation sediments coincidently with the activity of the factor. Figure 2 shows a tracing of optical density at 280 mμ from a

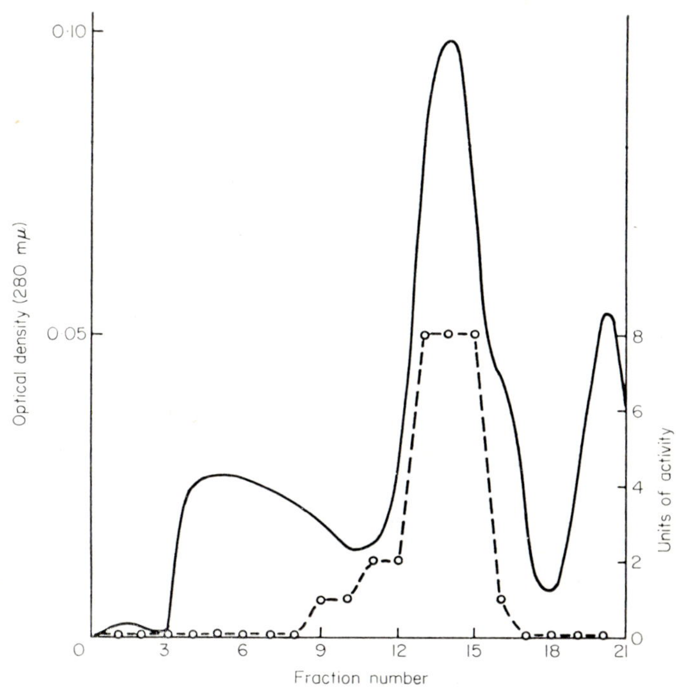

FIG. 2. Sedimentation of aggregation factor. A purified preparation of factor was centrifuged for 6 h at 25 000 × ***g*** on a sucrose gradient as previously described (Humphreys, 1965a). The optical density at 280 mμ in the gradient was recorded as the gradient was pumped through a recording spectrophotometer cell (solid line). Fractions of the effluent from the cell were collected and assayed for aggregation factor (circles) as previously described (Humphreys, 1965a).

zonal sedimentation of the factor preparation on a sucrose gradient. A peak of optical density coincides with a peak in activity suggesting that the material physically present in the peak is the active material. However, it must be stressed that this correlation is not yet experimentally demonstrated and thus remains an assumption upon which the significance of other experiments must rest. The experiments consist of treating the factor with various agents or enzymes to determine what effect they have on the physical particle as well as on the activity.

Removal of calcium ions destroys irreversibly the activity of the factor (Humphreys, 1963, 1967). One way of interpreting this requirement for calcium ions is that the large particles are made up of smaller subunit macromolecules (Margoliasch *et al.*, 1965) which associate by bonds involving calcium ion bridges to form the large structure. These subunits separate when the calcium ions are removed and cannot cause aggregation individually. Alternatively, calcium ions could be necessary for the active configuration of macromolecular groups on the factor particle which function in cell adhesion. Sedimentation experiments with factor inactivated by calcium ion removal showed that the physical particle remained intact. This result appears to rule out the first hypothesis and supports the second, although it is possible that the calcium ions hold small reactive molecules to the surface of a large particle.

Sodium dodecyl sulfate is a strong ionic detergent which generally separates all macromolecules not held together by covalent bonds. For example, it dissolves the ribosome into its individual RNA and protein macromolecules (Girard, Penman and Darnell, 1964). When the factor particles were treated with a 0·5% of solution of the detergent, they remained active and physically intact. Thus, the suggestions that the factor is made of smaller subunits appears invalid. Instead, the large particle seems to be held together by covalent bonds and thus may be considered to be one large macromolecule.

Proteases destroy the activity of the factor (Gasic and Galanti, 1966). If the factor is treated with high concentrations (1 mg/ml) of pronase which destroy activity, the physical particle remains intact. The proteins which are important for activity are thus external components of the particle and do not form any integral part of its structure. One might be tempted to suggest that protein is not a structural component at all; however, the structural proteins, as opposed to the active proteins, may simply be protected from pronase because they are inside the particle.

Polysaccharides appear also to be a component of the factor particle and may provide the structural core of the particle. This is suggested by the considerable portion of polysaccharide in the active preparations. It is confirmed by measuring the density of the factor in CsCl. The activity bands at about 1·5 to 1·55 g/cm^3. Since proteins rarely have a density greater than 1·35, the particle must contain some other constituent. The failure of the factor to show an absorption peak at 260 mμ indicates the presence of less than a few percent of nucleic acid which thus need not be considered as a major component. In addition, nucleic acids are not detectable by other means and the factor is not susceptible to ribonuclease or deoxyribonuclease. Since polysaccharides constitute presently about 70% of the material in the most pure preparations, and are the only other usual cellular constituents which could readily account for the high density, it seems most likely that they are a component of the active particle.

Together, these results suggest that the factor particle is one large macromolecule made up of polysaccharide, possibly as a structural element, with proteins involved in the functional activity of the particle and possibly attached to the surface of the polysaccharide molecule. However, there are still many important questions concerning this interpretation which these experiments do not answer and so additional researches should prove to be very interesting.

Acknowledgements

This work was supported by N.S.F. Grant # GB 7216. I wish to acknowledge the expert and cheerful assistance of Mrs. Pamela Lindstrom.

References

Curtis, A. S. G. (1962). Pattern and Mechanism in the Reaggregation of Sponges. *Nature Lond.* **196**: 245–248.

Galtsoff, P. S. (1925). Regeneration after dissociation (An Experimental Study on Sponges) II Histogenesis of *Microciona prolifera* Verra. *J. exp. Zool.* **42**: 223–251.

Gasic, G. J. and Galanti, N. L. (1966). Proteins and Disulfide Groups in the Aggregation of Dissociated Cells of Sea Sponges. *Science N.Y.* **151**: 203–205.

Girard, M., Penman, S. and Darnell, J. E. (1964). The effect of Actinomycin on Ribosome Formation in HeLa Cells. *Proc. Natl. Acad. Sci., U.S.A.* **51**: 205–211.

Holtfreter, J. (1939). Gewebeaffinität, ein Mittel den embryonal Formbildung. *Arch. exp. Zellforsch.* **23**: 169–209.

Humphreys, T. (1963). Chemical Dissociation and *in vitro* Reconstruction of Sponge Cell Adhesions. *Devl. Biol.* **8**: 27–49; 8 figs.

Humphreys, T. (1965a). Cell Surface Components Participating in Aggregation: Evidence for a New Cell Particulate. *Exp. Cell Res.* **40**: 539–543.

Humphreys, T. (1965b). Aggregation of Chemically Dissociated Sponge Cells in the Absence of Protein Synthesis. *J. exp. Zool.* **160**: 235–240.

Humphreys, T. (1967). The Cell Surface and Specific Cell Aggregation: pp. 195–210 in *The specificity of cell surfaces* (Davis and Warren, Eds). Prentice-Hall, Inc.

Huxley, J. (1912). Some phenomena of regeneration of *Sycon* with a note on the structure of its collar cells. *Phil. Trans. R. Soc.* B, **202**: 165–190.

Kuroda, Y. (1968). Preparation of an Aggregation-Promoting Supernatant From Embryonic Chick Liver Cells. *Expl Cell Res.* **49**: 626–637.

Lillien, J. E. and Moscona, A. A. (1967). Cell Aggregation: Its Enhancement by a Supernatant From Cultures of Homologous Cells. *Science N.Y.* **157**: 70–72.

MacLennan, A. P. and Dodd, R. Y. (1967). Promoting Activity of Extracellular Materials on Sponge Aggregation. *J. Embryol. exp. Morph.* **17**: 473–480.

Margoliasch, E., Schenck, J. R., Hargie, M. P., Barokas, S., Richter, W. R., Barlow, G. H. and Moscona, A. A. (1965). Characterization of Specific Cell Aggregating Materials from Sponge Cells. *Biochem. biophys. Res. Commun.* **20** (4): 383–388.

Moscona, A. (1952). Cell Suspension from Organ Rudiments of Chick Embryos. *Expl Cell Res.* **3**: 535–539.

Moscona, A. (1961). Rotation-mediated Histogenetic Aggregation of Dissociated Cells. *Expl Cell Res.* **22**: 455–475.

Sara, M., Liaci, L. and Melone, N. (1966). Bispecific Cell Aggregation in Sponges. *Nature N.Y.* **210** (5041): 1167–1168; 1 fig.

Wilson, H. V. (1907). On Some Phenomena of Coalescence and Regeneration in Sponges. *J. exp. Zool.* **5**: 245–258.

Symp. zool. Soc. Lond. (1970) No. 25, 335–352.

PROBLEMS AND SOME SOLUTIONS IN THE STUDY OF CELLULAR AGGREGATION

A. S. G. CURTIS

Department of Cell Biology, University of Glasgow, Glasgow, Scotland

SYNOPSIS

The sources of error in assessing the adhesiveness and specificity of adhesiveness of cells by use of reaggregation techniques are discussed in terms of the biophysics of such systems. A new technique for obtaining an absolute measurement of the adhesiveness of reaggregating cells is described: this technique avoids sources of error inherent in earlier techniques. The particular use of this technique to investigate whether or not specific adhesion occurs is described in theory. This new method does not require a means of recognizing the species or tissue type of individual cells. Experimental results using this technique to study the specificity of adhesion of mixtures of cells from the sponges *Halichondria panicea* Pallas, 1766, *Suberites ficus* Gray, 1806 and *Microciona fallax* Bowerbank, 1866 are reported. No evidence of specificity of adhesion was found.

INTRODUCTION

Studies on the reaggregation of cells commenced with the demonstration by Wilson (1907) that sponges could reform from suspensions of their cells. He found that there appeared to be a specificity in the aggregation such that, if cells from two sponge species were mingled, the resulting aggregates were composed each of cells of a single species type. This very remarkable phenomenon has been reinvestigated on a number of occasions, in particular by Huxley (1921), Galtsoff (1925) and by Humphreys (1963). It has greatly influenced thought and experiment on the morphogenesis of metazoan animals and the concept of "specific adhesion" has arisen primarily from work on sponge cell aggregation.

However, it is worth considering the nature of the evidence on which this concept of specific adhesion is based. Aggregates have been prepared by two main techniques. In the first method, the cells are allowed to sediment under gravity onto the bottom of a culture dish (during the sedimentation there may be an appreciable number of collisions between the cells). The cells then move around on the dish until they collide with one another, perhaps adhere and build up into aggregates. One of the defects of this technique is that it is quite possible that cells of like type are specifically attracted to move towards one another, with the consequence that any specificity seen is due to some process such as specific chemo-taxis or -kinesis and not to specific adhesion. The second method depends upon agitating a suspension of cells so that they collide and adhere to one another at collision to build up aggregates. This technique was introduced by Gerisch (1960) and was used by Humphreys (1963) to prepare aggregates of sponge cells. This technique avoids the criticism made above of the earlier technique. However, other criticisms can be advanced against the shaker technique and these will be considered shortly.

The investigator requires a technique to recognize whether or not individual aggregates are composed solely or mainly of cells of a single species or tissue type or whether each aggregate appears to contain contributions from each cell type present in the suspension in approximately the proportions they were present in the original suspension. Although a variety of techniques of a fairly reliable nature are available for the recognition of the type of an individual cell (e.g. radioactive labelling (Weston, 1963), immunofluorescent labelling (Okada, 1965), these techniques have never been used in the study of sponge aggregation as far as I know. Nearly all workers have used the natural colour of the different species of sponge as a means of recognizing the species type of cells in aggregates. The method of using the staining reactions of the cells as a means of their recognition was tried by Curtis (1962). Use of either of these techniques appears to be fraught with possibilities of error, and the probability of recognizing a single cell of one type in an aggregate otherwise composed of cells of the other type is very low when such techniques are used.

For these reasons, I have set out to devise a technique to test for the occurrence of specific adhesion which avoids the errors of existing methods. The resulting technique described in this paper has the advantage also that it provides a means of measuring the absolute adhesiveness of the cells: the magnitude of this value may reveal much about the mechanisms of the adhesion (see Curtis, 1966). The technique is based on several biophysical considerations which in themselves suggest that there may be other possible sources of error in existing methods of preparing and studying aggregates. Since these considerations are generally applicable to the whole field of aggregation studies, there follows an extended treatment of these points.

THEORETICAL CONSIDERATIONS

General

The two main techniques of preparing aggregates are, as has been mentioned, (i) sedimentation and cellular movement in a still culture medium and (ii) shaker aggregation. The first of these systems has never been analysed in biophysical terms but it seems that two separate phenomena which aid aggregation take place in such a system. During the sedimentation of the cells to the bottom of the culture dish, collisions occur between the cells. A proportion at least of these collisions lead to the cells involved adhering together. After the cells are on the bottom of the culture dish they move around, either randomly or in an oriented manner, and as a result of their movements make collisions with one another and these also help to build up aggregates.

The shaker system also leads to the collision of cells with one another. Humphreys (1963) used this technique to obtain aggregation of the sponge *Haliclona oculata* Pallas,1766. A variety of shaking motions have been used by different workers, e.g. end-over-end shaking of a tube (Gerisch), reciprocating motion (Curtis and Greaves, 1965) and gyratory shaking (Moscona, 1961). Shaking will produce gradients of velocity of flow of the medium inside the culture flasks but the physical processes which take place as a result of such shaking are most easily understood by considering collision processes amongst particles in suspension in a laminar flow. In this discussion, it will be assumed that cells are equivalent to particles which are spheroids and of a density inappreciably different from that of the suspending medium. These two assumptions are often approximately correct for suspensions of cells. With refinements the theory could be made to apply to non-spheroidal cells. When particles are placed in a laminar shear flow (a shear flow is one in which a velocity gradient exists in a direction normal to the direction of flow), collisions will take place between particles in suspension. The reason for this is that particles whose centres lie on different streamlines will be moving at different velocities from particles in other streamlines. Thus, whenever two particles lie on streamlines less than one particle diameter in separation they must inevitably collide if the faster moving particle starts behind the more slowly moving one. The instantaneous collision rate b_{ij} in such a case is given by the relationship

$$b_{ij} = \tfrac{4}{3}\mathbf{G}n_i n_j(r_i + r_j)^3 \qquad (1)$$

where $\mathbf{G}$ is the shear rate, n_i and n_j the number of particles per unit volume of types i and j of radii r_i and r_j, respectively. This relationship is due to Smoluchowski (see Swift and Friedlander, 1964).

Mason and his co-workers (Mason and Bartok, 1959 and Karnis Goldsmith and Mason, 1966) have shown both experimentally and theoretically that particles which do not possess any force of attraction for one another will not form aggregates as a result of such collisions.

If an additional force of attraction exists between the particles, however, then a proportion of all of the collisions will lead to the capture of one particle by the other. In this way, a two-member aggregate is built.

These aggregates may undergo further collisions either with single particles or aggregates and if adhesions form at a proportion of the collisions large aggregates will form.

This idealized situation is approached very closely indeed in a Couette flow system in which a rotational laminar shear flow is set up. (Mason and Bartok, 1959.) However, as aggregates build up in such a system, they will tend to a limiting size. The reason for this is that the shear forces acting on an aggregate tending to pull it apart increase as the square of the radius of the aggregate (Albers and Overbeek, 1960). Thus, small particles (i.e. single cells) will be brought into collision by the action of shear and may adhere and form aggregates as a result but, as the aggregates increase in size, shear effects will act to limit the size of individual aggregates. The equilibrium size of an aggregate will depend upon the average adhesiveness of its component particles and the rates of breakup due to shear effects versus the rate of building up of aggregates under the action of shear. Shear flow conditions will be produced in any type of shaking system and this accounts for the effectiveness of so many types of shaker in producing the collisions between cells by which their adhesive properties become manifest by the formation of aggregates.

There are, however, two defects inherent in the techniques of preparing aggregates currently in use. In shaker systems of either the reciprocating or gyratory type the flow is such that the shear rate varies greatly both with time and in different parts of the culture container. This large standard deviation in the shear rate will in turn affect the collision rate and thus lead to a degree of irreproducibility in the measurement of aggregation using conventional shaker systems. An unfortunate consequence of the variance in shear rate is that aggregates will be limited in size by the maximum transient shear force while the rate of formation of aggregates will be proportional to the average shear rate. Clearly, this is not an efficient method of preparing aggregates. A related defect applies particularly to the widely used technique of gyratory shaking. In such systems, the flow rate will be approximately zero in the centre of the flask. The marked velocity gradient in such flasks leads to the migration of aggregates towards the centre of the flask. (By comparison, in a Couette viscometer the minimum velocity of flow is only slightly less than the maximum velocity.) Consequently, the aggregates which have already formed tend to separate out from the unaggregated cells. This diminishes the rate of build up of aggregates since the collision rate is proportional to the square of the population density. As a result, the use of the gyratory shaker may give a misleading impression of the relative adhesiveness of two cell types, greatly accentuating small differences in adhesiveness. The chief defect of all existing techniques is that it is impracticable to obtain any realistic measurement of the shear rate **G**. Thus it is impossible to standardize collision rates or to obtain any reasonable idea of their value.

A new technique for investigating cellular adhesiveness: theory

These difficulties can be resolved by the use of an instrument in which laminar shear flow conditions are attained. The Couette viscometer (see Fig. 1) is suitable for this. The instrument consists of a concentric pair of stainless steel cylinders separated radially by a gap of ca. 1·5 mm. The outer cylinder can be rotated by an electric motor, situated underneath it, at speeds between 0 and 0·4 rev/sec. The inner

FIG. 1. A general view of the Couette viscometer. The inner bob is shown on the right hand side of the apparatus.

cylinder or bob hangs freely from a torsion suspension while the instrument is in use. The space between the cylinders is filled with a cell suspension and the outer cylinder is set in motion. If a suitable design of this Couette-type viscometer has been chosen (see Van Wazer, Lyons, Kim and Colwell, 1963), the flow conditions in the suspension are extremely close to ideal Couette laminar shear flow and the shear rate is practically of constant value at all radial positions in the gap. The value of the shear rate, **G**, is calculated from the geometry of the instrument and the rate of relative rotation of the two cylinders. It is important to realize that in this work the viscometer is not being used to measure viscosity but rather that it is being used to provide constant shear conditions of known value. It is hoped to publish a full description of the instrument (A. S. G. Curtis, in preparation).

Use of this instrument provides a technique for studying aggregation which has several particular advantages. These are:

(1) a constant shear is applied to the cell suspension. As a result of the extremely small standard deviation in the shear rate, the possibility of discrepancies between the maximum size of the aggregates and the rate of aggregation will be minimized.

(2) The small range and gradient of flow velocities in the gap will have the effect of diminishing the separation of aggregates from unaggregated cells, a phenomenon which is one of the main defects of the gyratory shaker.

(3) The shear rate is known so that Eq. (1) can be evaluated. It is extremely valuable to have a knowledge of the value of **G** because it can be used to obtain a quantitative measurement of cellular adhesiveness. The theory and technique of such a measurement is explained in the following paragraph.

When collisions take place between particles such as cells, there is a certain probability that these collisions will be effective in producing adhesions between the particles. This probability α is the probability that two particles will adhere at a collision: it is often termed the stability ratio and is a measure of the effectiveness of collisions in forming aggregates. If there is no adhesive force between the particles, we may write $\alpha = 0$. The greater the adhesive force between the particles, the greater the value of the probability α. Fuchs (1934) developed a technique for deriving the adhesive energy between two particles from the value of the stability ratio. A. S. G. Curtis and L. M. Hocking (in preparation) have shown this technique to be erroneous, because assumptions are made about the rate of approach of two particles which are unjustifiable in most cases, and have developed a new theory for deriving the absolute value of adhesive forces between particles from values of α. Swift and Friedlander (1964) have shown that, in a Couette flow system, the total number of particles of all classes (i.e. single particles, 2, 3, 4, ..., n membered particles) per unit volume $N_{\infty t}$ at time t is related to the initial number of particles per unit volume (if they were monodisperse) $N_{\infty o}$ by the relationship

$$\ln \frac{N_{\infty t}}{N_{\infty o}} = - \frac{4\mathbf{G}\,\phi\alpha t}{\Pi} \tag{2}$$

where $\mathbf{G}$ is the shear rate and ϕ the volume fraction occupied by the particles. This relationship is derived by Swift and Friedlander by a suitable though complex integration of Eq. (1). Application of Eq. (2) immediately allows the experimental determination of values of the stability ratio (or effective collision rate) for any type of suspension by use of the Couette viscometer to provide a known value for $\mathbf{G}$ and measurement of $N_{\infty o}$ and $N_{\infty t}$ by a particle counting technique, and of ϕ by a total particle volume measurement. Such a technique does not appear to have been available previously and indirect methods of obtaining values of α had to be used (see Wilkins, Ottewill and Bangham, 1962), or use has had to be made of unjustified values (see Steinberg and Roth, 1964).

Two points should be made concerning this new technique of measuring the stability ratio for the adhesion of particles (and thus of their absolute adhesiveness). First, it should be appreciated that the collisions produced by the action of Brownian motion in such systems will form only a small proportion of all collisions providing that $\mathbf{G}$ has a value appreciably greater than zero, the size of single particles be fairly large (say 5 μ diameter), and the population density of single particles such that they occupy only a small fraction of the volume of the suspension (see Steinberg and Roth, 1964). Thus, no correction for the action of Brownian motion needs to be made in such measurements provided the above conditions are maintained.

A second point, which is of especial application to studies on sponge reaggregation, arises from the consideration of the effect of the viscosity of the suspending medium on cellular aggregation. In shaker systems, whether gyratory or reciprocating, if two flasks are shaken at the same rate but have media of different viscosity in each of them then the average shear rate (and collision rate) will be lower in the flask with the more viscous medium. This has certain consequences for the aggregation of cells in the two flasks. Since the average shear rate is lower in the flask with the more viscous medium, the rate of aggregation (other things being equal) will be slower in that flask. The maximal shear rate will also be lower in the flask with the more viscous medium, however, and thus larger aggregates will survive against break-up than in the other flask so that larger aggregates will be formed eventually in the more viscous medium. Therefore, the paradox may occur that cells which are equally adhesive will form larger aggregates in the flask with the higher viscosity culture medium, even though their aggregation will be slower. An even more unsatisfactory result is possible, namely that cells of type A, which are less adhesive than those of type B, would form larger aggregates if the viscosity of the medium in their flask was higher than that in the flask containing B type cells. Since the size of aggregates has been widely used as a measure of cellular adhesiveness and since no measures have been taken to control the viscosity of culture media in such experiments, it is apparent that misleading results may have been obtained. In addition, several workers (e.g. Humphreys, 1963) have claimed that certain macromolecular materials specifically or non-specifically increase the adhesiveness of aggregating cells as judged by the size of the aggregates formed. It may be that these substances have done no more than increase the viscosity of the medium with the consequent effect on aggregate size. Since much of the evidence for specific adhesion of cells, promoted by macromolecular materials, is based on experiments of this type, it seemed desirable to develop new techniques which would avoid these sources of error. Use of the Couette viscometer for aggregation of cells entirely avoids these sources of error. If the viscosity of the suspending medium placed in the viscometer is increased, then it is necessary to raise the power input to the machine in order to run it at a constant rotational speed. Provided the speed of rotation is kept constant, viscosity changes have no effect on shear rate and thus in turn no effect on collision rate, aggregation rate or break up rate.

It can now be appreciated that existing shaker techniques for investigating the aggregation of cells are subject to several sources of serious error and that, in particular, experiments involving the introduction of materials which may alter the viscosity of the suspension medium, such as various macromolecular fractions, are potentially misleading. It is clear, however, that the use of the Couette technique for studying aggregation provides a superior method for investigating problems of cellular aggregation. With these considerations in mind, I have designed a new means of testing for the occurrence of specific adhesion of cells.

A new test for the specific adhesion of cells: theory

Examination of Eq. (1) shows that the collision rate for any class of particle is proportional to the square of the population density if we consider only one type of particle (i.e. i particles are the same as j particles). Suppose that cells of two types A and B of equal adhesiveness are being aggregated in a mixture and that there is no specificity in their adhesion, then their aggregation rate will be proportional to $(n_A+n_B)^2$, where n_A and n_B are the population densities for A and B type cells, respectively. Consider, however, the sitation in which adhesions are specific so that cells will only adhere to other cells of their own type. Then there are in effect, two entirely separate populations of aggregating cells and their aggregation rates are respectively proportional to $(n_A)^2$ and $(n_B)^2$. The combined aggregation rate is proportional to $(n_A)^2+(n_B)^2$. Thus, if specific adhesion occurs, the aggregation rate of a mixture of two types of equally adhesive cell will be reduced by a factor $[(n_A)^2+(n_B)^2]/[(n_A+n_B)^2]$, relative to the situation in which there is no specific adhesion. The effect of this can be appreciated easily by considering the aggregation of two cell types, which both have the same adhesiveness (stability ratio) when considered as single cell types, mixed in equal proportions. If there is no specificity in their adhesions the measured stability ratio for the mixture will be exactly the same as for the pure cell types, but if there is specificity of adhesion, the measured stability ratio for a 50 : 50 mixture will be exactly 50% of the value for the pure cell types. The expected relative stability ratio for two cell types mixed in differing proportions and where one cell type has the same adhesiveness (stability ratio), or 50%, or 25%, of the stability ratio of the other type is shown in Fig. 2. Both the case of non-specificity of adhesion and the case of specificity of adhesion are represented here.

Thus, the proposed technique for the investigation of the question of the specificity or otherwise of cell adhesion involves the measurement of the stability ratios of two pure cell types and of mixtures of the two types in varying proportions and then the calculation of whether the measured stability ratios for the mixture are of lower value than would be expected if adhesion was non-specific. There are several advantages to this technique. First, there is no problem concerning the identification of the species or tissue type of cells in the aggregation. It is not necessary to know the type of any individual cell whether in an aggregate or unaggregated. Second, this technique allows the recognition of the degree of specificity that may operate; for example only half the cells of one type may show specific adhesion while the other half show non-specific adhesion. No technique so far introduced has allowed an answer to the question of the completeness of the specificity that has been claimed. Third, the technique avoids the problems associated with traditional methods of studying aggregation; this point has been discussed above. For these reasons, this new technique has been applied to the question of the specificity of sponge cell reaggregation.

MATERIALS AND METHODS

Cell suspensions were prepared from three species of sponge, *Halichondria panicea*, *Suberites ficus* and *Microciona fallax*. The last of these species occurs as a commensal on the shells of *Aequipecten opercularis*. All were obtained in the vicinity of Millport, Isle of Cumbrae, Buteshire, U.K., by shore collection or dredging. The suspensions were made by first cleaning small fragments of the sponges of detritus and of other organisms with a pair of fine forceps. The small pieces of sponge were then washed quickly twice in an artificial sea water medium containing no divalent cations (of composition sodium chloride 35 g/l, potassium chloride 7·0 g/l, 2-amino-hydroxymethyl 1,3-propanediol 3 g/l, buffered to pH 7·6) and transferred to a medium containing the chelating agent diamino-ethane tetracetate 0·001 M pH 7·8 (in the calcium- and magnesium-free artificial sea water). After exposure of the sponge fragments to this medium for 5 min, the medium was replaced with two changes of the calcium- and magnesium-free artificial sea water (CMF sea water) and the sponge fragments were dispersed into a cell suspension by suction several times through a fine pipette orifice. The cell suspensions were allowed to stand for 5 min so that spicules and other inorganic matter could sediment. They were then centrifuged at ca. 300 ***g*** for 5 min while cooled to 1°C to sediment the whole cells. The supernatant, which contains bacteria, protozoa, cell fragments, etc., was discarded. The cells were resuspended in the same Ca- and Mg-free artificial sea water and the centrifugation repeated. After the second centrifugation, the cells were resuspended in 5 ml of this saline. This centrifugation process reduced the bacterial and protozoal contamination of the cell suspensions below 1000 organisms/ml. The suspension was then diluted with sterile natural sea water (sterilization was effected by use of a sterile filtration technique) to a final volume of 30 ml and single cell population densities of accurately known value within the range $0{\cdot}6 \times 10^6 - 2 \times 10^6$ cells/ml. All these operations except the centrifugation, were carried out at room temperature (20–25°C). Microscopic examination of the cell suspensions showed that they were in all cases composed of more than 85% single cells. Viability tests on the suspensions were carried out by plating out the cells in Falcon Petri dishes and measuring the percentage of cells that adhered and spread on the culture dish surface. Although this is only an approximate measure of viability, it appears that at least 75% of the cells were alive.

Cell suspensions containing cells of two species types in known proportions were prepared by mixing known volumes of the suspensions of each species type. The population density of cells in pure and mixed species suspensions was measured by the use of phase-contrast microscopy of haemocytometers to obtain cell counts per unit volume. The same technique was used to obtain total particle population density counts $N_{\infty t}$ at intervals after aggregation started. Immediately after the pure species type or mixed species type suspensions had been prepared and a sample removed for measurement of $N_{\infty 0}$ the suspension was placed in the Couette viscometer. The outer cylinder of the viscometer was then set in motion at a known speed of rotation. Samples of the suspension were removed at intervals and the value of $N_{\infty t}$ measured by haemocytometry. After about one hour, during which time an appreciable amount of aggregation normally takes place, the measurement was completed with five values for $N_{\infty t}$ being obtained at known values of t. The suspension was then removed from the viscometer and examined for the presence of visible aggregates (these were invariably present after one hour of aggregation).

RESULTS

The stability ratios (i.e. the probability that a collision between two cells or aggregates results in the formation of an adhesion) were measured for the aggregation of suspensions of cells from *Microciona fallax*, *Suberites ficus* and *Halichondria panicea*. Measurements of the stability ratio were made for suspensions of single and of mixed species types. The measurements are, of course, average measurements even for cells of a single species type and do not reveal whether the various types of cell in a sponge body differ in their adhesiveness. The experiment was designed so that a majority of measurements of the stability ratio of a pair of species in mixed suspension was obtained for mixtures containing less than 50% of the more adhesive cell type. The reason for this is illustrated in Fig. 2 from which it can be seen that the deviation between the curves for non-specific and specific adhesion are maximal for cell mixtures containing less than 50% of the more adhesive type. The mean values and standard deviations for the stability ratio measurements for the pure cell types and for their mixtures in pairs in various proportions are given in Table I.

Each mean value for the stability ratio of a mixture or pure species given in Table I or plotted in Fig. 3 is the mean of a set of successive measurements obtained from the aggregation of one cell suspension over the period of approximately one hour. The small values of the

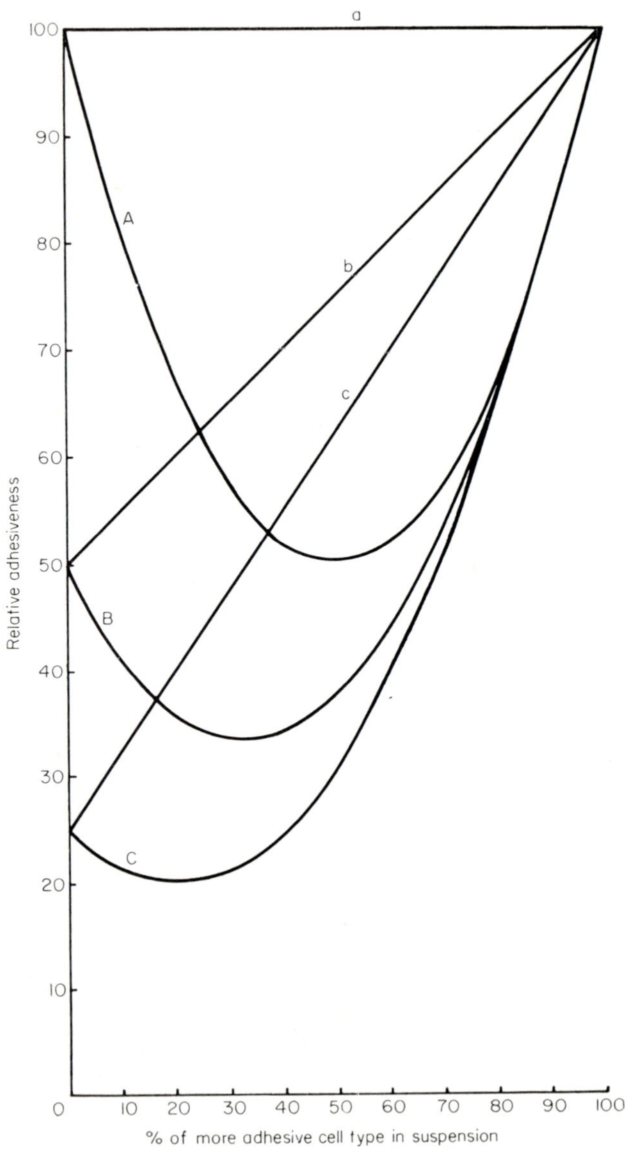

TABLE I

Stability ratios for mixtures of cells of pairs of sponge species

	Value of stability ratio α	standard deviation of α
A. *Halichondria panicea* and *Suberites ficus*		
% of *H. panicea* cells		
0	0·158	0·022
22	0·168	0·036
33	0·177	0·006
54	0·179	0·010
75	0·181	0·016
92	0·185	0·003
100	0·188	0·018
B. *Microciona fallax* and *Halichondria panicea*		
% of *M. fallax* cells		
0	0·188	0·018
12	0·195	0·008
26	0·387	0·036
42	0·420	0·010
100	0·720	0·098
C. *Microciona fallax* and *Suberites ficus*		
% of *M. fallax* cells		
0	0·158	0·022
15	0·215	0·018
27	0·450	0·031
45	0·465	0·007
62	0·590	0·074
100	0·720	0·098

$n = 5$, for each value of α.

standard deviations of these means (see Table I) suggest that the adhesiveness of the cells remains constant over the period of measurement—there is neither a decrease nor an increase of adhesiveness. The relationships between stability ratio and the proportions of cell types

FIG. 2 (facing page). This graph shows the theoretical relationships between relative adhesiveness (stability ratio) and the proportions of two cell types in a reaggregating suspension for the two conditions of specific and non-specific adhesion. Curve A shows the relative adhesiveness that would be expected if specific adhesion operates for two cell types which have quantitatively the same adhesiveness when this is measured for either cell type alone. Curve a shows the relative adhesiveness that would be expected for two cell types of equal adhesiveness (when this is measured for either cell type alone) if there is no specificity of adhesion. Curves B and b refer, respectively, to the specific and non-specific adhesion of mixtures of two cell types, one of which has 50% of the adhesiveness of the other. Curves C and c refer respectively to the specific and non-specific adhesion of mixtures of two cell types, one of which has 25% of the adhesiveness of the other.

in the three mixtures of pairs of the three species are plotted in Fig. 3. The linear regressions for these plots are shown as lines in Fig. 3 and are of the equations

a. *Halichondria panicea – Suberites ficus*

$$y = 0{\cdot}162 + 0{\cdot}00026x \qquad s_b < 1 \times 10^{-5}$$

where x is the % of *Halichondria panicea* cells in the mixture and s_b the standard deviation of the regression coefficient.

b. *Halichondria panicea – Microciona fallax*

$$y = 0{\cdot}185 + 0{\cdot}0054x \qquad s_b < 1 \times 10^{-5}$$

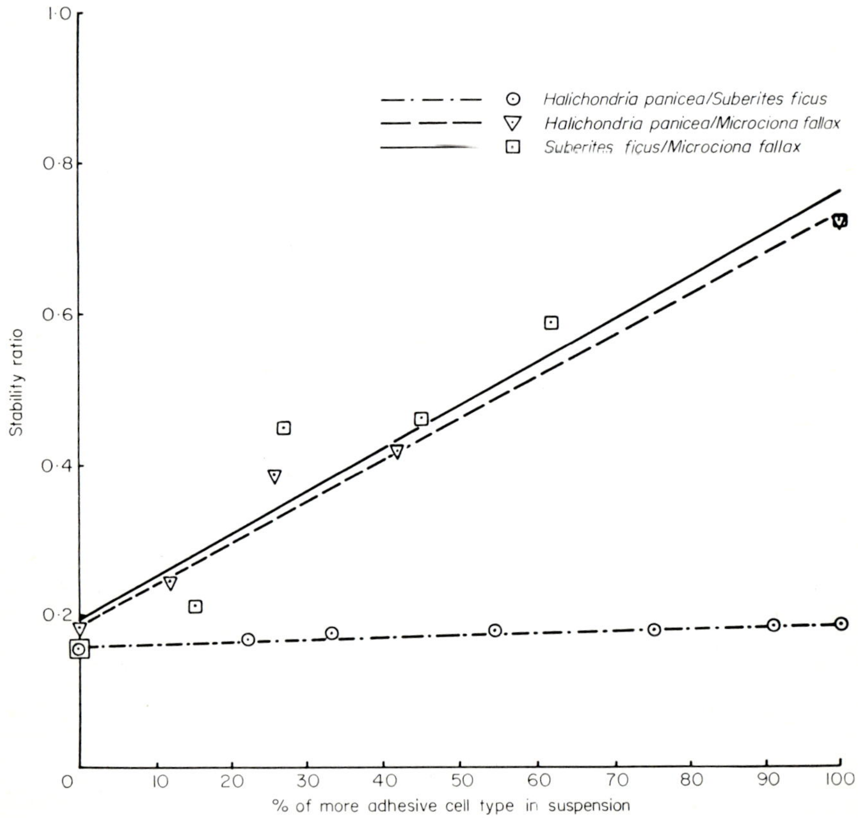

FIG 3. This graph shows stability ratio ("relative adhesiveness") as ordinate and the proportions of cells of two species of sponge as abscissa plotted for the experimental measurements on three pairs of species of sponge. The lines are the respective linear regressions.

and

c. *Suberites ficus – Microciona fallax*

$$y = 0{\cdot}195 + 0{\cdot}0057x \qquad s_b = 0{\cdot}00085$$

In each of cases b and c, x is the % of *Microciona fallax* cells in the mixed suspensions.

Examination of the regressions plotted in Fig. 3 and of the values of the standard deviations of the regression coefficients shows that the linear regressions fit the data extremely well for each pair of species. This result means that there is no evidence of specificity of adhesion of the cells of any one species amongst any of the three pairs of species whose aggregation was examined in this work. An estimate of the precision of the experimental results can be obtained by examining the question of the amount of specificity of adhesion which might occur but which would not be detected by these results. Considering the combination *Suberites ficus: Microciona fallax*, the results for which have a greater variance from the regression than the other combinations, the greatest difference between the values of the stability ratio for specific and non-specific adhesion would occur at $x = 25\%$ approx. Using the test given by Snedecor and Cochran (1967, p. 154), to give the 95% confidence limits in the value of the stability ratio at $x = 25\%$, it is found that there is a less than 0·05 probability of a value of the stability ratio being obtained such as would indicate that 20% of the cells adhere specifically, the remainder non-specifically. Obviously, the probability of obtaining a value of the stability ratio indicating a higher proportion of specifically adhering cells is even lower. The results for the other combinations of species show even lower probabilities of the occurrence of even a small degree of specific adhesion.

DISCUSSION

In general, it can be said that the experimental results clearly do not fit the hypothesis that cells of these species of sponge show specificity of adhesion under the conditions of reaggregation used in this work. They also do not fit the hypothesis that a small proportion of the cells show specificity of adhesion, unless the proportion be of very small size. It may be argued that the particular conditions of cell dispersion (dissociation) and reaggregation were such as to destroy or damage mechanisms for specific adhesion. However, conditions of dispersion and reaggregation were in many respects similar to those used by Humphreys (1963) in which he found evidence of specificity of adhesion. When Humphreys took the sponges *Microciona prolifera* and *Haliclona oculata*, dispersed them in calcium- and magnesium-free artificial sea water and washed them in this medium, he found that their cells reaggregated rapidly at temperatures between 18 and 25°C. Practically all single cells were reported to have entered aggregates within one hour of the start of aggregation. As a result of experiments on aggregation at lower temperatures, Humphreys suggested that an adhesive substance was removed from the sponges by the method of dispersion and that it could only reform when aggregation was carried out at moderately high temperatures, e.g. 18–25°C. If this is the case, it might be expected that the cells would grow more adhesive during the course of aggregation. In the experiments reported here, the final washings of the cells in CMF sea water were carried out at low temperature so that the cells would not be expected to synthesize such an adhesive substance until rewarmed in the Couette viscometer, if Humphreys' interpretations are generally applicable. No evidence has been found in the present work to suggest that the adhesiveness of the cells changed during the first hour of aggregation. This result suggests that it is not necessary to put forward the hypothesis that these cells produce a substance which aids adhesion in order for them to aggregate. Perhaps the main difference in procedure between this work and that reported by Humphreys, other than the use of different species, was that he exposed the cells to calcium- and magnesium-free sea water for 6–9 h in order to complete dispersion, while I exposed the cells to media lacking Ca or Mg ions for only 25 to 30 min. Since Humphreys makes no report of viability tests on the cells, it may be wondered whether his cells were relatively in the same state as those used in this work.

At the moment, it is impossible to decide whether or not specific adhesion of cells occurs in other sponge species, such as those which were used by Humphreys, or even under special conditions in the species used in this work. However, it can be pointed out that the new technique reported in this paper, which is superior to earlier techniques in avoiding problems of the identification of species type of cells, in providing a measure of the degree of specificity, and in avoiding a number of possible errors discussed on p. 338, yields no evidence for specificity of adhesion in three species. In addition, the technique has led to results which suggest that there is no synthesis of an adhesive substance required for cell adhesion during aggregation and that there are no other metabolic interactions between two cell populations which might affect their adhesiveness. In the discussion of the theory of aggregation studies, a number of reasons were given to suggest how experimental results might incorrectly suggest that adhesive cements or other substances were active in producing an adhesive state in the cell.

The values which have been obtained for the stability ratios of cells of these species and their mixtures are high by comparison with measured values for avian and mammalian cells (A. S. G. Curtis, in preparation). A possible explanation of this can be advanced in the light of the theory (Curtis, 1966) that adhesion between cells is due to a balance between the action of electrostatic forces of repulsion and the London force of attraction. In media such as sea water, the high value of the ionic strength means that electrostatic forces of repulsion will tend to be relatively diminished by comparison with those in media of ionic strength such as are found in mammalian and avian body fluids. Consequently, other things being equal, the adhesiveness of cells in sea water will tend to be higher than that of avian or mammalian cells.

Acknowledgements

I am most grateful to the Director, Deputy Director, and many other members of the staff of the Marine Station of the Scottish Marine Biological Association for providing facilities to carry out this work. In particular, I would like to thank Mr. W. S. Finlayson for providing supplies of the sponge species used.

References

Albers, W. and Overbeek, J. Th. G. (1960). Stability of emulsions of water in oil III. Flocculation and redispersion of water droplets covered by amphipolar monolayers. *J. Colloid. Sci.* **15**: 489–502.

Curtis, A. S. G. (1962). Pattern and mechanism in the reaggregation of sponges. *Nature, Lond.* **196**: 245–248.

Curtis, A. S. G. (1966). Cell adhesion. *Sci. Progr., Oxford.* **54**: 61–86.

Curtis, A. S. G. and Greaves, M. F. (1965). The inhibition of cell aggregation by a pure serum protein. *J. Embryol. exp. Morph.* **13**: 309–326.

Fuchs, N. (1934). Uber die stabilitat und aufladung der aerosole. *Z. Physik.* **89**: 736–743.

Galtsoff, P. S. (1925). Regeneration after dissociation. (An experimental study on sponges). I. Behaviour of dissociated cells of *Microciona prolifera* under normal and altered conditions. *J. exp. Zool.* **42**: 183–222.

Gerisch, G. (1960). Zellfunktionen und zellfunktionswechsel in der entwicklung von *Dictyostelium discoideum*. I. Wilhelm Roux *Arch. Entw. Mech. Org.* **152**: 632–654.

Humphreys, T. (1963). Chemical dissolution and *in vitro* reconstruction of sponge cell adhesions—I. Isolation and functional demonstration of the components involved. *Devl. Biol.* **8**: 27–47.

Huxley, J. (1921). Differences in viability in different types of regenerates from dissociated sponges, with a note on the entry of somatic cells by spermatozoa. *Biol. Bull. mar. biol. Lab.*, Woods Hole **40**: 127–129.

Karnis, A., Goldsmith, H. L. and Mason, S. G. (1966). The kinetics of flowing dispersions. I. Concentrated suspensions of rigid particles. *J. Colloid Interface Sci.* **22**: 531–553.

Mason, S. G. and Bartok, W. (1959). The behaviour of suspended particles in laminar shear. In *Rheology of disperse systems:* 16–48; Mill, C. C. (Ed.). London: Pergamon Press.

Moscona, A. (1961). Rotation mediated histogenetic aggregation of dissociated cells. *Exp. Cell Res.* **22**: 455–475.

Okada, T. S. (1965). Immunohistological studies on the reconstitution of nephric tubules from dissociated cells. *J. Embryol. exp. Morph.* **13**: 299–307.

Snedecor, G. W. and Cochran, W. G. (1967). *Statistical methods.* Ames, Iowa: Iowa State University Press. 6th ed.

Steinberg, M. S. and Roth, S. A. (1964). Phases in cell aggregation and tissue reconstruction. An approach to the kinetics of cell aggregation. *J. exp. Zool.* **157**: 327–338.

Swift, D. L. and Friedlander, S. K. (1964). The coagulation of hydrosols by Brownian motion and laminar shear flow. *J. Colloid Sci.* **19**: 621–647.

Van Wazer, J. R., Lyons, J. W., Kim, K. Y. and Colwell, R. E. (1963). *Viscosity and flow measurement. A laboratory handbook of rheology.* New York & London: Interscience Publishers.

Weston, J. A. (1963). A radioautographic analysis of the migration and localization of trunk neural crest cells in the chick. *Devl. Biol.* **6**: 279–310.

Wilkins, D. J., Ottewill, R. H. and Bangham, A. D. (1962). On the flocculation of sheep leucocytes—II. Stability studies. *J. theor. Biol.* **2**: 176–191.

Wilson, H. V. (1907). On some phenomena of coalescence and regeneration in sponges. *J. exp. Zool.* **5**: 245–258.

CELLS, INTEGRATION AND MORPHOGENESIS

Symp. zool. Soc. Lond. (1970) No. 25, 353–364.

LES CELLULES DES ÉPONGES

CLAUDE LÉVI

Laboratoire de Biologie des Invertébrés Marins
Muséum National d'Histoire Naturelle, Paris, France

SYNOPSIS

Studies on the structure and morphogenesis of Porifera demonstrate clearly that the simultaneous presence of choanocytes and pinacocytes is a characteristic attribute of the group. Choanocytes and pinacocytes form, respectively, a choanoderm and a pinacoderm; the former controls the nutrition of the sponge, while the latter controls the interrelations between the mesohyl and the external medium.

Second, increasing thickness of the mesenchyme induces an early differentiation of the internal collencytes. Pinacocytes and collencytes are apt to differentiate to ultimate states such as porocytes, contractile cells, etc. In this way they perform vital homeostatic functions and at times they may assume the role of phagocytes.

Study of sponge cells suggests that the three following properties—differentiation of a ciliary apparatus, ability to phagocytose and synthesis of the precursors of intercellular proteins—are characteristic of the primary cell types and are not specific signs of an ultimate state of functional differentiation.

INTRODUCTION

Il est courant d'envisager la structure d'un métazoaire supérieur adulte en termes d'organes, et de considérer son fonctionnement comme la résultante d'interrelations organiques. L'étude des Métazoaires inférieurs et des embryons de Métazoaires supérieurs a longtemps reposé sur la notion de feuillets, et c'est l'analyse des relations de l'ectoderme, de l'endoderme, de l'endomésoderme ou de l'ectomésenchyme qui sert de base à la compréhension biologique de ces animaux.

C'est dans une conception similaire de feuillets que l'organisation des Spongiaires a d'abord été considérée, et à la fin du XIXe siècle; les embryologistes ont cherché les homologies de ces feuillets avec ceux des autres Métazoaires; mais il n'a jamais été certain que cette homologie soit vérifiable. Il semble actuellement plus intéressant de reconsidérer la structure de l'éponge en termes de relations cellulaires et de rechercher la signification fonctionnelle des diverses catégories cellulaires des Spongiaires et de leurs larves en distinguant les types fondamentaux et leurs variantes fonctionnelles. L'objectif de cette analyse sera seulement atteint lorsqu'il aura été possible de séparer en cultures pures les diverses cellules de l'éponge pour pouvoir les réunir à volonté et en proportions définies. L'une des conditions de cette analyse est l'identification préalable des cellules de l'éponge basée notamment sur l'examen de leur comportement in vivo et sur celui de leur structure fine en microscopie électronique.

Des listes plus ou moins longues de types cellulaires de Spongiaires ont été publiées, dont les plus classiques sont celles de Sollas (1888), de von Lendenfeld (1889) et de Minchin (1900). Ces classifications sont fondées sur l'examen de la forme des cellules, de leur position dans l'éponge et de l'apparence et la nature de leurs inclusions cytoplasmiques endogènes ou exogènes.

Il n'est pas dans mon intention de redécrire ici en détail chaque type cellulaire des Spongiaires. La composition cellulaire des Éponges appartenant aux différents ordres, comme celle de leurs larves, se révèle cependant assez variable pour qu'il soit utile de rechercher les cellules fondamentales et d'en distinguer les cellules différenciées secondaires. Nous verrons ainsi qu'il existe chez les éponges trois familles cellulaires qui s'apparentent aux trois types cellulaires fondamentaux signalés par Willmer (1960) chez tous les Métazoaires.

LES TYPES FONDAMENTAUX SOMATIQUES: CHOANOCYTES ET PINACOCYTES

Parmi tous les types cellulaires ainsi décrits, quels sont les types fondamentaux dont peuvent dériver tous les autres? Il est certain que le choanocyte est l'un d'eux. On caractérise une éponge grâce à ses choanocytes et une éponge sans choanocytes est inconcevable. Seules les involuties n'ont pas de choanocytes et jamais les formes de résistance, telles que les involuties ou les gemmules n'ont été considérées comme des éponges fonctionnelles.

Les pinacocytes sont également essentiels: ils limitent le mésohyle même si celui-ci reste mésogléen et l'éponge peut parfaitement fonctionner à la suite du couplage de l'activité des pinacocytes et des choanocytes. Cette structure primitive au sens anatomique du terme existe presque chez les *Oscarella* Vosmaer,1887 où les sclérocytes n'existent pas. Il convient de constater qu'elle est indépendante de la complexité d'organisation du système aquifère déjà leuconoide dès la fin de la métamorphose. Chez *Oscarella*, les porocytes n'existent pas si ce n'est au niveau prosopylaire et les cellules du mésohyle sont rares. Quelques archéocytes de structure analogue à celle des pinacocytes et des choanocytes, exception faite des appareils cinétiques et des villosités cytoplasmiques, circulent dans la mésoglée et leur rôle est celui de gonocytes primordiaux; chez les jeunes *Oscarella* après métamorphose, ces cellules mésohylaires peuvent manquer totalement. Il est probable, mais non prouvé, que ces cellules amoeboides dérivent des pinacocytes.

Il n'existe pas d'éponges sans pinacocytes et sans choanocytes. Ce sont bien les deux types cellulaires essentiels au maintien de la structure et du fonctionnement de l'animal; chez *Oscarella*, les différences de structure entre pinacocytes et choanocytes sont faibles. Les cellules larvaires dont ils dérivent sont semblables en dépit de leur détermination précoce et de leur séparation par des cellules à bâtonnet nucléaire, qui deviendront d'ailleurs des pinacocytes après métamorphose. En faible quantité, les fibres de collagène existent cependant dans la larve et chez l'adulte, et il est vraisemblable que les pinacocytes d'*Oscarella* ont une faible activité fibrogénétique. Il est encore bien difficile de déterminer si la structure histologique des *Oscarella* est le reflet d'un état primitif ou le signe d'une simplification évolutive, mais il est clair que du point de vue cellulaire, elle traduit la primauté des choanocytes et des pinacocytes dans le fonctionnement somatique de l'éponge.

Chez les Calcaires, le mésohyle est également peu cellularisé, excepté par les sclérocytes et les gonies, mais on y trouve néanmoins plusieurs types cellulaires secondaires, même chez les espèces asconoïdes. Les porocytes y sont superficiels.

Chez toutes ces éponges, les volumes nucléaires des choanocytes et des pinacocytes sont comparables, et la plupart des noyaux ont un nucléole bien visible.

Aux deux types cellulaires fondamentaux de l'éponge: choanocyte et pinacocyte correspondent deux types de strates cellulaires épithéliales en contact avec le milieu extérieur et le milieu interne, mais la strate choanocytaire est surtout concernée par le milieu extérieur et la strate pinacocytaire par le milieu interne qu'elle délimite.

COLLENCYTES, PINACOCYTES ET DERIVES

Brien (1943) a très nettement indiqué comment l'évolution des Éponges se traduisait dans l'embryogénèse par une accélération de plus en plus rapide de la différenciation des cellules du mésenchyme liée à l'augmentation pondérale de la mésoglée, caractéristique de la plupart des espèces acalcaires.

Les larves parenchymella des éponges acalcaires incubantes ont en effet un massif cellulaire interne qui peut être diversifié et les cellules baignent dans une mésoglée riche en fibres de collagène. Dans cette larve, les pinacocytes peuvent manquer totalement ou se différencient au pôle postérieur à la surface d'une calotte polaire. Cette différenciation n'affecte pas uniquement les cellules embryonnaires strictement superficielles, mais également et à moindre degré, d'autres cellules sous-jacentes qu'on peut identifier comme collencytes. Il est certain que presque toutes les cellules du massif larvaire interne ont une activité fibrogénétique, mais cette activité s'atténue ou cesse lorsque la cellule se trouve placée en position superficielle où elle prend les caractères de pinacocyte. Il existe ainsi, en fonction de la position, une série morphologique et fonctionnelle qui relie le collencyte au pinacocyte. Les premiers collencytes se différencient à la suite de la prolifération des cellules internes de la larve et de l'isolement de ces cellules du milieu extérieur. Au cours de la métamorphose, des cellules internes peuvent être placées en position superficielle et deviennent alors des pinacocytes. Cette même transformation s'observe fréquemment chez les Spongiaires acalcaires dont les collencytes sont susceptibles de remplacer les pinacocytes disparus. Il reste à savoir s'il s'agit d'une transformation réversible et si des pinacocytes, isolés du milieu extérieur et replacés dans le mésohyle deviennent des collencytes. Ce phénomène ne paraît pas se produire naturellement, et ne semble pas avoir de signification biologique. L'expérience reste à faire. Le remplacement des pinacocytes par des collencytes ou des archeocytes montre au contraire l'importance fonctionnelle primordiale des pinacocytes et la parenté étroite des deux catégories cellulaires.

Remarquons que les collencytes, contrairement aux pinacocytes, n'existent pas obligatoirement chez toutes les éponges; Simpson (1963) n'en signale pas chez *Microciona prolifera* Ellis and Solander, 1786 où existent des archeocytes.

Il est certain que les porocytes, les cellules contractiles, les lophocytes appartiennent à la famille cellulaire des pinacocytes-collencytes, et leur différenciation traduit des exigences mécaniques locales de souplesse ou de rigidité liées à l'organisation et au fonctionnement du système aquifère.

Les Spongiaires, encore nommés Porifera, ont des porocytes qui permettent la pénétration de l'eau dans le système aquifère, mais on ignore encore si les porocytes, qui dérivent clairement des pinacocytes, sont des cellules différenciées spécialisées. On possède peu de descriptions précises de porocytes. Brien (1943) a décrit chez les Spongillidae le développement des pores intracytoplasmiques se perçant dans les pinacocytes au-dessous de l'assise pinacocytaire superficielle. Le pore peut être apparemment intracellulaire, mais il est certain qu'il est toujours extracellulaire (Annandale, 1907; Kirkpatrick, 1910; Borojević, 1966b), et qu'il résulte d'un repli annulaire d'un pinacocyte formant une sorte de sphincter unicellulaire autour d'un petit canal qui paraît être intracellulaire. Les porocytes jouent le rôle d'un diaphragme, et certains endopinacocytes prosopylaires ont la structure des porocytes. Il peut arriver chez les *Clathrina* Gray, 1867 que 2 ou 3 porocytes s'associent pour former un canal porocytaire rudimentaire.

Dans la majorité des pinacocytes ou collencytes d'éponges adultes, on n'observe aucune différenciation fibrillaire ou tubulaire intracytoplasmique. Mais dans les rhagons, des cellules manifestement déformables, et qui ont les caractères ultrastructuraux de pinacocytes et de collencytes contiennent de fines fibrilles allongées parallèlement à la surface.

Les cellules contractiles appartiennent à la famille des pinacocytes-collencytes, mais ces cellules internes n'interviennent pas dans la constitution de la surface d'échange et doivent être plutôt considérées comme des pinacocytes ou des collencytes modifiés en fonction de leur position particulière autour des ouvertures ou du trajet des gros canaux aquifères. Bien que ces cellules n'existent que dans les régions de l'éponge manifestement contractiles, il reste un doute sur la nature de ces fibrilles internes. On a discuté à maintes reprises la signification des fibrilles et des tubules cytoplasmiques, et l'école de Porter (cf. Buckley et Porter, 1968) considère que les filaments sont liés aux activités dynamiques de la cellule alors que les microtubules participent au développement et au maintien des formes cellulaires très asymétriques, et à l'activité cinétique des cils et flagelles.

Cette double interprétation permet d'expliquer les divergences quant à la présence unique de filaments ou la présence simultanée de filaments et de tubules dans les divers types de cellules contractiles d'éponge. On peut considérer que la synergie des filaments et des microtubules facilitent la permanence de la différenciation morphologique de la cellule dans des régions dynamiques de l'éponge où, par exemple, des sphincters se constituent.

Les lophocytes représentent une nouvelle variante de la différenciation des collencytes. La strate souspinacocytaire des lophocytes des Spongillides offre peu d'analogies avec le cortex ou les travées radiaires à lophocytes des Tetractinellides et des Clavaxinellides (*Tethya* Lamarck, 1815, *Axinella* O. Schmidt, 1862). Cependant, il s'agit bien dans les deux cas de collencytes qui participent à la consolidation périphérique axiale ou radiaire de l'éponge. La polarité apparente de ces cellules, soulignée par un sillage de collagène, peut s'expliquer aisément par une certaine mobilité individuelle, mais on peut imaginer aussi l'existence d'un flux cellulaire continu, mais sans doute très visqueux car les lophocytes se situent dans les zones relativement fermes de l'éponge. Notons enfin que les lophocytes existent chez les éponges à différenciation cellulaire larvaire précoce ou tardive.

ARCHAEOCYTES ET COLLENCYTES

Nous savons que les archaeocytes sont des cellules mobiles, caractérisées par un noyau vésiculaire à gros nucléole et sont capables de phagocytose. Or, chez diverses éponges, notamment chez les Hadromérides et Clavaxinellides où le collagène existe en très grande abondance, il est difficile de distinguer archaeocyte et collencyte. La plupart des lophocytes sont chargés de phagosomes et leur noyau est de type archaeocytaire. Chez *Microciona prolifera*, où Simpson (1963) n'a pas distingué de collencytes, l'éponge est néanmoins riche en collagène; on est donc conduit à relier très étroitement les collencytes aux archaeocytes. La parenté générale des archaeocytes, collencytes et pinacocytes, est encore attestée par l'étude histologique de diverses Keratosa qui révèle une remarquable uniformité de structure cellulaire, la différenciation des pinacocytes étant ellemême particulièrement peu accentuée.

A cette même grande famille cellulaire se rattachent les cellules squelettogènes de toute nature, et qu'il est impossible d'identifier avant qu'apparaissent les ébauches squelettiques.

Rappelant maintenant le rôle physiologique primordial des choanocytes et des pinacocytes, il faut noter que la différenciation des cellules mésohylaires que nous venons de passer en revue est fonction de leur position dans l'éponge, et répond à des besoins locaux dépendant de l'organisation du système aquifère et corrélatifs de la croissance et du renouvellement cellulaire normal ou accidentel. Dans les éponges à mésohyle développé, c'est l'archaeocyte qui fonctionne comme cellule souche, tout en assurant sa fonction phagocytaire.

ARCHAEOCYTES ET CELLULES À INCLUSIONS

Il est généralement affirmé que toutes les cellules mobiles du mésohyle contenant des inclusions dérivent des archaeocytes. Bien que la preuve absolue n'en ait pas encore été apportée, c'est hautement probable, et cette différenciation qui débute dès le stade larvaire se manifeste par la genèse des inclusions caractéristiques et par la réduction du volume nucléaire et nucléolaire. La signification des inclusions est totalement inconnue. Des éponges apparemment primitives comme les *Halisarca* Dujardin, 1838 en ont 3 ou 4 types différents; d'autres, beaucoup plus complexes en ont seulement 1 ou 2. Un gros effort reste à faire pour comprendre leur nature et leur rôle; même chez les éponges à mésohyle réduit, ces cellules existent; elles peuvent dériver de pinacocytes devenus internes.

Ainsi, distinguerons-nous pour terminer dans l'ensemble cellulaire des Éponges, trois sous-ensembles:

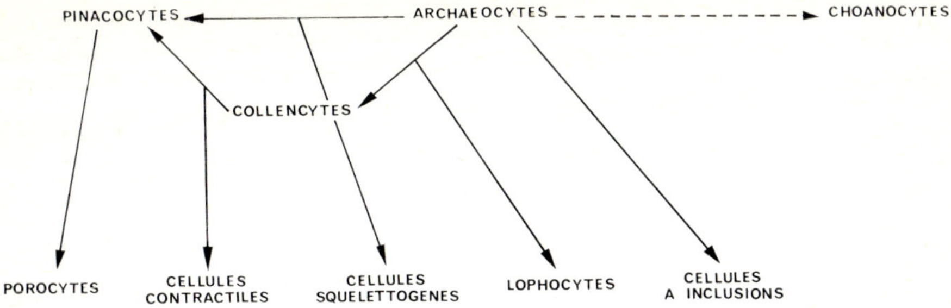

TABLEAU I

ORGANISATION DES CELLULES DE L'ÉPONGE

L'éponge est un être organisé. Les cellules mobiles circulant dans la mésoglée traversent souvent une trame cellulaire de type collencytaire elle-même interposée entre les feuillets choanocytaires et pinacocytaires. Quelquefois les cellules de la trame et les cellules mobiles ne constituent qu'une seule catégorie comme chez certaines Cornées. Dans la mésoglée intercellulaire, le collagène est très généralement abondant et même lorsqu'il est rare, les pinacocytes reposent sur une mésoglée fibrillaire. Toutes les modifications anatomiques adaptatives du système aquifère aux conditions écologiques retentissent sur la population cellulaire, mais affectent surtout la répartition topographique des divers types de cellules et la différenciation des catégories les plus exceptionnelles et les plus spécifiques.

On a beaucoup écrit à propos du problème des corrélations cellulaires chez les Éponges, et ce sujet est encore traité à ce Symposium. Du seul point de vue morphologique, nous nous bornerons à examiner l'organisation du choanoderme et du pinacoderme. Ces deux feuillets essentiels de l'éponge sont tout à fait dissemblables. Pour Brien (1967), l'éponge est formée d'un feuillet "gastrique" (qui délimite une cavité digestive et les choanocytes qui le constituent ont la structure et la physiologie des choanoflagellates) et d'un ectoderme original, ectomésenchyme périphérique délimité au contact de l'eau par un mésenthélium pinacocytaire.

Les deux notions d'un alignement de cellules flagellées et d'un mésenthélium périphérique correspondent assez bien aux observations récentes de ces deux ensembles cellulaires. Ni les choanocytes, ni les pinacocytes ne réalisent vraiment un épithélium. Il n'existe aucune membrane basale, à moins qu'on ne considère toute la mésoglée comme une vaste basale. Il n'existe pas de desmosomes. L'adhésivité réciproque des choanocytes et la cohésion des cellules d'une chambre flagellée sont cependant très fortes. Mais les surfaces d'adhésion des cellules voisines sont restreintes et les ponts adhésifs les plus constants sont situés dans le tiers basal des choanocytes. L'étendue normale des surfaces d'adhésion est difficile à apprécier.

Le choanoderme est moins organisé que l'enveloppe cellulaire ciliée larvaire dont il provient directement ou indirectement à la suite de la métamorphose. Les racines ciliaires des cellules larvaires ne se reconstituent généralement pas dans les choanocytes, et il est intéressant de signaler cette métamorphose de l'appareil cinétique d'une cellule larvaire ciliée se transformant directment en choanocyte également cilié. La cohésion des cellules larvaires est beaucoup plus forte au niveau des cinétosomes, ce qui n'est jamais le cas chez les choanocytes. Il serait intéressant de connaître les réactions immunologiques comparées des choanocytes à l'action d'antisérums préparés à partir de cellules ciliées larvaires ou de choanocytes.

Le pinacoderme n'est pas un épithélium de surface comparable aux épithéliums cubo-cylindriques des autres métazoaires; il représente un type d'organisation épithélioide (Willmer, 1960, p. 1137) qui évoque les endothélium et mésothélium des métazoaires. Les cellules forment une lame très mince et sont très étroitement contigues. Cette lame limite un milieu liquide et une mésoglée conjonctive, et on ignore à peu près tout du rôle physiologique de cette barrière cellulaire. Toutes les expériences, assez grossières, d'alimentation de l'éponge montrent que les choanocytes interviennent seuls dans les conditions normales, dans la première phase de la nutrition. Le pinacoderme pourrait intervenir dans la régulation de la composition du milieu extérieur du mésohyle.

Ces diverses considérations nous permettent maintenant d'examiner les notions de différenciation et d'indifférenciation des cellules d'éponge en évoquant à ce propos le schéma d'évolution cellulaire proposé par Willmer (1960) qui a luimême longuement considéré les Éponges et leurs larves comme une sorte de modèle de la situation primitive de l'organisation multicellulaire du Métazoaire, et qui se représente le métazoaire primitif comme un organisme composé de 2 types de cellules: flagellées saprobiontes et amoeboides holozoiques, en état de coopération symbiotique.

LES TROIS GROUPES CELLULAIRES DES ÉPONGES: LEUR DIFFERENCIATION

L'étude du comportement des cellules de Vertébrés, mises en culture, a conduit Willmer à distinguer chez les Métazoaires trois types cellulaires fondamentaux: épithéliocytes, mécanocytes et amoebocytes dont il résume les principaux caractères (1960, p. 118). Analysant les types cellulaires de l'éponge, en tenant compte de cet essai de systématisation cellulaire, mais en se basant surtout sur la connaissance du développement et de l'histologie des Calcaires Clathrinidae, Willmer considère qu'il existe chez ces éponges:

(a) des épithéliocytes: les choanocytes,

(b) des mécanocytes issus des cellules ciliées de la larve: les scléroblastes, les porocytes et les pinacocytes,

(c) des amoebocytes dérivés des cellules polaires postérieures de la larve.

En nous reportant au Tableau I, nous pouvons effectivement distinguer trois groupes cellulaires correspondant aux trois types définis par Willmer, mais cette constatation appelle quelques remarques:

Epithéliocytes

Les choanocytes sont capables de microphagocytose et de pinocytose, et captent notamment toutes sortes de particules, digestes ou indigestes, entraînées par le courant d'eau jusqu'aux chambres choanocytaires. Ces cellules peuvent ainsi accumuler des quantités importantes de carmin ou de thorotrast qui pénètrent grain par grain à la périphérie distale de la cellule, et qui s'associent en boulettes basales par fusion des vacuoles qui les contiennent.

Ce type de comportement cellulaire s'observe dans diverses cellules de Métazoaires et notamment dans les cellules de Kuppfer du foie des Vertébrés qui, d'après Willmer, forment une sorte d'endothélium dérivé de cellules amoebocytaires. Cette absorption distale de particules n'est pas comparable à la phagocytose amoebocytaire typique telle qu'on peut l'observer chez les archaeocytes.

Amoebocytes

Les amoebocytes à inclusions semblent analogues aux cellules qui circulent à travers les tissus conjonctifs ou les fluides internes de nombreux Métazoaires. Leur rôle reste malheureusement inconnu.

Contrairement aux autres Métazoaires, les Spongiaires ont une couverture cellulaire externe homogène; il n'existe notamment pas de cellules à mucus externes. Or les mucus sont souvent excrétés en abondance par certaines éponges; ils sont secrétés par des cellules amoeboides de type rhabdifère qui sont situées dans le mésohyle et ne sont généralement émis qu'après rupture de la paroi pinacocytaire. Le rôle des mucus dans la physiologie normale de l'éponge mériterait d'être soigneusement étudié. Il n'a aucune fonction de lubrification externe, et intervient plus certainement dans la régulation de la composition de la mésoglée. La répartition des cellules rahbdifères, qui existent même dans les larves, est souvent périphérique et chaque type d'amoebocyte semble distribué dans l'éponge d'une façon caractéristique qu'il faudra analyser. Les amoebocytes des éponges peuvent assurer certaines fonctions dévolues chez les autres Métazoaires à certains épithéliocytes.

Mécanocytes

Il est clair que les mécanocytes des éponges sont analogues à ceux des autres Métazoaires; les collencytes sont fibroblastes et les pinacocytes s'organisent en une sorte de membrane "ectothéliale". Les cellules contractiles, avec ou sans différenciation de fibrilles ou de tubules cytoplasmiques, font partie de ce même ensemble de myoblastes-myxoblastes défini par Willmer.

L'analyse de la structure fine des cellules d'éponge appartenant aux trois catégories montre une assez remarquable uniformité de composition et de répartition des organelles cytoplasmiques. Les mitochondries sont toujours peu nombreuses, de dimensions analogues ($2\ \mu/0{\cdot}25\ \mu$) et à crêtes espacées. L'appareil de Golgi dictyosomial est semblable dans les pinacocytes, les choanocytes et les archaeocytes à quelques détails près; les ribosomes sont dispersés et quelques sacs ergastoplasmiques s'observent parfois à la base des choanocytes ou dans la périphérie de certains archaeocytes et pinacocytes.

Le cytoplasme contient des vésicules variées qui ne caractérisent en particulier aucun type cellulaire fondamental. Dans les cellules à inclusions, il y a régression et condensation périnucléaire du cytoplasme en relation avec l'augmentation de volume des inclusions.

Il est classique de considérer l'archaeocyte comme une cellule indifférenciée et totipotente, seule capable de se différencier, et donc souche de tous les autres types cellulaires de l'éponge, y compris le cas échéant, des choanocytes (Borojević, 1966a). Si cette affirmation est valable pour les Acalcaires à mésohyle développé, elle est plus discutable pour les acalcaires à mésohyle réduit et pour les Calcaires dont les choanocytes ont été généralement considérés comme totipotents et souche de diverses cellules mésohylaires, notamment des gonies.

Que peut-on appeler cellule indifférenciée chez les Eponges? C'est par définition une cellule capable de subir une différenciation. Seuls les archaeocytes et, à beaucoup moindre degré, les pinacocytes et les collencytes, semblent actuellement susceptibles de différenciation somatique chez les Calcaires et les Acalcaires en fonction du développement quantitatif du mésohyle. La question reste ouverte en ce qui concerne la différenciation goniale ou pseudogoniale (cellules nourricières, trophocytes, etc. . .), puisqu'il est admis par de nombreux auteurs que les choanocytes sont capables de se transformer en gonies mâles et femelles.

Il est important de constater que toutes les cellules capables de différenciation assurent cependant une fonction physiologique dans l'éponge, indépendamment de leur évolution potentielle. Elles différent ainsi des cellules interstitielles des Cnidaires (Lentz, 1965) ou des néoblastes des Planaires (Le Moigne, Sauzin et Leinder, 1966; Sauzin, 1966) dont l'organisation cytoplasmique est nettement moins complexe que celle des cellules différenciées et dont le rapport nucléo-cytoplasmique est plus élevé. Elles en diffèrent surtout en conservant une fonction somatique que n'ont pas les cellules interstitielles, les néoblastes ou d'autres types de cellules souches de lignées cellulaires de Métazoaires, qui ne constituent qu'une réserve cellulaire. Restant totipotent comme un blastomère embryonnaire, l'archaeocyte montre une aptitude à la phagocytose, ce qui lui permet de jouer un rôle dans l'équilibre général de l'éponge, mais cette aptitude n'est pas signe de différenciation.

Ainsi, la différenciation d'un appareil ciliaire ou flagellaire, et l'aptitude à la phagocytose qui servent à caractériser divers groupes d'organismes unicellulaires ou deux phases évolutives d'un même individu unicellulaire sont deux manifestations morphologiques qui ne permettent pas de s'assurer de la différenciation définitive et achevée des cellules d'organismes multicellulaires.

En ce qui concerne les mécanocytes, on peut se demander si la synthèse de fibres glycoprotéiques extracellulaires est le signe d'un état différencié ou si c'est la conséquence la plus visible d'une activité de cellules encore capables de différenciation. La transformation de collencytes en cellules contractiles ou de pinacocytes en porocytes montre que la spécialisation fonctionnelle des mécanocytes reste possible chez les éponges. Il est difficile d'affirmer que les fibrilles mésogléennes ne peuvent se structurer qu'en présence de collencytes. Il suffit de noter l'abondance de collagène chez des éponges calcaires où les cellules mésohylaires sont rares; sans doute, les fibres de collagène sont-elles surtout abondantes chez les espèces à collencytes, mais il est souvent très difficile de distinguer alors collencytes et amoebocytes, notamment chez les Keratosa et la fibrogénèse mésogléenne résulte sans doute de l'activité plus ou moins spécifique de l'ensemble des cellules fondamentales du mésohyle, pinacocytes compris.

Ainsi la différenciation ciliaire, l'aptitude à la phagocytose et la sécrétion de précurseurs indispensables à une fibrillogénèse extracellulaire sont-elles trois manifestations de la vie cellulaire qui semblent caractériser les trois grands ensembles cellulaires de Willmer, et qui ne sont pas, en tous les cas chez les Éponges—le signe d'un état différencié définitif.

References

Annandale, N. (1907). The nature of the pores in *Spongilla*. *Rec. Indian Mus.* **1**: 267–273.

Borojević, R. (1966a). Etude expérimentale de la différenciation des cellules de l'éponge au cours de son développement. *Devl Biol.* **14**: 130–153.

Borojević, R. (1966b). Éponges Calcaires des Côtes de France. II. Le genre *Ascandra* Haeckel *emend. Archs Zool. exp. Gén.* **107**: 247–367.

Brien, P. (1943). L'embryologie des éponges. *Bull. Mus. r. Hist. nat. Belg.* **19**: 1–20.

Brien, P. (1967). Les Éponges. Leur nature métazoaire—Leur gastrulation—Leur état colonial. *Annls Soc. r. zool. Belg.* **97**: 197–235.

Buckley, I. K. et Porter, K. R. (1968). Cytoplasmic fibrils in living cultured cells. *Protoplasma* **64**: 349–380.

Kirkpatrick, R. (1910). On a remarkable Pharetronid Sponge from Christmas Island. *Proc. Roy. Soc. Lond.*, B **83**: 124–133.

Kirkpatrick, R. (1911). On *Merlia normani*, a sponge with a siliceous and calcareous skeleton. *Quart. Jl. Microsc. Sci.*, **56**: 657–702.

Le Moigne, A., Sauzin, M. J. et Lender, T. (1966). Comparaison de l'ultrastructure du néoblaste et de la cellule embryonnaire des Planaires d'eau douce. *C.r. hebd. Séanc. Acad. Sci. Paris* **263**: 627–628.

von Lendenfeld, R. (1889). A monograph of the Horny Sponges. 936 pp. London.

von Lendenfeld, R. (1889). Experimentelle Untersuchungen über die Physiologie der Spongien. *Z. wiss. Zool.* **48**.

Lentz, T. L. (1965). The fine structure of differentiating interstitial cells in *Hydra. Z. Zellforsch. mikrosk. Anat.* **67**: 547–560.

Minchin, E. A. (1900). Porifera in *Treatise on Zoology*. **2**. 1–188. Lankester, R. (Ed.) London: A. et C. Black.

Sauzin, M. J. (1966). Etude au microscope électronique du néoblaste de la Planaire *Dugesia gonocephala* (Turbellarié, Triclade) et de ses changements ultrastructuraux au cours des premiers stades de la régénération. *C.r. hebd. Séanc. Acad. Sci. Paris* **263**: 605–608.

Simpson, T. L. (1963). The biology of the marine sponge *Microciona prolifera* (Ellis & Solander). I. A study of cellular function and differentiation. *J. exp. Zool.* **154**: 135–151.

Sollas, W. J. (1888). Report on the *Tetractinellida* collected by H.M.S. Challenger, during the years 1873–1876. *Rep. H.M.S. Challenger Scient. Results Zool.* **25**: 1–458.

Willmer, E. N. (1960). *Cytoloy and Evolution*. London–New York: Academic Press.

Symp. zool. Soc. Lond. (1970) No. 25, 365–397.

ERGEBNISSE DER ENTWICKLUNGSPHYSIOLOGIE BEI SPONGILLIDEN

HANS MERGNER

Abteilung für Entwicklungslehre des Zoologischen Institutes der Justus Liebig-Universität, Giessen, West Germany

SYNOPSIS

Young individuals of Spongillidae are highly suitable for investigations of cytophysiology and developmental physiology because of their transparency. The techniques which have been used by various workers for the culture of spongillids are reviewed. Although the tissues of young spongillids contain cells of only a few types, these cells have a great variety of potencies. The morphogenesis of spongillids, from the initial behaviour of cells within the gemmules to the completion of development of the functional sponge, presents many problems of general interest to students of developmental biology.

The endo- and exopinacoderms of spongillids may be regarded as epithelia and it is suggested that these epithelia have a physiological polarization which is matched by a polarization of their inductive capacities. This hypothesis is supported by the results of experimental transplantation and implantation, particularly so in the case of experiments on the induction of oscules. Tests of the compatability of different epithelia indicate that the inductive agents are highly specific.

EINLEITUNG

Unter den vielzelligen tierischen Organismen wurde den Poriferen schon immer eine Sonderstellung eingeräumt. Ihre Organisation stellt nämlich den einmaligen Versuch der Natur dar, einen Weg der Vielzelligkeit zu beschreiten, für den vor allem der hohe Freiheitsgrad der Einzelzelle kennzeichnend ist (Ankel, 1964). Diese vagilen Einzelzellen werden aber unter das Gesetz der Zuordnung zu einer gemeinsamen morphologischen und funktionellen Leistung gebracht. Trotz des fast ständigen Ortswechsels der zellulären Bauelemente bleiben also die Konstanz der Formelemente und die reibungslose Funktion eines Strudelorganismus gewahrt. Für uns ist gerade eine derartige Strudlerorganisation, wie sie sich in unseren Versuchsobjekten, den Spongilliden, darstellt, ein großer Vorteil: Gibt sie uns doch die Möglichkeit, anhand von Überlegungen allgemein gültiger Art wie auch spezifischer Fragestellungen gezielte Eingriffe vorzunehmen.

Süßwasserschwämme zeigen als dreidimensional gewachsene, noch durchsichtige Jungschwämme den einfachen und leicht überschaubaren Aufbau ihrer Organisation (Abb. 1). Die Anzahl der in ihnen vorkommenden differenzierten und für eine bestimmte Leistung determinierten

Zellsorten ist gering, die Vielfalt der Zellpotenzen dagegen groß. Durch ihren räumlich und zeitlich lockeren Zusammenschluß zu Epithelien bilden bestimmte Zellen wasserdichte Bauelemente, deren Ausformung zu verschiedenen Leistungssystemen, zu Organen, in einfacher, aber wirkungsvoller Weise die Strudel-Filterfunktion des Schwammorganismus ermöglicht.

Wegen dieser Eigenschaften und wegen der Möglichkeit, die verschiedenen Epithel-Einzellamellen zu jeweils neuen Doppellamellen zu formieren, eignen sich Jungschwämme in besonderer Weise für zell- und entwicklungsphysiologische Untersuchungen. Dies gilt vor allem für die Klärung von Fragen, welche im Zusammenhang mit Determination und Potenzumfang der Zellen auftreten und damit sogar bis in das Geschehen innerhalb der Gemmula zurückgreifen. Alle bisherigen Arbeiten, die seit Erscheinen der ersten Mitteilung von Ankel (1948) durchgeführt wurden, waren daher auf Potenzprüfungen an determinierten Zellen von Süßwasserschwämmen ausgerichtet, mit dem Ziel, den Determinationsweg von der Gemmula bis zum fertigen Jungschwamm vollständig aufzuklären. Hierzu wurden immer wieder Verträglichkeitsprüfungen mit den einzelnen Epithelien vorgenommen, in deren Verlauf auch induktive Eigenschaften isolierter Epithelverbände festgestellt werden konnten (Mergner, 1964).

VERSUCHE UND BEOBACHTUNGEN

Zum Verständnis der experimentell ausgelösten morphologischen und funktionellen Veränderungen im Schwammorganismus erscheint es notwendig, zunächst einen Überblick über das Geschehen in der Gemmula zu geben, über die Normogenese des Jungschwammes von seinem Auskeimen aus dieser bis zum Erreichen der vorläufigen Ordnung und über die Funktionsprinzipien seines Strudel-Filtersystems. Dabei wird jeweils auf die wichtigsten experimentellen Eingriffe zur Aufklärung unbekannter Vorgänge während der einzelnen Entwicklungsabschnitte hingewiesen. Anschließend wird über Potenzprüfungen an determinierten Zellen berichtet und eine Hypothese zur physiologischen Orientierung der Epithelzellflächen entwickelt, welche das oft unterschiedliche Ergebnis dieser Potenzprüfungen deutet. Schließlich sollen die Untersuchungsergebnisse über die induktiven Eigenschaften isolierter Epithelkugeln und über die Natur des Induktionsstoffes dargestellt werden.

Untersuchungen an der Gemmula

Süßwasserschwämme bilden gegen Ende ihrer sommerlichen Wachstumsperiode beschalte Dauerstadien aus, die sog. Gemmulae. Diese enthalten eine größere Anzahl zunächst einkerniger primärer Archaeocyten; das sind teilungsfähige, mit Dottereinschlüssen beladene Zellen. Mit Fertigstellung der Gemmulaschale werden die primären Archaeocyten nach einem bestimmten Gradienten durch normale Mitosen zweikernig, zuletzt in der Umgebung des Porus (Brien, 1932; Leveaux, 1939; Wintermann, 1951). Sie werden damit zu sekundären Archaeocyten, als welche sie in der reifen Gemmula überwintern. Sowohl primäre als auch sekundäre Archaeocyten können während der winterlichen Diapause als totipotent angesehen werden. Innerhalb ihrer Reaktionsnorm liegen also noch alle Möglichkeiten für die Ausdifferenzierung sämtlicher den Jungschwamm aufbauenden Zellsorten (Mergner, 1964).

Faktoren der Gemmulabildung

Während über die Morphogenese der Gemmula dank der Untersuchungen von Leveaux (1939) und Rasmont (1955) genaue Vorstellungen bestehen, sind die sie auslösenden Faktoren noch kaum bekannt. Wahrscheinlich liegen sie aber teils in der Verschlechterung der standortgebundenen und jahreszeitlichen Milieubedingungen, teils in der Summierung der negativen Auswirkungen des Zellstoffwechsels begründet. Die Gemmulabildung darf daher nach Kilian (1964) als "Alterserscheinung" angesehen werden, die durch keine optimalen Außenbedingungen verzögert werden kann, infolge ungünstiger Milieueinflüsse jedoch auch schon früher einzusetzen vermag. Ihr Abschluß bezeichnet zumeist auch das Ende des individuellen Lebens des Schwammes.

Doch lassen sich *Ephydatia fluviatilis* Linnaeus, 1759 und *Spongilla lacustris* Linnaeus, 1759 in mit Fischen besetzten Aquarien in mehreren aufeinanderfolgenden Generationen züchten. Dabei keimen die jeweils gebildeten Gemmulae regelmäßig schon aus, während der Mutterschwamm noch für längere Zeit daneben weiterexistiert (Mergner, 1965, nicht veröffentlicht). Kilian (1964) hat derartige "perennierende" Formen sogar in Freiland beobachten können.

Faktoren der Diapause

Neu gebildete Gemmulae müssen vor dem Auskeimen eine Diapause durchmachen, die je nach Art verschieden lang ist und bei den einheimischen Arten normalerweise mit der winterlichen Ruhezeit zusammenfällt. Bei *Ephydatia fluviatilis* beträgt sie nur 1–2 Wochen, bei anderen Arten dagegen 6–8 Wochen. Nach Rasmont (1954) stellt die Diapause von *Ephydatia fluviatilis* und der algenhaltigen Gemmulae von *Spongilla lacustris* nur eine kältebedingte Entwicklungshemmung dar, die sich durch Temperaturerhöhung jederzeit wieder aufheben läßt. Kilian (1964) konnte sogar die 6–8 wöchige obligatorische Ruhezeit der algenfreien Gemmulae von *Spongilla lacustris* durch Einfrieren für 2–3 Tage auf den Zeitraum von einer Woche einschränken.

Mit physiologischen Problemen der Diapause haben sich vor allem Zeuthen (1939) und Rasmont (1956, 1962) befasst. Zeuthen beobachtete einen osmotischen Druckanstieg in der Gemmula auf das 3–4 fache des normalen Wertes von 30 mM Natriumchlorid, der jedoch noch vor dem Auskeimen wieder auf den Normalwert absinkt und möglicherweise mit den in der Gemmula ablaufenden Kernteilungen zusammenhängt. Rasmont hat dagegen die Rolle der Temperatur und chemischer Einflüsse als Mittel zur Beendigung der Diapause untersucht, ferner die respiratorischen Abläufe und die Veränderungen der Reservestoffe während der Ruhezeit.

In natürlichen Gewässern wird die Ruhezeit der Gemmula durch das Eintreten günstigerer Lebensbedingungen beendet, d.h. in kühlen Klimaten mit steigender Erwärmung des Wassers im Frühjahr und in warmen Klimaten mit Eintreffen der Regenzeit oder der Überschwemmungsperiode. Die Gemmulae als Dauerstadien müssen daher über eine entsprechende Kälte- oder Trockenresistenz verfügen. Darüber liegen eine Anzahl Beobachtungen vor:

Nach Kilian (1964) können Gemmulae einheimischer Spongilliden in Wasser bei +2°C Temperatur je nach Art 4–6 Jahre keimfähig aufbewahrt werden, nach meinen Beobachtungen sogar bis zu 8 Jahren, wobei jedoch der Prozentsatz der noch auskeimenden Gemmulae von Jahr zu Jahr ständig absinkt. Eingefrorene Gemmulae verschiedener Arten keimten nach Weltner (1893) nach 17 Tagen zu 50% aus, solche von *Spongilla fragilis* Leidy, 1851 sogar noch nach 59 Tagen. Kilian (1964) ließ frisch gebildete Gemmulae für 4 Tage bei −20°C einfrieren, von denen bei stufenweiser Erhöhung der Temperatur 90% wieder ausschlüpften. Die Gemmulae unserer einheimischen Spongilliden sind also gegen alle im Bereich ihres Vorkommens normalerweise auftretenden Kältegrade geschützt. Ähnlich verhält es sich mit der Trockenresistenz: Von Lendenfeld (1889) gibt für *Spongilla lacustris* eine Trockenresistenz von 3 Jahren an, Wierzejski (1935) für eine nicht näher bezeichnete Art sogar eine solche von 30 Jahren. Kilian (1964) erzielte bei Gemmulae von *Ephydatia fluviatilis*, die bei einer Temperatur von +22°C lufttrocken gehalten wurden, eine Schlüpfrate von 95%. Demnach können also auch unsere einheimischen Spongilliden längere Trockenperioden überdauern.

Feinstruktur und chemischer Aufbau der Reservestoffe

Bis zur Fertigstellung eines funktionsfähigen Kanalsystems als Abschluß der Normogenese sind die frühen Entwicklungsstadien des Jungschwammes auf die Ernährung durch Reservestoffe angewiesen. Diese als "Dotter- oder Eiweißeinschlüsse" bekannten Reservestoffe finden sich in großer Zahl als stark lichtbrechende, schiffchenförmige Körper von 5–10 μ Länge innerhalb der Archaeocyten. Sie werden beim Aufbau des Jungschwammes allmählich aufgezehrt und sollen nach Kilian (1964) für etwa 2–3 Wochen ausreichen.

Elektronenoptische Untersuchungen (Ruthmann, 1965; Feige, 1969) haben die lichtoptisch gewonnenen Erkenntnisse über Aufbau, Feinstruktur und Abbau der Dotterkörner vertieft, histochemische Untersuchungen (Pourbaix, 1934, 1935; Kauffold und Spannhof, 1963; Tessenow, 1969) haben sie bestätigt. Danach werden die Schollenkappen von elektronenoptisch dichten, stark basophilen Nucleoproteinen gebildet, die eine weniger dichte, acidophile Medianplatte aus Proteinen einschließen. In Aussparungen der Schollenrandzone sind Lipidgranula eingelagert, deren Gesamtheit als Gürtelzone die Scholle umgibt, während eine mehrschichtige Membran die Außenhülle bildet.

Aus diesen Befunden geht hervor, daß die Dottereinschlüsse für ihre Verwendung als Reservestoffe ideal zusammengesetzt sind: Verschiedene wichtige Stoffe sind portionsweise in die Schollen eingelagert und lassen sich bei beginnender Differenzierung leicht verteilen als eine günstige Voraussetzung für die rasche Entwicklung des Gemmulainhaltes zum Jungschwamm (Kauffold und Spannhof, 1963). Nach den übereinstimmenden Befunden von Ruthmann (1965) und Feige (1969) zeigen weitgehend abgebaute Dotterschollen eine Zunahme der Vakuolenbildung und Granulierung in der Medianplatte sowie eine Vermehrung der konzentrischen Membransysteme, während die dichten Schollenkappen bereits abgebaut sind.

Morphogenese, Aufbau und Funktionieren des Jungschwammes

Kurz vor dem Auskeimen des Schwammkörpers im Frühjahr setzt sich der Differenzierungsprozess innerhalb der Gemmula fort. Die Archaeocyten werden, wiederum nach dem gleichen Gradienten wie bei ihrer Differenzierung zu sekundären Archaeocyten, vielkernig: Mitosewellen laufen ab, ohne daß sich sofort Plasmotomien anschließen. Die Vielkernstadien sind als erste Stufen der weiteren Zelldifferenzierung anzusehen. Sie zerfallen schließlich in einkernige Archaeocyten, von denen sich ein Teil, und zwar der in Porusnähe gelegene zuerst, in Histoblasten umwandelt (Brien, 1932; Leveaux, 1939;

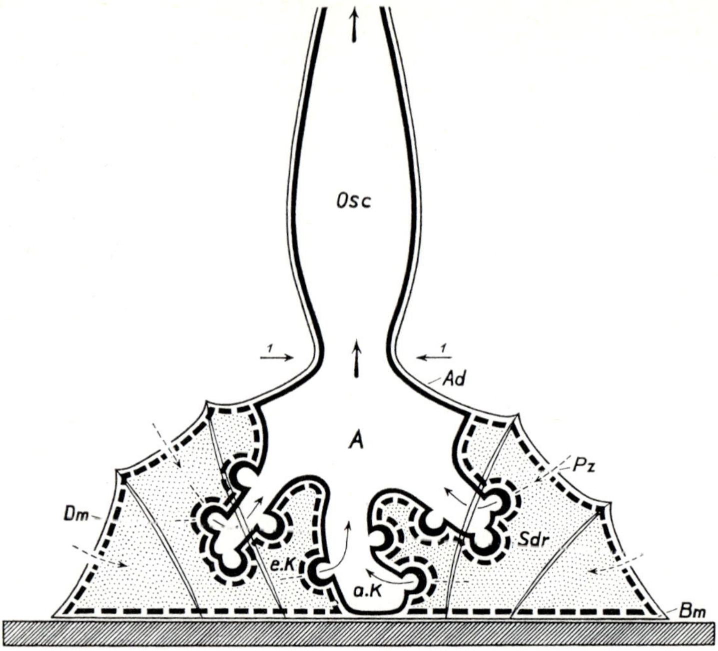

ABB. 1. *Ephydatia fluviatilis.* Vereinfachtes Schema eines frei gewachsenen Jungschwammes. Erläuterungen im Text. Gestrichelte Pfeile, Weg des einströmenden Wassers; Ausgezogene Pfeile; Weg des ausströmenden Wassers; Pfeile 1, Schnittstelle zur Gewinnung von Ocularrohrkugeln. (Nach Mergner, 1964.)

Verwendete Abkürzungen

A, Atrium; Ad, Atrialdach; a.K, ausführender Kanal; Bm, Schwammbasis; Dm, Dermalmembran; e.K, einführender Kanal; ö, künstlich gesetzte Implantationsöffnung; Osc, Oscularrohr; Osc.K, Oscularrohrkugel; Pz, Porenzelle; Sdr, Subdermalraum.

Verwendete Zeichen

Dünne Linien, Exopinacocyten der Dermal—u. Basalmembran; Dicke Linien, Ap-Endopinacocyten (ausführendes System); Dickgestrichelte Linien, Pros-Endopinacocyten (einführendes System); Weiße Flächen, Räume des ausführenden Systems; Lockerpunktierte Flächen, Räume des einführenden Systems; Dichtpunktierte Flächen, Fremdkugelmaterial; Schraffierte Flächen, Gemmulae und Unterlagen (Objektträger).

Wintermann, 1951). Diese Histoblasten, Vorläufer der verschiedenen epithelbildenden Pinacocytensorten, verlassen zusammen mit einkernigen Archaeocyten als erste die Gemmula und breiten sich um sie herum als Keimkörper aus.

Mit dem Entstehen der Vielkernstadien und ihrem Zerfall durch Plasmotomien sind vermutlich eine Potenzverteilung auf die Tochterzellen und damit eine Einengung ihrer Reaktionsnorm verbunden. Beide Vorgänge setzen sich während der weiteren Differenzierungsschritte in den auskriechenden Zellen fort, bis mit Erreichen der vorläufigen Ordnung des Jungschwammes sämtliche von ihm benötigten Zellsorten entstanden sind (über Einzelheiten der Zelldifferenzierung siehe Wierzejski, 1915, 1935; Brien, 1932; Wintermann, 1951). Diese Zellen sind dann nicht mehr totipotent; die ihnen verbliebenen Potenzen sind festgelegt. Gleichzeitig muß aber auch ihre Determinierung erfolgt sein. Sie zeigen nämlich nur noch die Reaktionsmöglichkeiten und das lagespezifische Verhalten von Zellen desjenigen Epithelsystems, dem sie angehören (Mergner, 1964).

Faktoren der Keimung

Mit steigender Erwärmung des Biotops im Frühjahr keimt bei ausreichendem Sauerstoffgehalt der Gemmulainhalt als Jungschwamm aus. Versuche von Brien (1932) und Kilian (1964) haben ergeben, daß der Temperaturbereich für das Auskeimen zwischen 15° und 30°C mit einem Optimum bei 23°C liegt, bei zusätzlicher Belüftung des Kulturmediums jedoch zwischen 11° und 35°C. Außerdem keimen Gemmulae in gut durchlüfteten Medien durchwegs 2–3 Tage früher als in solchen ohne Belüftung, nämlich bei 23°C bereits 3–5 Tage nach Ansetzen der Kultur. Alle diese Versuchsergebnisse bestätigen also die Abhängigkeit der Keimung von Temperatur und Sauerstoffgehalt.

Demgegenüber scheinen andere Faktoren wie der osmotische und hydrostatische Druck innerhalb der Gemmula oder die pH-Konzentration des Kulturmediums keine besondere Rolle beim Austreten des Gemmulainhaltes zu spielen. Nach Zeuthen (1939) gehen der osmotische Druckanstieg in der Diapause und der ihm entsprechende Anstieg des hydrostatischen Binnendruckes schon mehrere Wochen vor dem Auskeimen wieder zurück. Auch konnte Kilian (1964) zeigen, daß aus keimungsfähigen Gemmulae, die mit feinen Glasfäden angestochen werden, kein Zellmaterial austritt. Die Porusmembran wird also sicher nicht infolge erhöhten Binnendruckes geöffnet, sondern vermutlich auf chemischem Wege aufgelöst.

Auch die pH-Konzentration ist offensichtlich nicht für das Austreten der Zellen entscheidend: Nach Kilian (1964) erfolgt nämlich der Keimungsvorgang regelmäßig innerhalb eines pH-Bereiches von 5·5–8·0, während in den Grenzbereichen von pH 4·8–5·5 und 8·0–8·2 Verzögerungen von 2–3 Tagen und nur noch vereinzeltes Auskeimen beobachtet werden. Keimpfropfe, die sich sogar noch bei pH 9·0 gelegentlich bilden, dann aber in ihrer Entwicklung stagnieren, breiten sich nach ihrer Überführung in den optimalen pH-Bereich weiter aus. Diese Ergebnisse besagen eindeutig, daß Gemmulae in nahezu allen natürlichen Gewässern auskeimen können.

Morphogenese des Jungschwammes

Über das Ausbreiten der geschlüpften Zellen rings um die Gemmula und das Anheften des jungen Schwammkörpers auf seiner Unterlage hat Brøndsted (1943) eine genaue Darstellung gegeben. Wintermann (1951) hat die Bildung des Jungschwammes in flachen Räumen untersucht, Mergner (1964) dagegen das Enstehen der vorläufigen Ordnung am frei gewachsenen Jungschwamm. Hierzu interessiert uns vor allem die Frage, wie die Epithelsysteme in der Morphogenese aufgebaut werden, d.h. wie die Doppellamellen aus ihren beiden Einzellamellen entstehen, die jeweils aus einer anderen Pinacocytensorte zusammengesetzt sind (Abb. 2):

Zunächst verharren die aus der Gemmula geschlüpften Zellmassen, der Keimkörper, unter einer einschichtigen, noch lockeren Histoblastenbedeckung in einer mehrstündigen "Zellanarchie". In ihr entstehen durch scheinbar ungeordnetes Auseinanderweichen der Zellen eine Anzahl Binnenlakunen, die anschließend von zarten einschichtigen Endopinacocyten-Lamellen ausgekleidet werden. Dann beginnen sich diese Begrenzungslamellen vom Rand des Keimkörpers und von innen her nach und nach gegen die äußere Histoblastendecke zu legen und mit ihr gemeinsam das doppelschichtige Dermalmembranepithel zu bilden. Seine Außenlamelle wird mit dieser Unterlagerung durch Korrelation zur Exopinacocytenschicht, die innere zur Pros-Endopinacocytenschicht. Gleichzeitig hebt sich die damit fertiggestellte Dermalmembran, durch einregulierte Kieselnadeln gestützt, von den im Innern verbliebenen Zellmassen ab und überdacht den derart entstandenen Subdermalraum. Innerhalb der Dermalmembran formt sich ein Teil der Pros-Endopinacocyten nach einem bestimmten Muster zu Porenzellen um, durch deren intrazelluläre, verschließbare Poren das Wasser in den Subdermalraum strömt.

Die restlichen Binnenlakunen im Schwamminnern kleiden sich mit einer Ap-Endopinacocytenlamelle aus und verbinden sich untereinander zu den ausführenden Kanälen. Aus fortgesetzten Teilungen von Archaeocyten entstandene Choanocytenhaufen bilden hohlkugelige Epithelverbände und verbinden sich nunmehr als Geißelkammern mit den ausführenden Kanälen. Von diesen ausgesparte Zwischenräume werden zu einführenden Kanälen mit einer Begrenzung aus Pros-Endopinacocyten und gewinnen Anschluß an den Subdermalraum. Schließlich wächst ein ausführender Kanalabschnitt durch den Subdermalraum hoch und durchstößt als junges Oscularrohr die Dermalmembran.

Der Ablauf der vorstehend kurz dargestellten Normogenese von

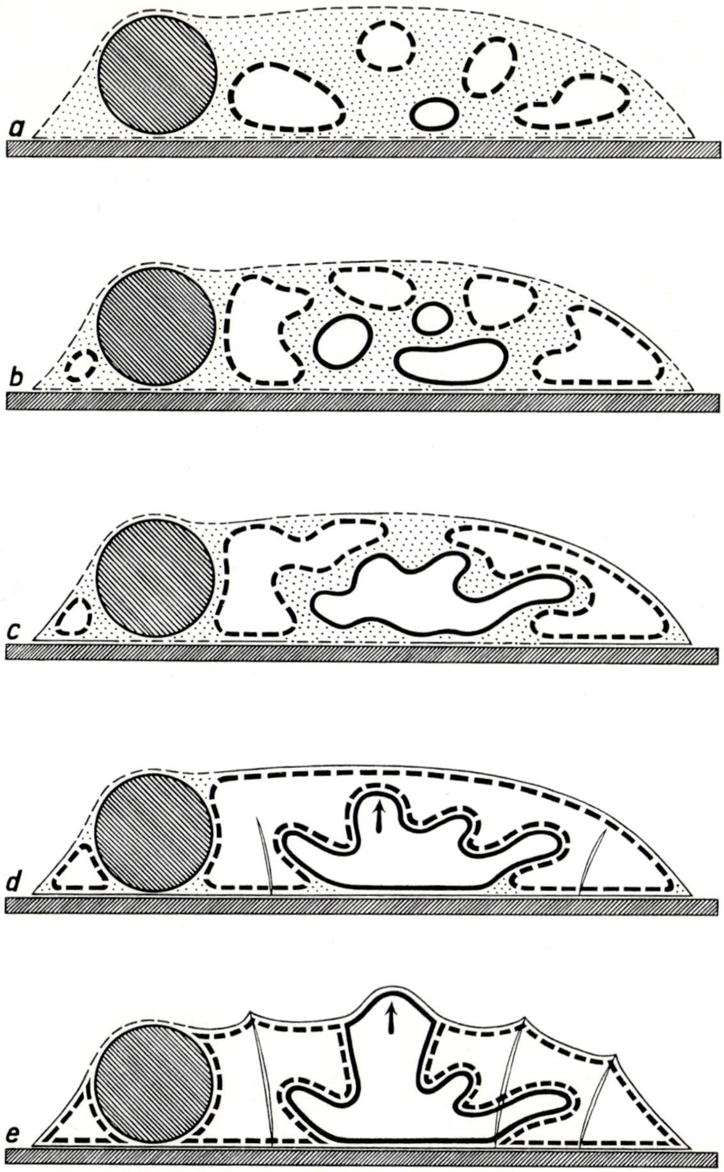

Abb. 2. Frühentwicklung des frisch geschlüpften Jungschwammes. Erläuterungen im Text. Geißelkammern, Porenzellen und absolute Größenzunahme nicht berücksichtigt; Punktiert, Zellbrei aus Archaeocyten und Histoblasten; Weiß, Lakunen und Kanalsystem; Dünne unterbrochene Linien, Histoblasten-Bedeckung und—Unterlage; Pfeile, Oscularrohrbildung. (Nach Mergner, 1964.)

Ephydatia fluviatilis wirft eine Anzahl Fragen auf, mit deren Lösung sich verschiedene Untersucher befasst haben: Schon Brøndsted (1936) hat erkannt, daß sich die aus der Gemmula schlüpfenden Zellen aktiv auf ihrer Unterlage ausbreiten. Daß diese Ausbreitung aber unabhängig von der Schwerkraft erfolgt, haben Ankel und Eigenbrodt (1950) nachgewiesen, indem sie Schwämme an senkrechten Flächen oder gar am Oberflächenhäutchen des Wassers auskeimen ließen. Kilian (1964) wiederum hat die Wachstumsgeschwindigkeit, die einer S-Kurve folgt, und die optimalen Endwerte des beim Aufbau erzielten Wachstums berechnet; er hat hierzu Flächen- und Volumenmessungen von Jungschwämmen miteinander verglichen, die aus einzelnen oder mehreren Gemmulae gekeimt waren. Danach werden Optimalwerte bei einem gemeinsamen Zuchtansatz von drei Gemmulae erreicht, während die Endwerte der Wachstumsgeschwindigkeit mit weiter steigender Gemmulazahl wieder abnehmen. Das Größenwachstum der Spongilliden folgt also offensichtlich den gleichen Wachstumsgesetzen, wie sie für Tierpopulationen gelten.

Über die Zelldifferenzierung während der Morphogenese geben Wierzejski (1915, 1935), Brien (1932), Meewis (1935, 1936) und Wintermann (1951) Auskunft. Über Determination und Zellverhalten sowie über Korrelationserscheinungen zur Herstellung der Doppelepithelien haben in erster Linie Ankel (1948, 1964), Ankel und Wintermann-Kilian (1952), Ankel, Wintermann-Kilian und Kilian (1954) und Mergner (1964) gearbeitet. Bei Mergner finden sich auch ausführliche Definitionen der hierbei verwendeten Begriffe. Brøndsted's Vermutung (1936), die Epithelien des Jungschwammes seien syncytial strukturiert, wurde bereits von Brien widerlegt; auch andere Untersucher konnten den zellulären Aufbau der Epithelien eindeutig nachweisen, zum Teil anhand von elektronenoptischen Befunden (Feige, 1969).

Schließlich muß in diesem Zusammenhang auch auf die Stellung der Poriferen im tierischen System eingegangen werden. Noch bis in die jüngste Zeit hinein wurde ihnen eine Sonderstellung als Parazoa eingeräumt, obwohl Tuzet (1948), Ankel (1948) sowie Wintermann-Kilian und Ankel (1954) die Frage nach dem Vorkommen echter Gewebe zumindest bei Spongilliden längst in positivem Sinne entschieden haben: Demnach bilden sämtliche Epithelien der Süßwasserschwämme feste Verbände mit funktionell polarer Differenzierung und der Fähigkeit zur Organbildung. Auch haben die Versuche von Mergner (1964) zur Oscularrohr-Neubildung diese Erkenntnisse noch vertieft und zur Deutung des funktionell differenten Verhaltens der einzelnen Epithelverbände untereinander beigetragen. Alle diese Ergebnisse beweisen also das Vorhandensein echter Gewebe bei

Poriferen und rechtfertigen deren Einordnung unter die niederen Metazoen.

Organisation und Funktionieren des Jungschwammes

Der in der Morphogenese entstandene, dreidimensional gewachsene Jungschwamm (Abb. 1) haftet auf seiner Unterlage mittels des von seiner basalen Exopinacocytenschicht ausgeschiedenen Spongins. Über dieser Bodenfläche erhebt sich, aus zarten, durchsichtigen Epithelien aufgebaut, der Schwammkörper, dessen gesamte freie Oberfläche ebenfalls von einer Exopinacocytenschicht bedeckt ist. Seine innere Organisation zeigt zwei streng voneinander geschiedene, epithelbegrenzte Hohlraumsysteme, deren eines das eingestrudelte Wasser einleitet, während das andere das gefilterte Wasser wieder ausleitet. Das "einführende System" ist mit Pros-Endopinacocyten ausgekleidet und umfasst den von der Dermalmembran gegen die Außenwelt abgeschirmten Subdermalraum und die einführenden Kanäle. Das "ausführende System" wird dagegen von Ap-Endopinacocyten begrenzt und besteht aus den ausführenden Kanälen, dem Atrium als deren Sammelraum und dem freistehenden Oscularrohr mit dem Osculum als Egestionsöffnung.

Die einzige Verbindung zwischen beiden Systemen führt durch halbkugelige, aus einer wechselnden Anzahl von Choanocyten zusammengesetzte Hohlorgane, die Geißelkammern. Sie sind mit einer weiten Öffnung, der Apopyle, fest in das Epithel der ausführenden Kanäle eingefügt, während ihre gewölbte Außenfläche den einführenden Kanälen zugekehrt ist. Kleine Lücken zwischen den Choanocyten, die Prosopylen, ermöglichen den Eintritt des Wassers in die Geißelkammern.

Diese zentralen Durchströmungsorgane stellen zugleich Pumpen und Filterapparat des Schwammkörpers dar. Der Geißelschlag ihrer Choanocyten erzeugt nämlich einen gerichteten Wasserstrom, aus welchem deren Zellkörper die lebensnotwendigen Stoffe herausfiltern. Infolge der Tätigkeit der Geißelkammern herrschen in beiden Bereichen des Wasserleitungssystems unterschiedliche Druckverhältnisse: Der Unterdruck im einführenden System fordert eine Stützung der Dermalmembran durch zeltstabartige Skelettelemente, zu Pfeilern verkittete Kieselnadeln, die den notwendigen Abstand der Dermalmembran zum Kanalsystem sichern. Der Überdruck im ausführenden System bedingt dagegen die Aufblähung der ausführenden Kanäle, des Atriums und des Oscularrohres und hält zugleich dessen Mündung offen.

Der Weg des Wasserstromes führt also durch die Poren der Dermalmembran in den Subdermalraum. Hier wird das Wasser auf die einführenden Kanäle verteilt und durch die Prosopylen in die Geißelkammern eingestrudelt, wo es von den Choanocyten abgefiltert wird. Durch deren Pumpwirkung wird es dann in die ausführenden Kanäle weiterbefördert und gelangt über Atrium und Oscularrohr ins Freie.

Diese knappe Darstellung der Schwammorganisation und des Funktionierens des Strudel-Filtersystems basiert vorwiegend auf den Ergebnissen von Untersuchungen, die Ankel, Wintermann, Kilian und Mergner in den Jahren 1948 bis 1964 durchgeführt haben. Dabei hat Kilian (1952) in erster Linie Probleme des Wassertransportes und der Nahrungsaufnahme bearbeitet, während Ankel, Wintermann-Kilian und Kilian (1954) gemeinsam Fragen der Wasserschüttung und der Sklerenbildung bei skelettlosen Fladenschwämmen untersucht haben. Derartige extrem flachgewachsene Schwämme entstehen als Fehlbildungen in kieselsäurefreien Kulturmedien. Über die Druckverhältnisse im Oscularrohr hat Mergner (1964) neue Einsichten vermittelt.

Erhaltung und Wiederherstellung der Schwammorganisation

Seit langem ist die besondere Befähigung der Spongilliden zur Reorganisation nach Dissoziation bekannt, etwa nach Durchpressen des Schwammgewebes durch ein Tuch. Nachdem bereits Brien (1929, 1935, 1936–1937) dieses Phänomen gründlich untersucht hat, haben sich auch neuere Arbeiten wieder eingehend mit offen gebliebenen Fragen befasst: Korotkova (1963) unterscheidet aufgrund ihrer Versuche an zahlreichen Schwammarten zwischen Regenerationsprozessen, die bei solitär wachsenden Formen überwiegen, und solchen der Reorganisation oder somatischen Embroygenese (Tokin, 1963), die vor allem bei weniger integrierten, kolonialen Formen wie bei den Spongilliden vorkommen. Dabei wird unter somatischer Embryogenese die Herstellung eines vollständigen Organismus aus Komplexen somatischer Zellen verstanden, unter Regeneration dagegen die Restitution eines Körper- oder Organteiles.

Einzelheiten über Fragen des Zellkontaktes dissoziierter Schwammzellen als einer wesentlichen Voraussetzung für Reorganisation und Regeneration bringen Ganguly (1960), Simons (1963) sowie Mookerjee und Ganguly (1964): Danach zeigen dissoziierte *Ephydatia*-Zellen beim Zellkontakt während ihrer Aggregation ein deutliches Unterscheidungsvermögen gegenüber gleichsortigen bzw. verschiedensortigen Zellen. Da sich aber bestimmte Zellsorten wie Pinacocyten, Skleroblasten und Choanocyten nach ihrer Dissoziierung stets nur wieder in ihren ursprünglichen Zelltyp redifferenzieren können, benötigen sie für ihr erneutes Zusammentreten zu Epithelverbänden dieses Unterscheidungsvermögen. Erst hierdurch werden also erfolgreiche Reorganisations- und Restitutionsprozesse ermöglicht.

Probleme der Erhaltung der Schwammorganisation hat Wintermann (1951) analysiert. Sie fand dabei als Minimalbedingung für das Fortbestehen eines Schwämmchens das Vorhandensein einer Geißelkammer, die von wenigen Archaeocyten und Pinacocyten umgeben ist. Derartige "Kleinstschwämme" sind in der Lage, sich zu größeren Exemplaren weiterzuentwickeln. Bei skelettlosen Fladenschwämmen, wie sie von Ankel, Wintermann-Kilian und Kilian (1954) in kieselsäurefreien Medien gezüchtet wurden, bildet sich bereits über je 2–3 Geißelkammern ein kleines Oscularrohr. Seine Entstehung wird wahrscheinlich durch den Wasserdruck als formbildenden Reiz veranlasst. Auch Mergner (1964) erhielt bei Induktionsexperimenten funktionierende Kleinstoscularrohre aus nur 30 Ap-Endopinacocyten, wenigen Archaeocyten und 3 Geißelkammern. Obwohl diese Kleinstoscularrohre nur der Dermalmembran des Unterlageschwammes und ohne irgendeine Verbindung zu dessen Kanalsystem aufsaßen, blieb ihre Mündung für längere Zeit geöffnet. Alle diese Versuchsergebnisse zeigen eindeutig, daß für die Erhaltung einer funktionsfähigen Schwammorganisation nur ein Minumum an Zellen erforderlich ist, unter denen aber unbedingt Choanocyten, Pinacocyten und Archaeocyten vorhanden sein müssen.

Regulationen des Kanalsystems

Im Verlauf ihrer Untersuchungen über die Erhaltung der im fertigen. Jungschwamm eingestellten Ordnung ist Wintermann (1951) auf Regulationsvorgänge aufmerksam geworden, welche das Wasserleitungssystem ständigen, mehr oder weniger tiefgreifenden Änderungen unterwerfen. Bei derartigen regulativen Umbauten können sich die ausführenden Kanäle erweitern und begradigen, aber auch verengen und dadurch das Oscularrohr zum Verschluß bringen. Dann wird an anderer Stelle ein neues Oscularrohr gebildet, das die Funktion des primären Rohres übernimmt. Häufig werden damit gleichzeitig die ausführenden Kanäle verlagert oder sogar völlig neu angeordnet und, hierdurch verursacht, die Wasserstromrichtung umgekehrt. Letzteres kann auch erfolgen, wenn zwei einzeln gewachsene Jungschwämme der gleichen Art miteinander verschmelzen und ihre beiden Kanalsysteme zu einem einheitlichen vereinigen. Niemals aber werden bei irgendwelchen regulativen Umbauten einführende Kanäle zu ausführenden und umgekehrt (Wintermann, 1951).

Für die Klärung derartiger und ähnlicher Fragen sind freilich dreidimensionale Jungschwämme nicht immer geeignet. Das Gewirr der zahlreichen Kanäle im Schwamminnern mit der Fülle der ihnen angelagerten Geißelkammern verwehrt nämlich häufig den Einblick in die tieferen Regionen des Schwammes, vor allem auf seine basale Anheftungsfläche. Hier hat nun Ankel mit seinen Mitarbeitern (1944–1950, 1951) durch die Methode der Unterdeckglaszucht praktisch

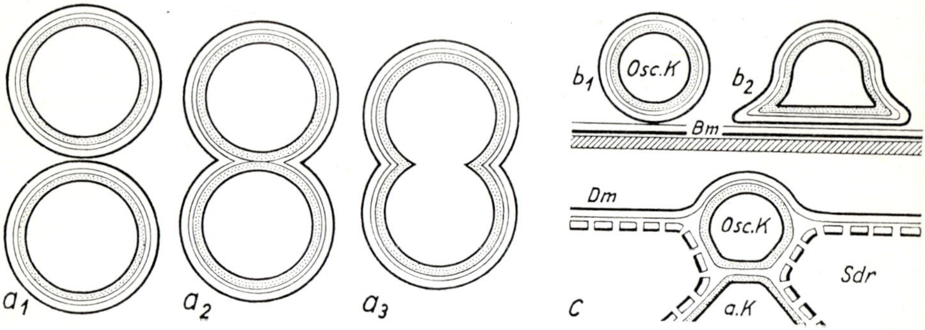

ABB. 3. Möglichkeiten der physiologischen Orientierung der Epithelflächen bei artgleichem Zellmaterial. Erläuterungen im Text. Dicke Linien, Außenflächen der Epithel-Einzellamellen; Dünne Linien, Innenflächen der Einzellamellen; Weiß, Exopinacocyten; Punktiert, Ap-Endopinacocyten; Gestrichelt, Pros-Endopinacocyten.
a, Vereinigung zweier Oscularrohrkugeln;
b, Auflegen einer Oscularrohrkugel auf isolierte Exopinacocytenmembran der Schwammbasis;
c, Vereinigung einer in den Subdermalraum eingesunkenen Oscularrohrkugel mit einem ausführenden Kanalabschnitt zur Oscularrohr-Neubildung. (Nach Mergner, 1966.)

zweidimensionaler Schwämme, also "lebender Flachschnitte", neue Einsichten in Bau und Verhalten eines Strudlers eröffnet. Er hat 1964 eindrücklich auf die mit Unterdeckglaskulturen durchgeführten Versuche und die dabei erzielten Ergebnisse hingewiesen. Es genügt daher, nur einige der darin liegenden Probleme aufzuzeigen:

Der Mechanismus der Durchströmung erfordert zwei absolut wasserdichte Kanalsysteme, das einführende und das ausführende, allein verbunden durch die hydromotorischen Zentren der Geißelkammern. Sie sind zugleich Sammelstellen für alle eingeströmten Partikeln und Garant für den lebensnotwendigen Gasstoffwechsel (Ankel, 1964). Jeder die Geißelkammern umgehende Kurzschluß kann daher den Effekt des Gesamtsystems ernsthaft gefährden. In diesem Sinne haben alle am Aufbau des Schwammkörpers beteiligten Zellen die Aufgabe, in korreliertem Zusammenspiel das Leitungssystem dicht zu halten. Daher erscheint die Unterordnung der vagilen Einzelzelle mit ihrem hohen Freiheitsgrad unter den Zwang der funktionellen Stabilität als das Entscheidende. Erst die Anwendung dieses Prinzips erlaubt das Fortbestehen der Schwammfunktionen auch bei noch so tiefgreifenden Umbauten.

Dies gilt gleichermaßen für die bereits erwähnte regulative Neugestaltung des Kanalsystems während längerer Beobachtungszeiten wie auch für kurzfristig reversible Umbauten. So kann die Modulation der Durchströmungsintensität sogar bis zum zeitweisen Stop der Pumpleistung führen, indem sich die Geißelkammern kontrahieren und die ausführenden Kanäle bis zum zwangsläufigen Verschwinden der einführenden erweitern (Wintermann, 1951). Noch ungeklärt sind dagegen die Ursachen für die plötzlichen Starrezustände in der steten "Peristaltik" des Kanalsystems, bei welchen alle beteiligten Zellelemente für kurze Zeit zu absoluter Bewegungslosigkeit gelangen (Ankel, 1964). Offensichtlich sind diese Zellen in ihrem Verhalten gut funktionierenden Korrelationen unterworfen.

Am eindrucksvollsten erweist sich dies bei Reorganisationen, wie sie Wintermann (1951) durch Eingriffe in das Kanalgefüge von Unterdeckglasschwämmen herbeigeführt hat . Durch Druck auf das Deckglas lassen sich nämlich alle Grade von Schädigungen setzen, von geringen Störungen bis zur totalen Dissoziation der Schwammepithelien und bis zur Aufhebung jeglicher Ordnung im Kanalsystem. Trotzdem aber erscheint dieses System wie in einer Kopie des Ausgangsmodells bereits nach wenigen Stunden wieder in den ursprünglichen Grenzen restituiert.

Untersuchungen über den Potenzumfang der Epithelzellen

Die oben gegebenen Beispiele zeigen deutlich, wie entscheidend die Unterordnung der so vagilen und physiologisch selbständigen Einzelzellen unter das Ziel der Aufrechterhaltung eines geschlossenen Durchströmungssystems ist. Ankel und Mitarbeiter (Eigenbrodt, Wintermann, Kilian und Mergner, 1948–1966) haben sich daher schon frühzeitig mit diesen Einzelzellen und ihren Eigenschaften auseinandergesetzt, in erster Linie natürlich mit den drei Pinacocytensorten als den Epithelzellen. Sie haben deren morphologische und funktionelle Differenzierungen untersucht, ihre Ansätze zu echter Gewebebildung und ihre Fähigkeit, durch Zusammenschluß zu Epithelmembranen wasserdichte Bauelemente herzustellen als Voraussetzung für das Funktionieren der Durchströmungseinrichtungen des Schwammes. Ankel und Wintermann-Kilian (1952) haben dabei von den beteiligten Zellen eine Anzahl Eigenschaften, welche die Epithelbildung erst ermöglichen, gefordert und in Versuchen auch nachgewiesen:

Hierhin gehören vor allem das Bestreben isolierter Epithelzellen, sich in Plattenepithelien zu vereinigen (Wintermann, 1951), und die Bildung von Epithel-Doppellamellen aus jeweils zwei einschichtigen Einzellamellen (Wintermann, 1951; Ankel und Wintermann-Kilian, 1952). Ferner befassten sie sich mit den spezifischen Eigenschaften der freien Zellkanten, mit dem differenten Verhalten der inneren und äußeren Epithelzellflächen (Wintermann-Kilian und Ankel, 1954) und mit den Auswirkungen des Verhaltens der Einzelzellen gegenüber dem von ihnen gebildeten Verband. Zum Verständnis des Folgenden müssen die Ergebnisse dieser Untersuchungen etwas näher betrachtet werden, zumal sie eine Grundlage für die nachfolgenden Feststellungen über die induzierenden Eigenschaften des Oscularrohrepithels bilden.

Verträglichkeitsprüfungen mit Epithelien

Abgeschnittene Osculartohre von *Ephydatia fluviatilis* und anderen Spongilliden schließen sich sofort an der Schnittkante und am Osculum und wandeln sich innerhalb von 24 Stunden in flüssigkeitserfüllte, pralle Hohlkugeln um. Ihre Wandung besteht wie beim Osculartohr aus einer äußeren Lage von Exopinacocyten und einer inneren von Ap-Endopinacocyten. Das physiologisch differente Verhalten beider Epithelschichten bestimmt durch Sekretion von Flüssigkeit in den Hohlraum hinein die Herstellung des notwendigen Binnendruckes (Wintermann und Ankel, 1951).

Zwei oder mehrere dieser "Osculartohrkugeln" haften bei Berührung aneinander und verschmelzen derart miteinander, daß eine größere, wiederum zweischichtige Kugel mit polygonalem Zellmuster entsteht (Abb. 3a). Dabei bilden sich an der Haftstelle durch Auflösen der Kantenbindungen benachbarter gleichsortiger Epithelzellen und Neubindung der freien Zellkanten zunächst eine eng umschriebene Kommunikation und damit ein gemeinsamer Innenraum. Durch Umordnen der Einzelzellen rundet sich dieser Innenraum schließlich gleichmäßig ab.

Beide Schichten, Exopinacocyten und Ap-Endopinacocyten, stellen nun zwar aneinandergelagerte, jedoch in sich geschlossene und untereinander zu keiner Vereinigung befähigte Epithelsysteme mit verschiedenen Leistungen und Eigenschaften dar. Unter diesen Eigenschaften fällt die hohe physiologische Selbständigkeit der beiden Zellsorten auf. Sie zeigt sich schon in ihrer Dehnbarkeit und Kontraktilität sowie in der damit verbundenen Fähigkeit zur Formveränderung des von ihnen gebildeten Verbandes. Noch mehr beeindruckt ihre Befähigung, Zellkanten kurzfristig zu lösen oder neu zu verbinden. Voraussetzung hierfür sind das Ausschalten bzw. Aktivieren der entsprechenden Bindekräfte und das Auflösen bzw. Neuformieren der histologisch nachweisbaren Interzellularschicht. Erst hierdurch ergibt sich die Möglichkeit zur Umordnung der Epithelien, wie sie sich bei der Vereinigung zweier Oscularrohrkugeln vollzieht.

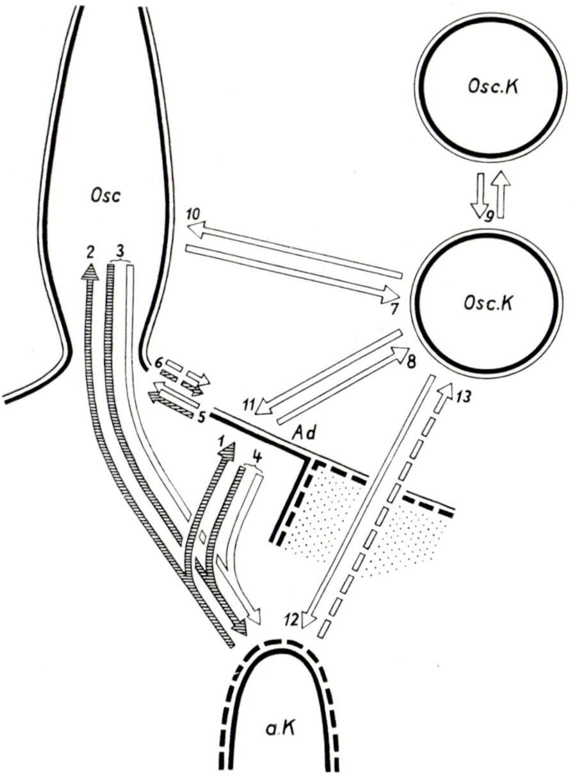

Abb. 4. Die möglichen strukturellen und funktionellen Zustandsänderungen der Ap-Endopinacocyten als Beweis für das "lagespezifische Verhalten" dieser Zellen und die Erhaltung ihrer Gesamtpotenz im Bereich des ausführenden Systems. Erläuterungen im Text. 34. Schraffierte Pfeile, Natürliche, bei Normogenese und Regulationen spontan auftretende Zustandsänderungen; Weiße Pfeile, Künstlich hervorgerufene Zustandsänderungen; Ausgezogene Pfeile, Nachgewiesene Zustandsänderungen; Gestrichelte Pfeile, Theoretische mögliche Zustandsänderungen. (Nach Mergner, 1964.)

Wintermann und Ankel (1951, 1952, 1954) haben diese regulativen Potenzen dazu benutzt, die Leistungen der einzelnen am Aufbau des Schwammkörpers beteiligten Epithelzellen experimentell nachzuprüfen, um einen Einblick in das lagespezifische Verhalten der betreffenden Epithelien zueinander zu gewinnen. Auch Mergner (1959, 1964, 1966) hat seine Versuche zur Oscularrohrinduktion auf diesen Potenzprüfungen aufgebaut. Aus der Vielzahl der durchgeführten Versuche über die Verträglichkeit der einzelnen Epithelien untereinander seien hier nur einige wenige aufgeführt:

Ebenso wie zwei oder mehrere Oscularrohrkugeln miteinander zu einer größeren Kugel verschmelzen, vereinigen sich auch aus Dermalmembranschläuchen gezüchtete Kugeln. Ihre Außenlamelle weist wiederum Exopinacocyten auf, während die Innenlamelle nunmehr aus Pros-Endopinacocyten besteht (Ankel und Wintermann-Kilian, 1952). Wegen dieser unterschiedlichen Determinierung der beiden Innenschichten ist es auch unmöglich, eine Oscularrohrkugel mit einer Dermalmembrankugel zu verschmelzen, obwohl beide Außenschichten aus Exopinacocyten bestehen und daher den gleichen Potenzgehalt haben.

Im Gegensatz hierzu lassen sich Oscularrohrkugeln in Atrialdachepithel und in funktionierende Oscularrohre völlig einordnen (Abb. 4; Mergner, 1964). Alle in diesem Versuch verwendeten Innenschichten, also die des Atrialdaches, des Oscularrohres und der Oscularrohrkugel, stellen nämlich Derivate des ausführenden Kanalsystems dar und bestehen daher aus Ap-Endopinacocyten. Sie besitzen also den gleichen Potenzgehalt. Da auch die jeweiligen Außenschichten gleich determiniert sind, lassen sich die Transplantate mit den Unterlagen vereinigen.

Legt man jedoch eine Oscularrohrkugel auf isolierte lebende Exopinacocytenmembran der Schwammbasis (Abb. 3b), dann haftet sie zwar unter Bildung einer Kalotte auf dieser, aber die beiden aufeinander liegenden Exopinacocytenschichten ordnen sich nun nicht mehr zu einem einheitlichen Epithel zusammen, wie dies bei der Vereinigung zweier Oscularrohrkugeln geschieht (Ankel und Wintermann-Kilian, 1952). Trotz ihrer nachweislich gleichartigen Determinierung und des gleichen Potenzumfanges bleiben die beiden Schichten daurend getrennt.

Zur physiologischen Orientierung der Epithelzellflächen

Die Ergebnisse der oben gewählten Beispiele zeigen in ihrer Gesamtheit, daß eine Vereinigung zweier Epithelsysteme den gleichen Potenzgehalt ihrer Einzellamellen voraussetzt. Die Diskrepanz zwischen den einzelnen Versuchsergebnissen macht jedoch deutlich, daß für eine dauerhafte Verschmelzung auch noch andere Faktoren hinzutreten müssen. Mergner (1966) hat daher die Frage der "physiologischen Orientierung" der Epithelzellflächen aufgegriffen. Er hat hierzu das differente Verhalten der Innen- und Außenflächen der Epithellamellen, also ihre Polarität, untersucht und folgende Gesetzmäßigkeiten gefunden:

Beim Verschmelzen zweier artgleicher Oscularrohrkugeln zu einer größeren Kugel (Abb. 3a) kommen zunächst die physiologischen Außenflächen der beiden Exopinacocytenschichten aufeinander zu liegen und sie werden auch tatsächlich miteinander verbunden. Genauso vereinigen sich anschließend die beiden Ap-Endopinacocytenschichten, nur sind hier beim Anlagern ihre physiologischen Innenflächen gegeneinander gerichtet. Prinzipiell die gleiche Orientierung der Epithelzellflächen und die gleiche Reihenfolge in der Vereinigung der Einzellamellen finden sich beim Verschmelzen zweier Dermalmembrankugeln und beim Einordnen einer Oscularrohrkugel in Atrialdachepithel oder Oscularrohr.

Ganz anders ist jedoch die Orientierung der Epithelzellflächen beim Auflegen der Oscularrohrkugel auf die isolierte Exopinacocytenschicht der Schwammbasis (Abb. 3b). Hier kommt nämlich die physiologische Außenfläche der Kugel-Exopinacocyten auf die Innenfläche der Basalmembran-Exopinacocyten zu liegen. Eine dauernde Vereinigung der beiden Schichten kann daher in diesem Versuch nicht erfolgen.

Aus diesen und ähnlichen Beobachtungen hat Mergner (1966) folgende Hypothese abgeleitet: Zu einer dauernden Vereinigung sind artgleiche und gleichsortige Epithellamellen nur befähigt, wenn sie mit ihren beiderseitigen Außen- oder Innenflächen gegeneinander liegen, nicht aber, wenn Außen-auf Innenflächen treffen. Daraus wiederum läßt sich folgern, daß die beiden verschiedensortigen Einzellamellen jedes Doppelepithels im Schwammkörper stets mit ihren Innenflächen gegeneinander liegen und mit ihren Außenflächen voneinander abgewendet. Diese Hypothese hat sich auch bei den später noch zu behandelnden Versuchen zur Oscularrohrinduktion bestätigt.

Ohne ihre Kenntnis bestand jedoch ursprünglich die Auffassung, die Exopinacocyten verschiedener Schwammregionen besäßen auch verschiedene physiologische Eigenschaften. Der vergebliche Versuch, die Exopinacocytenschicht einer Oscularrohrkugel mit der einer Dermalmembrankugel oder den Exopinacocyten der Schwammbasis zu vereinigen, scheint diese Auffassung auch zu bestätigen. Begründet wurde sie aber damit, daß das lagespezifische Verhalten der Pinacocyten und der von ihnen gebildeten Verbände Mißbildungen verhindere, die zu Funktionsstörungen führen könnten (Ankel und Wintermann-Kilian, 1952). Vom biologischen und funktionsmorphologischen Standpunkt aus erscheint diese Deutung auch durchaus sinnvoll. Und doch sprechen einige Beobachtungen und Überlegungen gegen sie (Mergner, 1964):

So bildet ein bestimmter Prozentsatz in Kulturen gezüchteter Jungschwämme regelmäßig mehrere Oscularrohre aus. Andere Schwämme können zusammenwachsen und unter Aufgabe ihrer bisherigen Individualität ein einheitliches Kanalsystem mit mehreren funktionierenden Oscularrohren bilden. Alle diese "überzähligen" Oscularrohre wären also im Sinne der eben gegebenen Deutung "Mißbildungen", die von der Natur nicht vorgesehen sein könnten.

Ferner können neue Oscularrohre entweder als normale Organbildungen während der Normogenese oder als natürliche Regulationsphänomene infolge einer Umkehrung der Wasserstromrichtung im bereits funktionierenden Schwammkörper entstehen (Wintermann, 1951). In beiden Fällen werden jedoch die aus dem ausführenden Kanalsystem vorwachsenden Oscularrohre beim Durchstoßen der Dermalmembran von deren Exopinacocyten besiedelt. Die Außenlamellen von Dermalmembran und Oscularrohr sind also cytologisch einheitlich und sollten schon deshalb ein Aneinanderheften gestatten.

Versuche zur Oscularrohrinduktion

Diese Überlegungen haben Mergner (1959, 1964) veranlasst, im Rahmen von Untersuchungen über Potenzumfang und Determination der Pinacocyten neue Versuche zur lagespezifischen Verträglichkeit von Oscularrohrepithel gegenüber Dermalmembran anzusetzen. Es sollte also das Verhalten eines Doppelepithels aus Exopinacocyten und Ap-Endopinacocyten gegenüber einem solchen aus Exopinacocyten und Pros-Endopinacocyten geprüft werden. Zugleich sollte aber auch die Induktionsfähigkeit determinierter Zellen nachgewiesen werden. Außerdem musste sich zwangsläufig zeigen, wie der Unterlageschwamm mit dem anders zusammengesetzten Kugelmaterial zurechtkommen würde, vor allem, wie er sein wichtigstes Funktionsprinzip, die scharfe Trennung von ein- und ausführendem Wasserleitungssystem, wahren würde. Diese Induktionsversuche wurden teils unter Mitverwendung des Kugelmaterials, teils ohne dieses, also als rein stofflich bedingte Induktionen, durchgeführt. Später wurden sie durch Untersuchungen über die Artspezifität des Induktionsstoffes ergänzt (Mergner, 1966).

Oscularrohrinduktionen unter Einbau des Kugelmaterials

Legt man eine Oscularrohrkugel auf die Dermalmembran eines Unterlageschwammes, dann zeigt sich mit hoher Wahrscheinlichkeit, nämlich mit etwa 77·5 % Erfolgsquote, folgender Vorgang (Abb. 5): Die Kugel haftet auf der Dermalmembran fest und breitet sich auf ihr aus. Unter Umordnungserscheinungen wie bei der Vereinigung zweier Oscularrohrkugeln fügt sich das beiderseitige Exopinacocytenepithel ineinander. Dadurch kommen die Ap-Endopinacocyten der Kugel unmittelbar auf die Pros-Endopinacocyten der Dermalmembran zu liegen. Die beiden verschiedensortigen Epithellamellen können natürlich nicht miteinander verschmelzen. Die Kugel sinkt mehr und mehr in den Subdermalraum ein und tritt mit dem nächstliegenden Abschnitt des ausführenden Kanalsystems in Verbindung. Dieser kann sogar der Kugel entgegenwachsen, sich dabei vom ausführenden Hauptkanal abtrennen und seine Wasserstromrichtung umkehren. Die beiderseitige Annäherung wird oft durch eine Kontraktion des Subdermalraumes beschleunigt. Beide Ap-Endopinacocytenlamellen, nämlich die von Kugel und ausführendem Kanal, verbinden sich miteinander. Dann bedarf es nur noch des Wasserüberdruckes des ausführenden Kanalabschnittes, um die Kugel aufzublähen. Sie bildet am freien Ende eine Egestionsöffnung und wird damit nach etwa 20 Stunden zum funktionsfähigen Oscularrohr, das je nach Wasserdruck kürzer oder länger bestehen bleibt.

Bei einer Analyse des Versuchsablaufes (Mergner, 1964) läßt sich zunächst eine durch die offensichtliche Verträglichkeit der beiden genäherten Exopinacocytenschichten bestimmte Anheftungsphase erkennen. Diese Verträglichkeit war wegen der morphogenetisch gleichen Herkunft und des gleichen Potenzgehaltes der beiden Außenlamellen zu erwarten. Sie entstammen nämlich beide der ersten Histoblastenbedeckung des ausgeschlüpften Keimkörpers. Die anschließende Ausbreitungsphase bringt Zell- und Epithelverschiebungen rings um die breite Anheftungsfläche der Kugel. Die Zellen des beiderseitigen Exopinacocytenepithels vereinigen sich nun miteinander und ziehen sich aus der gemeinsamen Haftfläche zurück.

In der Einsenkungsphase wie auch in der nachfolgenden Anschlußphase werden infolge des Tiefersinkens der Kugel mehrfach hintereinander und kurzfristig jeweils neue Doppelepithelien zusammengestellt. Dabei vereinigen sich die beiden Pros-Endopinacocytenschichten des Subdermalraumes im unmittelbaren Umkreis der Kugel, ebenso auch die Ap-Endopinacocyten der Kugel und des ausführenden Kanalsystems. Alle diese Vereinigungsvorgänge werden wiederum mit dem Aneinanderlegen der physiologischen Außen- oder Innenflächen gleichsortiger Epithelien eingeleitet. Hier stimmen also jeweils physiologische Orientierung und Potenzgehalt der sich vereinigenden Lamellen überein (Abb. 3c). Beim kurzfristigen Zusammenstellen der Doppelepithelien aus verschiedensortigen Einzellamellen verhindert dagegen deren unterschiedlicher Potenzgehalt trotz richtiger Orientierung eine Verschmelzung dieser Lamellen.

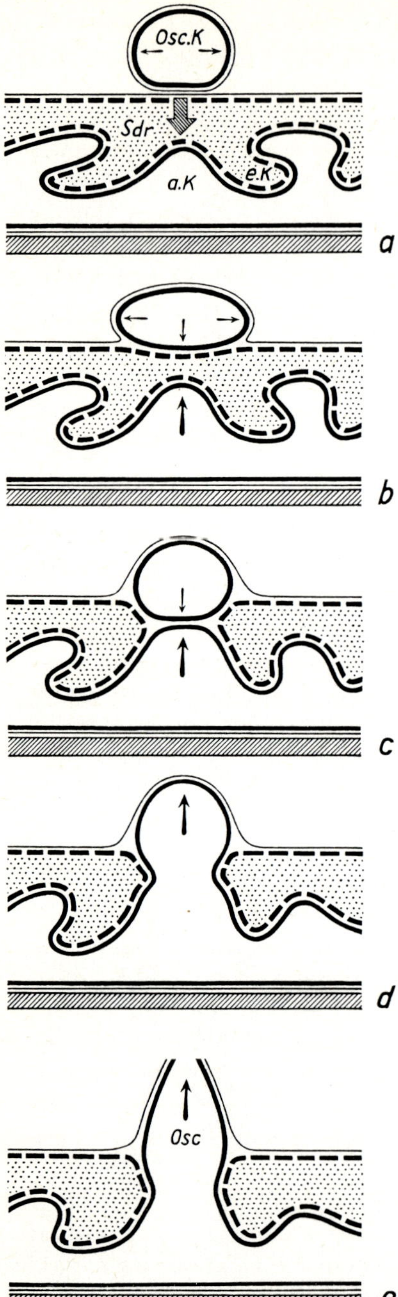

Abb. 5.

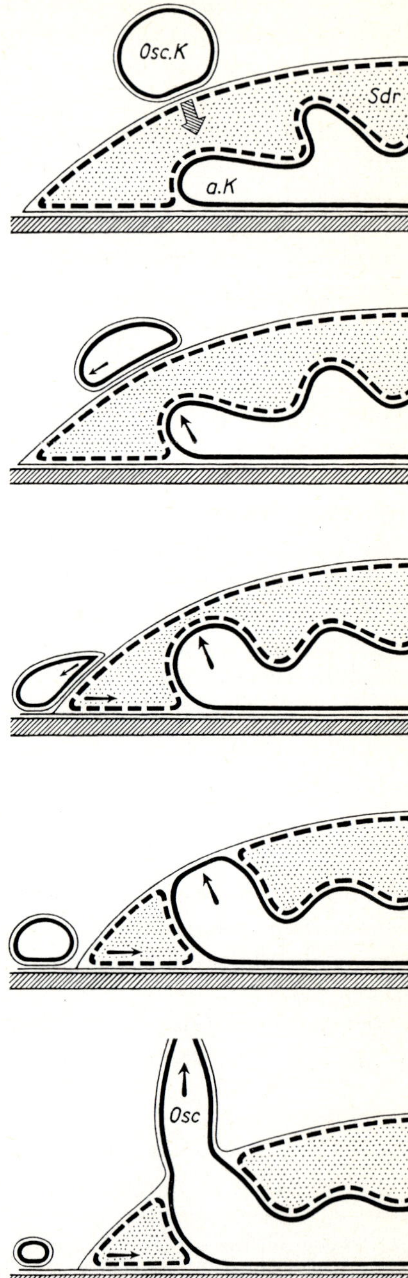

Abb 6.

Die im vorhergehenden Abschnitt begründete Hypothese läßt sich also nunmehr, wie folgt, erweitern: Zur Bildung eines Doppelepithels aus verschiedensortigen Einzellamellen genügt deren richtige physiologische Orientierung, während ihre Determinicrung unterschiedlich sein kann. Zum Ineinanderordnen zweier gleichsortiger Epithellamellen müssen dagegen Determinierung wie auch physiologische Orientierung übereinstimmen.

Die ganze Einsenkungsphase wird durch die Fähigkeit der Pinacocyten bestimmt, in kürzester Zeit ihre Kantenbindungen zu lösen und wieder neu zu formieren. Ebenso überrascht die Leichtigkeit, mit welcher sich die beiden Einzelschichten jeder Doppellamelle voneinander trennen und wieder neu zusammentreten. Hingewiesen sei auch auf die Art und Weise, wie die in den Subdermalraum eindringende Kugel durch dessen Pros-Endopinacocyten abgeschirmt wird. Das für die Schwammfunktion notwendige Prinzip der scharfen Trennung beider Wasserleitungssysteme wird wiederum von Anfang an verfolgt und nur deshalb kann die Oscularrohr-Neubildung erfolgreich durchgeführt werden. Der Schwamm ist also befähigt, selbst bedeutenden Eingriffen in seinen Organismus ohne wesentliche Störung seiner Durchströmung zu begegnen und das zusätzliche Epithelangebot in sein System einzubauen.

Ferner tritt hier deutlich die unterschiedliche Determinierung von Pros- und Ap-Endopinacocyten hervor. Erstere können grundsätzlich nur die Innenauskleidung des einführenden Systems herstellen, letztere nur die des ausführenden Systems. Schließlich zeigen das Einsinken der Kugel und das gleichzeitige Entgegenwachsen eines ausführenden Kanalabschnittes mit seiner Abtrennung vom Hauptkanal und Umkehrung seiner Wasserströmung induktive Einflüsse, die nur vom Kugelmaterial ausgehen können.

In der folgenden Anschlußphase bilden Kugel und ausführender Kanal durch Verbinden ihrer beiderseitigen Ap-Endopinacocytenschichten ein einheitliches System. Die Kugel gleicht nunmehr einer beginnenden Oscularrohr-Neubildung. In der abschließenden Aufblähungsphase bestimmen Wasserdruck, Angebot an Zellmaterial und stoffliche Wirkung das Ausmaß des letzten Induktionsschrittes. Dabei

---

ABB. 5. Induktion eines Oscularrohres durch eine auf die Dermalmembran aufgelegte Oscularrohrkugel unter Mitverwendung des kugeleigenen Zellmaterials. Erläuterungen im Text. Schraffierter Pfeil, Induzierender Reiz des Kugelmaterials; Kleine schwarze Pfeile, Bewegungen der Kugel; Große schwarze Pfeile, Vorstoß des ausführenden Kanals und des späteren Oscularrohres. (Nach Mergner, 1964.)

ABB. 6. Rein stofflich bedingte Induktion eines Oscularrohres ohne Mitverwendung kugeleigenen Zellmaterials. Erläuterungen im Text. (Nach Mergner, 1964.)

wirken Wasserdruck und Beinflussung durch die Wasserschüttung des primären Oscularrohres gegeneinander. Außerdem hat sich erwiesen, daß größere Kugeln mit höherer Wahrscheinlichkeit neue Oscularrohre ergeben als kleine Kugeln. Den ortstreuen Einbau des Kugelmaterials hat Mergner (1964) durch Verwenden neutralrot gefärbter Kugeln nachweisen können, den Anschluß der Kugel an das Kanalsystem dagegen mittels graphischer Methoden.

Die Bildung der Oscularrohrmündung führt wiederum zu der Frage, inwieweit dabei determinierte Zellen ein lagespezifisches Verhalten zeigen: Beim Abschneiden des Oscularrohres schließt sich regelmäßig sein Osculum. Seine Mündungszellen unterscheiden sich dann durch nichts mehr von den übrigen polygonal geformten Zellen der fertigen Oscularrohrkugel. Es ist also unmöglich, etwa absichtlich die Kugel in ihrer ursprünglichen Ausrichtung, die sie als Oscularrohr hatte, auf die Dermalmembran aufzulegen. Trotzdem aber bildet sie nach ihrem Anschluß an das Kanalsystem stets eine neue Mündung. Es muß daher jede andere Kugelzelle genauso die Funktion einer Mündungsgrenzzelle übernehmen können. Sie zeigt dann nur ein dementsprechendes lagespezifisches und jederzeit reversibles Verhalten; ihr Potenzgehalt erlaubt ihr diesen Funktionswechsel.

Noch deutlicher wird dies, wenn die soeben geöffnete Mündung etwa von der Seite zum freien Ende des Oscularrohres wandert. Dann wird der Rand der Egestionsöffnung in gleitender Ablösung von immer neuen Zellen dargestellt. Die Lochbildung, die hier paradoxerweise geradezu als ein morphologisches Element erscheint, wird durch aktive Leistungen zwischen den sie begrenzenden Zellen weitergereicht (Ankel, 1964). In wieder anderen Versuchen wechselte die Lage des Osculum auch ohne Wanderung mehrmals zeitlich und örtlich voneinander unabhängig oder es entstanden mehrere Mündungen gleichzeitig an einem Oscularrohr. Diese Ergebnisse dokumentieren eindeutig das kurzfristig reversible, lagespezifische Verhalten determinierter Zellen und ihre Befähigung zur raschen Umordnung des von ihnen gebildeten Epithelzellmusters (Mergner, 1964).

Der Potenzumfang der Ap-Endopinacocyten und die ganze Breite ihrer strukturellen und funktionellen Modifikabilität lassen sich sogar in einem einzigen, aber sehr instruktiven Versuch darstellen (Abb. 7): Biegt man nämlich das distale Ende eines großen, neu induzierten Oscularrohres gegen das Atrialdach des primären Hauptoscularrohres herab, dann ordnet es sein Epithel in das Atrialdachepithel derart ein, daß nunmehr der Wasserstrom des induzierten Oscularrohres in den Atrialraum umgeleitet wird. Die Ap-Endopinacocyten der aufgelegten Kugel haben sich also zunächst dem ausführenden Kanal-

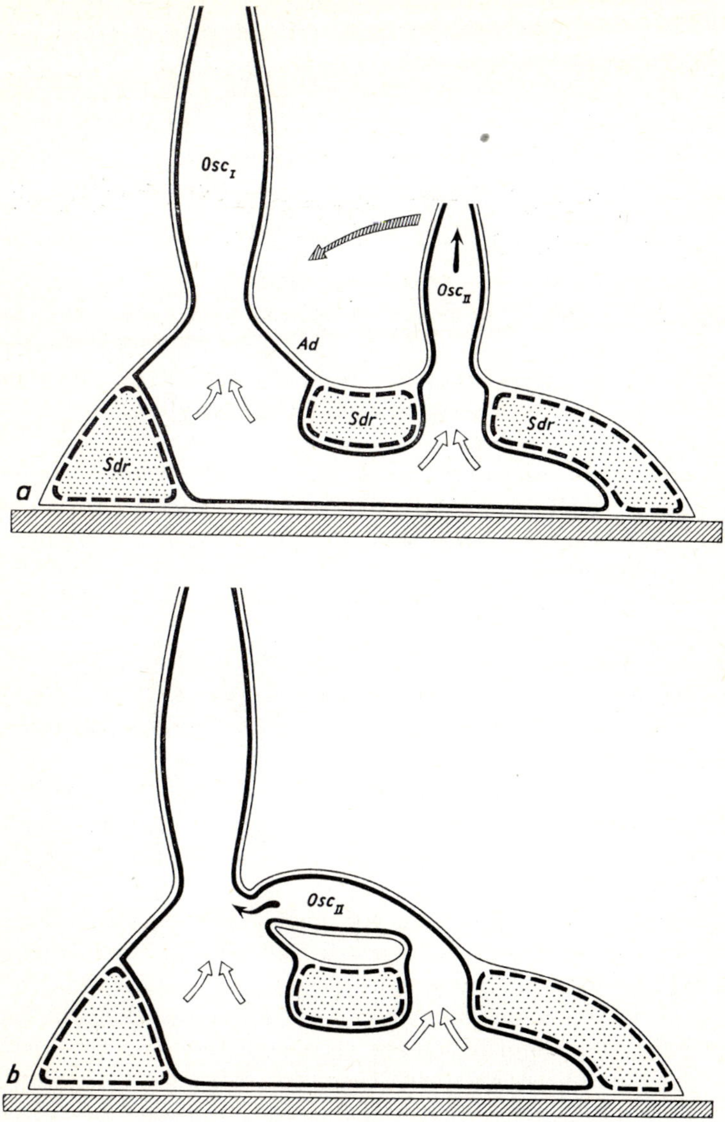

ABB. 7. Nachweis des Potenzumfanges der Ap-Endopinacocyten. Erläuterungen im Text. Weiße Pfeile, Weg des auströmenden Wassers; Schwarze Pfeile, Weg des ausströmenden Wassers im neugebildeten Oscularrohr (Osc$_{II}$); Schraffierter Pfeil, Herabbiegen des neugebildeten Oscularrohres zum Einordnen seiner Spitze in das Atrialdach. (Nach Mergner, 1964.)

system angeschlossen und ein neues Oscularrohr gebildet und dann diese Neubildung teilweise wieder in das Atrialdach eingefügt. Die gleichen Zellen sind demnach zum Anschluß an das ausführende System befähigt, zum Ausüben der Oscularrohrfunktion und zum Einordnen in das Atrialdachepithel (Abb. 4). Sie können also jegliche Funktion im Rahmen des gesamten ausführenden Systems übernehmen als des überhaupt möglichen Bereiches ihres Vorkommens (Mergner, 1964).

Rein stofflich bedingte Oscularrohrinduktionen

Schon beim Einordnen des Kugelmaterials in den Unterlageschwamm und bei der Neubildung des daraus hervorgehenden Oscularrohres ließen sich mehrfach induktiv bedingte Phänomene erkennen. Doch wird dies noch viel deutlicher, wenn die Kugel nur eine begrenzte Kontaktzeit mit der Unterlage hat und frühzeitig von der Haftstelle abrutscht, so daß keinerlei Kugelmaterial für den Einbau oder für eine weitere Reizaussendung zurückbleibt. Dann wächst nämlich der nächstgelegene ausführende Kanalabschnitt gegen die ursprüngliche Auflagestelle der Kugel hoch und bildet genau hier ein neues Oscularrohr (Mergner, 1964). Dessen Zellmaterial muß also ausschließlich dem Unterlageschwamm entstammen (Abb. 6). Damit ist bewiesen, daß zum Induzieren eines neuen Oscularrohres der Einbau des Kugelmaterials in den Unterlageschwamm gar nicht erforderlich ist.

Eine derartige Neubildung, die als Indukt zu kennzeichnen ist, kann demnach nur die Antwort sein auf den Induktionsreiz der aufgelegten Kugel als des Induktors. Die strenge Lokalisierung der Reizwirkung auf die vorbestimmte Stelle und der hohe Wirkungsgrad über eine relativ große Entfernung hinweg lassen aber den Schluß zu, daß es sich hierbei um einen Induktionsreiz stofflicher Natur handeln muß.

Wiederum werden die Regeln der physiologischen Orientierung der Epithelzellflächen durch die Beobachtungen während des Ablaufes der rein stofflich bedingten Oscularrohrinduktion bestätigt. Nur treten in diesem Fall die Verschmelzungsvorgänge gegenüber der wiederholten Bildung neuer Doppelepithelien aus verschiedensortigen Einzellamellen zurück. Erstaunlich ist auch die Tatsache, daß eine Induktion selbst dann noch erfolgt, wenn sie für die Gesamtfunktion des Unterlageschwammes sinnlos oder gar störend wird, wenn etwa sein ausbalanziertes Strömungssystem durch das neue Oscularrohr irgendwo angezapft wird. Das bedeutet, daß die Informationsübertragung des Induktionsstoffes unausweichlich ist und daher zwangsläufig zu Störungen des Schwammorganismus führen muß (Ankel, 1964).

Aber auch hier zeigen sich wieder die in den Schwammzellen

vorhandenen regulativen Potenzen, die dem Ziel der Aufrechterhaltung eines geschlossenen Durchströmungssystems zugeordnet sind: Der Subdermalraum wird gegen die sinnlose Neubildung abgeschirmt und damit die Trennung beider Wasserleitungssysteme gesichert. Schließlich werden auch die durch das Induktionsexperiment gesetzten Störungen vom Schwamm aktiv wieder ausgeglichen. Die induzierten Oscularrohre werden nämlich nach einiger Zeit des Funktionierens wieder eingeschmolzen.

Über die Natur des Induktionsstoffes

Die Zwangsläufigkeit der Informationsübertragung beim Induktionsvorgang spricht also für einen rein stofflichen Induktionsreiz. Über seine Natur lassen sich vorerst jedoch nicht einmal Vermutungen anstellen. Für die vorgesehene chemische Analyse wird eine große Menge des zarten, nur 1–2 μ dicken Kugelepithels benötigt und dadurch ein bedeutender technischer Aufwand erforderlich. Zudem geht der Induktionsreiz ausschließlich vom lebenden Zellmaterial aus; Versuche mit abgetöteten Kugeln verliefen bisher stets negativ. Möglicherweise lassen sich aber beide Schwierigkeiten, nämlich der Einsatz einer großen Materialmenge und zugleich noch induktiv wirksamen Zellmaterials, durch sukzessives Einfrieren von Kugeln beheben.

Zunächst aber wurde damit begonnen, nach anderen Methoden zur erfolgversprechenden Präzisierung der Stoffnatur des Reizes zu suchen. Die eine wäre mit der Zucht von Ap-Endopinacocyten in vitro gegeben, die einerseits als "Sender" des Induktionsreizes der Kugel und andererseits als dessen "Empfänger" im Epithelverband eines ausführenden Kanalabschnittes angesehen werden müssen. Gegen eine Gewebezucht dieser Zellen spricht jedoch die Tatsache, daß ausdifferenzierte Pinacocyten nicht mehr teilungsfähig sind. Ein weiterer Weg könnte darin liegen, die Reaktion des Unterlageschwammes bei Auflage von Fremdkörpern zu beobachten, wie etwa von kleinen Agarblöckchen, welche für eine Zeitlang der Einwirkung von Oscularrohrkugeln ausgesetzt waren, oder von Hühnerplasma und ähnlichem. Mergner (1966) hat jedoch als nächstliegende Methode Versuche angesetzt, durch welche die Artspezifität des Induktionsstoffes festgestellt werden sollte.

Über Fragen des artspezifischen Verhaltens bei Spongilliden bestehen bereits verschiedene Untersuchungen: Wierzejski (1935) ließ Gemmulae verschiedener Arten dicht nebeneinander keimen. Er konnte beobachten, daß die ausgekeimten Jungschwämme niemals miteinander verwuchsen, ja daß sie sogar in ihrem Wachstum gehemmt wurden. Ankel und Wintermann-Kilian (1952) schlossen aus ähnlichen Beobachtungen auf das Vorhandensein eines die Keimung hemmenden

Stoffwechselproduktes. Wintermann (1951) wiederum versuchte vergebens, durch Zerquetschen von zwei dicht nebeneinander gewachsenen, artverschiedenen Unterdeckglasschwämmen "bispezifische Konglomerate" (de Laubenfels 1926/27, 1928) zu erzielen. Die Zellen wanderten jedoch regelmäßig wieder zu ihrer Ursprungsart zurück. Wintermann führt dieses Verhalten auf die artspezifischen Unterschiede der beiden basalen Sponginplatten zurück, die eine Besiedlung mit artfremden Zellen nicht ermöglichen.

Die Annahme, daß auch der Induktionsstoff artspezifisch sein könnte, lag also durchaus im Bereich des Wahrscheinlichen. Methodisch hat Mergner (1966) sog. "xenoplastische Induktionen" erprobt, also Osculärrohr-Neubildungen mit artfremdem Kugelmaterial, von der Überlegung ausgehend, daß bei erfolgreicher Neubildung die Existenz eines Induktionsstoffes oder zumindest seine Artspezifität in Frage gestellt werden müßten. Als Unterlageschwämme bzw. "Xenotransplantate" und "Xenoimplantate" hat Mergner wechselweise drei Arten verwendet, nämlich *Ephydatia fluviatilis*, *Ephydatia mülleri* und *Spongilla lacustris*.

Bei den insgesamt 80 Versuchen kam es zwischen den verschiedenen Spongilliden-Arten niemals zu auch nur kurzfristigen Anheftungs- oder Verschmelzungsvorgängen oder gar zu einer Oscularrohr-Neubildung wie bei den Kontrollversuchen mit arteigenen Kugeln. Im Gegenteil, die beiden Arten trennten sich regelmäßig und in kürzester Zeit wieder voneinander, allerdings je nach der Ausgangslage des Fremdmaterials zum Unterlageschwamm in sehr verschiedener Weise (Abb. 8): Die Kugeln konnten sowohl über das Kanalsystem der Unterlage wie auch durch deren Epithelien ausgestoßen werden, vor allem aber durch das Dermalmembran-Epithel (Abb. 9). Dabei tritt das Xenoimplantat entweder als vollständige Kugel ohne stärkere Deformierung oder in einzelne kleine Zellaggregate zerlegt nach außen, um sich hier wiederum miteinander zu vereinigen. In beiden Fällen kann das artfremde Material interzellulär zwischen den Grenzen der Epithelzellen hindurchdringen oder intrazellulär durch die Dermalmembranporen. Oft wird sogar eine ganz bestimmte Öffnung bevorzugt, auf die sich bereits an anderen Stellen durchtretendes Material wieder zurückzieht, um hier gemeinsam nach außen zu treten. Die Tendenz der Zellen, Aggregate von einer gewissen Mindestgröße zu bilden, ist also offensichtlich. Alle diese Vorgänge werden regelmäßig mit einer örtlich begrenzten Kontraktion des Unterlageschwammes in der Umgebung des Xenoimplantats eingeleitet. Ihre räumliche Ausdehnung hängt dabei vom Wirkungsbereich des Implantats ab, ihre zeitliche Begrenzung von dessen Entfernung.

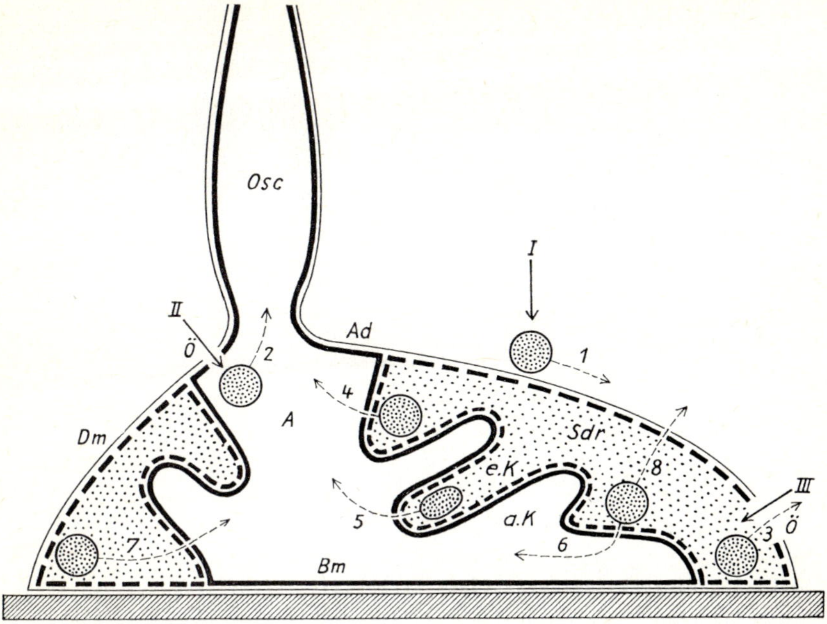

Abb. 8. Möglichkeiten der Trans- und Implantation von Fremdkugeln in den Unterlageschwamm. Erläuterungen im Text. Durchgezogene Pfeile I—III: Trans- und Implantationen. I, Auflegen auf Dermalmembran; II, Implantieren in das Atrium; III, Implantieren in den Subdermalraum; Gestrichelte Pfeil 1–8: Entfernen des Fremdkugelmaterials. 1, Abrutschen von der Dermalmembran; 2, Auswerfen aus dem Atrium; 3, Ausstoßen aus der Implantationsöffnung; 4, Ausstoßen aus dem Atrium; 5, Entfernen über einführende Kanäle; 6, Entfernen über ausführende Kanäle; 7, Aktives Wandern zur Geißelkammerschicht; 8, Entfernen durch die Dermalmembran (s. auch Abb. 9). (Nach Mergner, 1966.)

Natürlich liegt die Frage nahe, ob bei den Trennungsvorgängen beide Schwammarten beteiligt sind? Beobachtet man daraufhin das Implantat, dann findet sich sein ursprünglich kugelig geformtes Material häufig in langausgezogene und weitverzweigte Zellstränge gegliedert, deren Randzellen Pseudopodien zeigen und die ihre Lage ständig verändern. All dies beweist die Fähigkeit des Fremdmaterials zur aktiven Fortbewegung im Hohlraumsystem der Unterlage, zum Aufsuchen einer geeigneten Durchtrittsstelle und zum Eindringen zwischen den Zellgrenzen des Unterlageepithels. Dieses Epithel wiederum ermöglicht das Durchtreten des Implantats durch die Befähigung seiner Zellen zum kurzfristigen Lösen und Wiedereingehen ihrer epithelialen Bindungen und zur raschen Formveränderung.

Zwingt man nun in Kontrollversuchen die Unterlage, implantierte tote Materie, wie etwa kleine Stanniolstückchen, zu entfernen, dann

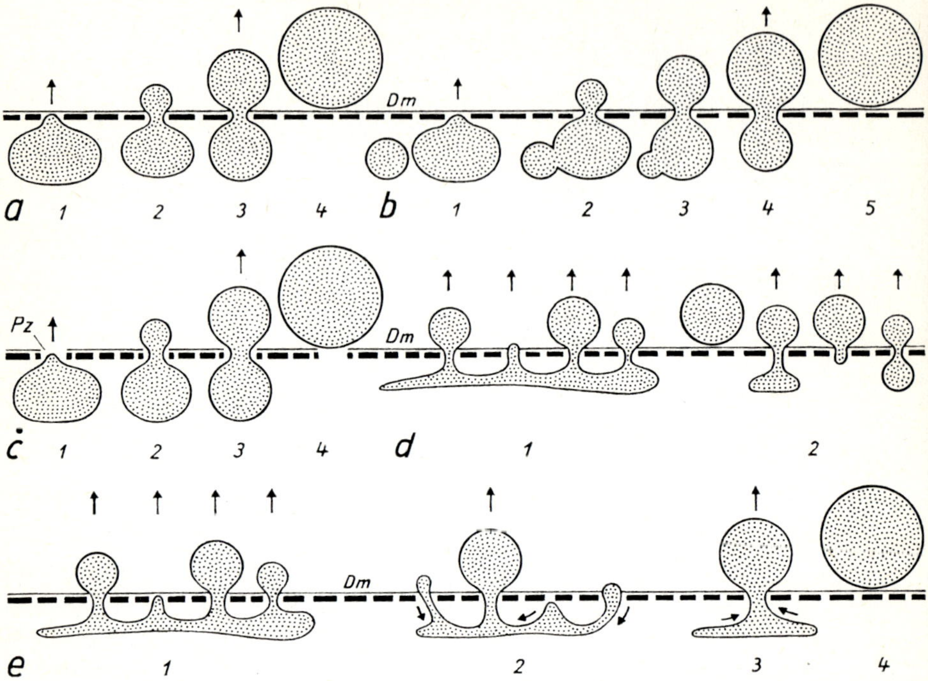

ABB. 9. Möglichkeiten des Durchtretens der Fremdkugel durch die Dermalmembran der Unterlage. Erläuterungen im Text. Pfeile, Richtungen des Durchtretens. (Nach Mergner, 1966.)

vermag sie zwar auch dies, aber der Zeitbedarf hierfür erhöht sich auf das 3 bis 10-fache. Setzt man ferner die Zuchttemperatur vom Optimalwert um 25°C auf nur 18°C herab, dann erhöht sich der Zeitbedarf sogar auf das 16-fache.

Die vorstehenden Versuchsergebnisse beweisen also, daß sowohl Implantat wie auch Unterlageschwamm aktiv an der Trennung der beiden verwendeten Arten beteiligt sind. Das implantierte artfremde Kugelmaterial gelangt aber in allen Fällen nach außen. Diese Tatsache spricht für eine gegenseitige Unverträglichkeit der beiden Zellmaterialien, die ihrerseits wiederum aus deren unterschiedlicher Artspezifität resultiert. Die Unmöglichkeit, "xenoplastische Induktionen" neuer Oscularrohre zu erzielen, während doch Kontrollversuche mit arteigenen Kugeln regelmäßig zu Oscularrohr-Neubildungen führen, beweist daher auch die Artspezifität des Induktionsstoffes. Wir sind damit einen Schritt weiter auf dem Weg zur Aufklärung seiner Natur.

ZUSAMMENFASSUNG

Spongilliden eignen sich als Jungschwämme in besonderer Weise für zell- und entwicklungsphysiologische Untersuchungen. Der Aufbau ihrer einfachen Organisation ist dank der Durchsichtigkeit leicht zu überblicken, die Anzahl der in ihnen vorkommenden Zellsorten ist gering, die Vielfalt der Zellpotenzen dagegen groß.

Entwicklungsphysiologische Probleme führen vom Zellgeschehen innerhalb der Gemmula über den Auskeimvorgang aus dieser und die Morphogenese des Jungschwammes bis zum Einsetzen seiner Funktion.

Damit ist auch die Frage nach dem Potenzgehalt der differenzierten und für eine bestimmte Leistung determinierten Zellen gestellt. Sie wird anhand von Verträglichkeitsprüfungen mit verschiedenen Epithelien angegangen und durch eine Hypothese zur physiologischen Orientierung der Epithelzellflächen ihrer Lösung näher gebracht.

Im Zuge der Potenzprüfungen, die letztlich dem Verständnis des gesamten Determinationsweges von der Gemmula bis zum fertigen Jungschwamm dienen sollen, wurden an isolierten Epithelverbänden induktive Eigenschaften festgestellt: Auf Unterlageschwämme aufgelegte Oscularrohrkugeln induzieren, teils unter Einbau des Kugelmaterials, teils durch rein stoffliche Wirkung, neue Oscularrohre.

Untersuchungen über die Natur des Induktionsstoffes mittels xenoplastischer Induktionen erbrachten den Nachweis seiner Artspezifität und zeigten das Verhalten von Xenoimplantaten im Hohlraumsystem der Unterlageschwämme.

Die Ergebnisse der mehr als zwanzigjährigen Untersuchungen des Gießener Laboratoriums für Spongillidenforschung wurden in besonderem Maße berücksichtigt.

Literatur

Ankel, W. E. (1948). Über Fragen der Formbildung und der Zelldetermination bei Süßasserschwämmen. *Verh. dt. zool. Ges.* **1948**: 58–66.

Ankel, W. E. (1964). Der Süßwasserschwamm *Ephydatia fluviatilis* (Einleitende Worte zur Vorführung des Films). *Verh. dt. zool. Ges.* **1964**: 426–444.

Ankel, W. E. und Eigenbrodt, H. (1950). Über die Wuchsform von *Spongilla* in sehr flachen Räumen. *Zool. Anz.* **145**: 195–204.

Ankel, W. E. und Wintermann-Kilian, G. (1952). Über den Nachweis von morphologischen und funktionellen Differenzierungen bei den Epithelzellen der Spongillidae. *Verh. dt. zool. Ges.* **1951**: 248–255.

Ankel, W. E., Wintermann-Kilian, G. und Kilian, E. F. (1954). Fehlbildungen bei *Ephydatia fluviatilis* und ihre Bedeutung für das Verständins der normalen Korrelationen. *Verh. dt. zool. Ges.* **1954**: 361–371.

Brien, P. (1929). Biologie de la régénération des Éponges. *Annls Soc. r. Sci. méd. nat. Brux.* **1929**: 9–16.

Brien, P. (1932). Contribution à l'étude de la régénération naturelle chez les Spongillidae, *Spongilla lacustris, Ephydatia fluviatilis. Archs. Zool. exp. gén.* **74**: 461–506.

Brien, P. (1935). Régénération et Régulation. 2me *Congrès National Sci.* **1935,** *Comtes Rendus*, **2**: 1086–1091.

Brien, P. (1936/37). La réorganisation de l'Éponge aprés dissociation par filtration et phénomènes d'involution chez *Ephydatia fluviatilis. Archs. Biol., Paris* **48**: 185–268.

Brøndsted, H. V. (1936). Entwicklungsphysiologische Studien über *Spongilla lacustris L. Acta zool. Stokh.* **17**: 75–172.

Brøndsted, H. V. (1943). Formbildungsprozesse bei einem sehr primitiven Metazoon, dem Süßwasserschwamm *Spongilla lacustris L. Protoplasma* **37**: 244–257.

Feige, W. (1969). Die Feinstruktur der Epithelien von *Ephydatia fluviatilis. Zool. Jb.* (*Anat.*) **86** (21): 177–237.

Ganguly, B. (1960). The differentiating capacity of dissociated sponge cells. *Wilhelm Roux' Arch. Entw. Mech. Org.* **152**: 22–34.

Kauffold, P. und Spannhof, L. (1963). Histochemische Untersuchungen an den Reservestoffen der Archaeocyten in Gemmulen von *Ephydatia Mülleri Lbk. Naturwissenschaften* **50**: 384.

Kilian, E. F. (1952). Wasserströmung und Nahrungsaufnahme beim Süßwasserschwamm *Ephydatia fluviatilis. Z. vergl. Physiol.* **34**: 407–447.

Kilian, E. F. (1964). Zur Biologie der einheimischen Spongilliden. Ergebnisse und Probleme. *Zool. Beitr. N. F.* **10**: 85–159.

Korotkova, G. P. (1963). On the types of restoration processes in Sponges. *Acta biol. hung.* **13**: 389–406.

de Laubenfels, M. W. (1926/27). Bi-specific conglomeration of sponges. *Carnegie Instn. Wash. Yb.*, **26**: 219–222.

de Laubenfels, M. W. (1928). Interspecific grafting using sponge cells. *J. Elisha Mitchell scienc. Soc.* **44**: 82–86.

von Lendenfeld, R. (1889). Experimentelle Untersuchungen über die Physiologie der Spongien. *Z. wiss. Zool.* **48.**

Leveaux, M. (1939). La formation des gemmules chez les Spongillidae. *Annls Soc. r. zool. Belg.* **70**: 53–96.

Meewis, H. (1935). Contribution à l'histogénèse de *Spongilla lacustris. C. r.* 11e *Congr. Nat. Sci. Bruxelles,* **1935**: 1062–1071.

Meewis, H. (1936). Contribution à l'étude histologique des Éponges de l'eau douce: *Spongilla lacustris, Ephydatia fluviatilis. Mém. Mus. r. Hist. nat. Belg.* (2) **3**: 519–537.

Mergner, H. (1959). Über die Induktion neuer Oscularrohre bei Süßwasserschwämmen. *Naturwissenschaften* **46**: 632.

Mergner, H. (1964). Über die Induktion neuer Oscularrohre bei *Ephydatia fluviatilis. Wilhelm Roux' Arch. Entw. Mech. Org.* **155**: 9–128.

Mergner, H. (1966). Zum Nachweis der Artspezifität des Induktionsstoffes bei Oscularrohr-Neubildungen von Spongilliden. *Verh. dt. zool. Ges.* **1966**: 522–564.

Mookerjee, S. and Ganguly, B. (1964). Contact reaction in sponge aggregation. *Wilhelm Roux' Arch. Entw. Mech. Org.* **155**: 525–534.

Pourbaix, N. (1934). Etude histochimique des substances de réserve au cours de la reproduction asexuée. *Annls Soc. r. zool. Belg.* **65**: 41–58.

Pourbaix, N. (1935). Formation histochimique des gemmules d'Éponge. *Annls Soc. r. zool. Belg.* **66**: 33–35.

Rasmont, R. (1954). La diapause chez les Spongillides. *Bull. Acad. r. Belg. Cl. Sci.* **40**: 288–304.

Rasmont, R. (1955). La gemmulation des Spongillides. (III). Rupture de la diapause par des agents chimiques. *Annls Soc. r. zool. Belg.* **85**: 173–181.

Rasmont, R. (1956). La gemmulation des Spongillides. (IV). Morphologie de la gemmulation chez *Ephydatia fluviatilis et Spongilla lacustris. Annls Soc. r. zool. Belg.* **86**: 349–387.

Rasmont, R. (1962). The physiology of gemmulation in fresh-water sponges. *Symp. Dev. Growth* **20**: 3–25.

Ruthmann, A. (1965). The fine structure of RNA-storing archaeocytes from gemmules of fresh-water sponges. *Q. Jl. microsc. Sci.* **106**: 99–114.

Simons, J. R. (1963). The adhesion and coalescence of sponge cells after dissociation. *Proc. Linn. Soc. N.S.W.* **88** (1): 28–34.

Tessenow, W. (1969). Lytic processes in development of fresh-water sponges in *Lysosomes in biology and pathology* **1**: 392–405.

Tokin, B. P. (1963). Regeneration and somatic embryogenesis. *Symp. biol. Hung.*, **3**: 11–45.

Tuzet, O. (1948). La place des Spongiaires dans la classification. *Int. Congr. Zool. Paris* **13,** Rapport, Section VI: 429–432.

Weltner, W. (1893). Bemerkungen über den Bau und die Entwicklung der Gemmulae der Spongilliden. *Biol. Zbl.* **13**: 119–126.

Wierzejski, A. (1915). Beobachtungen über die Entwicklung der Gemmulae der Spongilliden und des Schwammes aus den Gemmulis. *Bull. int. Acad. pol.*, (B) **2**: 45–77.

Wierzejski, A. (1935). Süßwasserspongien. Monographische Bearbeitung. *Mém. Acad. pol. Sci.* (B) **9**: 1–242.

Wintermann, G. (1951). Entwicklungsphysiologische Untersuchungen an Süßwasserschwämmen. *Zool. Jb.* Anat. **71**: 427–486.

Wintermann, G. und Ankel, W. E. (1951). Über das Verhalten reiner Epithelverbände bei Spongilliden. *Naturwissenchaften* **38**: 71–72.

Wintermann-Kilian, G. und Ankel, W. E. (1954). Über das Verhalten der Zellen bei der Vereinigung isolierter Epithelkugeln von *Ephydatia fluviatilis. Wilhelm Roux' Arch. Entw. Mech. Org.* **147**: 171–200.

Zeuthen, E. (1939). On the hibernation of *Spongilla lacustris. Z. vergl. Physiol.* **26**: 537–547.

Symp. zool. Soc. Lond. (1970) No. 25, 399–413.

PROLIFERATION ACTIVITY AND SYNTHESIS OF PROTEIN IN THE CELLS OF FRESH-WATER SPONGES DURING DEVELOPMENT AFTER DISSOCIATION

S. M. EFREMOVA

Department of Embryology, Leningrad State University, Leningrad, U.S.S.R.

SYNOPSIS

Generation time, duration of mitosis, proliferation activity and protein synthesis have been studied in various groups of cells in the fresh-water sponge *Ephydatia fluviatilis* after their dissociation and during their subsequent development.

An increase in proliferation activity and a corresponding decrease in generation time and duration of mitosis in nucleolate amoebocytes was observed. The reverse was observed in anucleolate amoebocytes and collencytes, in which these phenomena remained constant during development. During the early stages of aggregate development, protein synthesis is much more intensive in nucleolate amoebocytes than in other cells. At later stages the radioactivity of the protein synthesized in these cells diminishes, while it becomes higher in anucleolate amoebocytes. The changes in intensity of the synthetic processes, in the behaviour as well as in the morphology of various groups of cells, and in the duration of their life cycles, are correlated with the disintegration and reintegration of the biological systems which they comprise.

INTRODUCTION

The process of development of sponges from complexes of somatic cells after their dissociation has been studied extensively, not only from the morphological aspect but also from the biochemical. Details of this morphogenesis are discussed in the works of Wilson (1907, 1910), Müller (1911), Galtsoff (1925), Brien (1937), Ganguly (1960), Tuzet and Connes (1963), Borojević and Lévi (1964) and others. Many scientists have studied the nature of specific interactions between aggregating cells in heterogeneous suspensions (e.g. Spiegel, 1954; Moscona, 1963; Humphreys, 1965; Yazikov, 1965, 1967).

My interest in the processes of development of sponges from dissociated cells is connected with the study of various types of regenerative morphogenesis. Some of these processes, named "somatic embryogenesis" by Tokin (1959, 1964) and Korotkova (1963, 1968), are connected with profound disintegration of the mechanisms responsible for the integration of the organism. Others—regeneration proper—involve local disintegration of tissues without disturbance of the general integration mechanisms responsible for the symmetry and polarity of the organism. According to Tokin (*loc. cit.*), in conditions of disintegration, cells become free from the integrating influences of tissues, organs, and of the organism as a whole, and become capable of revealing their maximum potentialities, i.e. of giving rise to a whole organism. This capability is well known in the plant kingdom. In conditions of disintegration changes must occur, not only in the morphological characteristics of the nucleus and cytoplasm of the cells but also in the duration of their individual lives.

According to Tokin's theories on the ontogeny of cells (Tokin, 1934, 1942) division of the cell may be considered as the crowning stage of its development and the daughter cells must be most labile in their formative potency in the first moments after fission.

Development of sponges from dissociated cells presents a convenient model for studying one type of somatic embryogenesis. The studies have been based on considerations of the changes in the state of the cells and of their life cycles in a state of disintegration of their original biological system. I have attempted to determine whether or not there occur changes in proliferation activity, in generation length and in the duration of mitosis in the cells at various stages of development of cell aggregates. Müller (1911) and Brien (1937) refer to mitoses occurring in the cells of sponge aggregates. Sarà (1965) described large numbers of dividing cells of the sponge *Petrosia ficiformis Poiret*, 1789 during early stages of aggregation and attributed great importance to mitotic division as the factor which, together with adhesion, produces an increase in the number of cells in aggregates. No quantitative evaluation of this process has been made yet, however, nor is it yet clear which cells divide mainly mitotically.

Another set of phenomena which merit investigation are the metabolic processes, such as protein synthesis, in the cells of sponges developing after dissociation.

MATERIAL AND METHODS

The experimental material was the fresh-water sponge *Ephydatia fluviatilis* Linnaeus, 1759, collected from the Starii Peterhof ponds in the vicinity of Leningrad. The sponges were rubbed through three-layered bolting cloth with 150 μ meshes. The resulting suspension was put into a Petri dish filled with filtered pond water. Aggregate formation took place in 3 h. In order to standardize the conditions of development, only aggregates of 1mm diameter were taken for further experiment. The culture temperature was kept between +18 and +19°C.

Proliferation activity was studied by blocking mitoses at the metaphase with colchicine. This method of studying proliferation was suggested by Dondua, Efremov, Krichinskaya and Nikolaeva (1966) for the investigation of proliferation activity in embryonic systems. It is a preferable technique to that involving counts of the mitotic index, as the latter does not show the true proliferation activity when the duration of mitosis varies.

Proliferation activity (**P**) is considered to be the percentage of metaphases blocked by colchicine in unit time and is calculated by the formula

$$\mathbf{P} = \frac{[\mathrm{NB}]}{t},$$

where **P** is proliferation activity, NB the index of blocked metaphases and t the time of colchicine action. The duration of a mitosis is calculated by the formula

$$t_m = \frac{[m] \times \mathrm{t}}{[\mathrm{NB}]},$$

where t_m is the duration of a mitosis and m the mitotic index. The length of generations was calculated by the formula

$$\mathbf{T} = \frac{t}{[\mathrm{NB}]} \times \ln 2.$$

At various intervals after dissociation (7 h, 24 h, 48 h, 72 h, 96 h) the aggregates were placed for 4 h in a 10^{-4} concentration of colchicine solution in pond water.

The first blocked metaphases appeared after 0·5 h, thus showing that 3·5 h is the colchicine action time. Blocked metaphases differed from normal metaphases in the appearance of a compact clump of chromatin which absorbed dyes strongly. The blocked metaphases were counted in granulated nucleolate amoebocytes and in anucleolate hyaline amoebocytes and collencytes. The statistical reliability of differences between "colchicine-mitosis" values at each of the stages of development was determined by the Mann–Whitney test.

In the autoradiographic investigations the amino acid phenyl-alanine-C^{14} was used as a precursor of protein. Table I shows the scheme of experiments.

The specific activity of the phenyl-alanine-C^{14} was 51 mCu/g and its concentration in the medium in all experiments was 3 μCu/ml. All the aggregates, in all the experiments, were exposed to this medium for 3 h.

In the first series of experiments the suspension of freshly dissociated cells was placed immediately in the C^{14} medium. After 3 h the aggregates

TABLE I

Scheme of autoradiographic experiments

No. of Series	Age of Aggregates	Time from end of phenyl-alanine-C^{14} treatment to fixation of material
1	0 h	0 h, 24 h, 48 h, 72 h, 7 days
2	24 h	0 h, 24 h, 48 h, 72 h
3	72 h	0 h, 24 h
4	5 days	0 h

formed in the medium were removed and replaced in filtered pond water. Some were fixed immediately, while the others were left to develop and were fixed successfully after 25 h, 48 h, 72 h and seven days.

In the second series of experiments the forming aggregates were put into the solution of amino acid tracer 24 h after dissociation. As in the first series of experiments, they were removed from the tracer solution and fixed immediately and after a further 24 h, 48 h and 72 h of development.

In the third series the aggregates were placed in the isotope solution 72 h after their dissociation, when they had already begun to flatten, and they were fixed immediately (0 h) or 24 h after their removal from the phenyl-alanine-C^{14} medium.

In the fourth series of experiments the sponges, five days after dissociation, had developed functional aquiferous systems.

The sponges used as controls were of sizes up to 1 cm diameter. They were placed, attached to their substrates, in a phenyl-alanine solution for 3 h, removed and fixed in Bouin's solution.

Sections of the sponges, 5 μ thick, were covered with liquid photographic emulsion of the "P" NIKFI type, and autoradiographs prepared. Tracks of radioactivity were counted in nucleolate amoebocytes, collencytes, and choanocytes, using an eyepiece micrometer grid. The mean tracks per grid-square (40 μ^2) were calculated from counts of tracks in 500 squares. Five to eight aggregates of developing sponges were fixed at a time for each set of autoradiographic counts. The significances of differences between mean counts were determined by the Mann–Whitney test.

RESULTS

Proliferation activity

The data on the calculated proliferation activity are shown in Fig. 1. Seven hours after dissociation the proliferation activity (**P**) was 0·73 % mitoses per hour, which corresponds to that in a normal sponge. Within 24 h it had increased to reach 1·04 % mitoses per hour. At the end of 48 h **P** had returned to the level of the 7-h stage. At the stage of flattening of the aggregates, and during further development of the sponge (represented at 72 and 96 h), the **P** of nucleolate amoebocytes increased sharply, reaching 2·40 % mitoses per hour at the 72-h stage and 3·64 % mitoses per hour at the 96-h stage.

The proliferation activity of anucleolate amoebocytes and of collencytes (Fig. 1) remained virtually unchanged at all stages of development. The rate of division of anucleolate cells during development after dissociation departed little, if at all, from that in normal sponges.

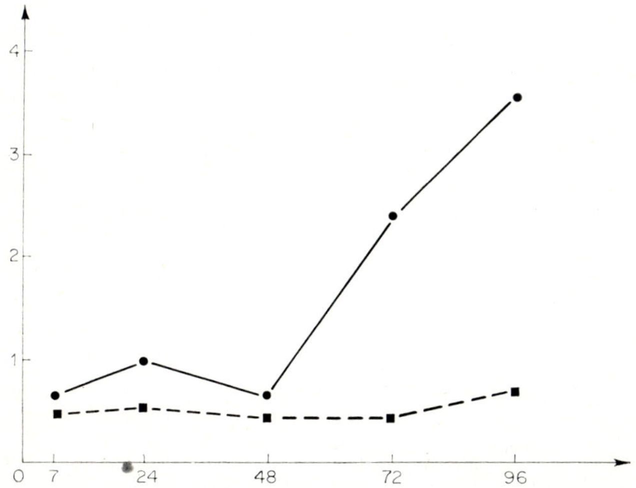

FIG. 1. Change in proliferation activity of sponge cells developing after dissociation. Abscissa, time after dissociation (hours); ordinate, proliferation activity (**P**) in per cent of mitoses per hour; unbroken line, nucleolatea moebocytes; broken line, anucleolate amoebocytes and collencytes.

During the development of sponges from dissociated cells, a considerable decrease in length of generations is observed in nucleolar cells, in accordance with the increase in their proliferation activity. The duration of generations (**T**) in nucleolate amoebocytes decreases from

97·02 h at the 7-h stage to 18·8 h at the 96-h stage. At the 24-h stage **T** = 67·22 h, at 48 h after dissociation **T** = 105·5 h, and 72 h after dissociation **T** = 28·8 h.

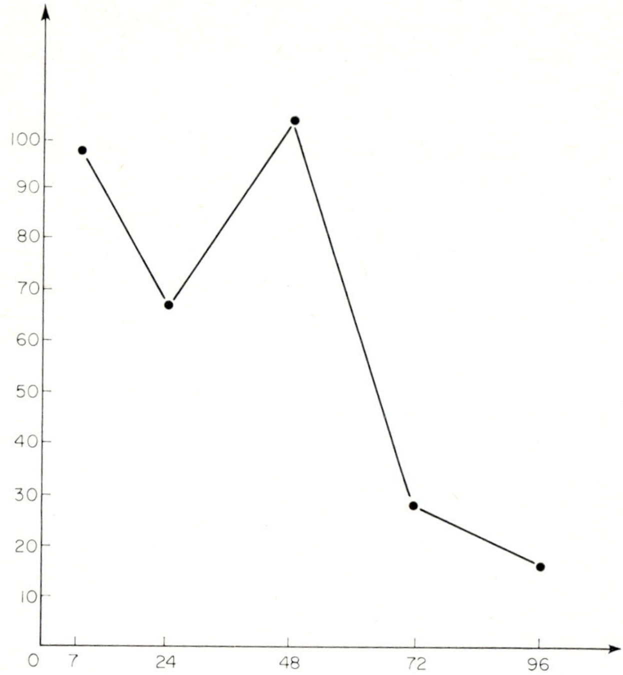

FIG. 2. Change in generation time in nucleolate amoebocytes. Abscissa, time after dissociation (hours); ordinate, generation time (**T**) in hours.

The duration of mitosis in nucleolate amoebocytes decreases also from 2·24 h at the 7-h stage to 0·44 h at the 96-h stage (Fig. 3). Nevertheless, despite the change in mitosis duration being similar to that in length of generations, the mitosis duration is too short in relation to the generation length for it to be the sole cause of the change in generation length. The decrease in length of generations presumably occurs at the expense of other phases of the mitotic cycle.

A decrease in the generation length for nucleolate amoebocytes following dissociation (by a factor of nearly five in 96 h), compared to that of a normal sponge, is a substantial shift in the life cycle of a cell. The stimulus of this shift is probably the actual dissociation of the cells. It is well known that dissociation of cells in tissue culture leads to a dedifferentiation of the cells and to an increase in the percentage of

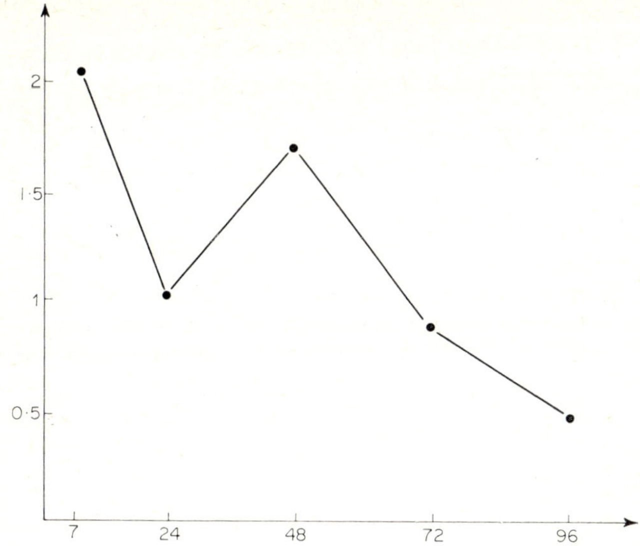

FIG. 3. Change in mitosis duration in nucleolate amoebocytes. Abscissa, time after dissociation (hours); ordinate, duration of mitosis ($t6$) in hours.

their nuclei synthesizing DNA. This is especially noticeable in the case of trypsinization of cells in single-layer cultures. In the experiments described above, an increase in proliferation activity was induced by removing cells from an integrated organism. Unfortunately, it is not possible to comment upon the dedifferentiation of nucleolate amoebocytes as there do not exist any data on the morphological expression of dedifferentiation in cells in a suspension.

An increase in the proliferation rate and a decrease in some parameters of the mitotic cycle allow us to assume that there must occur some activation or acceleration of synthetic processes which promote cell division.

Seventy-two hours after dissociation the aggregates fix themselves to the substrate and begin to flatten rapidly. At that stage some nucleolate amoebocytes start dividing and give rise to choanocytes. These in turn give rise to choanocyte chambers, while rudiments of spicules appear in the cytoplasm of some amoebocytes. A marked increase in the number of anucleolate amoebocytes and collencytes occurs as nucleolate amoebocytes transform themselves into these cell types. These processes evidently follow upon an intense proliferation of nucleolate amoebocytes. A dedifferentiation of anucleolate amoebocytes and collencytes from the mother sponge probably also occurs in the aggregates. However, dedifferentiation of these cells does not go far, as is shown by the fact that their proliferation rate remains steady throughout development, and does not deviate from the rate found in the normal sponge.

An essential proportion of these cells is supplied by granulated nucleolate amoebocytes (Efremova, 1968).

Intensity of protein synthesis

Figure 4 shows changes in the radioactivity of protein in various types of cells depending upon the age of the aggregates at the moment of their immersion in the isotope solution. The most active absorption of phenyl-alanine is seen in the protein of nucleolate amoebocytes (average 5·3 tracks per 40 μ^2) in aggregates formed in the first hours after dissociation. Choanocytes from the mother sponge show fewer tracks (average 4·2 per 40 μ^2), while the average number of tracks recorded for collencytes is 2·2 per 40 μ^2—2·5 times less than that in nucleolate amoebocytes.

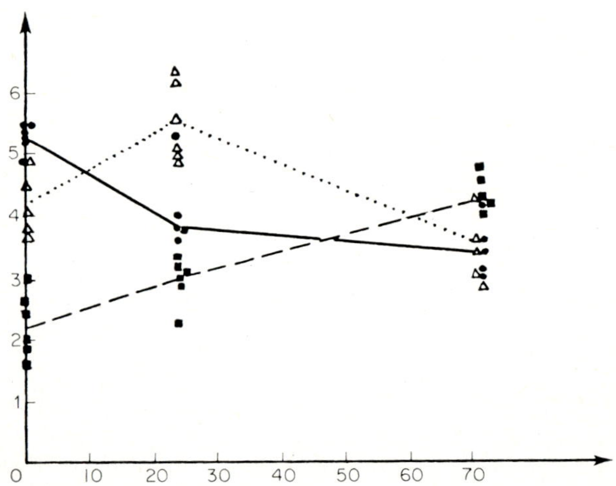

FIG. 4. Change in protein radioactivity in nucleolate amoebocytes, collencytes, and choanocytes at various stages of development from dissociated cells. Abscissa, time after dissociation (hours); ordinate, mean number of tracks per 40 μ^2 of cells; unbroken line, nucleolate amoebocytes; broken line, collencytes; dotted line, choanocytes.

The control sponges showed the following average numbers of tracks in their cell types: nucleolate amoebocytes 3·0, choanocytes 3·5 and collencytes 2·6 tracks per 40 μ^2.

At the next stage (24 h after dissociation) the C^{14} content in the nucleolate amoebocytes is lower than at the previous stage, while it is higher in choanocytes and in collencytes. In flattening aggregates (72 h after dissociation) the intensity of C^{14} absorption by the collencytes increases, while the intensity in the nucleolate amoebocytes remains constant. Inclusion of C^{14} by the choanocytes is similar to that shown by the amoebocytes. At this stage a very heavy inclusion of phenyl-alanine-C^{14} is clearly detectable in the basal pinacocytes. A comparison of the autoradiographs of early aggregates with those of control sponges shows that incorporation of C^{14} into choanocytes and anucleolate amoebocytes scarcely differs at all in the two developmental forms, while the C^{14} incorporation in nucleolate amoebocytes is much higher in the aggregates than in the control sponges. After five days, the aquiferous system and isolated spicules appear in the flattened sponges. At this stage the amount of C^{14} incorporated in the basal pinacocyte layer exceeds considerably that observed in all other cell elements, indicating an intense synthesis of protein in that layer (Fig. 5).

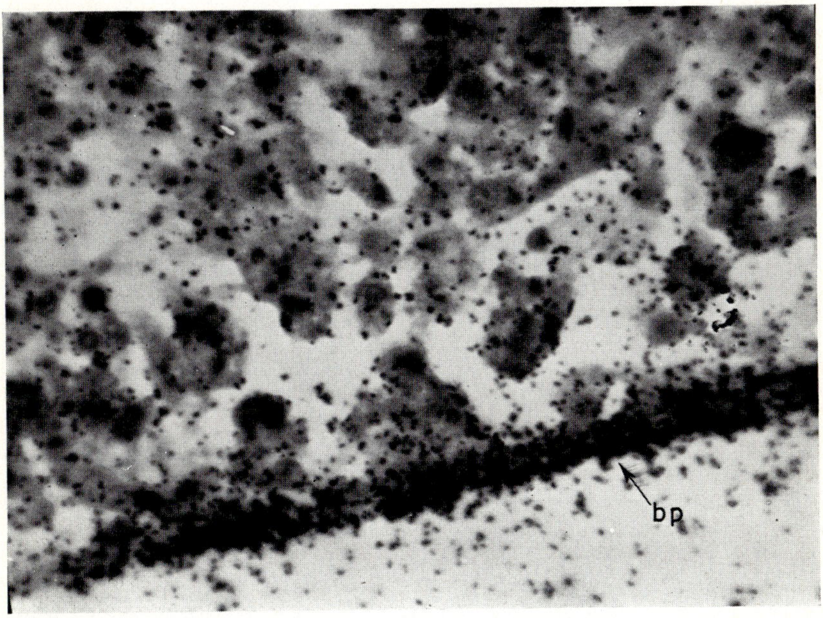

FIG. 5. Increased phenyl-alanine-C^{14} inclusion in basal sponge pinacocytes 5 days after dissociation. bp, basal pinacocytes. Magnification, $\times 16 \times 7$.

The dynamics within aggregates of accumulation and discharge of the phenyl-alanine tracer are shown in Figs 6 and 7 as a function of time after the aggregates' immersion in the isotope solution. A comparison of the autoradiographs of the aggregates and of the sponges in the first and second series of experiments shows that in both cases an accumulation of C^{14} continues in all types of cells studied for 24 h after their transfer from the isotope solution. In the next 24 h the number of tracks decreases and continues to decrease slightly up to 72 h. In the first series of experiments, however, this decrease is distinct, while in the second series (Fig. 7) the amount of C^{14} inclusion hardly differs from that observed 48 h after immersion in the isotope solution.

In the first series of experiments the phenyl-alanine uptake by a cell does not continue to diminish during the full 7 days following exposure to the isotope (the apparent increase in Fig. 7 in the mean values of the tracks for the three cell types is not statistically reliable). Thus it is obvious that a longer period of time is needed for a full exchange of C^{14} in the synthesized protein.

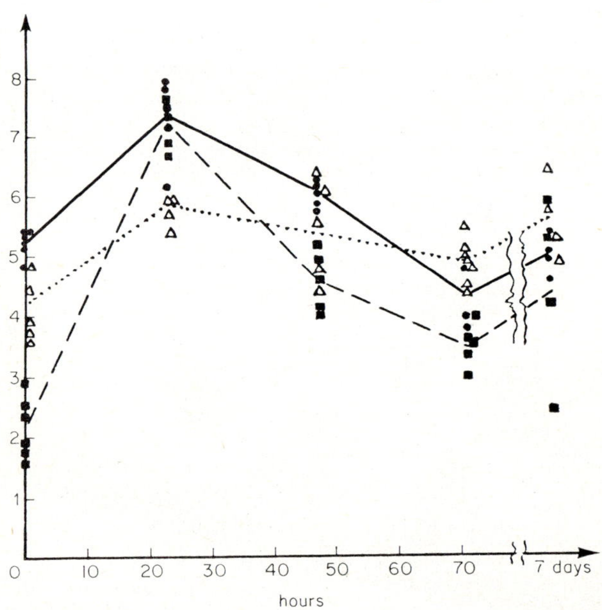

FIG. 6. Change of protein radioactivity in nucleolate amoebocytes, collencytes, and choanocytes as a function of time after aggregate immersion in phenyl-alanine C^{14} solution. Time of C^{14} addition, 0 h after dissociation; abscissa, time of material fixation; ordinate, mean number of C^{14} tracks on 40 μ^2 of cells; solid line, nucleolate amoebocytes; broken line, collencytes; dotted line, choanocytes.

It is interesting that a sharp increase in the C^{14} inclusion (3·5 times, on average) occurs in the anucleolate amoebocyte aggregates during the 24 h following their removal from the isotope solution and that a corresponding decrease in C^{14} inclusion occurs in the following 24 h (Fig. 6).

In the second series of experiments the change in intensity of C^{14} inclusion in anucleolate amoebocytes during the 48 h following exposure to the isotope is far less marked (Fig. 7). Without any doubt this indicates a difference between cells absorbing labelled amino acid while in a dissociated state and cells exposed to the isotope when beginning to arrange themselves into an integrated system.

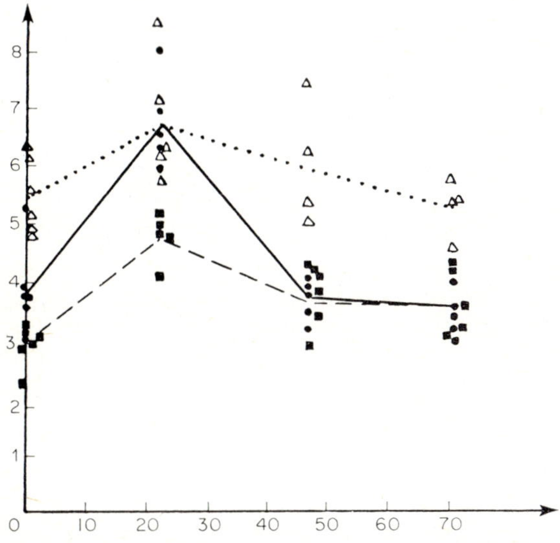

FIG. 7. Change of protein radioactivity in nuclear amoebocytes, collencytes, and choanocytes as a function of time of the immersion of the aggregate in phenyl-alanine-C^{14} solution. Aggregate immersed in the phenyl-alanine solution 24 h after dissociation. Conventions as in Fig. 6.

In aggregates exposed to the phenyl-alanine-C^{14} 72 h after dissociation, the mean rates of accumulation of C^{14} in the three types of cell were the same during the 24 h following immersion in the isotope solution (Fig. 8).

When undertaking the series of autoradiograph experiments, I expected the development of a sponge from aggregates of dissociated cells to be characterized by a general increase in the metabolism of the whole cell system and especially amongst the nucleolate amoebocytes.

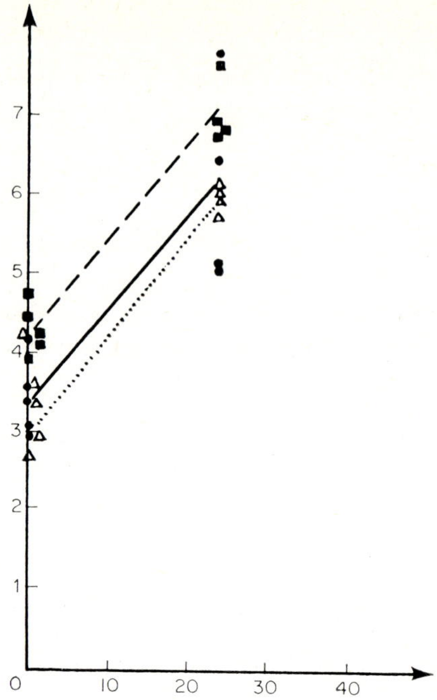

FIG. 8. Change of protein radioactivity in nucleolate amoebocytes, collencytes, and choanocytes as a function of time of the aggregate immersion in the phenyl-alanine-C^{14} solution. Aggregates immersed in the phenyl-alanine solution 72 h after dissociation. Conventions as in Fig. 6.

The investigations, however, did not confirm this expectation. They showed that more complicated relations exist between the cells at various stages of development. Cell aggregation depends upon an increased rate of synthesis of protein in nucleolate amoebocytes, these being the basic and most active elements uniting and forming the aggregates.

Within 24 h of dissociation, when the aggregates are covered with an epithelial layer and some lacunae are beginning to form, the intensity of phenyl-alanine-C^{14} inclusion, and, therefore, of protein synthesis, falls sharply and approaches that of an intact adult sponge. A connection between aggregation of cells and protein synthesis by the cells was indicated by Guidice (1965), who investigated the mechanism of reaggregation of cells of sea-urchin larvae, and by Mookerjee, Ganguly and Gouri (1965), who studied aggregation of dissociated cells of *Ephydatia fluviatilis* in a penicillin solution.

These events take a different pattern in the anucleolate amoebocytes, protein synthesis being less conspicuous in the few cells originating from the dissociated mother sponge. At the stage of aggregate flattening, however, the C^{14} inclusion in these cells increases. The increase is presumably associated with a proliferation of nucleolate amoebocytes and their transformation into anucleolate amoebocytes.

Synthetic activity is especially conspicuous in the basal epithelium of a sponge. Borojević and Lévi (1967) have demonstrated, by electron microscopy, the existence of a layer of spongin fibres between the substrate and the basal epithelium, the spongin fibres being produced by the basal epithelial cells. This discovery is highly relevant to the preferential inclusion of phenyl-analine-C^{14} in the basal membrane of a forming sponge. We do not know the extent to which phenyl-alanine is characteristic of the composition of spongin. Saper and White (1958), in a paper chromatography study of the amino acid content of a Mediterranean sponge, found that the phenyl-alanine spot was one of the brightest, together with that of hydroxyproline, leucin and proline. Such an analysis has not been carried out on fresh-water sponges. It is clear, nonetheless, that changes in the processes of synthesis in a forming cell system can be revealed by use of labelled phenyl-alanine.

CONCLUSIONS

1. In the course of development from dissociated cells, an increase in proliferation activity takes place in nucleolate amoebocytes. This is associated with a decrease in the length of generations and the duration of mitosis in these cells. Proliferation activity remains unchanged in nucleolate amoebocytes and collencytes.

2. In the process of cell aggregation, protein synthesis proceeds more rapidly in nucleolate amoebocytes than in other cells and exceeds that in the nucleolate amoebocytes of intact sponges. The radioactivity of protein synthesized in these cells decreases during later development.

3. The increased inclusion of C^{14} in collencytes, and especially in basal pinacocytes, unlike that in nucleolate amoebocytes, indicates the active part that the two former cell types play in synthetic processes in general and in spongin formation in particular.

4. A thorough disintegration of a sponge results not only in a change in its cells' behaviour and morphology, but also in their life cycle duration and in the intensity of their various synthetic activities. The extent of these changes is related to the state of differentiation achieved prior to dissociation.

References

Borojević, R. and Lévi, C. (1964). Morphogenèse expérimentale d'une éponge à partir de cellules de la larve nageante dissociée. *Z. Zellforsch. microsc. Anat.* **68**: 57–69.

Borojević, R. and Lévi, P. (1967). Le basopinacoderme de l'éponge *Mycale contarenii* (Martens). Technique d'étude des fibres extracellulaires basales. *J. Microscopie*, **66**: 857–862.

Brien P. (1937). La réorganisation de l'éponge apres dissociation par filtration et phénomenes d'involution chez *Ephydatia fluviatilis. Archs Biol.* **48**: 185–268.

Dondua, A. K., Efremov, V. I., Krichinskaya, E. B. and Nikolaeva, I. P. (1966). Mitotic index, duration of mitosis, and proliferation activity in the early phases of the chick embryo. *Acta biol. hung.* **17**: 127–143.

Efremova, S. M. (1968). Character of the proliferating activity of different types of cells of the sponge *Ephydatia fluviatilis* in the course of development after dissociation. *Arkh. anat. gistol. embriol.* **54**: 96–101.

Galtsoff, P. S. (1925). Regeneration after dissociation (an experimental study on sponges). *J. exp. Zool.* **42**: 183–222.

Ganguly, B. (1960). The differentiating capacity of dissociated sponge cells. *Roux' Arch. Entw. Mech.* **152**: 22–34.

Guidice, G. (1965). The mechanism of aggregation of embryonic sea urchin cells: a biochemical approach. *Devel. Biol.* **12**: 233.

Humphreys, T. (1965). Aggregation of chemically dissociated sponge cells in the absence of protein synthesis. *J. exp. Zool.* **160**: 235–239.

Korotkova, G. P. (1963). Dependence of restorative processes in sponges on the character and level of their integration. *Zh. Obshch. Biol.* **24**: 445–453.

Korotkova, G. P. (1968). The classification and evolution of regulatory processes. *Zh. Obshch. Biol.* (in press).

Mookerjee, S., Ganguly, B. and Gouri, C. V. (1965). Action of penicillin on cell aggregation of sponges. *Indian J. exp. Biol.* **3**: 1.

Moscona, A. A. (1963). Studies of cell aggregation: demonstration of materials with selective cell binding activity. *Proc. nat. Acad. Sci. U.S.A.* **49**: 742–747.

Müller, H. (1911). Das Regenerationsvermögen der Süsswasserschwämme insbesondere Untersuchungen über die bei ihnen Vorkommende Regeneration nach Dissociation und Reunition. *Roux' Arch. Entw. Mech.* **32**: 397–446.

Saper, J. and White, W. E. (1958). Amino acid composition of sclero-protein of the sponges. *Nature Lond.* **181**: 285–286.

Sarà, M. (1965). Mitotic activity during cell reaggregation of sponges. *Naturwissenschaften* **52**: 503.

Spiegel, M. (1954). The role of specific surface antigens in cell adhesion. Part 1. The reaggregation of sponge cells. *Biol. Bull. mar. biol. Labs Woods Hole.* **107**: 130–148.

Tokin, B. P. (1934). Problems of the ontogenesis of cells. I. Regeneration in the light of ontogenesis of cells. *Biol. Zh.* **3**: 279–293.

Tokin, B. P. (1942). Ontogenesis of cells and blastomatic growth. *Zh. Obshch. Biol.* **3**: 262–282.

Tokin, B. P. (1959). Regeneration and somatic embryogenesis.

Tokin, B. P. (1964). Regeneration and somatic embryogenesis. In *Regeneration and wound healing. Symp. biol. hung.* **3**: 11–45.

Tuzet, O. and Connes, R. (1963). Recherches histologiques sur le reconstitution de *Sycon raphanus* O.S. à partir des cellules dissociés. *Vie Milieu* **13**: 20–26.

Wilson, H. V. (1907). On some phenomena of coalescence and regeneration in sponges. *J. exp. Zool.* **5**: 245–258.

Wilson, H. V. (1910). Development of sponges from tissue cells outside the body of the parent. *Bull. Bur. Fish., Wash.* **28**: 1265–1271.

Yazikov, A. A. (1965). Aggregation of dissociated cells in sponges *Reniera cinerea* (Grant), *Halichondria panicea* (Pallas) and *Ephydatia fluviatilis* (Lamarck) and the capacity of formed multicellular complexes for restoration to tho complete organism. *Zh. Obshch. Biol.* **26**: 690–693.

Yazikov, A. A. (1967). The effect of polyvalent and monovalent ions on intercellular reaction during aggregation of cells of the sponge *Ephydatia fluviatillis* (Lamarck). *Zh. Obshch. Biol.* **28** (2): 231–236.

Symp. zool. Soc. Lond. (1970) No. 25, 415–422.

SOME NEW ASPECTS OF THE PHYSIOLOGY OF FRESH-WATER SPONGES

R. RASMONT

Laboratoire de Zoologie générale
Université libre de Bruxelles, Belge

SYNOPSIS

Gemmule formation in *Ephydatia fluviatilis* is found to be inhibited by continuous illumination. Respiration rate is found to be increased by illumination. Variation in duration of culture temperatures and illumination have interacting effects upon the rate of gemmule production.

The phenomenon of extrusion of spicules during early morphogenesis is strictly limited by time and related temporally to the morphogenesis of the aquiferous system. Spicules are the first structures to be differentiated after hatching and, although they act as supports for the aquiferous system, the relationship between total sponge mass and the mass of skeleton required for support of an adequate aquiferous system is complex. The number of erratic spicules increases as the square of the number of gemmules that have contributed to the developing sponge. Expulsion of spicules is less, but onset of gemmulation earlier, in sponges formed from gemmules hatching in sequence than from sponges formed from gemmules hatching contemporaneously.

Morphologically and physiologically distinct strains of *Eph. fluviatilis* have been discovered. These show inability for fusion of gemmule-products between strains but ready fusion of gemmule-products within strains. The independence, as shown by gemmule physiology and behaviour, is persistent whether the sponges be cultured artificially or in the field.

INTRODUCTION

Much of the research activity at the "Laboratoire de Zoologie générale", at the University of Brussels, is centred on the physiology of fresh-water sponges. Many results collected recently by individual research workers relate to a collective research project on the physiology of cell differentiation during reproduction by means of gemmules and so these results have not yet been published individually. However, as some of them, despite their present incompleteness, might interest other students in the field, we thought they should be reported upon briefly. This article should not, therefore, be regarded as a thorough review of the present state of the question but as a preliminary report on some contributions by our laboratory to sponge physiology.

THE INFLUENCE OF LIGHT ON THE ONSET OF GEMMULATION

In previous articles, I have reported how sponges can be induced to gemmulate *in vitro* (Rasmont, 1962, 1963).

These sponges themselves had been cultivated in Petri dishes, from gemmules gathered from a brook near Brussels. In order to obtain sponges of definite size, the gemmules are arranged carefully in appropriate clusters. The dishes must then be kept undisturbed on the laboratory table for about one week until the gemmules hatch. In the experiments cited above, the sponges were further kept in an incubator, at constant temperature (20°C) and darkness. At constant intervals, they were removed briefly from the incubator for observation in the laboratory. On each of these occasions, the sponges underwent a change in temperature and illumination. However, as these changes were the same in each experiment for all the dishes, they do not render invalid the general conclusions drawn from the whole range of experiments.

In a further set of experiments, it occurred to us (Rasmont and Schmidt, 1967) that the respiration of sponge gemmules is light sensitive. This made us fear that light might have interfered in some unsuspected way with our previous experiments.

Therefore, we had a constant temperature cabinet installed, large enough to allow cultivation and observations to be made without any change of the environmental conditions of the sponges. We first tried to put this room under constant illumination, the light source being fluorescent tubes (Osram, Universal white), the illumination being 400 lux on the cultivation bench. In these experimental conditions, we tried unsuccessfully to repeat our gemmulation experiments. Under every set of conditions in which we had previously obtained gemmulation, we failed to observe the phenomenon again.

Even under fluorescent illumination reduced to one-half its previous value, or under the much dimmer light of incandescent bulbs (25 lux), we did not obtain a single case of gemmulation as long as the sponges were kept under constant illumination. We then performed an experiment in which some sponges were kept under constant light, while others were permanently or periodically masked from the light with black cardboard boxes. The results were striking—although the gemmulation frequency varied from one illumination condition to another, they were consistently nil in constant light.

In another experiment, sponges were kept in constant darkness although in various cycles of two temperatures: 20°C and 23°C. Here, the gemmulation frequency varied from one set to another but gemmulation occured in every set of conditions.

The conditions and results of these experiments can be summarized as follows.

Conditions

The sponges were of the species *Ephydatia fluviatilis* Linnaeus, 1759, type α. They were cultivated in disposable polystyrene Petri dishes, in **M** medium (Rasmont 1961). Before being incubated, the gemmules had been cleaned in hydrogen peroxide. Each Petri dish contained 120 gemmules, arranged in 10 groups of 12, that gave rise to 10 sponges. (For further technical data see Rasmont, 1961, 1963.)

Results

The results can be expressed in numbers of gemmules produced by the 20th day per 100 sponges; they are summarized in Table I.

TABLE I

Conditions (in hours per 24 h period)					
Illumination		Temperature		Number of sponges in each condition	Number of gemmules per 100 sponges
0	25 lux	20°C	23°C		
—	24	24	—	60	0
8	16	24	—	20	90
16	8	24	—	30	135
24	—	24	—	50	68
24	—	16	8	30	73
24	—	8	16	30	127
24	—	—	24	30	20

Analysis of these experimental results shows clearly that light and temperature act upon gemmulation, different cyclic conditions resulting in different gemmule yields. Further experiments are presently being made to acquire a more comprehensive understanding of the action of various light cycles on gemmulation. Our very first results indicate that the relations are intricate and different from one strain of sponges to another, even within the same species.

However, the results summarized in Table I, as well as those obtained from further experiments, are consistent with our previous fruitless attempts to obtain gemmulation, in that they show that the phenomenon is completely inhibited by constant illumination, at least in strain α of *Ephydatia fluviatilis*.

These findings, together with the results of our previous experiments on the respiration of incubated gemmules (Rasmont and Schmidt, 1967) made it the more desirable that the respiratory metabolism of adult sponges be measured under varying illumination.

THE EFFECT OF LIGHT ON THE RESPIRATION OF ADULT SPONGES

Fully developed sponges are difficult to handle in experiments involving Warburg apparatus. To obtain manometric readings large enough to be accurate, it is necessary to use rather large sponges in small flasks. However, it is impossible to cultivate sponges in these very conditions, this being probably due to the accumulation of noxious, or at least inhibitory, substances within too small a volume.

With the aid of one of our co-workers, A. Weisman, we overcame this difficulty in the following way.

We hatch the gemmules in conditions that are as close as possible to our standard conditions (Rasmont, 1961) except for the fact that, within the Petri dishes, the gemmules are placed on narrow strips of dialysis membrane. On this substrate the sponges develop very well and they can be handled with ease and transfered to small cylindrical Warburg flasks with wide necks, where they survive in good condition for 24 h. After measurements have been made the sponge-bearing strips can be taken back to Petri dishes for further observations.

Using this technique, the respiration of full-grown sponges in various environmental conditions is presently being studied in the laboratory.

Preliminary results were obtained from sponges, eight days old, of the *Ephydatia fluviatilis* species, strain α, each sponge being hatched from 200 gemmules and being devoid of symbiotic algae. They showed an oxygen consumption of 2·8 μl/h in total darkness, against a consumption of 4·6 μl/h under a 1500 lux illumination.

It is not yet possible, in the present state of our experiments, to formulate any hypothesis on the metabolic mechanisms that relate respiration to illumination.

In our previous work, we had been driven to the hypothesis that the onset of gemmulation is related to an imbalance in numbers between the reserve-containing archaeocytes and the differentiated cells within the sponge. All that can be stressed, presently, is that this hypothesis does not conflict with the discovery that gemmulation is inhibited by light, insofar as the differentiation of archaeocytes and the utilization of their reserves are respiration-dependent.

THE EARLY REGULATION OF SPICULATION

In a previous article (Rasmont, 1963), I reported that in sponge cultures, under some conditions and during a short phase of the growth of the organisms, many spicules appeared on the bottom of the Petri dishes, outside the sponges themselves.

One of our co-workers, Miss A. Mawet, has undertaken a closer study of this phenomenon and the results she has obtained so far allow us to formulate the following provisional conclusions.

1. The spicules that appear outside the sponges get there by expulsion through the oscula of the sponges. Therefore, they should be called "erratic" rather than "ectopic" as I suggested originally.

2. The total number of spicules, both functional and erratic, produced by each sponge is approximately proportional to the number of gemmules contributing to the sponge. On the other hand, the larger the sponge, the earlier does spicule expulsion begin, and the greater is the final proportion of discarded spicules. In fact, the number of erratic spicules expelled by each sponge increases as the square of the number of gemmules that have contributed to the sponge.

The expulsion of megascleres is strictly limited in time and coincides with the gradual organization of the aquiferous systems. It should be remembered that the spicules are among the first structures that differentiate in the hatching sponge, many hours before the aquiferous system or even the isolated choanocyte chambers appear. Therefore, the rate of formation of the first spicules cannot be controlled by any feed-back from the aquiferous system they are destined to support. It is easily understood that the number of spicules that are formed is closely correlated with the total number of cells involved in the building of the sponge, i.e. to the number of gemmules contributing these cells.

Once the construction and regulation of the aquiferous system sets in, the superfluous spicules are disposed of in the same way as the other debris that are normally discharged by the exhalant current. Although we do not know the laws that govern the pattern of the skeleton of the fully differentiated sponge, it is probably reasonable to assume that the number of spicules needed is less than directly proportional to the size of the sponge. This very simple assumption would account for the fact that a larger proportion of spicules are expelled by larger sponges.

The fact that spicule expulsion ceases once the aquiferous system is completed probably implies that, by that time, some feed-back mechanism sets in that correlates the differentiation of new spicules to the architectonic requirements of the sponge.

3. Feeding by the sponges interferes strongly with the expulsion of spicules. It retards the phenomenon and, to some extent, reduces the total number of erratic spicules. This relationship is the more striking since the expulsion of megascleres is strictly contemporaneous with the organization of the aquiferous system, whereas feeding is

impossible or ineffective as long as the aquiferous system is not at least partially functional. Therefore, it is within a short time, in fact within two days, that the first nourishment inhibits the shedding of spicules.

We still lack evidence that would allow us to choose between several hypotheses that might account for this interaction. However, it is possibly significant that feeding is a process that appears to enhance gemmulation formation (Rasmont, 1963) and reduces altogether at least one manifestation of precocious differentiation.

4. In a previous set of experiments, we had noted that sponges hatched from a group of gemmules, one whose members had been added to the batch two or three days late, form gemmules more readily than sponges hatched from the same total number of gemmules put together all at once. We explained the fact by assuming that in the first instance the differentiation of the cells of the "late" gemmula within an already differentiating sponge was at least partly inhibited, giving rise to an accumulation of undifferentiated cells ready for the formation of a new gemmule. If this hypothesis is valid, one might expect that in such cases, the shedding of erratic spicules would be lessened.

Recent experiments by Miss A. Mawet, do indeed show that, $3 + 1$ (late) gemmule sponges give rise to significantly less erratic spicules than 4 gemmule sponges do, at least for the strain α of *Ephydatia fluviatilis*. As for strain β, there is no significant difference in the erratic spiculation of the sponges with or without late addition of a gemmule. Our latest experiments show that there are also differences between the two strains' gemmulation.

INTRA-SPECIFIC DIFFERENCES BETWEEN STRAINS

From our previous work (Rasmont, 1956), we knew that within the species *Ephydatia fluviatilis*, gemmules gathered at the same moment, in the same place, but from different sponges, can either fuse when hatching close together or give rise to sponges that are unable to merge even on close contact. Our first assumption was that the incompatibility between two such types was attributable to sex difference. However, the further discovery that other gemmules, gathered from different places, would fuse neither with each other nor with any one of the previously found types made us abandon this hypothesis and suspect the existence of genetically different strains. In addition, several inconsistencies between the results of parallel physiological experiments run with different gemmule stocks made us aware of the possibility that the different strains might have different physiological characteristics.

Therefore, we devised techniques that allowed us to gather and characterize pure stocks of gemmules, i.e. stocks within which all the gemmules would give rise to sponges able to fuse together.

We can define a strain as the collection of all the individual sponges, the gemmules of which give rise to sponges that are able to fuse together.

One of our co-workers, Dr. G. Van de Vyver, is presently investigating more thoroughly the problem of the origin and implications of strain differences with the species *Ephydatia fluviatilis*.

Careful testing of gemmules coming from five different places in Belgium has already made evident the existence of eight strains and this brings us to suspect that a plurality of strains is the rule, at least in *Ephydatia fluviatilis*.

In 1967, we collected most of our *Ephydatia fluviatilis* gemmules from two sites, near Brussels, which we may call A and B. Each of these places harboured two strains: α and β in location A, γ and δ in B.

What made this discovery the more striking was the fact that A, a small stream flowing from a pond, was literally crowded with a great many individual sponges.

In March 1968, Miss Van de Vyver prepared with each of the four strains a set of glass plates, each one bearing a sponge hatched from several hundred gemmules of the same strain.

After having been incubated in the laboratory, these plates were suspended side by side in the outflow of a small pond that had been chosen because of its suitable characteristics and because it was devoid of any indigenous sponge.

The sponges grew very well, underwent no sexual reproduction and started gemmulating in mid-July. They were gathered in the Autumn. Within the sponges themselves, as well as in the gemmules they had built, there are morphological differences between strains. What is more significant is that gemmules of any one of the four strains are unable to fuse on hatching with hatched or hatching gemmules of any of the three other strains, although all the gemmules have been formed under exactly the same ecological conditions. On the other hand, gemmules of each strain give rise to sponges which can merge, not only with one another, but also with sponges originating from gemmules of the same strain but one year older, i.e. coming from different places and formed under different ecological conditions.

Therefore, it seems difficult to think of the strain differences within *Ephydatia fluviatilis* as being anything but genetic, although full proof of this cannot be given unless we succeed in obtaining sexual reproduction under controlled conditions. To our knowledge, this has never been done. One hope arises from the fact that, in the open, the sponges reproduce sexually around the summer solstice and that in the laboratory their physiology depends very strongly on illumination. Experiments are being started in our laboratory to induce sexuality in sponges by submitting them to various light cycles.

Other experiments should allow us to circumscribe the anatomical and physiological differences between the strains. Preliminary results in this field confirm our initial suspicion that there are differences—at least between α and β—regarding the conditions of gemmulation and of erratic spiculation.

Besides its intrinsic interest, the fact that there exist different strains within the same species has one obvious implication: it is unwise to expect repeatable results in any physiological investigation on fresh-water sponges if one does not make sure of working consistently with a definite strain or at least of separating the strains with which one is working.

References

Rasmont, R. (1956). La gemmulation des Spongillides. *Ann. Soc. r. zool. Belg.* **86**: 349–387.

Rasmont, R. (1961). Une technique de culture des Éponges d'eau douce en milieu contrôlé. *Ann. Soc. r. zool. Belg.* **91**: 147–156.

Rasmont, R. (1962). The physiology of gemmulation in fresh-water Sponges. *In Regeneration, 20th Growth Symposium*: 3–25, Rudnick, D. (Ed.). New York: The Ronald Press Cy.

Rasmont, R. (1963). Le rôle de la taille et de la nutrition dans le déterminisme de la gemmulation chez les Spongillides. *Devl. Biol.* **8**: 243–271.

Rasmont, R. and Schmidt, I. (1967). Mise en évidence du caractère photosensible de la respiration des gemmules de Spongillidae (Porifera). *Comp. Biochem. Physiol.* **23**: 959–967.

Symp. zool. Soc. Lond. (1970) No. 25, 423–436.

REGENERATION AND SOMATIC EMBRYOGENESIS IN SPONGES

G. P. KOROTKOVA

Department of Embryology, State University of Leningrad, Leningrad, U.S.S.R.

SYNOPSIS

The reactions of sponges to injuries vary greatly. Sponges can redevelop from aggregates of somatic cells after dissociation and from small fragments of the body. Also, regeneration phenomena like the healing of wounds, restoration of lost structures, etc. are highly characteristic of the group.

The first type of process, which Tokin and I have called "somatic embryogenesis", differs in its genesis and in the details of its formative processes from the second type, which is regeneration proper.

According to Tokin's hypothesis, there exists a close reciprocal relationship between these two types of restorative processes; the way in which restoration proceeds depends upon the degree of integration of the organism and on the type of injury inflicted. Experiments on sponges show that it is possible to induce experimentally either regeneration proper or somatic embryogenesis.

One of the distinctive features of somatic embryogenesis is the change of polarity and body symmetry of the animal during restoration. This is especially obvious during development of the sponges *Sycon lingua* and *Halichondria panicea* from small fragments.

Investigation of restorative morphogeneses in sponges with different types of organization, and subjected to similar injuries, enables us to draw the following conclusions:

1. The more integrated the form (e.g., unioscular sponges with distinct outline plan of body symmetry), the more often are experimental injuries repaired by regeneration proper and the rarer are restorations by somatic embryogenesis.
2. The less well integrated the sponge (e.g., multioscular forms without distinct body outline), the more often are experimental injuries repaired by somatic embryogenesis, while regeneration proper occurs less often.

The details of restorative morphogenesis of a lost part of the body and of the growth of a new organism in a trauma spot depend upon specific properties of the tissue organization of the particular sponge, e.g., on its epithelia being but slightly or strongly pronounced, on its differentiation being stable or loose, on the type of its mesogloea, etc.

INTRODUCTION

Most biologists apply the term regeneration to the whole spectrum of restorative morphogenetic phenomena. However, it is clear that genetically widely different processes are covered by this term. Such are the healing of wounds, the restoration of a lost part of the body, compensatory processes, heteromorphoses and additions, as well as a variety of processes leading to the development of a whole new organism from groups of somatic cells or from single somatic cells.

Tokin's distinction (1959, 1964) between the two main types of restorative and regulatory morphogenesis—regeneration proper and somatic embryogenesis—is most valuable. Tokin defines *regeneration* as the phenomena of restoration of lost parts of the body, organ, tissue or cell. A precondition for regeneration is, in Tokin's view, preservation of the normal mechanisms of integration in the rest of the organism, organ, tissue or cell.

He defines *somatic embryogenesis*, on the other hand, as the processes of development of whole new organisms from complexes of somatic cells. A pre-requisite for somatic embryogenesis is a complete reorganization of structures and functions of the former tissue and organ systems. This implies that groups of cells and single cells isolate themselves from the formative influences of adjoining cells and tissues and, revealing their own potentialities to the full, develop into a new organism.

An intermediate position between these two main categories of restorative phenomena is occupied by such phenomena as heteromorphoses and additions.

The necessity for distinguishing between the processes of regeneration and somatic embryogenesis arises from the fact that without this distinction we could neither make a proper analysis of the evolution of morphogenetic processes, nor assess the restorative morphogenesis and level of integration of any particular living system.

Since regeneration proper and somatic embryogenesis are initiated by two extremes of the degree of disintegration of a system, one cannot expect the evolution of these two types of processes to proceed in parallel. Rather, one would expect fundamentally different and diverging paths of evolution of the various types of restorative morphogenesis, depending on increases or decreases in the level of integration achieved by particular groups of organisms.

Accordingly, as the complexity of ontogeny or phylogeny increases, the capabilities for development of whole new organisms from somatic cells must diminish. This means that powers of asexual and vegetative reproduction must diminish, or disappear completely. At the same time somatic embryogenetic power, the potential for development of embryos from isolated blastomeres and the production of additive growths must diminish.

The reverse should also be true. With increased integrity of organization, regeneration proper will remain intact, growing weaker with the lessening of integrating mechanisms and the simplification of organization.

In view of the importance of the theoretical implications of this classification of restorative morphogenetic processes through the reciprocal dependence of their realization on the integration of living systems, the hypotheses mentioned above have been tested experimentally in our laboratories.

Since living systems differ not only in the levels of their organization, but also have qualitative peculiarities of structure, it is difficult to equate organisms by their level of organization. Consequently, we have restricted our choice of experimental material to those animals in which we can compare reactions to injuries inflicted on homologous organs, organ systems or parts of the body.

Three groups of animals with the simplest kind of organization were chosen for the experiments—sponges, coelenterates and lower worms. These forms are especially suitable because in the adults both regeneration proper and somatic embryogenesis can be induced in all three groups by similar experimental injuries.

At present, the most complete picture of correlations between regeneration proper or somatic embryogenesis and the level of integration of the organism has been obtained for sponges.

RESTORATIVE MORPHOGENESIS IN UNIOSCULAR AND MULTIOSCULAR SPONGES

Although sponges are, in general, very poorly organized multicellular organisms, they do show a range of degrees of integration. There seems to be no doubt that colonial forms are less well integrated than solitary forms. This has led us to compare responses to injuries in unioscular sponges with responses to injuries in multioscular sponges.

The question of whether or not multioscular sponges are colonies or individuals is still open to discussion (*vide* Beklemishev, 1964; Tuzet, Pavans de Ceccatty and Paris, 1963; Hadži, 1966). We consider multioscular sponges to be colonies. However, whatever the view of their organization may be, a looser functional integration of their multioscular apparatus, together with their more amorphous general body shape, must indicate a less well developed degree of integration than occurs in sponges with more definitive body symmetry.

Amongst multioscular sponges there are several species in which individuals are only partially separated from each other and there are other species in which separate zooids are quite distinct. In the latter group the individuality of separate animals is not suppressed by the integration of the colony. Such sponges can be considered as occupying an intermediate position between the unioscular forms and those multioscular forms with only imperfect boundaries between individuals—

i.e., in which colony integration suppresses the individuality of the component zooids.

Unfortunately, we have been unable to work with such an ideal range of forms for while the range of degrees of integration is readily accessible, it must include forms with aquiferous systems of different types.

The majority of our experiments were performed on three species from the littoral of the Barents Sea—*Leucosolenia complicata* Montagu, 1818, *Sycon raphanus* O. Schmidt, 1862 and *Halichondria panicea* Pallas, 1766. *S. raphanus* is unioscular; the other two species are multioscular.

We may compare the responses to experimental injury of two species of *Sycon* which have similarly organized aquiferous systems. *Sycon raphanus* is normally a solitary sponge with a well defined and stable body symmetry. *S. lingua* Haeckel, 1872, on the other hand, after achieving a certain size, transforms itself into a colony with only indistinct boundaries between individual zooids. Young specimens of these species, after being cut into two or four fragments by parallel longitudinal cuts, follow two different paths of morphogenesis. *S. raphanus* knits together the cut edges in the plane of the baso-apical axis, preserving its original polarity. In this manner a new tube arises, with only minimal transformations in the region of reunition (Fig. 1a). The strips cut from *S. lingua*, however, bend inwards in the direction of the paragastral cavity and a thin regeneration membrane grows between the approaching edges of the body wall. While the process of approach proceeds, apical spicules of the marginal fringe disappear, and so does part of the oscular rim, the original polarity is lost and a new osculum develops (Fig. 1b).

When the two species of *Sycon* are cut transversely into two or three parts, the fragments of *S. raphanus*, as in the previous experiments, preserve their original polarity and restore the oscular and basal parts of the body on the corresponding planes of their wounds. In *S. lingua*, on the other hand, 10–15% of the fragments develop two oscula in the plane of the cut (Fig. 1c).

Comparing the reactions of these two species of sponge with very similar tissue organization and canal systems, but differing in their degrees of integration, we observe two fundamentally different types of restorative morphogenesis. In the less highly integrated *S. lingua*—a potential colonial sponge—very often the healing process proceeds as somatic embryogenesis.

The next series of investigations was performed on *L. complicata*—a colonial sponge with well outlined individual zooids—and on *H. panicea*—a colonial sponge with poorly marked borders between

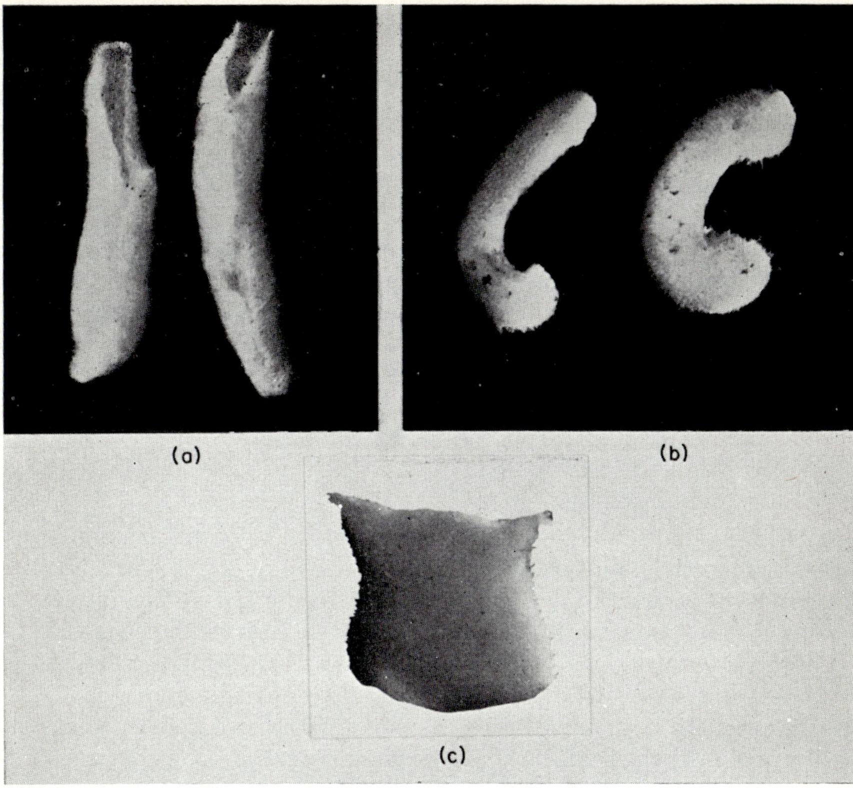

FIG. 1. Regenerative processes in *Sycon* after various operations.
(a) Knitting of the wound edges in *Sycon raphanus* after longitudinal bisection. (b) Formation of "closed" sponges after longitudinal bisection of *S. lingua*. (c) Formation of two oscula in a fragment of *S. lingua* after transverse section.

separate zooids. Experiments showed that in both species post-traumatic morphogenesis readily takes the course of somatic embryogenesis. For example, when small fragments are cut out of the body wall of *L. complicata*, or when scattered burns are inflicted on its oscular tubes, diverticuli arise from the wound sites. Very often a membrane covers the original osculum and a new colony develops, with new oscular tubes. Various instances of regeneration proper in this sponge have been described in earlier papers (Korotkova, 1961; Korotkova and Gelikovskaya, 1963).

In *H. panicea*, when its oscular mounds are damaged, a slow and very often a typical regeneration occurs. Sometimes the opening of the old osculum is sealed up and an oscular bud arises near the trauma spot (Korotkova, 1962a).

Comparing the responses of different species of sponge to similar injuries, we find a striking demonstration of the important role played by integration in these animals. Multioscular, less well integrated sponges respond much more often by morphogenesis resulting in restoration of the lost structures (Korotkova, 1963a).

Our researches revealed a variety of types of somatic embryogenesis, the type depending in each case upon the structure of the animal and the kind of asexual reproductive mechanisms peculiar to the species under investigation. Our experiments also suggest that one or other kind of asexual reproduction is imitated to some degree by the sponge during restoration but the exact degree of imitation depends upon the experimental conditions. The researches also show that the kind of somatic embryogenesis induced depends upon many factors, such as the physiological state of the animal, the nature of the experimental injury and the culture conditions.

The balance between regeneration proper and somatic embryogenesis can be altered in favour of the desired type of morphogenesis by changing the conditions of the experiment. Every factor that tends to increase the local or general disintegration of the sponge tends also to direct the formative processes towards somatic embryogenesis. For instance, operating with a blunt instrument, inflicting additional neighbouring wounds or application of carcinogens leads to the transformation of the oscular tubes of *L. complicata* into closed, stolon-like, bodies which change polarity and produce numerous buds (Korotkova, 1961; Korotkova and Gelikovskaya, 1963, 1966).

HISTOGENIC PROCESSES IN REGENERATION

Histogenic processes have been studied most thoroughly in *L. complicata* (Jones, 1957; Korotkova, 1962b). The restoration of an injured oscular tube is not accompanied by the formation of a regeneration blastema. Instead, a two-layered pinacocyte membrane arises initially by the growth and stretching of the superficial layer of exopinacocytes and adjacent parts of the mesogloea (Fig. 2, r.m.). Then a choanocyte epithelium spreads along the inner surface of the membrane (Fig. 2, ch.e.). The cells of all layers of the body retain their original histogenic potentials. In the body of the sponge the epithelial type of tissue predominates, and the peculiarities of its growth and stretching are easily observed during the process of regeneration and the restoration of the missing parts by proliferation of cells of the corresponding layers.

Restoration of the injured parts of the body in sponges of leuconoid organization is marked by activation of the nucleolate amoebocytes; their activity is essential in the restoration of the surface membranes and of some other structures (Brøndsted, 1953; Korotkova and Nikitin, 1968).

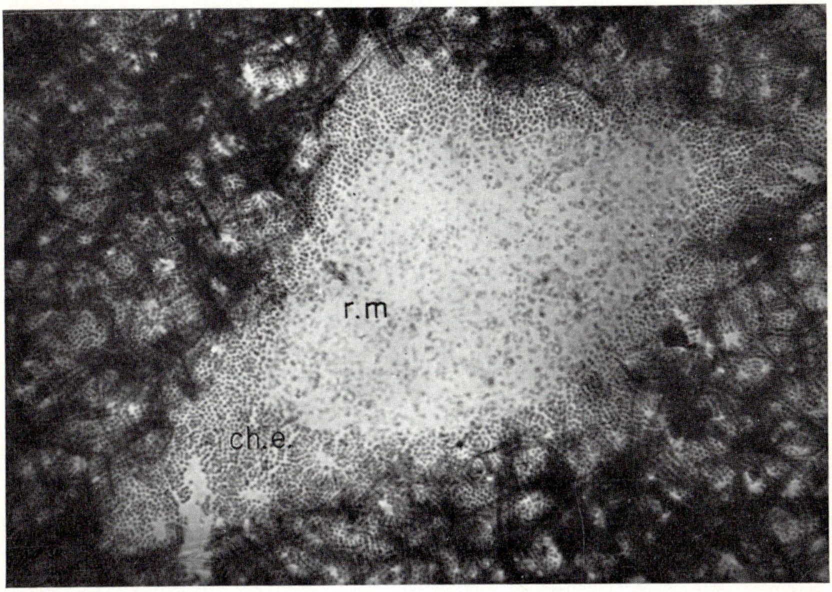

FIG. 2. Spreading of choanocyte epithelium (ch.e.) on the inner surface of the regeneration membrane (r.m.) on the 4th day of regeneration of the body wall of *Leucosolenia complicata*.

Regeneration processes in *Sycon* also have their especial characteristics, expressed initially in a destruction of choanocyte chambers in the vicinity of the wounded surface and in an accumulation of amoebocyte aggregates of various origins (Tuzet and Paris, 1963).

Thus it can be seen that analogous restoration processes of small areas of the body, for instance, differ amongst species and are accompanied by differing degrees of morphallactic reconstruction around the wound surface. On the other hand, the types of restorative morphogenesis resulting from comparatively small injuries are not well known and we can only surmise the degrees of species specificity which they may show.

It is probable that, as the canal system and mesogloea increase in complexity, there is an increase in destructive processes and in the extent of degeneration of certain cell complexes near the wound surface following any given type of injury. This is probably associated with the detailed nature of protective reactions which in sponges take the form of production of antibiotic substances and of encapsulation of foreign bodies by amoebocytes (Nigrelli, Jakowska and Calventi, 1959; Tuzet and Paris, 1964; Connes, 1968 and others).

HISTOGENIC PROCESSES IN SOMATIC EMBRYOGENESIS

Somatic embryogeneses in sponges occur as a great variety of types of processes: development from conglomerates of cells after dissociation of tissues, from small fragments of the body, by experimentally induced budding, by reconstruction of large pieces of the body together with change of polarity and formation of a new colony and by many other similar processes. Development after dissociation is the phenomenon which has been studied to the greatest extent (Wilson, 1907; Huxley, 1911; Müller, 1911; Galtsoff, 1923, 1925a,b; Brien, 1937; Sivaramakrishnan, 1951; Ganguly, 1960; Curtis, 1962; Tuzet and Connes, 1962; Borojević and Lévi, 1964a,b 1965; Sindler and Burnett, 1967 and others).

Yet there remain some controversies over the mechanisms leading to adhesion of cells, the extent of dedifferentiation of various cell types and the share that each type has in the formative process. Probably, the difficulties of somatic embryogenesis, like those of regeneration proper, can be overcome in a number of ways and the particular method employed depends upon the degree of integration and organization of the sponge before dissociation. The cultivation of sponge cells in artificial media, under controlled standard conditions, will probably be of great assistance in solving the problems connected with the morphogenetic potentials of various homogeneous and heterogeneous cell complexes. At present the data we possess are derived from experiments on dissociated cells cultured in ordinary sea water or in fresh water.

Development of sponges from small fragments of the body has been studied in several species of sponge and the development is found to proceed in a number of ways. For instance, fragments of *L. complicata* of about 1 mm diameter develop into whole sponges in much the same way as do larger fragments (Jones, 1957). This development does not involve any breakdown of the three-layer organization of an asconoid sponge—pinacocyte epithelium, mesogloea and choanocyte epithelium. These layers, by bending towards the inner part of the sponge and growing, close off the inner cavity (Fig. 3). The greatest amount of

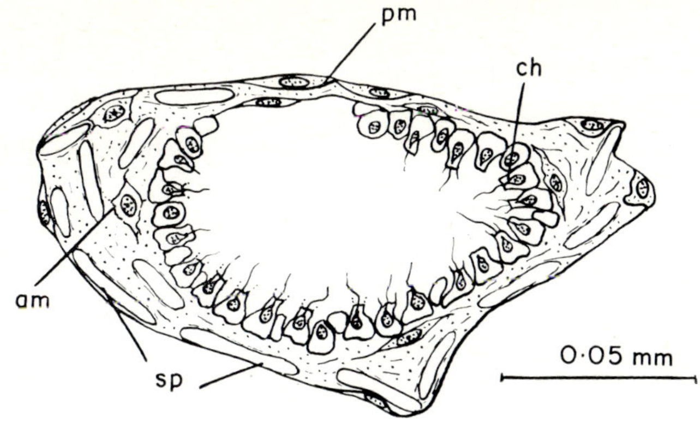

FIG. 3. Cross section of *Leucosolenia complicata* developing from a small piece of the body. pm, pinacocyte membrane; ch, choanocytes; sp, spicules; am, amoebocytes.

reorganization and reconstruction apparently involves the middle layer of the body because spicule orientation becomes less regular.

Sycon lingua, when developing from small fragments of the body about 2 mm in diameter, undergoes complicated destructive changes in the parts of the canal system adjoining the wound surface. A new apical section arises by the reconstruction of the lining of the paragastral cavity of the marginal portions of the fragment (Fig. 4). The polarity of the new sponge is usually perpendicular to the original axis of polarity.

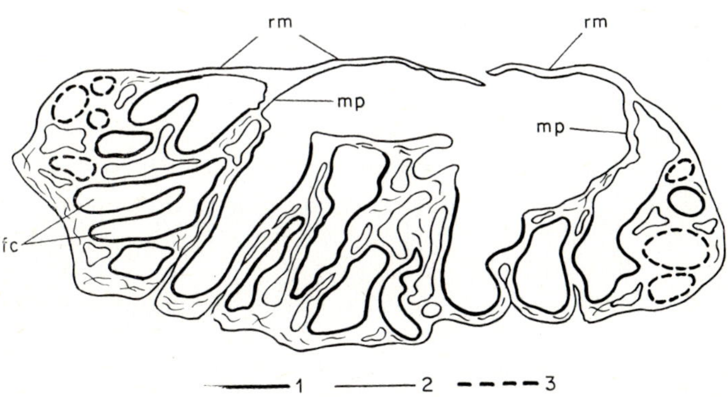

FIG. 4. Cross section of *Sycon lingua* developing from a small fragment of the body at the stage of the regeneration membrane formation (future apical section). Schematic drawing from a preparation. 1, choanocyte epithelium; 2, pinacocyte epithelium; 3, injured choanocyte chambers; rm, regeneration membrane; mp, mesenchyme partitions; fc, fusing canals.

A redeveloping sponge passes through a stage when the old aquiferous system stops functioning. This period of growth of a "closed" sponge coincides with a transformation, and in some cases a degeneration, of the choanocyte lining of some of the old choanocyte chambers (Korotkova, Efremova and Kadantseva, 1965).

Development of leuconoid sponges from small fragments of the body also involves profound reconstruction of the original organization (Maas, 1910; Connes, 1966; Korotkova and Nikitin, 1968). For instance, in a redeveloping *H. panicea*, the original apico-basal orientation of structures characteristic of the species (Fig. 5a) is gradually replaced by a radial orientation (Fig. 5b) and this, in turn, is gradually replaced by a baso-apical orientation after the sponge is attached (Fig. 5c). *Halichondria panicea*, like other sponges developing from small pieces of the body, passes through a closed stage which coincides with the period of maximal reorganization of the canal system. The process of redevelopment of a leuconoid sponge from a small body fragment is accompanied by active migration and redifferentiation of granular nucleolate amoebocytes (Fig. 5a, b, c).

Thus it can be seen that the original type of tissue organization of the sponge determines the type of mechanism which will predominate in the somatic embryogenesis. The degrees of dedifferentiation of certain comparable groups of cells, and their subsequent share in analogous formative processes, vary with the complexity of the mesogloea and of the aquiferous system of the sponge.

Histogenic processes involved in experimentally induced budding, or in reorganization of large areas of the colony involving change of polarity, have not been sufficiently well studied for us to relate precisely the details of this kind of morphogenesis to the level of their integration and to species-specific characters of their aquiferous systems and mesogloeae.

A comparative analysis of several kinds of restorative morphogenesis in relation to their dependence on varying integration and the changing organization of the aquiferous system reveals that the level of integration affects the ratio of regeneration proper to somatic embryogenesis. Potentially, either process can occur when similar injuries have been inflicted on sponges of different species and either process can proceed in a variety of ways. Yet, as a rule, the lower the level of integration of the animal, the more often the formative process turns to somatic embryogenesis. The mechanisms involved in a particular type of somatic embryogenesis are determined, not so much by the level of integration, as by the specific properties of organization inherent in the asconoid, syconoid and leuconoid type. This accounts for the fact that

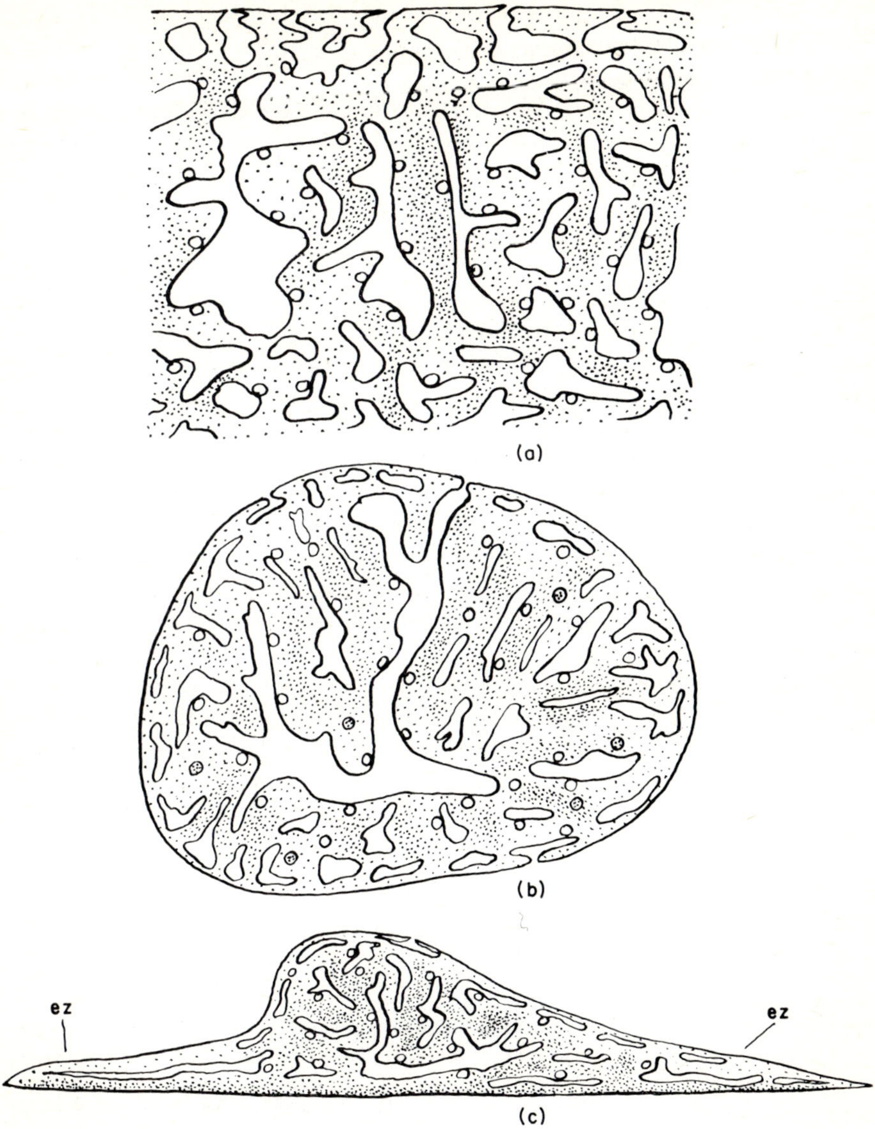

FIG. 5. Change in organization of *Halichondria panicea* in the course of its development from a small fragment of the body. Schematized drawings from a preparation. (a), Cross section of the sponge immediately after operation; (b), Section of the sponge at the stage of maximal rounding of the fragment; (c), Cross section of the sponge after its fixation to the substrate.
The continuous line borders surfaces lined by pinacocyte epithelium. Circles show choanocyte chambers. (Degenerating choanocyte chambers have dots inside the circles.) Density of dots shows the distribution of nucleolate amoebocytes in the mesogloea. ez, edge zone.

the poorly integrated, multioscular sponges of the asconoid type, with their amorphous body shape, redevelop by somatic embryogenesis as readily as do multioscular sponges of the syconoid or leuconoid type.

The most commonly occurring feature of somatic embryogenesis, which distinguishes it from regeneration proper, is, apparently, the change in polarity and pattern of symmetry. These changes provide the two main indications of the level of integration of the organism. Change in polarity of asconoid sponges does not lead to any breakdown of the aquiferous system or of the choanocyte epithelium lining the gastral cavity. This is confirmed by Jones' experiments (1958) as well as by our own (Korotkova and Gelikovskaya, 1963). The change of polarity of oscular tubes is first revealed by the change in the state of the mesogloea and by the spicule orientation. Similar processes in the change of polarity and body symmetry in syconoid, and especially in leuconoid, sponges are accompanied by cell migrations and destructive changes in the canals and choanocyte chambers. This indicates a more precisely determined morphallactic reconstruction process. In other words, one and the same form of somatic embryogenesis, or of regeneration proper, may be brought about by a variety of mechanisms, each being characterized by a variety of types of behaviour of similar groups of cells.

Much research remains to be carried out to determine the range of types of restorative morphogenesis, their species-specificity and the contributions made by different cell groups in each type, and to determine the extent to which these phenomena are controlled by the degree of disintegration of the original tissue and organism and by the culture conditions.

References

Beklemishev, V. N. (1964). Bases of comparative anatomy of invertebrates. Moscow publishing house "Science". I and II.

Borojević, R. and Lévi, C. (1964a). Ultrastructure des cellules d'éponge en suspension cellulaire. *European regional Conference on Electron Microscopy* **3,** P. Czech. Acad. Sci.: 561–562.

Borojević, R. and Lévi, C. (1964b). Métamorphose artificielle des larves d'Eponges après dissociation et réagregation des cellules larvaires. *C. r. Hebd. Séance. Acad. Sci. Paris* **259**: 4364–4366.

Borojević, R. and Lévi, C. (1965). Morphogénèse experimentale d'une éponge à partir de cellules de la larve nageante dissociée. *Z. Zellforsch. mikrosc. anat.* **68**: 57–69.

Brien, P. (1937). La réorganisation de l'Eponge après dissociation par filtration et phénomènes d'involution chez *Ephydatia fluviatilis*. *Arch Biol.* **48** (2): 185–268.

Brøndsted, H. V. (1953). The ability to differentiate and the size of regenerated cells after repeated regeneration in *Spongilla lacustris*. *Q. Jl. microsc. Sci.* **94** (2): 177–184.

Connes, R. (1966). Aspects morphologiques de la régenération de *Tethya lyncurium* Lamarck. *Bull. Soc. zool. Fr.* **91**: 43–53.

Connes, R. (1968). Réactions de défense de l'éponge *Tethya lyncurium* Lamarck, vis-à-vis des micro-organismes et de l'amphipode *Leucothoe spinicarpa* Abildg. *Vie Milieu* **18** (2): 281–289.

Curtis, A. S. G. (1962). Pattern and mechanism in the reaggregation of sponges. *Nature, Lond.* **196**: 245–248.

Galtsoff, P. S. (1923). The amoeboid movement of dissociated sponge cells. *Biol. Bull. mar. biol. Lab. Woods Hole* **45**: 153–161.

Galtsoff, P. S. (1925a). Regeneration after dissociation (an experimental study on Sponges). I. Behavior of dissociated cells of *Microciona prolifera* under normal and altered conditions. *J. exp. Zool.* **42**: 183–222.

Galtsoff, P. S. (1925b). Regeneration after dissociation (an experimental study of Sponges). II. Histogenesis of *Microciona prolifera* Verr. *J. exp. Zool.* **42**: 223–256.

Ganguly, B. (1960). The differentiating capacity of dissociated sponge cells. *Arch. EntwMech. Org.* **152**: 22–34.

Hadži, J. (1966). Vprasanje individualitete pri spuzvah. *Razpr. slov. Akad. Znan. Umet.* (med. vede) **9**: 165–204.

Huxley, J. S. (1911). Some phenomena of regeneration in *Sycon*, with a note on the structure of its collar-cells. *Phil. Trans. R. Soc.* (B) **202**: 165–202.

Jones, W. C. (1957). The contractility and healing behaviour of pieces of *Leucosolenia complicata*. *Q. Jl. microsc. Sci.* **98**: 203–217.

Jones, W. C. (1958) The effect of reversing the internal water-current on the spicule orientation in *Leucosolenia variabilis* and *L. complicata Q. Jl. microsc. Sci.* **99**: 263–278.

Korotkova, G. P. (1961). Regeneration and somatic embryogenesis in the calcareous sponge *Leucosolenia complicata* Mont. *Acta biol. hung.* **11**: 315–334.

Korotkova, G. P. (1962a). Regeneration and somatic embryogenesis in the colonial horny silica sponge *Halichondria panicea* (Pallas). *Vest. leningr. gos. Univ.* **15**: 33–45.

Korotkova, G. P. (1962b). Behaviour of the cell elements in the calcareous sponge *Leucosolenia complicata* Mont. during regeneration. *Acta biol. hung.* **13**: 1–30.

Korotkova, G. P. (1963a) Regeneration and somatic embryogenesis in calcareous sponges of the sicon type. *Vest. leningr. gos. Univ.* **16**: 34–47.

Korotkova, G. P. (1963b). On the types of restoration processes in sponges. *Acta biol. hung.* **13**: 389–406.

Korotkova, G. P. and Gelikovskaya, M. A. (1963). Recherches expérimentales sur le phénomène de la polarité chez les Eponges calcaires du type ascon. *Cah. Biol. mar.* **4**: 47–59.

Korotkova, G. P. and Gelikovskaya, M. A. (1966). Effect of carcinogenic substances on calcareous sponges. *Vest. leningr. gos. Univ.* **9**: 33–43.

Korotkova, G. P., Efremova, S. M. and Kadentseva, A. G. (1965). Peculiarities of morphogenesis during development from small pieces of the body (*Sycon lingua*). *Vest. leningr. gos. Univ.* **21**: 14–30.

Korotkova, G. P. and Nikitin, N. S. (1968). Peculiarities of morphogenesis during the development of the horny silica sponge *Halichondria panicea* from a small fragment of the body. *Trudy murmansk. morsk. biol. Irat.* In press.

Maas, O. (1910). Über Nichtregeneration bei Spongien. *Arch. EntwMech. Org.* **30**: 356–378.

Müller, K. (1911). Das Regenerationsfermögen der Süsswasserschwämme. *Arch. EntwMech. Org.* **32**: 397–446.

Nigrelli, R. F. Jakowska, S. and Calventi, I. (1959). Ectyonin, an antimicrobial agent from the sponge *Microciona prolifera* Verrill. *Zoologica, N.Y.* **44** (4): 173–176.

Sindler, W. F. and Burnett, A. L. (1967). A time-lapse photographic analysis of sponge cell reaggregation. *J. gen. Physiol.* **50**: 1089–1090.

Sivaramakrishnan, V. R. (1951). Studies on early development and regeneration in some Indian marine sponges. *Proc. Indian Acad. Sci.* **34**: 273–310.

Tokin, B. P. (1959). Regeneration and somatic embryogenesis. Leningrad University publishing house, Leningrad.

Tokin, B. P. (1964). Regeneration and somatic Embryogenesis. *Symp. biol. hung.* **3**: 11–45.

Tuzet, O. and Connes, R. (1962). Recherches histologiques sur la réconstitution de *Sycon raphanus* O.S. à partir des cellules dissociées. *Vie Milieu* **13**: 703–710.

Tuzet, O. and Paris, J. (1963). Recherches sur la régenération de *Sycon raphanus* O.S. *Vie Milieu* **14**: 285–291.

Tuzet, O. and Paris, J. (1964). Réaction tissulaires de l'Eponge *Suberites domuncula* (Olivi) Nardo vis-à-vis de ses commensaux et parasites. *Vie Milieu* Suppl. **17**: 147–155.

Tuzet, O., Pavans de Ceccatty, M. and Paris J. (1963). Les éponges sont-elles des Colonies? *Archs. Zool. exp. gen.* **102**: 14–19.

Wilson, H. V. (1907). On some phenomena of coalescence and regeneration in sponges. *J. exp. Zool.* **5**: 245–258.

Symp. zool. Soc. Lond. (1970) No. 25, 437–448.

LA POLARITÉ DE L'OEUF ET LA SYMÉTRIE DE LA LARVE DES ÉPONGES CALCAIRES

ODETTE TUZET

Laboratoire de Zoologie Générale, Université de Montpellier, Montpellier, France

SYNOPSIS

The ovoid egg of the calcareous sponges, lying beneath the choanocytes, is polarized as soon as its growth is completed. Its minor axis, being perpendicular to the epithelial base, is the chief axis, through which the first plane of cleavage passes. The animal pole is in contact with choanocytes. There is another axis, the major one. The pronuclei join together at the point of intersection of the two axes.

Heterocoela. In *Sycon raphanus*, *Grantia compressa*, *Leucandra aspera* and *L. gossei* the chief axis of the embryo coincides with the primary axis. The polar globules are projected towards the poles of the chief axis. The penetration area of the spermiokyste corresponds to the ectoblastic animal pole. The carrying cell and spermiokyste of *Leucandra nivea* may enter the oocyte surface at any point; this anomaly is correlated with this special mechanism of fertilization. The Heterocoela larva possesses a tetraradiate symmetry, as indicated by the "cross-patterned cells" which can be seen from the four-blastomere stage to the amphiblastula.

Homocoela. In *Leucosolenia botryoides*, the ovocyte polarity becomes plain when a chromidial emission gathers round the animal pole, under the choanocytes. The amphiblastula has the same aspect and polarity as that of the Heterocoela. The tetraradiate symmetry is determined by the "cross-patterned cells". The spermiokyste of *Clathrina coriacea* does not enter from the region of the choanocytes but from the opposite pole. The animal pole is located in the position which it occupies in other Calcarea after the surface inversion. It contains one or several ectoblastic cells.

Pharetronida. The polarity of the ovocyte of *Petrobiona massiliana* is determined from the beginning of its growth. The amphiblastula behaves as that of the other calcareous sponges.

INTRODUCTION

La polarité de l'oeuf est le facteur principal déterminant l'orientation de la premiere mitose de l'oeuf fécondé et des plans de clivage ultérieurs, entrainant la polarité et la symétrie de l'adulte.

Pour Caullery (1939) c'est le point de sortie des globules polaires qui définit un axe perçant l'oeuf en deux points: le pôle animal, correspondant à l'endroit où sortent les globules polaires et le pôle végétatif, ou antipôle, qui lui est opposé. C'est autour de ce dernier que se forme l'endoderme et autour du pôle animal l'ectoderme. Il semble, cependant, que la polarité de l'oeuf ne soit pas nécessairement liée à l'endroit où se forment les globules polaires.

Huxley et de Beer (1934) soulignent, chez les Amphibiens, l'importance de la polarité. L'extrémité frontale de l'adulte se forme à proximité du pôle animal de l'oeuf, l'extrémité postérieure au pôle végétatif. Ancel et Vintemberger (1942) montrent que le spermatozoïde n'est l'agent de symétrie bilatérale que dans le cas où l'on place l'oeuf le pôle animal en haut. Si on fait tourner cet oeuf après l'insémination, le plan de fécondation fait avec le plan de symétrie des angles variant de 0 à 180°. Chez le Cnidaire *Amphisbetia* (*Sertularia*) *operculata* Hincks, 1868 (Teissier, 1930; Teissier et Teissier, 1930, 1931) l'axe et les deux pôles de l'oeuf sont déterminés avant la fécondation. Un pôle est coloré par un pigment orangé et la blastula a son pôle antérieur, correspondant à l'ectoderme, blanc, et son pôle postérieur, qui par invagination polaire donnera l'endoderme, orangé. Le pôle de l'oeuf où se fait l'émission des globules polaires et où commence la segmentation correspond au pôle oral de l'hydranthe. La polarité de l'oeuf est la même chez les Cnidaires *Dynamena pumula* (Hincks, 1868 et *Hydractinia echinata* Hallman, 1871.

Gatenby (1920a, b) ne reconnait pas de différenciation régionale dans l'oeuf de *Grantia compressa pennigera* Fabricius 1780. Il s'est demandé si l'on pouvait parler de polarité pour l'ovocyte et l'embryon, mais sa discussion manque de base, car il n'avait pas observé la blastule à flagelles internes et l'inversion des surfaces de cette blastule.

D'après nos observations, les Eponges calcaires se comportent commes les autres Métazoaires, l'oeuf est polarisé dès la fin de sa croissance. Il a une forme ovoide et se trouve placé sous les choanocytes, de telle façon que son petit axe est perpendiculaire à la base épithéliale. Le petit axe AB est l'axe primaire, l'axe de polarité par lequel passe le premier plan de segmentation. Le pôle animal est au contact des choanocytes. Il y a un autre axe important, le grand axe CD, c'est près de l'un de ses pôles que sont émis les globules polaires. Les pronuclei s'unissent au croisement des deux axes (Fig. 1). De plus, la larve des Eponges calcaires possède un plan de symétrie bilatérale, auquel s'ajoute une symétrie tétraradiaire, en particulier pour l'amphiblastule.

ÉPONGES CALCAIRES HÉTÉROCOELES

Chez *Sycon raphanus* O. Schmidt, 1862 et *Grantia compressa* Fabricius, 1780, Duboscq et Tuzet (1937) ont montré que l'axe principal de l'oeuf et de l'embryon coincide avec le trajet d'entrée du spermatozoïde (Fig. 2). Cette coincidence n'est pas complète, le spermatozoïde peut pénétrer dans le voisinage de ce pôle. Les globules polaires sont émis près de l'un des pôles du grand axe CD, on les trouve soit du côté des choanocytes qui surmontent l'oeuf, soit du côté opposé (Fig. 3).

Ces axes de symétrie correspondent à ceux que l'on observe chez les Coelentérés, où le pôle de maturation se confond avec le pôle du grand axe et où l'axe du fuseau de la première mitose de segmentation contient l'axe principal, coïncidant plus ou moins avec le trajet d'entrée du spermatozoïde (Teissier, 1930; Teissier et Teissier, 1930, 1931). Les oeufs de *Sycon raphanus* et de *Grantia compressa* possèdent donc des territoires préformés représentés par la zone située du côté du point de pénétration du spermatozoïde qui donnera l'ectoblaste et par l'antipôle formant l'endoblaste, ainsi que le montre l'embryogenèse. On ne voit pas chez l'oeuf de différenciations régionales dans le vitellus, qui a une répartition uniforme, mais le pigment se localise au pôle végétatif. La pesanteur ne joue aucun rôle dans la répartition des substances de l'oeuf. On peut, en effet, trouver des oeufs fécondés placés aux quatre points cardinaux d'un tube radial et cependant tous parfaitement orientés par rapport au centre du tube.

Nous pouvons donc dire des oeufs de ces deux Eponges, ce que Dalcq (1938) disait de ceux des Amphibiens: "Il y a un méridien particulièrement favorable à la fécondation et une symétrie préalable dans l'ovocyte à la fin de sa croissance". Nous ne dirons pas, comme lui, dans l'oeuf vierge, car nous ne savons pas jusqu'à quand ce qualificatif peut être employé pour l'ovocyte de l'Eponge. Cesse-t-il d'être vierge après la pénétration du spermiokyste ou bien seulement lorsque la spermie est sortie de son kyste?

Chez *Sycon* et *Grantia*, la cellule charriante avec son spermiokyste se colle d'abord en un point quelconque de la paroi de l'ovocyte placée sous les choanocytes, puis elle migre à la surface de l'ovocyte et peut aller jusqu'à une extrémité du grand axe. Elle revient ensuite sous les choanocytes, jusqu'à ce qu'elle ait trouvé pour s'y fixer le point privilégié de pénétration.

Les *Leucandra* (e.g., *aspera* O. Schmidt, 1862, *nivea* Grant, 1826, *gossei* Bowerbank, 1862) montrent que le point de pénétration du spermatozoïde et le point de sortie des globules polaires ne sont pas nécessairement liés à l'établissement de l'axe de polarité. Chez *Leucandra aspera*, le spermiokyste pénètre souvent au pôle animal de l'oeuf, mais parfois aussi à une certaine distance de celui-ci. Chez *Leucandra gossei*, la cellule charriante se place toujours au pôle animal qu'elle déprime pour y faire pénétrer le spermiokyste. Enfin chez *Leucandra nivea*, la cellule charriante, qui entre toute entière dans l'ovocyte avec son spermiokyste, le fait en n'importe quel point de la surface, aussi bien au pôle animal (Fig. 4) que sur le flanc de l'oeuf, ou sur sa face végétative. Cette anomalie doit être liée à la pénétration de la cellule charriante dans l'ovocyte. L'émission des globules polaires a lieu, chez *Leucandra aspera*, au voisinage de l'un des pôles du grand axe (Fig. 5).

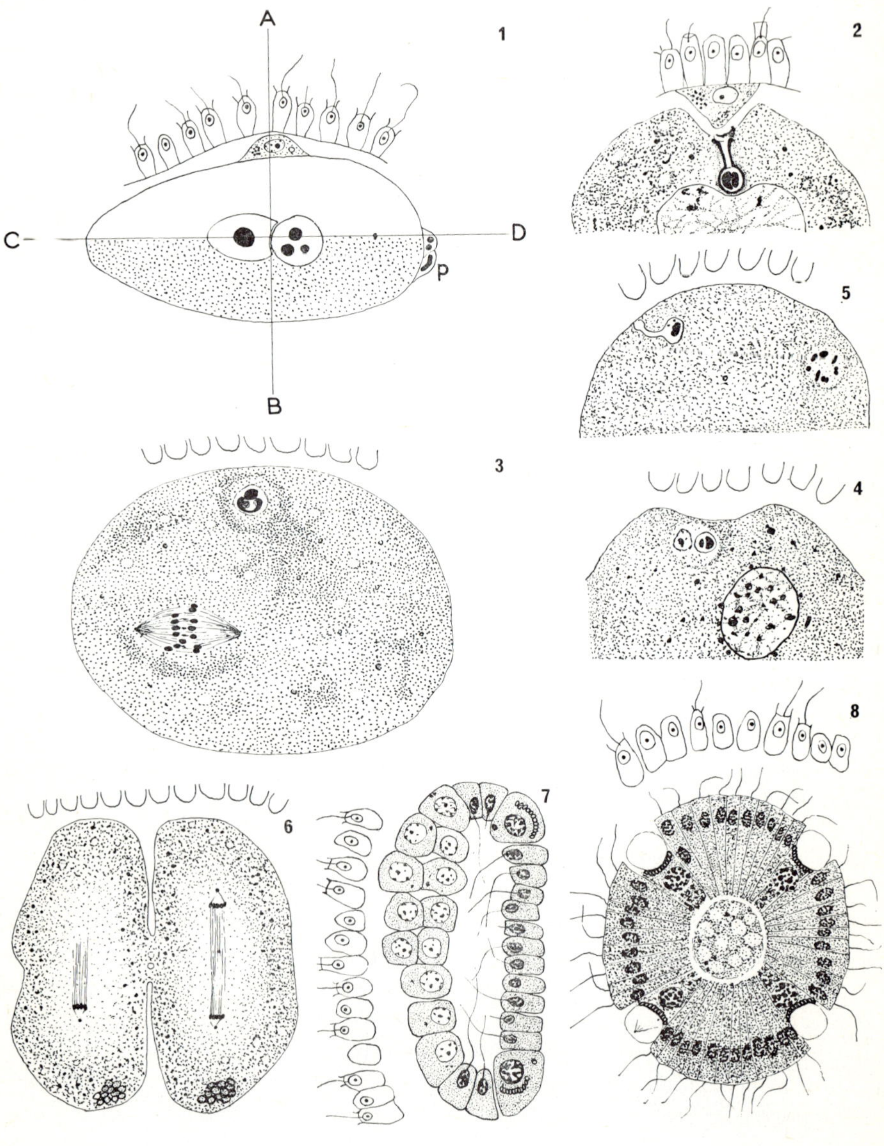

Chez les Hétérocoeles, les pronuclei subissent une rotation avant l'établissement de la première mitose de segmentation. L'axe du fuseau se confond avec le grand axe et le premier plan de segmentation contient l'axe principal AB. La deuxième division de segmentation est dans un plan perpendiculaire au premier (Fig. 6), la troisième passe à nouveau par le petit axe. C'est seulement la quatrième, avec son plan perpendiculaire aux trois premiers, qui détermine la placula diploblastique formée de huit cellules supérieures ectoblastiques et de huit cellules inférieures endoblastiques contenant le pigment mélanique.

La larve possède, en outre, une symétrie tétraradiaire, bien définie par la situation des cellules en croix, depuis le stade à huit blastomères jusqu'à l'amphiblastule. Les futures cellules en croix sont visibles dès le stade à deux blastomères où elles sont annoncées par une plage de cytoplasme spécial, située à leur pôle inférieur (Fig. 6). Au stade à huit blastomères, quatre de ceux-ci, disposés en croix, montrent la plage différenciée. Enfin au stade à seize blastomères, quatre des huit blastomères inférieurs endoblastiques montrent les cellules en croix. Elles resteront au nombre de quatre et localisées dans l'endoderme jusqu'à l'amphiblastule. Les cellules en croix ont été observées chez *Grantia compressa*, *Sycon raphanus*, *S. ciliatum* Fabricius, 1780, *S. elegans* Bowerbank, 1845, *Leucandra gossei*, *L. nivea*, *L. aspera* (Duboscq et Tuzet, 1941, 1942, 1944). La plage de protoplasme différencié va donner un arc d'olives chromatiques situé d'abord en arrière du noyau au stade de la stomoblastule (Fig. 7) puis, devenant latéral au moment de l'inversion des surfaces, il se place finalement au-dessus du noyau. Les cellules en croix ont, au début de leur développement, un flagelle comme les cellules endodermiques, elles le perdent assez vite.

Les cellules en croix ont une situation bien précise, il n'en existe que quatre par larve, situées à égale distance les unes des autres, de telle sorte qu'une coupe transversale heureuse les montre disposées en croix (Fig. 8). Elles ont la hauteur des cellules flagellées, mais sont plus larges et destinées à dégénérer. Chez *Sycon raphanus*, elles s'altèrent dans

---

FIG. 1 (Page ci-contre). Schéma d'un oeuf de *Sycon raphanus* au moment de la conjugaison des pronuclei. p: globule polaire; AB: axe principal; CD: grand axe. En pointillé, le territoire qui donnera les cellules flagellées (endoderme). FIG. 2. *Sycon raphanus*, le spermiokyste a pénétré dans l'ovocyte qui a formé la capsule et le canal du spermiokyste. FIG. 3. *Grantia compressa pennigera*. Ovocyte montrant la métaphase de la première mitose de maturation et le spermiokyste encore inchangé. FIG. 4. *Leucandra nivea*. Cellule charriante ayant pénétré avec son spermiokyste dans l'ovocyte. FIG. 5. *Leucandra aspera*. Oeuf au moment de la métaphase du premier globule polaire, spermiokyste avec canal. FIG. 6. *Sycon raphanus*. Stade à deux blastomères, télophase. Au pôle inférieur de chacun d'eux, la plage d'olives chromatiques. FIG. 7. *Sycon ciliatum*. Stomoblastule à flagelles internes montrant les deux cellules en croix. FIG. 8. *Sycon raphanus*. Coupe transversale montrant les quatre cellules en croix.

l'amphiblastule et ont disparu dans la jeune larve qui vient de se fixer. Celles de *Grantia compressa* dégénèrent dès l'amphiblastule lenticulaire et celles de *Sycon elegans* sont expulsées synchroniquement dans le tube radial de l'Eponge mère (Fig. 9). Ces observations tendent à n'attribuer aucun rôle aux cellules en croix, mais elles ne diminuent pas leur importance morphologique. C'est par elles que se trouve établie la symétrie tétraradiaire des amphiblastules, provenant d'une double symétrie binaire.

La symétrie fondamentale est établie par rapport à un plan passant par le petit axe de la coupe transversale qui correspond au plan de polarité de l'oeuf. L'autre plan de symétrie, perpendiculaire au premier, passe par le grand axe de la section, qui peut être presque circulaire chez *Sycon* et *Grantia* ou plus ou moins elliptique chez d'autres Hétérocoeles. Cet axe est moins important que le premier et n'existe pas dans tous les cas, en particulier chez *Leucandra gossei* où les cellules en croix ont des positions différentes par rapport aux axes de polarité. Nous avons même vu une fois, chez cette espèce, une cinquième cellule en croix (Fig. 10) qui n'avait pas la couronne habituelle de bâtonnets chromatiques, mais seulement deux corpuscules plus gros, disposition que l'on trouve chez *Sycon elegans* et que nous avons interprété comme un arrêt de développement. Cette cellule surnuméraire tendrait à établir une nouvelle symétrie pentaradiare, au lieu de la symétrie tétraradiaire qui est générale.

ÉPONGES CALCAIRES HOMOCOELES

Le jeune ovocyte de *Leucosolenia botryoïdes* Ellis and Solander, 1786 s'installe sous les choanocytes avec son petit axe, axe primaire, perpendiculaire à la base des choanocytes (Tuzet, 1948). Son gros noyau est d'abord très rapproché du pôle animal, c'est alors que se fait une première émission chromidiale. Puis l'ovocyte s'allonge et absorbe un certain nombre de cellules nourricières, son noyau est revenu au centre du cytoplasme. Bientôt une deuxième émission chromidiale change sa structure et contribue à déterminer deux territoires différents (Fig. 11). La plupart des chromidies s'étalent sous la membrane de l'ovocyte située sous les choanocytes, la pénétrent et doivent en modifier la perméabilité, tandis que le noyau se place au pôle animal. Désormais, le territoire supérieur de l'oeuf, destiné à fournir l'ectoderme, diffère du territoire inférieur végétatif, dont le protoplasme est fluide et limité par une membrane délicate qui permet la formation de pseudopodes ramifiés, capteurs de cellules nourricières.

La position de la cellule charriante est variable, le pôle animal reste,

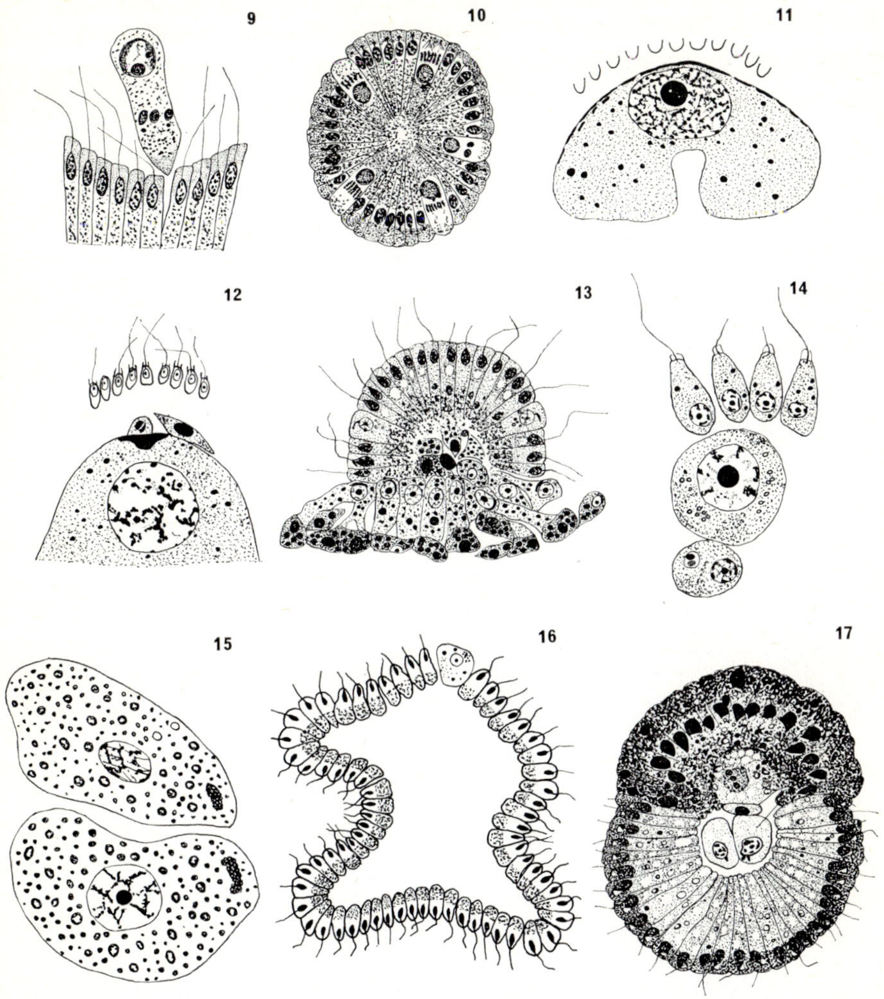

FIG. 9. *Sycon elegans.* Cellule en croix après l'expulsion. FIG. 10. *Leucandra gossei.* Amphiblastule avec une cellule en croix surnuméraire. (symétrie pentaradiaire). FIG. 11. *Leucosolenia botryoides.* Ovocyte montrant l'infiltration chromidiale de la membrane du pôle animal. FIG. 12. *Leucosolenia botryoides.* Ovocyte avec cupule et spermiokyste placés au pôle animal. FIG. 13. *Leucosolenia botryoides.* Amphiblastule incomplètement développée avec cellules granuleuses appliquées sur les choanocytes nourriciers dont quelques uns s'infiltrent vers le centre de la larve. FIG. 14. *Clathrina coriacea.* Petit ovocyte montrant la cellule charriante qui a migré à la surface de l'ovocyte. FIG. 15. *Clathrina coriacea.* Stade à deux blastomères. FIG. 16. *Clathrina coriacea.* Larve au stade préamphiblastule. FIG. 17. *Petrobiona massiliana.* Amphiblastule.

cependant, le point privilégié de pénétration du spermiokyste (Fig. 12), mais celui-ci, à la suite des mouvements de l'oeuf, peut entrer plus ou moins latéralement. Chez cette Eponge le point d'entrée du spermiokyste ne peut toujours servir à définir l'axe primaire et le lieu d'émission des globules polaires ne peut davantage contribuer à cette définition. Ils sont émis en un point situé entre le grand axe et le petit axe, dans le territoire végétatif.

L'amphiblastule de *Leucosolenia botryoïdes* a le même aspect et la même symétrie que celle des Hétérocoeles. Elle montre les mêmes cellules en croix qui y déterminent la même symétrie tétraradiaire (Fig. 13). Le pôle animal est bien le pôle ectoblastique, il continue à être le pôle nourricier de l'amphiblastule qui n'est pas ici entourée de cellules placentaires. On voit, par exemple, cette larve, logée dans la cavité du tube atrial tapissée d'endoderme dans lequel elle pénètre très tôt pour achever son développement, adhérant aux choanocytes qui bordent les tubes et sont devenus nourriciers et les suçant, grâce à ses cellules ectodermiques.

Chez *Clathrina coriacea* Ellis et Solander, 1786, le pigment dispersé dans l'ovocyte ne peut servir à distinguer le pôle animal, car il ne se localise pas (Tuzet, 1947). Le spermiokyste pénètre généralement après que la cellule charriante ait effectué une migration qui la conduit au pôle opposé à celui où elle s'était formée (Fig. 14). Le point d'entrée du spermiokyste semble déterminer ici le pôle animal, ectoblastique, inversé vis-à-vis des choanocytes par rapport à celui des Hétérocoeles et des Homocoeles à amphiblastules typiques.

Nous n'avons pu voir dans cette espèce l'émission des globules polaires. Le déterminant des cellules en croix s'observe, comme chez les autres Eponges calcaires, dès le stade à deux blastomères (Fig. 15), mais il ne nous a pas été possible de suivre son sort ultérieur.

La larve de *Clathrina coriacea* se forme très simplement, sans inversion des surfaces, les flagelles sont d'emblée externes. Toutes les cellules de la larve ne sont pas semblables, il existe au pôle animal, opposé aux choanocytes ayant fourni la cellule charriante, une ou plusieurs, selon les espèces, cellules sans flagelles, plus grosses que les éléments flagellés et à protoplasme granuleux qui correspondent aux cellules granuleuses ectoblastiques des Hétérocoeles et de *Leucosolenia* (Fig. 16). Minchin (1896, 1900) avait déjà décrit les larves blastulaires de *Clathrina* formées de cellules cylindriques flagellées entourant une grande cavité blastocoelienne et présentant à un pôle, chez *Clathrina blanca* Miklucho-Maclay, 1868, deux ou quatre grosses cellules granuleuses non flagellées.

Les larves de *Clathrina* Gray, 1867 n'auraient pas une polarité inversée des blastomères par rapport à celles des autres Eponges calcaires, c'est le point de pénétration du spermatozoïde qui serait inversé. La blastule de *Clathrina* ne montre pas d'inversion des feuillets, mais la polarité est inversée d'emblée dans l'ovocyte, le pôle ectoblastique est situé dans la position où il se trouvera chez les Hétérocoeles et les *Leucosolenia* Bowerbank, 1861 après le retournement aboutissant au démasquage des cellules flagellées. Cependant, à cause du petit nombre des cellules granuleuses, les cellules flagellées du pôle végétatif doivent contribuer aussi à la formation de l'ectoderme. La larve de *Clathrina* serait une blastule qui tendrait vers l'amphiblastule. Elle serait à un stade préamphiblastule, ainsi que le dit Brien (1943), le stade amphiblastule ne s'y réalise pas ou très imparfaitement, il est sauté et dépassé. Par migration des cellules blastodermiques dans la cavité blastocoelienne, la larve deviendra didermique atteignant le stade parenchymula.

ÉPONGES CALCAIRES PHARÉTRONES

La polarité de l'ovocyte de *Petrobiona massiliana* Vacelet et Lévi, 1956 est semblable à celle des Hétérocoeles (Vacelet, 1964). C'est le pôle animal. point de pénétration de la cellule charriante, qui donnera l'ectoderme larvaire. Le développement de *Petrobiona* est intéressant à deux points de vue: du fait de la réception précoce du spermiokyste et de la constance de son point de pénétration, la polarité est très nettement définie dès le début de la croissance de l'ovocyte; d'autre part, la destinée de ce pôle animal peut être suivie plus facilement que chez les autres Calcaires, car il est repérable par l'amas pycnotique dérivant du noyau de la cellule charriante, par les restes de la capsule du spermiokyste que l'on trouve encore au début de la segmentation et par l'induction précoce exercée par le pôle ectodermique sur les choanocytes. Comme chez les autres amphiblastules d'Eponges calcaires, après le retournement de la blastule, l'ectoderme sera postérieur (Fig. 17).

CONCLUSIONS

L'inversion des surfaces de la larve observée chez les Eponges calcaires Homocoeles (sauf les *Clathrina* où les cellules de la larve ont d'emblée leur position définitive), Hétérocoeles et Pharétrones, a une conséquence importante au point de vue de la polarité, le retournement déplaçant le pôle animal à l'antipode de sa situation première. Par le démasquage des cellules flagellées l'amphiblastule se trouve constituée et l'hémisphère flagellé endodermique est au voisinage des choanocytes.

On peut se demander si le pôle animal d'une amphiblastule après le démasquage reste au centre des cellules ectodermiques et s'il ne se trouverait pas au sommet de l'hémisphère flagellé qui dirige le mouvement de la nage. Korschelt (1936), en comparant l'amphiblastule de l'Eponge à la planule d'Hydraire, fait remarquer qu'on ne peut les homologuer qu'en supposant chez l'Eponge une inversion du sens de la nage et il ne recule pas à l'admettre. Nous pensons comme lui et nous voyons, selon la règle, le pôle animal placé au centre des cellules ectodermiques. Teissier (1930) a d'ailleurs vu au cours du développement d'*Amphisbetia operculata* deux inversions successives de la polarité physiologique et chez les Démosponges, la blastule flagellée d'*Oscarella* Vosmaer, 1887 (Meewis, 1939) se fixe par le pôle antérieur de la larve qui, comme dans l'amphiblatule des Calcaires, est le pôle endodermique qui s'invagine pour former ensuite les corbeilles vibratiles.

A première vue la répartition des cellules en croix détermine une symétrie quadriradiée qui se maintient rigoureusement tant qu'elle est régulière (cette symétrie tétraradiaire est d'ailleurs très fréquente chez les Invertébrés). Cependant, au lieu d'être symétrique par rapport à deux plans, certaines larves ne le sont que par rapport à un. Comme l'amphiblastule est généralement un peu aplatie, sa coupe transversale équatoriale au niveau des cellules en croix est plus ou moins elliptique et a deux plans de symétrie correspondant l'un à l'axe principal, l'autre au grand axe. Si, comme chez *Leucandra gossei*, la symétrie du deuxième plan est par trop approximative, elle crée un déséquilibre qui détermine l'apparition d'une cellule en croix surnuméraire et, par ce fait, d'une symétrie pentaradiaire. (Celle-ci n'implique en rien celle des Echinodermes.) Une seule chose est commune dans ces symétries, c'est qu'elles ont toutes à leur base un plan de symétrie bilatérale.

Ce que nous venons de dire de l'oeuf et des larves des Eponges calcaires peut s'étendre aux autres Eponges, en effet pour Brien (1943), malgré la diversité des modes de développement des Eponges, leur embryogenèse est uniforme, la larve type est une amphiblastule. On la retrouve chez les Eponges primitives, les Tetraxonides Carnosa, les Calcaires Homocoeles et Hétérocoeles. L'amphiblastule des Calcaires a fondamentalement la structure et le comportement de celle des Carnosa, elle en a la polarité, la différenciation selon l'axe antéro-postérieur en deux régions ecto-mésenchymateuse et endoblastique, enfin elle présente la gastrulation par invagination.

Bibliographie

Ancel, P. et Vintemberger, P. (1942). Sur le déterminisme de la symétrie bilatérale dans l'oeuf de Batracien. Rapports de la trainée spermatique avec le plan de symétrie bilatérale dans diverses conditions expérimentales. *C. r. Séanc. Soc. Biol.* **136**: 411–412.

Brien, P. (1943). L'embryologie des Éponges. *Bull Mus. r. Hist. nat. Belg.* **19**: 1–19.

Caullery, M. (1939). *Les progrès récents de l'embryologie expérimentale.* Paris: Edit. Flammarion.

Conklin, E. G. (1925). Cellular differentiation in *General cytology* Ed. Cowdry E. V.: 539–607, Univ. Chicago Press.

Dalcq, A. (1938). *Form and causality in early development.* Cambridge University Press.

Duboscq, O. et Tuzet, O. (1937). L'ovogenèse, la fécondation et les premiers stades du développement des Éponges Calcaires. *Arch. Zool. exp. gén.* **79**: 157–316.

Duboscq, O. et Tuzet, O. (1941). Sur les cellules en croix des *Sycon* (*Sycon ciliatum* Fabr., *Sycon coronatum* Ellis et Sol., *Sycon elegans* Bow.) et leur signification. *Arch. Zool. exp. gén.* **81**: 151–163.

Duboscq, O. et Tuzet, O. (1942). Recherches complémentaires sur l'ovogenèse, la fécondation et les premiers stades du développement des Éponges calcaires. *Arch. Zool. exp. gén.* **81**: 395–466.

Duboscq, O. et Tuzet, O. (1944). L'ovogenèse, la fécondation et les premiers stades du développement de *Sycon elegans* Bow. *Arch. Zool. exp. gén.* **83**: 445–459.

Gatenby, J. B. (1920a). The germ-cells, fertilisation and early development of *Grantia* (*Sycon*) *compressa. J. Linn. Soc.* Lond. **34**: 261–297.

Gatenby, J. B. (1920b). Further notes on the oogenesis and fertilization of *Grantia compressa. J. r. microsc. Soc. Lond.* **1920**: 277–282.

Huxley, J. et de Beer, J. R. (1934). *The elements of experimental embryology.* Cambridge: University Press.

Korschelt, E. (1936). *Vergleichende Entwicklungsgeschichte der Tier.* Ed. Jena Korschelt und Heider-Fischer.

Meewis, H. (1939). Contribution à l'étude de l'embryogenèse des Myxospongidae *Halisarca lobularis* Schmidt *Archs. Biol., Paris* **50**: 3–66.

Minchin, E. A. (1896). Note on the larva and the post-larval development of *Leucosolenia variabilis* with remarks on the development of other Asconidae. *Proc. r. Soc. Lond.* **60**: 42–52.

Minchin, E. A. (1900). Éponges calcaires. La Clathrine coriace. *Zool. descriptive Boutan*, I chap. V, Paris: Ed. Doin.

Teissier, G. (1930). Polarité morphologique et polarité physiologique de l'embryon des Hydraires. *C. r. Scéanc. Soc. Biol.* **105**: 671–673.

Teissier, G. et Teissier, L. (1930). Sur la polarité et les localisations germinales de l'oeuf des Hydraires. *C. r. Soc. Séanc. Biol.* **105**: 668–671.

Teissier, G. et Teissier, L. (1931). Sur l'origine germinale de la symétrie bilatérale de certains Hydraires. *C. r. Séanc. Soc. Biol.* **108**: 1124–1126.

Tuzet, O. (1947). L'ovogenèse et la fécondation de l'Éponge calcaire *Leucosolenia* (*Clathrina*) *coriacea* Montagu et de l'Éponge siliceuse *Reniera elegans* Bow. *Arch. Zool. exp. gén.* **85**: 127–148.

Tuzet, O. (1948). Les premiers stades du développement de *Leucosolenia botryoïdes* Ellis et Solander et *Clathrina* (*Leucosolenia*) *coriacea* Mont. *Annl. Sci. nat.* (*Zool.*) (11) **10**: 103–114.

Vacelet, J. (1964). Étude monographique de l'Éponge calcaire Pharétronide de Méditerranée *Petrobiona masisliana* Vacelet et Lévi. Les pharétronides actuelles et fossiles. *Recl. Trav. Sta. mar. Endoume* **50** (34): 1–125.

Symp. zool. Soc. Lond. (1970) No. 25, 449–466.

LES BASES ULTRASTRUCTURALES DES COMMUNICATIONS INTERCELLULAIRES DANS LES OSCULES DE QUELQUES ÉPONGES

MAX PAVANS DE CECCATTY, YVETTE THINEY et ROBERT GARRONE

Laboratoire d'Histologie et Biologie Tissulaire, Faculté des Sciences, Université de Lyon, Lyon, France

SYNOPSIS

The ultrastructures of histological patterns and cell junctions in the oscula of *Hippospongia communis*, Lamarck, 1816 *Euspongia officinalis* Linné, 1759, *Haliclona rosea* O. Schmidt, 1870 and *Verongia cavernicola* Vacelet, 1959 are described with a view to elucidating mechanisms of intercellular communication in sponges. Five main classes of cells were found: exo- and endopinacocytes, archaeocytes, spheriferous cells and contractile cells. Junctions do occur between all the cell types but the majority of junctions involve superficial pinacocytes and deep contractile cells. The oscula are constructed according to simple or complex patterns but all possess two epithelial layers enclosing, and connected with, an intermediate network.

Intercellular connections involve the basic apposition of plasma membranes, where desmosome-like and nexus-like junctions can be observed. Evidence of exchanges or transfer of material from cell to cell is illustrated.

Sponges clearly possess the ability to circulate substances or signals between cells, not only by the undirected spreading in extracellular spaces (the internal milieu), but also by means of directional transmission along circuits of connected cells.

INTRODUCTION

Dans un système pluricellulaire complexe, à partir du moment où un des éléments acquiert une spécialisation fonctionnelle quelconque, différente de celle des éléments voisins, une régulation s'impose, hors de laquelle l'organisation ne peut se maintenir en tant que telle. Cette régulation est assurée par une circulation intercellulaire de substances et de signaux (les substances pouvant elles-mêmes avoir valeur de signaux, comme dans le cas des hormones). Or, l'organisme d'un Spongiaire accomplit les fonctions essentielles à son développement, à sa survie et à sa reproduction. On doit donc rechercher si, à côté des mécanismes *cellulaires* généraux de croissance, de différenciation et de division, il met aussi en jeu des mécanismes *intercellulaires* tout aussi fondamentaux pour la régulation de ses diverses activités.

Chez les Métazoaires, la circulation interne de substances et de signaux s'opère de deux manières, qui ne s'excluent nullement l'une l'autre. On connait, d'une part, un mode de diffusion non directionnelle dans les espaces extracellulaires qui aboutit à une imprégnation locale

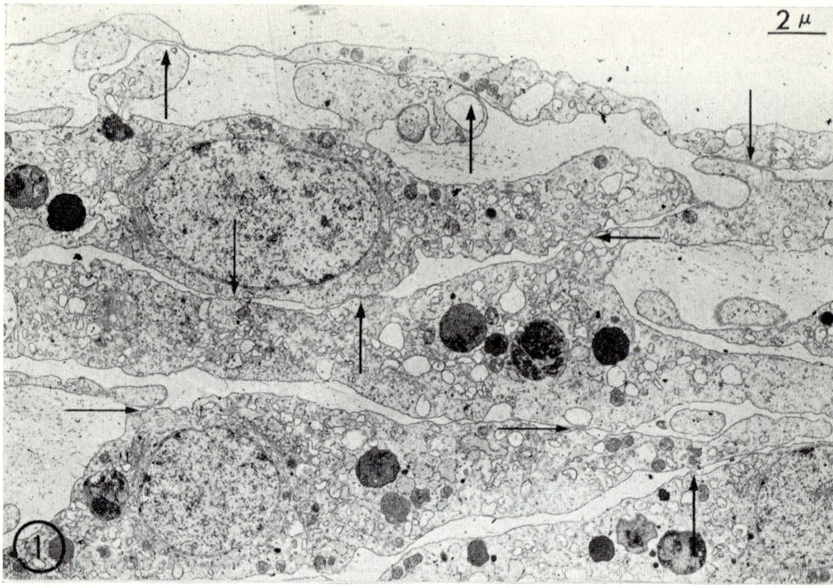
2 μ
1

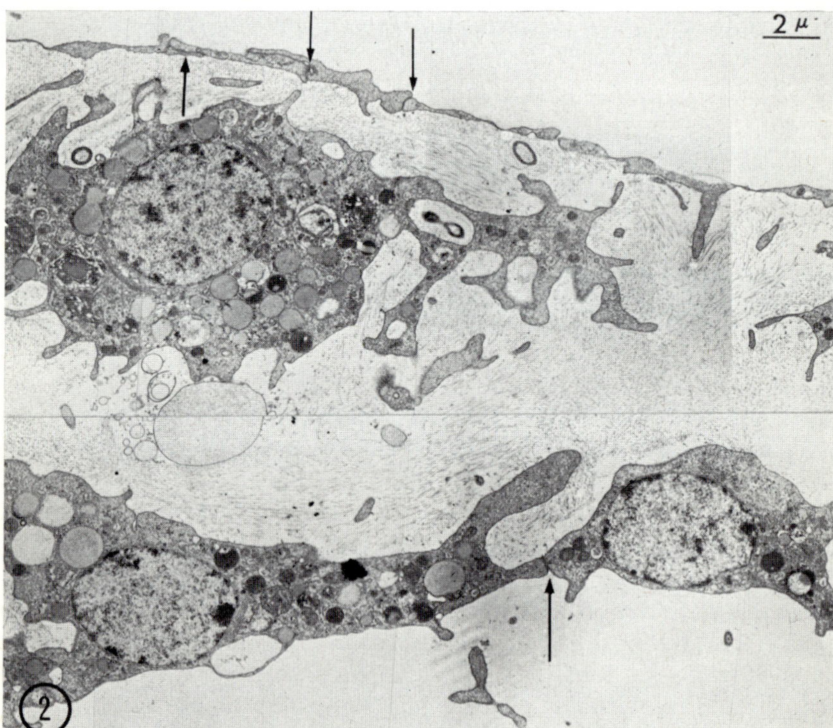
2 μ
2

ou générale de l'organisme. Dans un territoire donné, ou dans tout l'individu, les cellules sont atteintes et réagissent selon leur état. L'efficacité de cette diffusion dans les espaces dépend d'un milieu intérieur qui, chez les Eponges, pourrait être constitué par le "mésohile" (Hadži, 1949). On connait d'autre part, la transmission directionnelle le long de circuits constitués par des cellules en contact les unes avec les autres. Ne sont alors atteints par la circulation de substances ou de signaux que les éléments connectés. Sans que leur rôle soit bien connu, de telles connexions ont déjà été mises en évidence, chez les Eponges, au niveau des épithéliums et de certaines cellules mésenchymateuses (Pavans de Ceccatty, 1966a, b; Borojević et Lévi, 1967).

La compréhension des Spongiaires parait ainsi soumise, aujourd'hui, à l'analyse de leur milieu intérieur comme de leurs contacts intercellulaires qui font l'objet essentiel de ce travail.

Il est évident que cette étude ne peut être entreprise qu'en respectant le contexte histologique naturel des cellules. Nous avons donc cherché à surmonter les difficultés particulières aux Spongiaires, à leur plasticité cellulaire et à leur diversité anatomique, en choisissant des modèles histologiques stables et comparables: les oscules de Démosponges. Ils offrent l'avantage d'une localisation discernable, facilitant aussi bien la précision du prélèvement que l'observation *in situ* et l'intervention expérimentale. C'est ainsi que nous avons étudié les oscules d'*Hippospongia communis*, *Euspongia officinalis*, *Haliclona rosea* et *Verongia cavernicola*. Ils se présentent tous comme des membranes iridiennes parfois très fines et transparentes, parfois épaisses et plus ou moins pigmentées, toujours contractiles et faisant preuve enfin d'une grande capacité de régénération après ablation partielle ou totale.

Les techniques microscopiques utilisées sont celles des méthodes habituelles. Nous avons ainsi pratiqué la fixation directe simple ($0s0_4$) ou double (glutaraldéhyde + $0s0_4$) généralement à 0°C. Les prélèvements furent inclus dans l'Epon, coupés à l'ultramicrotome Reichert, et contrastés à l'acétate d'uranyle + plomb. Les observations et les photographies furent enfin réalisées avec les microscopes Hitachi HU 11 A et HU 11 C.

LES MODÈLES HISTOLOGIQUES OSCULAIRES

Deux types d'iris osculaires peuvent être observés. Le premier (chez *Haliclona* et *Euspongia*) est transparent, mince et extrêmement

(Page ci-contre) FIG. 1. *Haliclona*. Coupe de l'oscule. Entre les exopinacocytes (en haut) et l'épithélium interne (en bas), diverses cellules indifférenciées. Nombreux contacts (flèches. FIG. 2. *Hippospongia*. Coupe de l'oscule, près de l'orifice central. En bas endopinacocytes de l'épithélium interne. En haut, épithélium lamellaire externe, issu d'un exopinacocyte sous-jacent. Nombreux contacts (flèches).

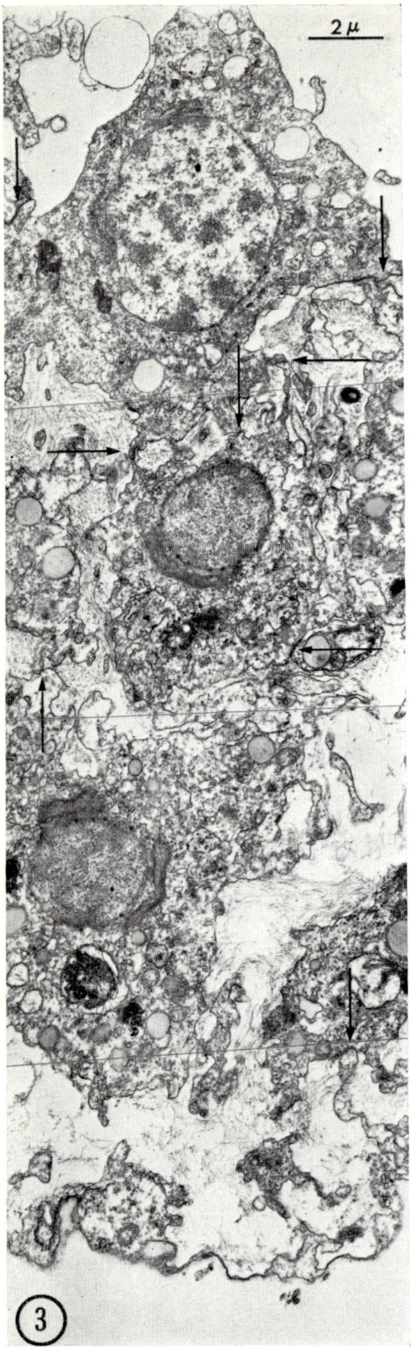
2 μ
3

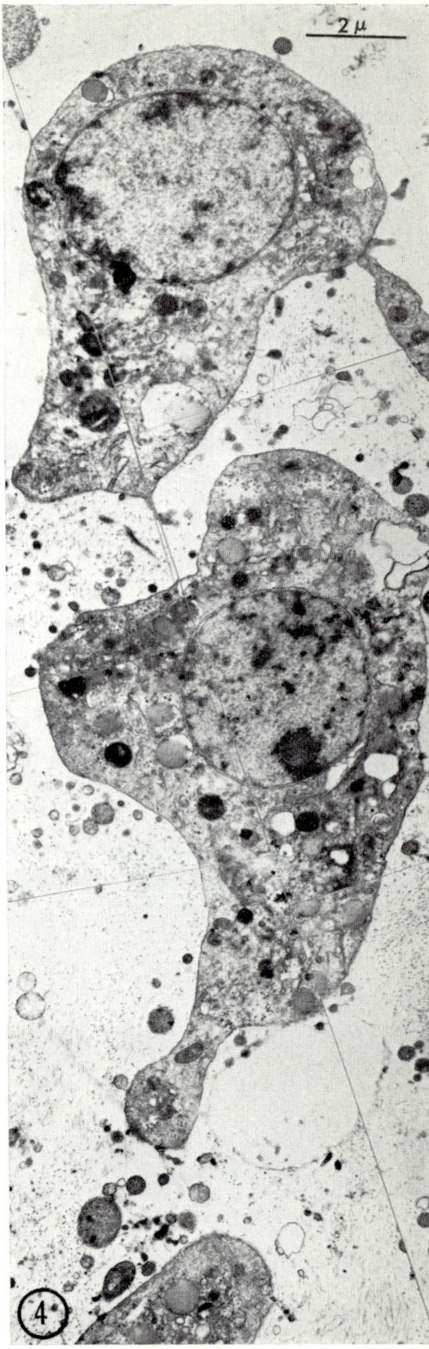
2 μ
4

délicat. Sa finesse permet une observation aisée sur le vivant, en microscopie photonique à contraste de phase qui décèle la superposition d'un nombre restreint de couches cellulaires. Le second type (*Hippospongia* et *Verongia*) diffère du premier par son épaisseur. La densité des superpositions cellulaires rend impossible l'observation vitale en microscopie photonique.

La microscopie électronique montre que ces deux types ne correspondent nullement à des catégories tranchées mais plutôt à une variation graduelle du même modèle. Les deux surfaces annulaires de l'oscule sont formées par un epithélium aplati. Du côté extérieur, les *exopinacocytes* sont en continuité avec ceux de l'épithélium de revêtement périphérique de toute l'éponge. On est frappé de n'y trouver jamais de noyau affleurant la surface, et marquant leur emplacement par un soulèvement local. Toute la couche épithéliale est mince, assez régulière. En fait, elle est formée par des expansions planes et jointives de cellules plus profondément situées dans le mésohile et qui correspondent aux exopinacocytes (Figs 1, 2 et 3).

Du côté interne, l'épithélium est en continuité avec celui qui tapisse le canal osculaire. Les *endopinacocytes*, parfois ciliés, sont étroitement associés les uns aux autres par leur bord. Leur noyau renfle la partie centrale de chaque corps cellulaire provoquant ainsi une ondulation régulière de la surface interne. Ils émettent, en direction du mésohile, des prolongements variés et font ainsi preuve d'une nette polarisation (Figs 1 et 3).

Cette différenciation des pinacocytes, leur polarité et leur manière de participer à l'édification de l'épithélium apparait très nettement au cours des phénomènes de régénération que nous avons pu observer chez *Hippospongia* (Figs 4 et 6).

Quoi qu'il en soit, tous les oscules examinés présentent une asymétrie de leurs bordures épithéliales extérieure et intérieure.

Dans les zones osculaires les plus fines—proches de l'ouverture centrale—les endopinacocytes superficiels, et les exopinacocytes profonds avec leurs expansions lamellaires de surface, suffisent à construire la totalité du modèle histologique (Fig. 2). Dans le mésohile, les espaces laissés libres entre les corps cellulaires des exopinacocytes sont comblés par du collagène.

(Page ci-contre) Fig. 3. *Hippospongia*. Coupe de l'oscule. Asymétrie des épithéliums externe (en bas) et interne (en haut). Nombreux contacts superficiels et profonds (flèches). Comparer avec la Fig. 4. Fig. 4. *Hippospongia*. Oscule en régénération de trois jours. Rupture des contacts intercellulaires et évolution des cellules (endopinacocytes, en haut et cellules mésohilaires, au milieu) vers un type amoeboide commun. Comparer avec la Fig. 3.

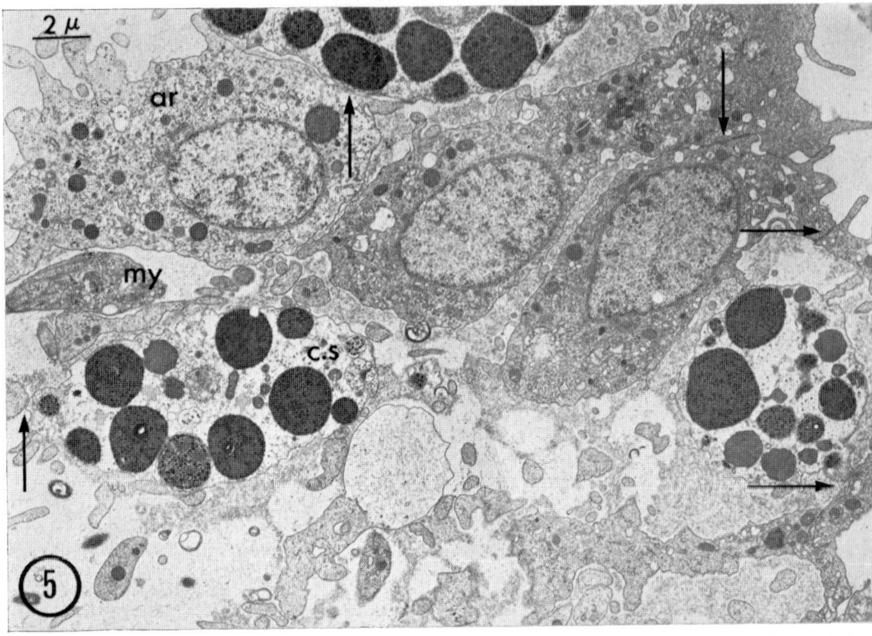
2 μ
ar
my
c.s
5

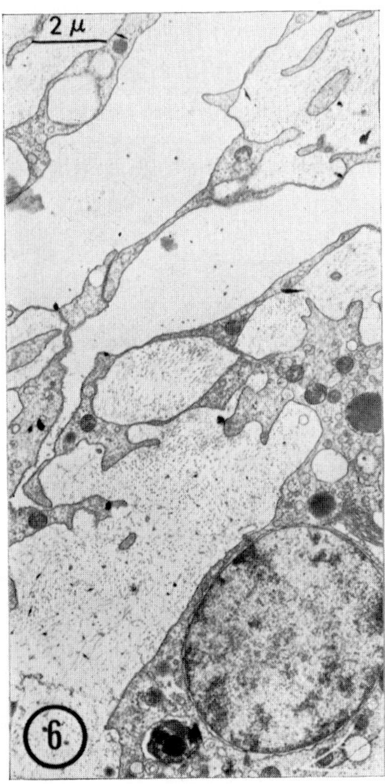
2 μ
6

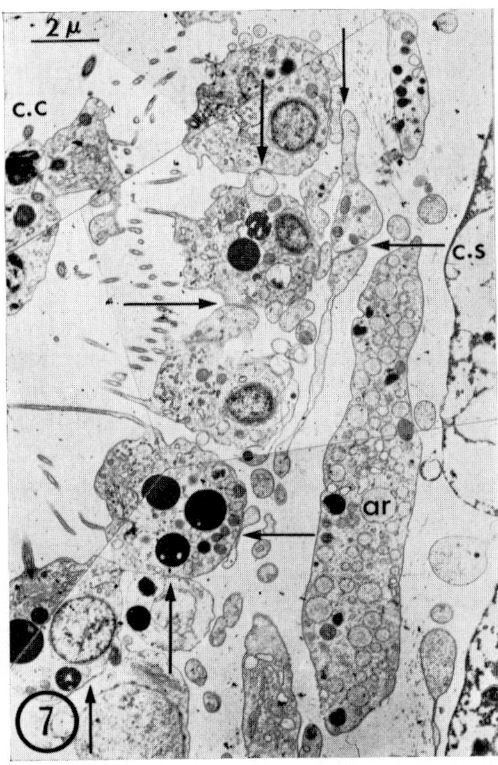
2 μ
c.c
c.s
ar
7

A partir de cette structure élémentaire fondamentalement bistratifiée, les autres modèles histologiques se compliquent par le développement du mésohile où l'on observe l'addition en nombre et en variété de types cytologiques différents (Figs 1 et 3). Nous y trouverons ainsi les principales catégories cellulaires des Eponges, sauf les choanocytes dont la proximité peut être cependant très grande (*Haliclona*) si bien qu'il n'est pas rare d'observer les corbeilles incluses à la périphérie de l'oscule (Fig. 7).

On relève la présence d'éléments peu différenciés (Figs 1 et 5), pourvus d'inclusions diverses et inconstantes, au nucléole souvent très marqué, et que l'on peut rattacher aux polyblastes (Tuzet et Pavans de Ceccatty, 1955; Connes, 1967) ou encore *archaeocytes* (Borojević et Lévi, 1964; Borojević, 1966).

Chez *Haliclona*, on rencontre quelquefois des scléroblastes identifiables grâce à leur élaboration minérale interne, leur riche vacuolisation et leurs microtubules.

Disséminés parmi les éléments précédents, les *cellules sphéruleuses* sont très reconnaissables malgré leurs aspects variables. Peu nombreuses et claires chez *Hippospongia et Euspongia*, elles sont plus fréquentes et munies de sphérules plus soulignés chez *Haliclona* (Figs 7 et 13), tandis qu'elles se démultiplient et montrent des inclusions tout à fait particulières (Figs 5 et 9) chez *Verongia* (Vacelet, 1967).

On remarque enfin des cellules caractérisées par leurs faisceaux de microfilaments cytoplasmiques. Ces éléments sont assez semblables chez *Hippospongia*, (Fig. 8) *Euspongia* et *Verongia* (Fig. 9) ainsi que chez d'autres Eponges (Pavans de Ceccatty, 1966a, b; Vacelet, 1966; Bagby, 1966). Chez *Haliclona*, ils révèlent une organisation plus poussée où la régularité des faisceaux, et leur disposition, met en relief une structure plus nettement paracristalline (Fig. 10). Ces cellules sont les seules susceptibles de rendre compte de la contractilité osculaire. Tous les auteurs sont d'accord pour leur attribuer la signification de *cellules myoblastiques* tandis que—à la suite de nos études sur *Tethya* et *Hippospongia* (Pavans de Ceccatty, 1966a,b)—nous y reconnaissons aussi les *cellules de type nerveux* de nos études antérieures (Pavans de Ceccatty, 1959).

Au total, les modèles osculaires comportent donc cinq catégories cellulaires principales: exo- et endopinacocytes, archaeocytes, cellules

(Page ci-contre) FIG. 5. *Verongia*. Cellules osculaires, ar: archaeocyte, c.s.: cellule sphéruleuse, my: cellule contractile. Contacts (flèches). FIG. 6. *Hippospongia*. Régénération osculaire (7 jours). Pinacocyte émettant la lame épithéliale. FIG. 7. *Haliclona*. Périphérie de l'oscule. Corbeille de choanocytes avec une cellule centrale (c.c.) A proximité de la corbeille, archaeocytes (ar) et cellule sphéruleuse (c.s.). Nombreux contacts (flèches).

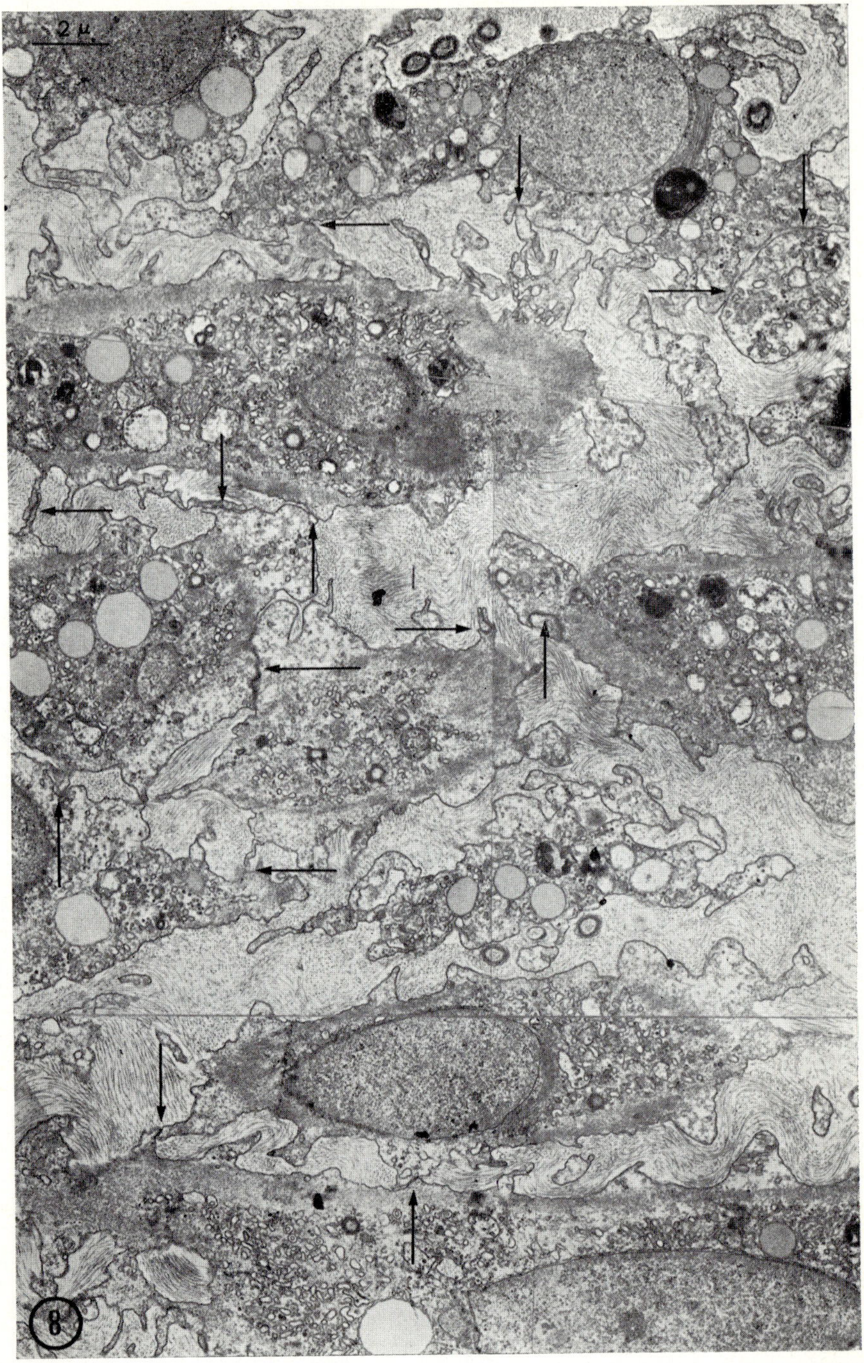
2 μ
8

sphéruleuses et cellules contractiles. Mais ces catégories n'expriment pas la totalité du système qui comprend de nombreuses formes ambigües. C'est ainsi que l'extrusion d'enclaves lipidiques se remarque dans les cellules contractiles (Fig. 11) comme dans bien d'autres types. Les phagosomes peuvent exister dans la plupart des différenciations. Les faisceaux de filaments apparaissent dans des exopinacocytes (Fig. 12) comme dans les myoblastes. Les tubules se distinguent dans des pinacocytes, des scléroblastes et dans les éléments contractiles. Que dire enfin de la fibrogenèse du collagène qui parait devoir se rattacher à un état fibroblastique transitoire de plusieurs types cellulaires?

La simple analyse des structures histologiques stables permet enfin une constatation intéressante quant à la disposition réciproque des cellules. Dans le cas d'*Haliclona* (Fig. 1), l'aspect du mésohile osculaire évoque celui d'un *parenchyme* aux espaces extracellulaires réduits. Chez *Hippospongia* (Fig. 8), les espaces sont plus développés, riches en collagène, et les cellules paraissent s'y déployer plus librement comme dans un *mésenchyme lâche* ou un tissu conjonctif très jeune.

LES CONNEXIONS INTERCELLULAIRES

Quelle que soit la complexité du modèle osculaire, l'importance des contacts intercellulaires apparaît dès le premier examen en microscopie électronique.

Chez *Haliclona* (modèle de type parenchymateux), les contacts entre les cellules sont nombreux et s'établissent souvent sur de larges surfaces aussi bien entre les cellules épithéliales qu'entre celles du mésohile, et à la jonction des unes et des autres (Fig. 1).

Chez *Hippospongia* (modèle de type mésenchymateux lâche), les larges jonctions intercellulaires continues se forment surtout au niveau des épithéliums de bordure. A l'intérieur, la texture moins serrée du tissu médian réduit la surface des zones de contact, mais pas leur nombre. Car les espaces intercellulaires plus développés sont alors parcourus de très nombreuses ramifications cytoplasmiques, qui dessinent un réseau et qui établissent des connexions soit en passant, soit par leurs terminaisons (Fig. 8).

Nous considérons comme le signe d'un contact intercellulaire le trajet parallèle de deux membranes plasmatiques, séparées par une

---

(Page ci-contre) FIG. 8. *Hippospongia*. Réseau de cellules contractiles interconnectées (flèches) dans le mésohile épais de l'oscule.

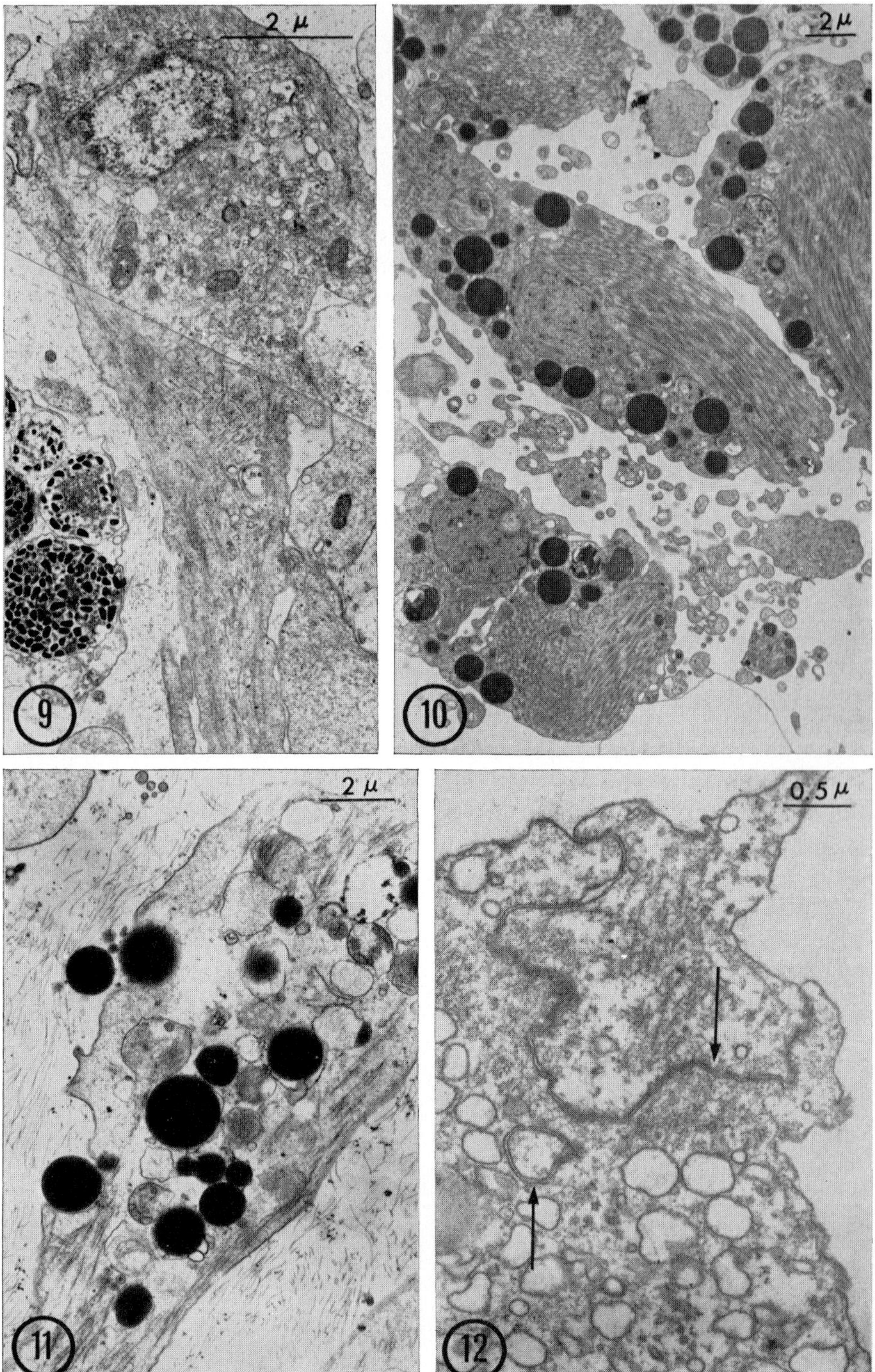
2 μ
9
2 μ
10
2 μ
11
0.5 μ
12

fente très souvent inférieure à 150 Å de large. A l'analyse, ces accolements révèlent trois types de connexions entre cellules identiques ou différentes.

On remarque le plus généralement, une *simple apposition des membranes* (Fig. 13) sans différenciation locale si ce n'est, parfois, l'organisation à peine marquée des substances de la fente intercellulaire en ponts irréguliers (et sans que de telles images puissent être confondues avec de véritables septa (Locke, 1965)). Ces contacts sont fréquents à tous les niveaux épithéliaux ou mésohilaires. La zone de jonction ainsi délimitée peut revêtir diverses formes: quasi-linéaire (aux bords des endopinacocytes aplatis, ou des lames superficielles des exopinacocytes), plane, ou en indentations dont certaines acquièrent une structure en "bouton-pression" couramment rencontrée au niveau des pinacocytes et des cellules contractiles (Fig. 14).

Le second aspect des contacts intercellulaires révèle une différenciation de la couche interne des membranes plasmatiques apposées (Fig. 15). Des épaississements, ou des dépôts de substance opaque aux électrons, suggèrent la *macula adhaerens* d'un desmosome (Farquhar et Palade, 1963); d'autant plus que chez *Hippospongia*, des tonofilaments cytoplasmiques viennent s'y insérer (Pavans de Ceccatty, 1966a). La présence de ces connexions a été relevée entre les pinacocytes principalement. Quand elles existent, ces *jonctions de type desmosomial* constituent toujours une partie, limitée à un millier d'Å, d'une simple apposition membranaire plus vaste. Souvent, elles encadrent un bref élargissement de la fente intercellulaire.

Enfin, la troisième catégorie de contacts n'apparait qu'après un examen utilisant une résolution élevée et un grandissement électronique direct de l'ordre de 60 à 80 000 diamètres, et portant sur les pinacocytes et les cellules contractiles. En effet, le long des contours suivis par les nombreuses appositions membranaires, il est possible de déceler, de place en place, un bref mais total accolement des membranes plasmatiques en présence (Fig. 16). Ces structurations évoquent les *zonula occludens* (Farquhar et Palade, 1963) mais sous forme de *contacts ponctuels du type nexus*. Toutes nos tentatives en vue de mieux conserver ce type de jonction, dont les images observées pourraient n'être que les vestiges de fusion plus étendue, se sont avérées vaines jusqu'à présent. Nos descriptions restent donc très réservées.

De très nombreux travaux ont été consacrés à la signification

---

(Page ci-contre) FIG. 9. *Verongia*. Cellule contractile (à proximité d'une cellule sphéruleuse, à gauche). FIG. 10. *Haliclona*. Cellules à filaments. FIG. 11. *Hippospongia*. Extrusion d'enclaves par une cellule contractile. FIG. 12. *Hippospongia*. Présence de microfilaments dans un pinacocyte. Contacts sinueux ou en indentation (flèches).

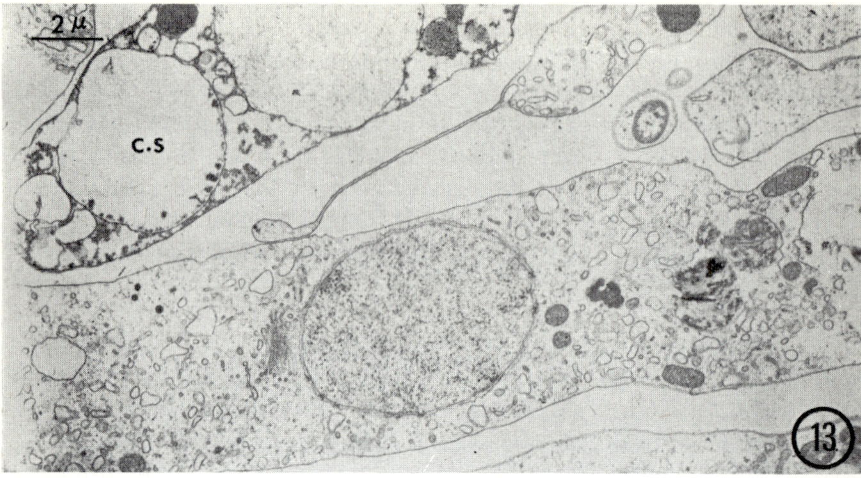
2 μ
C.S
13

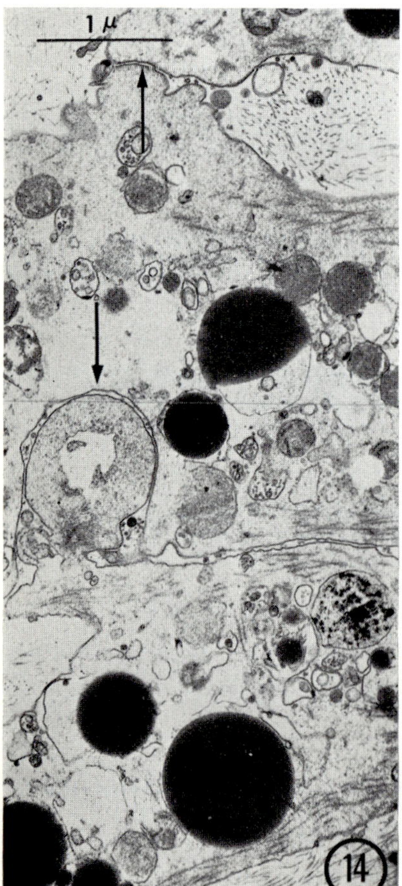
1 μ
14

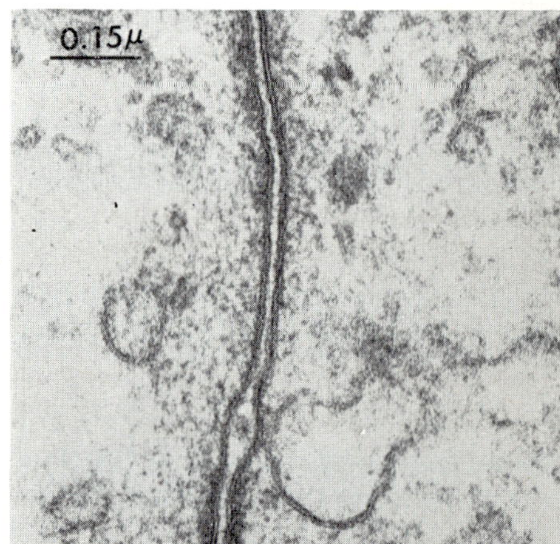
0.15μ
15

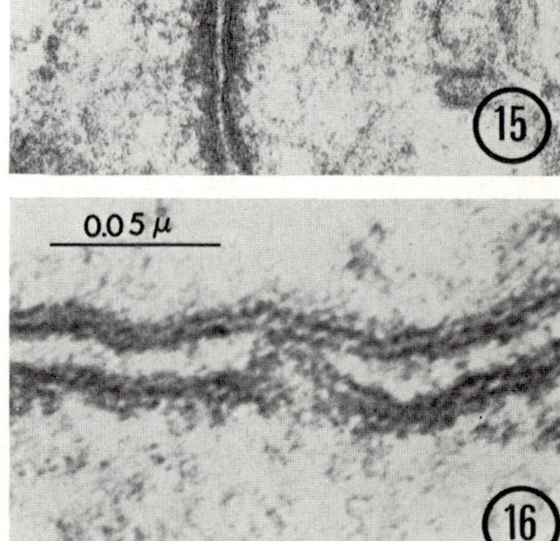
0.05 μ
16

physiologique des diverses formes de contacts intercellulaires chez les organismes animaux (Bennett, 1966; Dewey, 1965; Dewey et Barr, 1964; Kelly et Luft, 1966; Loewenstein, Socolar, Higashino, Kanno et Davidson, 1965; Loewenstein, 1967). En ce qui concerne les Eponges, les observations déjà accomplies ont été interprétées dans certains cas comme des modalités d'attachement mécanique entre cellules (Pavans de Ceccatty, 1966a, Borojević et Lévi, 1967) et, dans d'autres cas, comme des structures d'échange et de transfert (Pavans de Ceccatty, 1966a, b). Mais on doit souligner que, chez les Spongiaires, comme dans tout autre système pluricellulaire, ces deux interprétations ne sont ni contradictoires ni exclusives l'une de l'autre.

En effet, au niveau des connexions décelées dans les oscules, nous avons pu fixer des processus d'extrusion et de micropinocytose qui révèlent à l'évidence des *flux intercellulaires localisés*. Il n'est pas exceptionnel de noter le transfert, d'une cellule à une autre, d'une ou de plusieurs microvésicules (Figs 19 et 21) ou bien d'une substance moins nettement figurée (Figs 18 et 20). De telles observations ont été faites aussi bien sur les jonctions homogènes entre cellules du même type (pinacocytes ou cellules contractiles) que sur les contacts hétérogènes entre cellules de catégories différentes. Chez *Hippospongia*, une incubation de l'oscule, préalablement à sa fixation, dans une solution de réserpine a multiplié ces images d'extrusion microvésiculaire et, par voie de conséquence, de leur transfert par pinocytose au niveau des contacts entre deux cellules (Fig. 21).

Il est probable que ces échanges localisés n'ont pas la même signification physiologique. L'important est leur existence à l'endroit des jonctions intercellulaires.

Les pseudo-desmosomes ou les pseudo-nexus ne sont qu'une partie des surfaces d'apposition simple qui, à elles seules, signalent le contact et ses multiples possibilités fonctionnelles souvent associées les unes aux autres (Loewenstein, 1967). La condition essentielle paraît être un écartement minimal des membranes, conduisant parfois à des structures d'ancrage ou d'oblitération, et permettant des échanges trophiques ou informatifs.

Notre étude dégage la tendance dominante de trois catégories cellulaires à établir des relations par contacts, entre elles et avec les éléments

---

(Page ci-contre) FIG. 13. *Haliclona*. Jonction polarisée, par simple apposition des membranes entre deux cellules indifférenciées. Au dessus, une cellule sphéruleuse (c.s.) FIG. 14. *Euspongia*. Jonctions (flèches) par appositions membranaires en plan, ou en "boutonpression" entre trois cellules contractiles. FIG. 15. *Hippospongia*. Jonction de type desmosome entre deux pinacocytes. FIG. 16. *Hippospongia*. Jonction ponctuelle de type nexus entre deux pinacocytes.

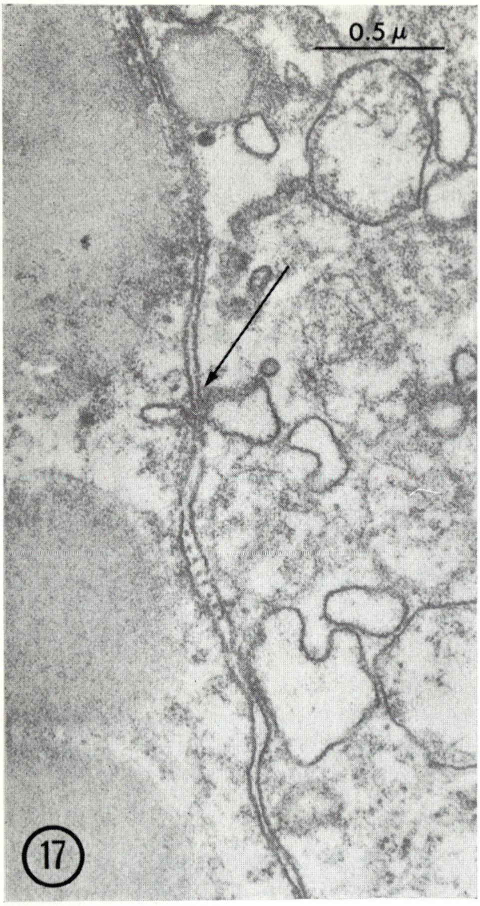
0.5 μ
17

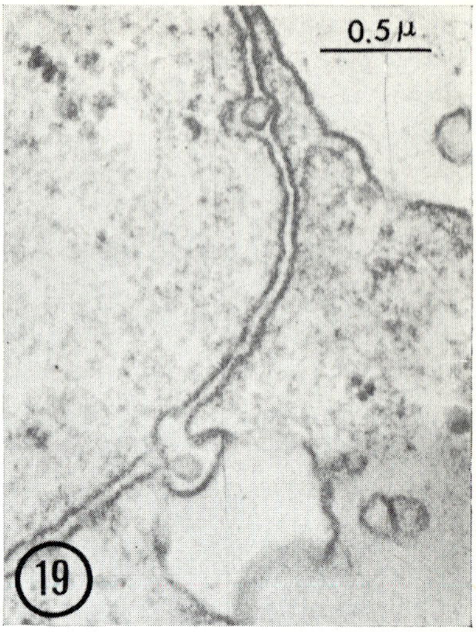
0.5 μ
19

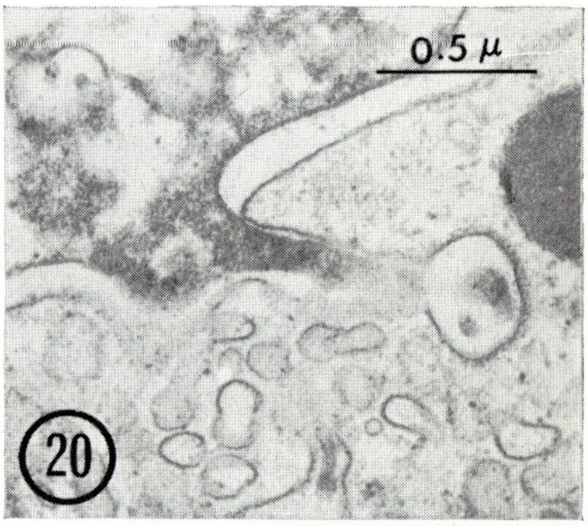
0.5 μ
20

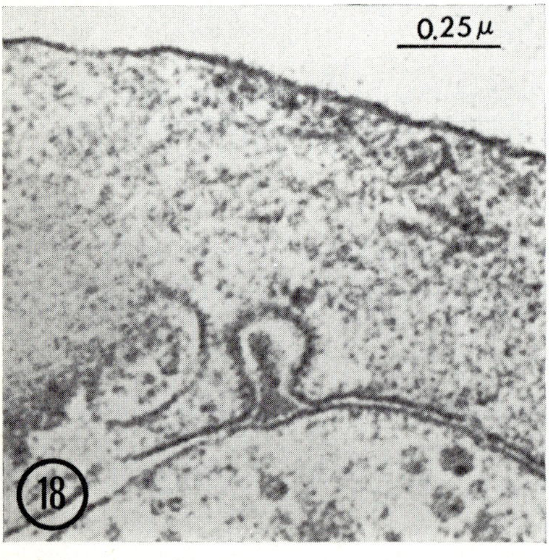
0.25 μ
18

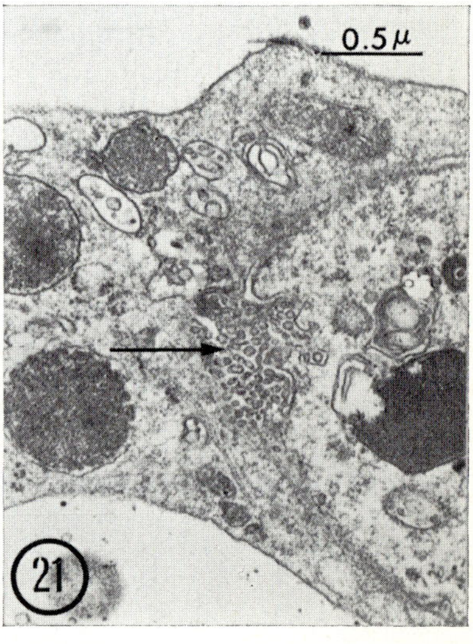
0.5 μ
21

voisins; ce sont les exopinacocytes, les endopinacocytes et les cellules contractiles. Nous considérerons à part les rares scléroblastes osculaires; les cellules sphéruleuses aux jonctions peu fréquentes; et enfin les choanocytes pourtant riches en connexions (Fig. 7) mais seulement présents à la limite des oscules.

En adoptant la terminologie de Willmer (1960) nous pouvons constater que les connexions intercellulaires au niveau des oscules sont essentiellement réalisées par des épithéliocytes dermiques et des mécanocytes dérivés des cellules flagellées embryonnaires. Ces éléments ont conservé ou rétabli leur capacité d'adhésion aux éléments voisins ainsi que le montrent les nombreux contacts établis dans le mésohile par les prolongements des pinacocytes et des cellules contractiles. Si bien que le modèle histologique osculaire révèle un ensemble d'éléments interconnectés constituant deux plans épithéliaux qui enserrent un réseau intermédiaire (Figs 3 et 8).

En ce qui concerne la spécificité de ces cellules, nous rappellerons qu'il faut les saisir "dans une phase morphologique quiescente pour les découvrir, mais les replacer ensuite dans leur dynamique propre pour les comprendre" (Pavans de Ceccatty, 1962). Comme le souligne Willmer: "au cours du déroulement de l'embryogenèse et de l'évolution, des gradients s'établissent presque aussi souvent que des dichotomies". Les cellules obéissent à une différenciation du type "plus ou moins"; et non seulement du type "tout ou rien". Pour nous, il est indiscutable que les Eponges constituent un des systèmes pluricellulaires les plus primitifs qui soit, où ces gradients paraissent de règle. Néanmoins, la différenciation fondamentale entre épithéliocytes, mécanocytes et amoebocytes, "est déjà présente quoique, peut-être, sous une forme plus réversible". Cette réversibilité est particulièrement mise en évidence par les processus régénératifs (Fig. 4) que nous avons pu observer sur *Hippospongia*. Au cours des premiers jours suivant l'amputation, tout le système interconnecté de la zone cicatricielle rompt ses attaches. Les cellules évoluent vers un type amoeboïde commun avant de reconstituer une structure normale.

(Page ci-contre) FIG. 17. *Hippospongia*. Contacts membranaires et échanges vésiculaires (flèches) entre deux cellules contractiles. FIG. 18. *Hippospongia*. Contacts membranaires et échanges de substance entre une cellule du mésenchyme (en bas) et un pinacocyte. FIG. 19. *Hippospongia*. Contacts membranaires et transfert de microvésicules entre deux pinacocytes. FIG. 20. *Haliclona*. Transfert de substance entre une cellule sphéruleuse (en haut, à gauche) et une cellule indifférenciée (en bas, à droite). FIG. 21. *Hippospongia*. Contacts membranaires et transfert de nombreuses microvésicules (flèches) entre un pinacocyte (à gauche) et une cellule du mésenchyme (à droite).

CONCLUSIONS

Le système pluricellulaire des Spongiaires met en évidence l'existence d'un milieu intérieur sous forme d'un mésohile limité par les épithéliums de surface. La polarisation des pinacocytes souligne la différence de phase entre les milieux extérieur et intérieur. Ce dernier, par ailleurs, offre plus d'une analogie avec les espaces conjonctifs lâches des Métazoaires supérieurs, par le développement de son collagène (Gross, Sokal et Rougvie, 1956; Gross et Piez, 1960; Pavans de Ceccatty et Thiney, 1964).

D'autre part, grâce aux relations intimes qui s'établissent entre les éléments épithéliaux et les éléments mésenchymateux réticulés, *les Eponges possèdent des circuits de communications cellulaires continus*, sur de plus ou moins grandes distances, et plus ou moins labiles. Dans la profondeur des tissus, les cellules contractiles—aussi considérées par nous comme de type nerveux (*vide* Pavans de Ceccatty, 1959, 1966b)—jouent un rôle fondamental. Elles participent aux échanges pratiqués directement de cellule à cellule tout au long du réseau.

Dans tous les cas, il est dès à présent possible de connaître les bases structurales des corrélations cellulaires chez les Spongiaires. Pour transmettre les substances et les signaux qui assureront l'intégration de tous les éléments du système, ces organismes primitifs peuvent mettre en jeu aussi bien des mécanismes de diffusion, dans les espaces de leur milieu intérieur, que des mécanismes de transfert le long des circuits constitués par des cellules interconnectées.

REMERCIEMENTS

Ces travaux ont été réalisés au sein de l'Equipe de Recherche Associée au CNRS n° 183—en partie au Laboratoire d'Histologie et au Centre de Microscopie Electronique appliquée à la Biologie, de la Faculté des Sciences de Lyon—et en partie à la Station Zoologique de Villefranche-sur-Mer. Ils ont bénéficié de la collaboration technique de Mme J. Villeneuve, M. P. Coulet et Mlle C. Baleydier. Les auteurs remercient tous ceux qui les ont ainsi aidés.

BIBLIOGRAPHIE

Bagby, R. M. (1966). The fine structure of myocytes in the Sponges *Microciona prolifera* (Ellis and Solander) and *Tedania ignis* (Duchassaing and Michelotti). *J. Morph.* **118**: 167–182.

Bennett, M. V. L. (1966). Physiology of electrotonic junctions. *Ann. N.Y. Acad. Sci.* **137**: 509–539.

Borojević, R. (1966). Étude expérimentale de la différenciation des cellules de l'éponge au cours de son développement. *Devl. Biol.* **14**: 130–153.

Borojević, R. et Lévi, C. (1964). Étude au microscope électronique des cellules de l'éponge: *Ophlitaspongia seriata* (Grant), au cours de la réorganisation après dissociation. *Z. Zellforsch. mikrosk. Anat.* **64**: 708–725.

Borojević, R. et Lévi, P. (1967). Le basopinacoderme de l'éponge *Mycale contarenii* (Martens). Technique d'étude des fibres extracellulaires basales. *J. Microscopie* **6**: 857–862.

Connes, R. (1967). Structure et développement des bourgeons chez l'éponge siliceuse *Tethya lyncurium* Lamarck. Recherches expérimentales et cytologiques. *Archs. Zool. exp. gén.* **108**: 157–195.

Dewey, M. M. (1965). The anatomical basis of propagation in smooth muscle. *Gastroenterology* **4**: 395–402.

Dewey, M. M. et Barr, L. (1964). A study of the structure and distribution of the nexus. *J. Cell Biol.* **23**: 553–585.

Farquhar, M. G. et Palade, G. E. (1963). Junctional complexes in various epithelia. *J. Cell Biol.* **17**: 375–412.

Gross, J. et Piez, K. A. (1960). The nature of collagen. I. Invertebrate collagens in: "*Calcification in Biological Systems*": 395–409. R. F. Sognaes édit. *Amer. Assoc. Adv. Sc.*

Gross, J., Sokal, Z. et Rougvie, M. (1956). Structural and chemical studies on the connective tissue of marine sponges. *J. Histochem. Cytochem.* **4**: 227–246.

Hadži, J. (1949). Problem mezoderma in celoma v luci turbelarijske teorije knidarijev. *Razpr. Razr. prirod. med. slove akad.* znanosti utmetn. Ljublj. 1949 Knjiga IV.

Kelly, D. E. et Luft, J. H. (1966) Fine structure, development and classification of desmosomes and related attachment mechanisms. *Int. Congr. Elec. micr. Kyoto* **6**: 401–402.

Locke, M. (1965). The structure of septate desmosomes. *J. Cell Biol.* **25,** 166–169.

Lowenstein, W. H. (1967). On the genesis of cellular communication. *Devl. Biol.* **15**: 503–520.

Loewenstein, W. H. Socolar, S. J., Higashino, S., Kanno, Y. et Davidson, N. (1965). Intercellular communication: renal, urinary bladder, sensory and salivary gland cells. *Science, N.Y.* **149**: 295–298.

Pavans de Ceccatty, M. (1959). Les structures cellulaires de type nerveux chez *Hippospongia communis* Lmk *Annls Sci. nat.* (Zool), (12) **1**: 105–112.

Pavans de Ceccatty, M. (1962). Système nerveux et intégration chez les Spongiaires. *Annls. Sci. nat.* (Zool.), (12) **4**: 127–137.

Pavans de Ceccatty, M. (1966a). Connexions cellulaires et jonctions polarisées du réseau intramésenchymateux, chez l'Éponge *Hippospongia communis* Lmk *C. r. hebd. Séanc. Acad. Sci. Paris,* **263**: 145–147.

Pavans de Ceccatty, M. (1966b). Ultrastructures et rapports des cellules mésenchymateuses de type nerveux de l'Éponge *Tethya lyncurium. Annls Sci. nat.* (Zool.), (12) **8**: 577–614.

Pavans de Ceccatty, M. et Thiney, Y. (1964). Essai d'interprétation de la fibrogenèse cellulaire du collagène, chez l'Éponge *Tethya lyncurium* LmK. *Vie Milieu,* Suppl. **17**: 129–146.

Tuzet, O. et Pavans de Ceccatty, M. (1955). La mobilisation en amoebocytes des cellules des *Halisarca* (Eponges siliceuses). Les polyblastes chez les Eponges. *C. r. Séanc. Soc. Biol.* **149**: 799–801.

Vacelet, J. (1966). Les cellules contractiles de l'Éponge cornée *Verongia cavernicola* Vacelet. *C. r. hebd. Séanc. Acad. Sci.* Paris **263**: 1330–1332.

Vacelet, J. (1967). Les cellules à inclusions de l'Éponge cornée *Verongia cavernicola* Vacelet. *J. Microscopie* **6**: 237–240.

Willmer, E. N. (1960). *Cytology and evolution*. New York and London: Academic Press.

Symp. zool. Soc. Lond. (1970) No. 25, 467–490.

DIFFÉRENCIATION CELLULAIRE DANS L'EMBRYOGENÈSE ET LA MORPHOGENÈSE CHEZ LES SPONGIAIRES*

RADOVAN BOROJEVIĆ

Laboratoire de Biologie des Invertébrés Marins, Muséum National d'Histoire Naturelle, Paris, France

SYNOPSIS

A comparative study of cellular differentiation during embryogenesis and morphogenesis in sponges reveals three quite distinct types of development.

1. Sponges with blastula larvae. Cleavage gives rise to a coeloblastula of equipotent blastomeres. At metamorphosis the first cellular differentiations, and separation of flagellate and amoeboid cell stocks, occur posteriorly and are controlled solely by the positions of the cells in the pupa. To this group belong sponges of the Calcarea Calcinea, *Polymastia* and *Raspailia* and *Tetilla*.

2. Sponges with parenchymella and amphiblastula larvae of the homosclerophorid type. Cleavage gives rise to a stereoblastula. Cellular differentiation takes place during embryogenesis. The principal types of cells of the young sponge are present in the larva. Metamorphosis is a reorganization of cells, known as the "renversement des feuillets". To the group with parenchymella larvae belong the Ceractinomorpha, the Hexactinellida and some Tectractinomorpha (*Tethya*). To the group with amphiblastula larvae belong the Homosclerophorida.

3. Sponges with amphiblastula larvae of the calcaronean type. The larva is derived from a distinctly polarized egg with definite presumptive regions. The posterior part of the larva corresponds to that pole of the egg designated by the penetration of the spermatozoan. Metamorphosis is accomplished by the invagination of the anterior pole region of the larva.

These three types of development are clearly distinct and, while it is possible to derive the development of the *Polymastia-Raspailia* and the *Tetilla* type from the parenchymella type (via, for example, the type of development found in *Halisarca*) by simplification and neotenic metamorphosis, the two types of development occurring in the Calcarea are so different that they must be considered as evolutionarily distinct from all developmental sequences in Demospongia.

The development of sponges is thus seen as a sequence of definite differentiations occurring at different times in embryogenesis and morphogenesis. The separation of flagellate and amoeboid stocks is the first sign of cellular differentiation. The amoeboid stock next divides into pinacocyte-collencyte and archaeocyte lines. Finally, the highly specialized cells are derived from archaeocytes. Amongst the majority of Demospongia, these three principal stocks are clearly distinct. Amongst the Calcarea, the distinction between the archaeocyte and the collencyte-pinacocyte stocks is not so clear.

INTRODUCTION

L'intérêt particulier que les zoologistes ont porté à l'embryologie des Spongiaires a deux raisons: son étude devait permettre de comprendre leurs relations avec les autres Métazoaires, et d'autre part, elle devait

* Cet article résume le travail presenté pour obtenir le grade de Docteur ès Sciences, Thèse n° AO. 2877, Faculté des Sciences, Université de Paris, 1968.

apporter des précisions sur les rapports entre les Métazoaires et les Protistes.

Les remarquables travaux de Schulze et Metschnikoff et ensuite de Delage, Maas et Minchin, ont démontrés que la distribution des régions différenciées de la larve des Eponges est inversée par rapport aux autres Métazoaires, et notamment que les choancytes, toujours comparés aux "cellules gastriques" ou à "l'endoderme", dérivaient de la région antérieure de la larve. Deux opinions s'en sont dégagées: l'une, défendue par Delage et Maas et plus récemment par Lévi, souligne cette inversion de l'embryon des Spongiaires, considérés comme des "Enantiozoaires"; l'autre, émise notamment par Brien, Duboscq et Tuzet, rapproche les Éponges des Eumétazoaires en soulignant les analogies de leurs gamétogénèse, de leur différenciation et de leurs modes de gastrulation.

Les Spongiaires comptent parmi les Métazoaires les plus simples, et nombreux sont les chercheurs qui les ont envisagés comme des colonies de Protistes. Leur embryologie paraissait particulièrement intéressante, car depuis les grands travaux d'embryologie descriptive de la fin du siècle dernier, il a été admis que les stades embryonnaires pouvaient donner des informations sur la genèse du phylum. L'étude de l'ontogénie des Eponges devait donc donner des précisions sur la passionante question du passage de l'état unicellulaire à l'organisation multicellulaire. Metschnikoff (1886) et Willmer (1960), ont discuté le développement des Spongiaires dans cette optique.

Nous avons remarqué que le développement de différents Spongiaires était plus diversifié qu'on ne le considère généralement. Nous avons donc voulu l'analyser dans l'optique des récents travaux sur la différenciation cellulaire des invertébrés et des vertébrés, afin de distinguer les principes de cette différenciation propres aux Éponges de ceux qui, d'autre part, sont communs à tous les animaux.

ANALYSE DU DÉVELOPPEMENT

Éponges à larve blastula

La diversité du développement des Spongiaires a été souvent soulignée; néanmoins, leurs larves pouvaient être comprises soit comme des parenchymellas, soit comme des amphiblastulas (cf. Brien, 1967a).

Nous avons eu l'occasion de constater que dans deux groupes très éloignés parmi les Spongiaires: les Calcinea et certaines Tetractinomorpha (*Polymastia* Bowerbank, 1866 et *Raspailia* Nardo, 1833) la larve est nettement différente. C'est une blastula, constituée de cellules équipotentes. (Nous rattachons ici *Tetilla* O. Schmidt, 1868, qui n'a pas de stade libre dans son ontogénie, mais dont la différenciation cellulaire est analogue à celle des Éponges à larve blastula). La différenciation des principales lignées cellulaires est, chez ces Éponges, postérieure à la fixation de la larve sur le support. La larve représente en réalité un stock de cellules indifférenciées dont le développement rappelle plus l'évolution d'une gemmule que la métamorphose d'une larve du type amphiblastula ou parenchymella.

Éponges calcaires calcinéennes

Chez les Calcinea, la segmentation donne naissance à une coeloblastula, dont les divisions cellulaires sont toutes tangentielles. A l'éclosion, la larve est une blastula constituée d'une couche régulière de blastomères flagellés. Chez certaines Calcinea, les larves possédent au pôle postérieur deux ou plusieurs cellules qui ont l'aspect de blastomères retardés, dépourvus de flagelles. Ce cas n'est pourtant pas un phénomène général, car ces cellules peuvent manquer ou être présentes en nombre variable chez différents individus (Hadži, 1916). On a attaché une grande importance à ces cellules, car Minchin (1900) les a décrites comme des archaeocytes qui donnent au cours du développement les petits amoebocytes hyalins dont dérivent les cellules reproductrices. Elles assureraient ainsi la continuité de la lignée germinale, en s'opposant en tant que "cellules primordiales" aux choanocytes et pinacocytes (Minchin, 1900). Cette opposition de deux lignées principales de cellules dès le début du développement permettrait d'établir une homologie entre ces larves et les amphiblastulas de type calcaronéen. Rien ne nous permet pourtant de confirmer cette opinion. Comme Hadži (1916) l'a déja souligné, la présence de ces cellules est fort instable; elles dégénérent et disparaissent d'ailleurs avant la métamorphose de la larve. D'autre part, la facilité avec laquelle la partie antérieure de la larve, séparée de la partie postérieure, se transforme en éponge complète, malgré l'absence des cellules postérieures non flagellées (Borojević, 1969a), indique que ces cellules ne représentent pas une lignée cellulaire comparable aux cellules postérieures des amphiblastulas de Calcaronea qui, elles, sont indispensables à la formation d'une éponge fonctionnelle (Maas, 1906).

Lors de la vie libre de la larve un petit nombre de cellules apparaît généralement dans la région postérieure du blastocoele. Ce sont des cellules de la paroi larvaire qui immigrent, perdent leur asymétrie, résorbent leurs flagelles, mais gardent néanmoins une morphologie de cellule larvaire. Cette immigration fut intérprétée depuis les travaux de Minchin (1896, 1900) comme la séparation de la lignée des choanocytes de celle des cellules amoeboides: pinacocytes et porocytes.

Il faut distinguer en réalité deux stades dans l'immigration des cellules larvaires. Le premier est une immigration individuelle de cellules localisée surtout dans la région postérieure de la larve. Elle est très lente; chez *Ascandra falcata* Haeckel, 1872, par exemple, à la fin de la troisième journée de la vie libre larvaire, la plus grande partie du blastocoele reste encore libre. Les cellules immigrées forment un petit groupe au pôle postérieur, et leur développement semble être momentanément arrêté. Mais, dès le quatrième jour, l'immigration prend l'aspect d'une véritable crise qui provoque très rapidement un effondrement et une désorganisation de toute la paroi larvaire. Elle débute au niveau d'un anneau dans la région équatoriale de la larve et se propage d'abord vers le pôle postérieur. Les cellules déjà immigrées sont englobées dans le mouvement général et ne se distinguent plus. Le blastocoele est bientôt comblé et la larve se transforme en une petite sphère dense, une pupa, qui se pose sur le fond et commence à s'étaler.

Nous tenons à souligner que la première phase d'immigration au pôle postérieur représente un phénomène passager sans conséquence sur le développement ultérieur. La deuxième phase n'est pas une véritable immigration, mais plutôt une réorganisation du matériel cellulaire. Elle n'a ni la signification d'une séparation des lignées cellulaires, ni celle de la mise en place de régions différenciées, car la pupa ainsi constituée est toujours composée de cellules morphologiquement identiques et équipotentes. A tout point de vue, cette pupa est identique à une diamorphie constituée à partir de cellules de la blastula dissociées (cf. Borojević, 1969a). La différenciation de ces cellules sera régie uniquement par des variations de conditions dues à la position de chaque cellule dans la pupa.

Après l'étalement de la pupa, des lacunes se forment et isolent d'abord la couche externe de cellules, puis apparaît une cavité centrale autour de laquelle s'organisera le jeune olynthus. Les cellules externes de la pupa se transforment progressivement en exopinacocytes: cette détermination sera fonction de leur localisation dans la pupa. Parmi les cellules sous-jacentes à la couche externe se différencient les premiers scléroblastes, tandis que la masse centrale donne naissance aux choanoblastes. A partir de ce stade, la pupa se transforme en jeune olynthus selon la voie décrite par Minchin (1900).

La différenciation cellulaire dépend donc pour chaque cellule de la période de son immigration et de déplacements cellulaires qui se produisent lors de l'étalement de la pupa. La destinée la plus probable des dernières cellules qui restent à la surface de la pupa est de devenir des pinacocytes, mais à ce stade leur sort peut encore être modifié si elles sont recouvertes par d'autres cellules, lorsqu'une larve par exemple vient se poser sur la jeune pupa, ce qui est assez courant (cf. Borojević, 1969a).

Polymastia *et* Raspailia

Les *Polymastia* et *Raspailia* sont ovipares. Les oeufs adhèrent sur le support durant la période de segmentation, englobés dans un mucus épais secrété par l'éponge-mère. Le clivage total et égal donne naissance à une morula qui se transforme en une blastula aplatie, dont le blastocoele est très réduit en raison de la forme "en semelle" de la larve. Les blastomères sont flagellés, la larve est benthique et rampante. Elle reste monoblastique jusqu'à la métamorphose. Après une longue période de vie libre (20 jours pour *Polymastia*), les larves se fixent sur le support; dans la pupa, constituée de cellules équipotentes, les cellules externes se différencient progressivement et sur place en exo- et basopinacocytes et les cellules internes en choanoblastes et cellules mésohylaires (cf. Lévi, 1956; Borojević, 1967).

Le mode de développement de ces éponges est ainsi analogue à celui des Calcinea, la seule différence importante réside dans l'absence d'une immigration cellulaire dans le blastocoele avant la fixation de la larve, ce qui s'explique sans doute par la forte réduction du blastocoele.

Tetilla

Tetilla est la seule éponge actuellement connue qui ne possède pas de stade libre. Ses oeufs, et plus tard ses blastomères, se fixent sur le support par des longs pseudopodes; les basopinacocytes semblent être les premières cellules différenciées, tandis que les choanocytes se forment seulement à la fin de la morphogénèse de la jeune éponge (cf. Watanabe, 1957, 1960). Ceci nous invite à rattacher *Tetilla* aux éponges à larve blastula, car le principe de la différenciation cellulaire est analogue, modifié uniquement par la fixation très précoce sur le support. Cette tendance se manifestait déjà chez *Polymastia*, dont l'oeuf est immobilisé par le mucus lors de la segmentation et dont la larve est peu mobile.

Éponges à larve parenchymella et amphiblastula (*type Homosclerophorida*)

La larve de la majorité des Spongiaires est une parenchymella; nous lui rattacherons l'amphiblastula des Homosclerophorida. La dissemblance entre ces deux types de larves réside essentiellement dans le fait que les deux régions différenciées sont emboîtées chez les parenchymella et placées bout à bout chez les amphiblastula, mais le développement de toutes ces larves peut être caractérisé par les trois points suivants:

1. La segmentation donne naissance à une sterroblastula.

2. La différenciation cellulaire y est très précoce. Elle est accompagnée par des migrations cellulaires lors desquelles les micromères

(futures cellules flagellées) occupent la périphérie et la partie antérieure de l'embryon, tandis que les macromères et leurs dérivés se regroupent à son intérieur et à son pôle postérieur.

3. A l'éclosion, la larve posséde les principales catégories cellulaires de l'éponge adulte; la métamorphose n'est donc qu'un réarrangement du matériel cellulaire présent dans la larve.

Parenchymella (*Ceractinomorpha, certaines Clavaxinellida* (Tethya), *Hexactinellida*)

Bien que toutes les parenchymellas soient caractérisées par les trois points mentionnés plus haut, il existe entre les formes primitives et les formes évoluées des différences qui portent essentiellement sur la durée des phases successives du développement.

L'embryogenèse de *Halisarca* Dujardin, 1838 peut être considérée comme particulièrement primitive (Lévi, 1956). La segmentation y est régulière et la morula est isoblastomérique. Une partie des blastomères s'insére ensuite à l'intérieur de l'embryon; les cellules externes se multiplient et se différencient en cellules flagellées, tandis que la masse interne évolue peu. Au moment de l'éclosion, les cellules internes se condensent au pôle postérieur; elles formeront le pinacoderme et le mésohyle lors de la métamorphose.

L'évolution à partir de ce type de développement est marquée par la différenciation cellulaire de plus en plus précoce et de plus en plus importante au cours de l'embryogenèse. Les divisions de segmentation deviennent inégales, la sterroblastula est composée de micro- et de macromères. La première période de l'histogenèse larvaire est caractérisée par la migration des micromères vers la périphérie et leur différenciation en cellules flagellées. Il s'agit là d'une véritable différenciation car ces cellules perdent leurs réserves nutritives et leurs capacités morphogénétiques deviennent limitées (Borojević, 1966). Les macromères se différencient plus lentement, et donnent naissance aux cellules mésohylaires et aux pinacocytes. Ces cellules se regroupent vers le pôle postérieur de la larve, et ce mouvement est bien visible grâce au regroupement des sclérocytes initialement dispersés dans l'embryon. Certains macromères conservent cependant leur aspect embryonnaire: après la métamorphose ils constitueront les archaeocytes. L'apparition des micromères est donc le premier signe de la séparation des principales lignées cellulaires. Leur mode d'apparition est cependant variable. Chez *Haliclona* Grant, 1841 et *Adocia* Gray, 1867, par exemple, les micromères apparaissent dans tout l'embryon, puis migrent vers sa périphérie (Meewis, 1939, 1941). Chez *Halichondria* Fleming, 1828 (Meewis, 1941) et chez *Potamolepis* Marshall, 1883 (Brien, 1967b), par contre, ils apparaissent directement à la périphérie de l'embryon. Ce site d'apparition des micromères est intéressant, car, s'ils se forment dans tout l'embryon, leur différenciation doit être imposée par les propriétés intimes des blastomères, qui, déjà engagés sur la voie de différenciation, vont migrer pour occuper la place qui leur convient. Si par contre la différenciation a lieu uniquement à la périphérie, elle pourrait être expliquée par des variations locales des conditions métaboliques (Brien 1967b). Les deux systèmes semblent exister chez les Eponges, et les particularités du développement d'un cas précis fait qu'un de ces systèmes sera prépondérant.

La différenciation cellulaire peut être très poussée. Chez les Hexactinellida, par exemple, la larve posséde un squelette particulier très élaboré (Okada, 1928). Chez les Spongillidae, les archaeocytes se différencient en choanoblastes et en choanocytes encore dans le mésohyle larvaire : la larve est en réalité une jeune éponge complète, entourée d'un épithélium flagellé qui a ici la signification d'un organe larvaire, disparaissant à la métamorphose.

La larve de *Tethya* Lamarck, 1815 mérite une attention toute particulière, car c'est la seule Clavaxinellida qui, d'après nos connaissances actuelles, posséde une larve parenchymella. Son embryogenèse est très semblable à celle de *Halisarca*, mais le stade blastula est encore plus régulier offrant une analogie avec la formation d'une blastula du type de *Polymastia*. La différenciation de cellules flagellées s'effectue à la fin de l'embryogenèse et malgré une vie larvaire prolongée, la différenciation mésohylaire est très faible. Lévi (1956) a déjà souligné combien il est étonnant qu'une éponge anatomiquement aussi complexe que *Tethya* ait une embryogenèse aussi simple. Nous ajouterons que deux autres Clavaxinellida : *Polymastia* et *Raspailia* possédent une larve blastula, encore plus simple. Il semble donc que si parmi les Ceractinomorpha, partant d'une forme primitive comme *Halisarca*, la complication de la structure de l'éponge adulte est accompagnée par une accélération et une augmentation de la différenciation cellulaire lors de l'embryogenèse, chez les Clavaxinellida, qui forment ainsi une lignée évolutive indépendante, la même complication morphologique s'accompagne d'une simplification du développement et d'un retardement de la différenciation cellulaire.

Amphiblastula (*Homosclerophorida*)

Chez les Homosclerophorida, la segmentation donne naissance à une sterroblastula typique, mais lors du développement ultérieur les cellules externes se divisent et s'organisent en une couche régulière, tandis que les blastomères internes entrent en cytolyse et sont résorbés par les cellules périphériques. Ce processus se termine par la formation d'une larve à blastoderme monostratifié, qui posséde des régions morphogénétiques différenciées nettement séparées. La métamorphose s'effectue par l'invagination du pôle antérieur de la larve dans le pôle postérieur. La larve se fixe par l'orifice d'invagination et l'oscule est une néoformation. Les cellules de la partie antérieure de la larve appartiennent à la lignée flagellée: elle donnent les choanocytes lors de la métamorphose. Les cellules postérieures, bien que munies de flagelles, ont la signification de la lignée amoeboide, car elles donnent naissance aux pinacocytes de l'éponge adulte (Meewis, 1938).

Comme Maas (1898) l'a indiqué, l'orientation et le développement des amphiblastulas des Homosclerophorida sont comparables à ceux des parenchymellas, ce qui nous permet d'établir une homologie entre ces larves. Chez les parenchymellas pourtant, toutes les cellules participent à l'histogenèse larvaire, tandis que les amphiblastula des Homosclerophorida sont caractérisées par le phénomène de lyse de blastomères internes. Ce phénomène est unique parmi les Spongiaires, mais analogue au processus de la formation de la cavité centrale chez les gastrulas de certains Hydraires (*Hydractinia*, cf. Van de Vyver, 1968a). La cavité centrale des larves des Homosclerophorida est donc une cavité secondaire et non un blastocoele, ce qui nous interdit de considérer cette larve comme primitive, dont les parenchymellas dériveraient par l'hypertrophie du mésohyle, opinion avancée par Brien (1943, 1967a).

Il nous reste à signaler que d'apres la description de Nassonow (1883), la larve de *Cliona* Grant, 1826 est une amphiblastula. Comme ni son embryogenèse, ni la différenciation de ses cellules ne sont connues, il nous est impossible d'établir ses relations avec les autres larves des Spongiaires.

Éponges à larve amphiblastula (*type Calcaronea*)

L'oeuf des Éponges Calcaires Calcaronéennes est polarisé dès le moment où l'ovocyte prend contact avec le système charriant chargé du spermiokyste. La segmentation est régulière: elle se termine par la formation d'une coeloblastula dans laquelle deux types cellulaires se distinguent de plus en plus nettement. Les cellules de la lignée flagellée forment leurs flagelles à l'intérieur du blastocoele, tandis que les cellules amoeboides, en nombre nettement inférieur, s'écartent et délimitent une ouverture dans le blastoderme: dès ce stade l'embryon est une stomoblastula. La phase suivante est caractérisée par le retournement de l'embryon, comparable au même processus dans le développement des Volvocales. Le pôle flagellé de l'embryon passe par l'ouverture de la stomoblastula et tout l'embryon se retourne en doigt de gant: l'axe antéropostérieur initial et l'orientation des cellules dans le blastoderme sont ainsi renversés de 180°. Le pôle antérieur de la larve amphiblastula est donc flagellé, son pôle postérieur est constitué de cellules de la lignée amoeboide; elle posséde une cavité centrale secondaire, plus ou moins remplie de cellules nourricières d'origine maternelle (Vacelet, 1964).

Pendant la vie libre de l'amphiblastula le nombre de cellules flagellées reste constant, tandis que les cellules postérieures continuent à se diviser, contournant et recouvrant progressivement la région antérieure de la larve. Ce processus prend l'aspect d'une véritable invagination du pôle antérieur lors de la métamorphose. Les cellules flagellées larvaires se transforment directement en choanocytes, tandis que les cellules amoeboides postérieures donnent les futures pinacocytes, les porocytes et les cellules mésohylaires.

L'oeuf des Calcaronea posséde donc des régions présomptives définies et, quoi que les premières divisions soient souvent égales, le sort de chaque cellule est déterminé. Ceci est particulièrement net pour les cellules en croix qui dérivent de plages cytoplasmiques aisément repérables par les colorations histologiques dès le début du développement embryonnaire. Le pôle de l'oeuf marqué par la pénétration du spermiokyste donne ainsi les cellules de la lignée amoeboide de la jeune éponge, tandis que le pôle opposé est à l'origine de cellules flagellées larvaires, dont dérivent les choanocytes.

La différenciation cellulaire est extrêmement précoce dans le cas des amphiblastulas des Calcaronea. Notons que les cellules flagellées sont les premières à se différencier comme c'est le cas dans l'embryogénèse des parenchymellas, mais que leur nombre se limite rapidement. La métamorphose correspond alors au moment où les cellules amoeboides sont suffisamment nombreuses pour manifester leur dominance morphogénétique.

Les cellules flagellées de l'amphiblastula se comportent dès le début du développement comme des cellules différenciées; elles ne sont pas comparables aux blastomères flagellés des Calcinea. Tandis que toutes les cellules d'une larve calcinéenne, notamment celles de la partie antérieure, sont capables de reconstituer une éponge adulte (Borojević, 1968b), les cellules flagellées d'une amphiblastula n'ont pas ces capacités: séparées des cellules amoeboides, elles périssent (Maas, 1906) tout comme les choanocytes de l'éponge adulte (Huxley, 1911). Les cellules postérieures de la larve amphiblastula au contraire sont capables de régénérer une éponge fonctionnelle. Elles sont donc totipotentes, et conjointement à la morphologie, elles conservent les capacités embryonnaires. Par leur

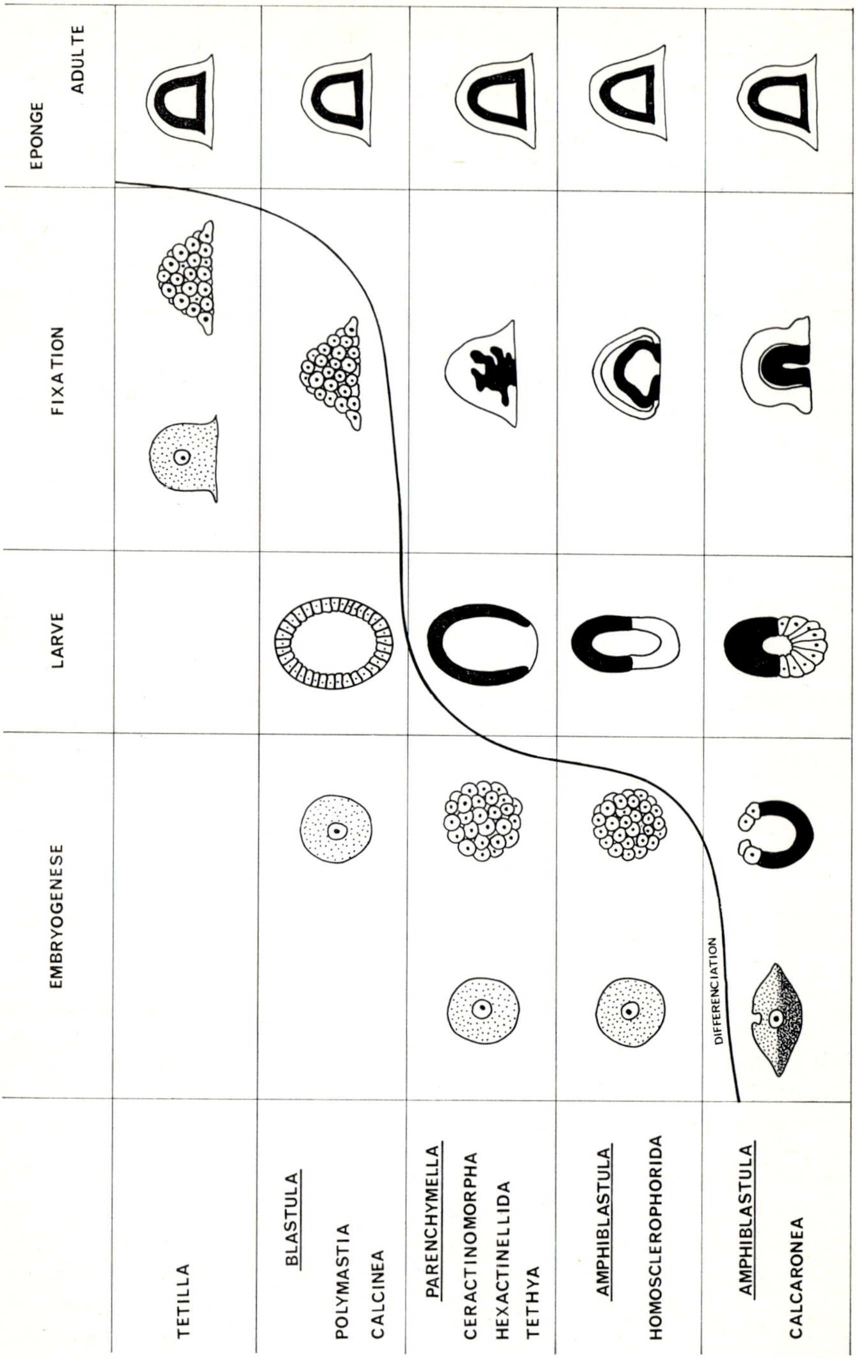

FIG. 1. Tableau recapitulatif des différents stades du développement chez les Spongiaires.

activité morphogénétique des pinaco- et des mécanocytes, elles assument le rôle de la lignée des collencytes; représentant en même temps une réserve de cellules indifférenciées elles assument également celle des archaeocytes.

DISCUSSION

Le développement des Spongiaires apparaît fort hétérogène. En réalité, cette hétérogénéité s'amenuise si on considère que le développement se résume en une série de transformations de blastomères en cellules différenciées telles qu'elles le sont chez l'adulte.

La différenciation entre les micromères (qui engendrent la lignée flagellée) et les macromères (point de départ de la lignée amoeboide) est la plus importante et la seule qui soit nette chez toutes les Éponges. Les changements de leur position relative vont provoquer, chaque fois que le déroulement du développement l'exige, des migrations, parfois très complexes. La première migration a lieu lors de la formation de la larve et correspond à une mise en place des micromères et à leur différenciation en un épithélium flagellé larvaire. La deuxième se produit lors de la métamorphose, quand la lignée flagellée passe à l'intérieur de l'éponge pour y former le choanoderme. Dans les deux cas, il s'agit d'une mise en place de cellules déjà engagées sur une voie de différenciation précise, et cette continuité de la lignée flagellée, qui constitue en réalité une réutilisation de matériel cellulaire différencié, est un phénomène particulier et caractéristique du développement des Spongiaires. C'est pourquoi les migrations cellulaires lors de développement embryonnaire chez les Spongiaires sont comparables aux migrations cellulaires et aux mouvements morphogénétiques observés lors de la reconstitution à partir de cellules d'éponge adulte dissociées, lors de processus involutifs et même lors des processus de croissance et de déplacement massif de certaines éponges. (Borojević, 1969b.)

Comme la fixation de la larve sur le support, c'est à-dire le passage de la vie libre à la vie sédentaire, coïncide avec la réorientation principale de la stéréotopie de l'éponge, le degré de différenciation cellulaire à ce moment est très important.

Si on considère l'ensemble des Spongiaires, on peut y découvrir une série complète de degrés de différenciation cellulaire au moment de la métamorphose; le phénomène de "renversement des feuillets", considéré comme le point cardinal dans l'ontogénie des Spongiaires n'est pas général. Le développement de *Tetilla* est un cas extrême, l'oeuf se fixant sur le support avant même le début de la segmentation. Chez les éponges à larves blastula, le clivage s'opére pendant la vie libre de l'embryon et la larve se fixe au stade de coeloblastula. La morphogénèse s'effectue sans "renversement des feuillets", la différenciation étant postérieure à la fixation. La majorité des Spongiaires possède une larve du type parenchymella ou amphiblastula; leur différenciation primaire se termine lors de l'histogenèse larvaire. Il s'est donc produit lors de la fixation une réorganisation complète du matériel cellulaire, appelée "renversement des feuillets". Enfin, chez les Spongillidae, où la différenciation cellulaire à l'intérieur de la larve est complète, les chambres choanocytaires apparaissent dans son mésohyle. L'épithélium périphérique est réduit à un organe larvaire qui disparaît lors de la métamorphose: il n'est plus possible, dans ce cas là, de parler de "renversement des feuillets".

Malgré ces différences dans la chronologie du développement, la morphogenèse d'un stade donné est toujours fonction d'un état d'équilibre entre les principales catégories cellulaires et nous avons déjà exprimé notre conviction (Borojević et Lévi, 1965) que toute la morphogénèse, larvaire et post-larvaire de l'éponge, est sous la double dépendance compétitive et complémentaire des deux catégories cellulaires: les cellules de la lignée flagellée et les cellules de la lignée amoeboide.

Lignée flagellée

Nous tenons à souligner que l'existence de flagelles ne constitue pas en soi un caractère de différenciation: il convient donc de distinguer les blastomères flagellés des cellules de la lignée flagellée. Comme la plupart des blastulas, celles des Spongiaires sont flagellées. La présence de flagelles n'implique pourtant pas une différenciation car leur totipotence est encore intégrale. Ceci est particulièrement net chez les Calcinea où toutes les cellules larvaires résorbent leurs flagelles avant la formation de la pupa, encore constituée de cellules équipotentes. Par contre, la formation de flagelles dans les cellules de l'épithélium larvaire des parenchymella va de pair avec leur différenciation. Ces cellules conserveront leur flagelle pendant toute leur vie, et, bien qu'elles soient capables de le résorber lors de la métamorphose ou de la reconstitution, elles sont flagellées à l'état fonctionnel.

Les cellules de la lignée flagellée sont donc les cellules typiques du kynoblaste (Metschnikoff, 1886), elles correspondent aux épithéliocytes de Willmer (1960), et sont caractérisées par le fait qu'elles sont en équilibre physiologique uniquement lorsqu'elles sont incluses dans un épithélium. Il est dès lors possible de les cultiver, soit à partir de la région antérieure de la larve amphiblastula (*Sycon* Risso, 1826: Maas, 1906, *Oscarella* Vosmaer, 1887: Lévi, 1956) soit à partir de fragments du choanoderme (*Sycon* Risso, 1826: Huxley, 1911; de Beer, 1922). Dans ce cas, elles forment des sphères creuses, flagellées à la surface externe. Ces sphères sont stables, elles ne manifestent aucune évolution. Elles simulent une blastula, sans en avoir les capacités morphogénétiques, ce qui montre une fois de plus que la morphogenèse de l'éponge résulte de la collaboration entre deux systémes morphogénétiques complémentaires. Elles représentent néanmoins (comme des structures semblables obtenues à partir des épithéliums des Polychètes (Thouveny, 1965) un modéle intéressant pour l'étude de l'état et de la stabilité de leur différenciation.

Les cellules de la lignée flagellée sont en équilibre lorsqu'elles séparent le milieu extérieur du milieu interne qui contient dans les conditions naturelles d'autres cellules, mais qui peut en être dépourvu dans les conditions expérimentales. Par contre, elles ne peuvent vivre à l'intérieur de l'éponge, sauf lors de déplacements morphogénétiques de durée limitée. Si elles ne sont pas rapidement incluses dans les chambres choanocytaires, elles sont phagocytées par des archaeocytes (Borojević et Lévi, 1965), ou à défaut de ceux-ci, meurent d'elles-mêmes, comme c'est le cas dans les cultures des cellules périphériques isolées d'une larve parenchymella (cf. Borojević, 1966). Ce comportement se retrouve dans toutes les cultures de cellules de la lignée flagellée (Maas, 1906; Huxley, 1911; Borojević, 1963).

Il est peu probable que les cellules flagellées soient capables de se redifférencier en cellules de la lignée amoeboide. Cette transformation a été décrite chez les éponges calcaires et chez *Halisarca* (Dendy, 1914; Jörgensen, 1917; Gatenby, 1920; Duboscq et Tuzet, 1937; Tuzet et Pavans de Ceccatty, 1955), mais ces observations devraient être revues et confirmées par l'expérimentation.

En tant que cellules différenciées, elles peuvent être utilisées par l'éponge comme matière de réserve, grâce à la capacité de phagocytose des archaeocytes; elles servent notamment de cellules nourricières lors de la croissance de l'ovocyte (Vacelet, 1964; Borojević, 1969a). Dans des conditions défavorables, naturelles (Weltner, 1893) ou artificielles (Fauré-Fremiet, 1932; Brien, 1937), lors de phénomènes d'involution ou de reconstitution, les choanocytes sont les premières cellules sacrifiées. Ceci pourrait expliquer le fait que les doses sublétales de rayons U.V. détruisent sélectivement les choanocytes (Fauré-Fremiet, 1932).

Lignée amoeboide

Tandis que la lignée flagellée est représentée chez toutes les Éponges adultes uniquement par les choanocytes, la lignée amoeboide est subdivisée chez la majorité des Spongiaires en pinacocytes-collencytes, archaeocytes, cellules spécialisées. Chez les formes les plus simples,

Halisarca, Éponges calcaires, les lignées des archaeocytes et des pinacocytes-collencytes ne sont pas séparées; les archaeocytes typiques apparaissent uniquement lors des phénomènes d'involution (Maas, 1910; Lévi, 1956). Chez la majorité des Spongiaires, les mécanocytes (pinacocytes, collencytes, sclérocytes) se différencient à partir des macromères lors de l'histogenèse larvaire; les archaeocytes sont en réalité les dernières macromères non différenciés.

Les relations entre les cellules sont ainsi plus complexes chez les Éponges à organisation simple car les cellules de la lignée amoeboide ont des capacités multiples, dont certaines prédominent à un moment précis de la vie. Cette prédominance déterminera leur morphologie et elles peuvent avoir successivement des caractères de pinacocytes, de porocytes ou d'archaeocytes typiques (cf. Maas, 1910; Borojević, 1969a). En quelle mesure une cellule est capable de prendre plusieurs de ces formes, l'une après l'autre, reste encore à être précisé. La différenciation des pinacocytes au moins ne semble pas être définitive, car leur transformation en phagocytes est un retour à une forme dont la morphologie est plus proche des cellules embryonnaires. Il reste encore à prouver que cette morphologie correspond effectivement à un état pluripotent. Le cas des phagocytes-cellules-nourricières qui apparaissent lors du développement des *Ascandra* Haeckel, 1872, par exemple, (Borojević, 1968b) n'est pas un cas de dédifférenciation, car ces cellules meurent, leur tâche de cellules nourricières ayant été accomplie. Une étude approfondie des processus d'involution chez les Éponges Calcaires, et notamment d'une éventuelle réorganisation à partir des réducties, devrait donner des informations intéressantes sur ce problème.

Chez les Éponges à mésohyle très différencié, tous les types cellulaires sont continuellement présents, et, d'après les résultats d'expériences effectuées jusqu'alors, la différenciation des cellules mésohylaires et des mécanocytes s'effectue à partir d'archaeocytes. Le retour des cellules différenciées à l'état d'archaeocytes n'a pas encore été prouvé (Borojević 1966).

Archaeocytes

Le rôle des archaeocytes dans une éponge semble être bien défini. D'une part, ils représentent une réserve de cellules totipotentes, d'autre part, ils sont les phagocytes. Par cette double activité, les archaeocytes agissent comme un régulateur de l'équilibre entre les différentes catégories cellulaires: ils épurent l'éponge et suppriment les cellules en excès, ils se différencient ensuite et forment, en réutilisant le matériel cellulaire, les types des cellules qui manquent.

Leur comportement dans l'éponge est typiquement celui d'amoebocytes de Willmer (1960). Ils se déplacent continuellement, et ne participent jamais à la constitution d'une structure stable de l'éponge. En culture, ils sont capables de constituer un revêtement monostratifié d'aspect épithélial, mais ceci s'accompagne en général par leur différenciation en basopinacocytes (Brøndsted, 1953).

Mécanocytes

Le rôle des mécanocytes est d'assurer l'intégrité de l'éponge et la stabilité de ses structures. Ce rôle complexe est assuré par la collaboration des mécanocytes de divers types et de comportements différents.

L'intégrité du milieu interne de l'éponge est préservé par la continuité du pinacoderme. Lors des processus morphogénétiques, le pinacoderme est la seule structure toujours présente, ce qui démontre que la diamorphie aussi bien que l'éponge fonctionnelle est une unité physiologique active qui tend à créer un milieu interne isolé et délimité. La désorganisation du pinacoderme est le premier signe de la dégénérescence irréversible d'une éponge.

La stabilité des structures est assurée d'une part, par l'adhésion du basopinacoderme au support et d'autre part, par la tension interne des structures de l'éponge et par l'édification d'un mésohyle lâche soutenu par le squelette. Tandis que les exopinacocytes manifestent une tendance à s'étaler sur les autres cellules et la majorité des cellules mésohylaires ont un comportement nommé "contact promoting", les basopinacocytes sont plus thigmotactiques vers les surfaces étrangères: ils s'étalent activement sur le support et s'y stabilisent en sécrétant une couche externe de collagène (Borojević et Lévi, 1967). Cette opposition de comportement des basopinacocytes et des autres cellules crée la tension interne qui oriente et organise les structures squelettiques et le système aquifère (Borojević, 1969b). Les collencytes constituent ensuite une trame cellulaire entre les deux surfaces de l'éponge, relativement stable dans son ensemble, bien que ces éléments soient mobiles, soutenus par le squelette rigide constitué par des sclérocytes et cellules sécrétrices de la spongine. Le terme de la stabilité des structures doit être compris dynamiquement car, si l'ensemble de l'éponge offre souvent une remarquable constance de la forme, les cellules y sont toujours en déplacement et même les collencytes se remodèlent et se déplacent continuellement. Les sclérocytes sont également mobiles, transportant les jeunes spicules, et à partir du moment où les spicules sont fixés par la spongine, les sclérocytes dégénèrent.

Différenciation

Nous avons indiqué que la présence des deux lignées cellulaires principales était indispensable à la constitution d'une éponge. Toute l'embryogenèse et la morphogenèse des Spongiaires se résument en une suite de rapports différents entre leurs deux tendances morphogénétiques.

Au cours de l'embryogénèse chez les Éponges, les cellules flagellées sont les premières à se différencier, et chez les Éponges à parenchymella, elles se séparent des archaeocytes et des collencytes centraux par migration centrifuge. Pendant cette période pré-larvaire, les collencytes en cours de différenciation, qui assument le rôle de mécanocytes, ne peuvent encore manifester leur dominance morphogénétique, et c'est l'épithélium flagellé qui représente le seul groupe cellulaire fonctionnel. A la fin de la période larvaire, cette situation est renversée au profit des collencytes, et cette inversion marque le début de la métamorphose. L'épithélium flagellé manifeste alors des signes de régression fonctionnelle, d'où sa fragmentation et sa désintégration une fois les collencytes devenus le groupe cellulaire dominant.

A partir de ce moment, les mécanocytes vont être les seuls chargés du maintien de la forme et de l'intégrité de l'éponge. On sait que Huxley (1911, 1921) a montré qu'une dissociation des cellules de *Sycon* aboutit à la reconstitution d'une éponge, mais qu'une suspension de choanocytes ou une suspension à dominance choanocytaire forme des organismes flagellés du type *Proterospongia*. Nous retrouvons ici l'opposition entre les deux tendances morphogénétiques, et seule la présence des mécanocytes en nombre suffisant peut assurer l'établissement des structures d'une éponge en tant qu'organisme multicellulaire.

Si donc on considère le comportement des principales catégories cellulaires des éponges, on est frappé par le fait que les choanocytes d'une part et les archaeocytes d'autre part se comportent exactement comme un assemblage d'unités indépendantes, et qu'il est difficile de voir une différence entre l'organisation d'une sphère choanocytaire et celle d'une colonie de flagellés, ou entre le comprotement des archaeocytes et celui des rhizopodes. Cette constatation est renforcée par le fait que les choanocytes, aussi bien que les archaeocytes, représentent sans doute des unités métaboliques indépendantes.

Par opposition à ces deux catégories cellulaires, les mécanocytes ont le comportement de cellules de Métazoaires typiques. Elles assurent la stabilité de la structure de toute l'éponge, formant un cadre sur lequel s'appuie le choanoderme et dans lequel circulent les archaeocytes et les cellules spécialisées. D'autre part, elles assurent l'intégrité du milieu interne en couvrant les surfaces libres de l'éponge par les pinacocytes.

Le comportement des choanocytes et des archaeocytes est donc celui de cellules individuelles, celui des mécanocytes est un comportement de cellules faisant partie d'une unité supérieure. Tandis que les choanocytes et les archaeocytes rapprochent les Spongiaires des Protistes, la présence et la différenciation des mécanocytes nous oblige à les considérer comme représentant des Métazoaires. Autrement dit, en comparant l'organisation de l'éponge à celle des Protistes, les différenciations des choanocytes et des archaeocytes sont des modifications de qualités que les Protistes possèdent déjà; la différenciation des mécanocytes est l'acquisition d'une qualité nouvelle, qui va de pair avec la constitution d'une unité supérieure possédant sa structure et son milieu interne défini.

En résumé, la différenciation principale d'une éponge est la séparation de la lignée flagellée et de la lignée amoeboide. Sans cette différenciation, on ne peut pas parler d'éponge, et une blastula ou une sphère choanocytaire ne représentent pas une éponge. Les deux tendances complémentaires doivent coexister pour former un organisme complet.

La lignée amoeboide assume le rôle de mécanocytes et d'amoebocytes tels qu'ils sont conçus par Willmer (1960), et cette double fonction est dans la majorité des cas matérialisée par la séparation des collencytes -pinacocytes des archaeocytes. Nous insistons sur ce fait car la conception classique de la différenciation cellulaire des Spongiaires formulée par Minchin (1900), et reprise par Willmer (1960), diffère de la nôtre. En effet, pour Minchin, les petits amoebocytes mis à part, les cellules mésohylaires et les pinacocytes dériveraient de la lignée flagellée, représentés par les cellules flagellées de la larve. La conception d'un transfert progressif des potentialités des cellules flagellées aux cellules amoeboides, développée surtout par Brien (1943) est aussi basée sur celle de Minchin. Comme nous l'avons expliqué, le rôle des deux lignées principales est également important chez toutes les éponges, le jeu de la chronologie du développement faisant ressortir la capacité morphogénétique de l'une ou de l'autre catégorie cellulaire, plus tôt ou plus tard dans le développement. Les amoebocytes et les mécanocytes ont une orgine commune et les cellules mésohylaires ne dérivent donc pas de deux types de cellules, constituant respectivement les deux régions, antérieure et postérieure de la larve. En ce sens, la différenciation des cellules des Spongiaires n'est pas comparable à celle des autres Métazoaires, chez qui Willmer (1960) a pu distinguer des cellules mésohylaires de deux origines embryonnaires: animale et végétative. Bien que la marche de la différenciation chez les Éponges soit particulière, le résultat final est le même, car les principes de la différenciation cellulaire des Éponges et d'autres Métazoaires sont superposables. Ce fait est particulièrement intéressant, car, comme

Willmer (1967) l'a souligné, les solutions des problèmes vitaux semblent être très peu nombreuses et la différenciation cellulaire en épithéliocytes, mécanocytes et amoebocytes semble être obtenue par évolution parallèle dans les groupes animaux différents.

Polarité des larves

Les larves de toutes les Éponges possèdent un axe antéro-postérieur, indiqué avant tout par le sens de leur nage. L'organisation de la larve autour de cet axe est très variable, et dépend directement du degré de différenciation de la larve.

Les cellules de la larve blastula sont équipotentes. Bien que l'immigration de cellules dans le blastocoele débute chez certaines Calcinea au pôle postérieur, l'axe antéropostérieur larvaire n'a aucune influence sur le développement ultérieur de l'éponge. La larve se transforme en sphérule dense avant la fixation sur le support, et toute trace de l'axe larvaire est ainsi perdue. De même, les larves de *Polymastia* se fixent sur le support par leur côté plat, et, dans les deux cas, la jeune pupa s'organise autour d'un axe perpendiculaire au support, comme c'est le cas dans tous les processus de monophogenèse à partir de diamorphies chez les Spongiaires (Borojević, 1969b). La polarité larvaire est ainsi purement fonctionnelle et concerne surtout la synchronisation et l'orientation de battements des flagelles.

Chez les parenchymella, la polarité est mieux marquée: chez les formes primitives, comme *Halisarca*, elle se définit tard dans l'embryogenèse, mais les premières divisions de clivage sont déja inégales chez les formes évoluées. La majorité des parenchymellas se fixent sur le support par leur pôle antérieur, mais ceci n'est pas sans exceptions, et les intenses déplacements cellulaires dans la pupa et les déplacements de l'ensemble de la pupa sur le support, qui suivent la fixation de la larve, font pratiquement disparaître l'axe primaire de la larve, et l'orientation de la nouvelle éponge est imposée par la position et la configuration du support.

Les amphiblastulas, notamment celles des Calcaronea, ont une polarité très tranchée, définie déjà dans l'oeuf, et comme leur métamorphose s'effectue par invagination du pôle antérieur, dans les cas typiques l'axe de symétrie du jeune olynthus recouvre celui de la larve et de l'oeuf. Il n'est pourtant pas rare que les conditions externes provoquent un ralentissement du stade pupa, pendant lequel les cellules continuent à se déplacer les unes par rapport aux autres, ce qui change profondément l'orientation de la jeune éponge.

La polarité de la larve et de l'éponge adulte ne se recouvrent donc que rarement et très imparfaitement. D'autre part, l'axe de la larve ne concerne souvent que le sens de la nage, et la différenciation du matériel cellulaire n'en dépend pas. L'importance théorique qu'on attribue à l'orientation de l'axe antéropostérieur est sans doute exagérée, et le fait que cet axe des parenchymellas est inversé si on le compare à celui d'autres Métazoaires, ou le fait qu'après deux retournements successifs lors du développement des Calcaronea cet axe est comparable à celui des Cnidaires, n'a pas plus de signification pour leur phylogenèse que les analogies de leurs développements avec celui des Volvocales. Il en est de même d'ailleurs lors du développement des Cnidaires, qui servent de point de référence dans les comparaisons des développements des Spongiaires et des autres Métazoaires. Les résultats récents de Belousov et Ostroumova (1966) et de Van de Vyver (1968) indiquent que les axes physiologiques de l'oeuf, de la larve et du jeune polype ne se recouvrent pas obligatoirement, comme le laissaient supposer les résultats des observations de Teissier (1931).

Conclusions sur l'évolution des Spongiaires

Eponges calcaires

Les genres *Clathrina* Gray, 1867 et *Leucosolenia* Bowerbank, 1861 possédent sans doute l'organisation la plus simple parmi les Éponges Calcaires. Leurs développements sont pourtant trés differents, et tout essai visant à faire dériver une forme de l'autre serait hasardeux. Chacun de ces genres est à la base d'un groupe évolutif complet, et ces groupes se développent parallèlement en passant par des stades comparables de complication morphologique (Borojević, 1968). Cette évolution morphologique n'est pas accompagnée obligatoirement par des modifications du développement: dans le groupe des Calcaronea, où nous connaissons l'ontogénie de nombreuses espèces, le développement est particulièrement homogène (Vacelet, 1964). Malgré l'organisation analogue de leurs représentants, les Calcinea et les Calcaronea représentent donc deux lignée évolutives parallèles et séparées depuis fort longtemps.

Les relations des Éponges Calcaires réunies actuellement sous le nom de Pharetronida avec les Calcinea et les Calcaronea doivent encore être précisées.

Démosponges

Tandis que le groupe d'Éponges Calcaires semble être au moins diphylétique, les Démosponges, a organisation infiniment plus variée, peuvent être dérivées d'une souche commune.

Si nous considérons le développement d'*Halisarca* comme primitif (*vide* Lévi, 1956), les parenchymellas typiques des Ceractinomorpha sont caractérisées par la différenciation de plus en plus précoce et de plus en plus importante de la lignée amoeboide et de ses dérivés. Cette tendance est poussée au maximum chez les Spongillidae où les archaeocytes donnent naissance aux choanocytes encore dans le mésohyle larvaire.

L'amphiblastula des Homosclerophorida dérive d'une parenchymella (Lévi, 1956). Sa segmentation se termine par la formation d'une sterroblastula, dont l'amphiblastula dérive par la lyse de cellules internes. C'est un mode d'embryogénèse tout particulier parmi les Spongiaires, et si l'organisation des Homosclerophorida est très simple, leur développement, loin d'être primitif, nous interdit de les placer à la base de l'évolution des Éponges. Nous les considérons comme une branche indépendante, séparée très précocément des autres Démosponges, sans suite dans l'évolution du groupe. Il faudrait sans doute les séparer des autres Tetractinomorpha dont elles se distinguent par le développement embryonnaire et par la cytologie des Éponges adultes.

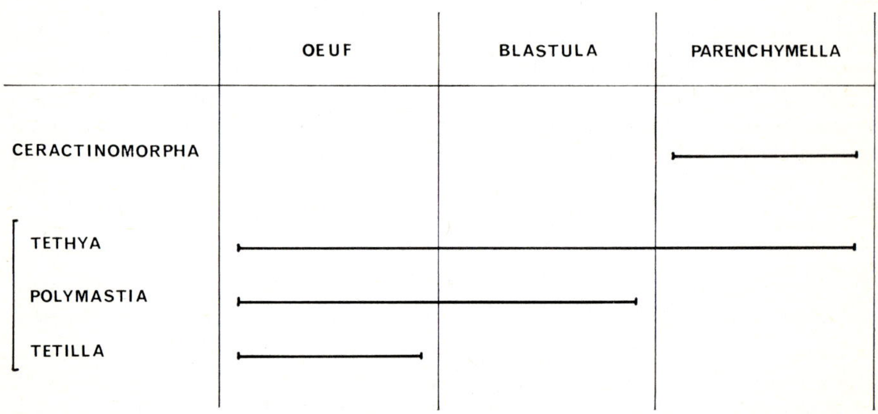

FIG. 2. Réduction progressive des stades libres dans la lignée évolutive des Tetractinomorpha.

Le développement des Tetractinomorpha (Homosclerophorida exclues) est caractérisé par la précocité croissante de la fixation des larves et par le retard de la différenciation cellulaire. Parmi les Clavaxinellida, *Tethya* passe par un stade libre blastula, qui se transforme avant la métamorphose en une parenchymella, selon un mode comparable à celui de *Halisarca*. Chez *Polymastia* et *Raspailia* la métamorphose intervient après le stade blastula, et le stade parenchymella est omis. *Tetilla* parmi les Tetractinellida, ne posséde plus de stade libre, la blastula et la pupa se confondent. Notons néanmoins que c'est le seul développement que nous connaissons actuellement parmi les Tetractinellida, et que les recherches complémentaires sont indispensables pour préciser si ce mode de développement, très particulier, est commun aux autres éponges du même groupe, ou s'il est propre à *Tetilla.*

Hexactinellides

Les Hexactinellida possédent un développement proche de celui des Ceractinomorpha. Leur structure très particulière nous autorise cependant à les isoler des autres Spongiaires, en une lignée évolutive indépendante. Une meilleure connaissance de leur biologie devrait permettre de mieux définir leurs relations phylogénétiques.

En résumé, nous voyons dans l'évolution des Spongiaires plusieurs lignées évolutives différentes. Celles des Calcinea, Calcaronea et des Hexactinellida sont homogènes et bien délimitées; celles des Démosponges est complexe. D'un type primitif, dont *Halisarca* est sans doute la plus proche, dérivent les Homosclerophorida caractérisées par la larve amphiblastula, les Ceractinomorpha caractérisées par l'importance de plus en plus grande de la différenciation cellulaire lors de l'embryogenèse et les Clavaxinellida auxquelles se rattachent probablement les Tetractinellida, caractérisées par la réduction progressive des stades libres et par la précocité de leur métamorphose.

Le développement des Calcinea offre certaine analogie avec celui de *Halisarca* (Borojević, 1969a). On peut donc être tenté de la considérer comme primitif, ce qui s'accorde bien avec la simplicité de l'organisation de certaines Calcinea. Mais l'extrême complexité du développement des Calcaronea, dont le niveau d'organisation est semblable à

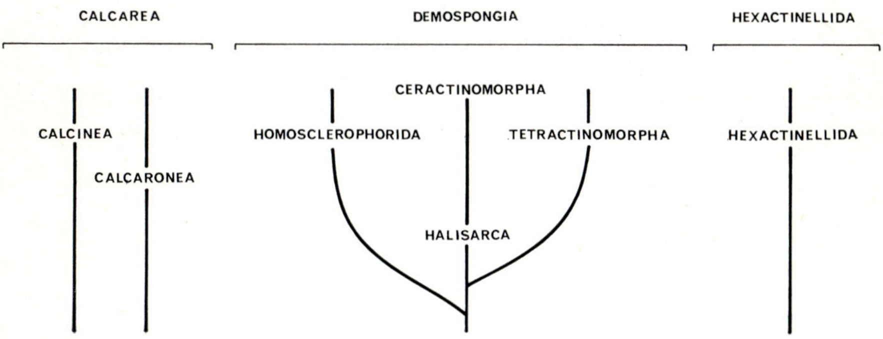

FIG. 3. Principales lignées evolutives des Spongiaires.

celui des Calcinea et la simplification secondaire du développement des Clavaxinellida, éponges sans doute évoluées, suffisent pour infirmer cette hypothèse.

Il nous paraît donc difficile de relier les lignées évolutives des Spongiaires à leur base et de les dériver toutes d'une forme primitive commune. Nous ne sommes pas opposés au principe de polyphylétisme des Porifera.

Bibliographie

Beer, G. R. de (1922). Notes sur les colonies artificielles de choanocytes. *Archs Zool. exp. gén.* **61**: 47–49.

Belousov, L. V. et Ostroumova, T. V. (1966). Gradients physiologiques et processus morphogénétiques chez les Hydraires. *Zh. obshch. Biol.* **27**: 336–348.

Borojević, R. (1963). Essai de reconstitution du choanoderme à partir d'une suspension choanocytaire d'Éponge Acalcaire. *C.r. hebd. Séanc. Acad. Sci. Paris* **257**: 961–962.

Borojević, R. (1966). Étude expérimentale de la différenciation des cellules de l'éponge au cours de son développement. *Devl. Biol.* **14**: 130–153.

Borojević, R. (1967). La ponte et le développement de *Polymastia robusta* (Démosponges). *Cah. Biol. Mar.* **8**: 1–6.

Borojević, R. (1968). Éponges calcaires des Côtes de France. IV. Genre *Ascaltis* Haeckel *emend. Archs. Zool. exp. gén.* **109**: 193–210.

Borojević, R. (1969a). Étude du développement et de la différenciation cellulaire d'Éponges Calcaires Calcinéennes (Genre *Clathrina* et *Ascandra*). *Annls. Emb. Morphogen.* **2** (1): 15–36.

Borojević, R. (1969b). Le comportement des cellules d'Éponge lors des processus morphogénétiques. *Archs. Biol. Liège.* Sous presse.

Borojević, R. et Lévi, C. (1965). Morphogénèse expérimentale d'une éponge à partir de cellules de la larve nageante dissociée. *Z. Zellforsch. mikroscop. Anat.* **68**: 57–69.

Borojević R. et Lévi, P. (1967). Le basopinacoderme de l'Éponge *Mycale contarenii* (Martens). Technique d'étude des fibres extracellulaires basales. *J. Microscopie* **6**: 857–862.

Brien, P. (1937). La réorganisation de l'Éponge après la dissociation par la filtration et phénomènes d'involution chez *Ephydatia fluviatilis*. *Archs Biol. Liège.* **48**: 185–268.

Brien, P. (1943). L'embryologie des éponges. *Bull. Museum Hist. Nat. Belg.* **19**: 1–20.

Brien, P. (1967a). Les Éponges. Leur nature métazoaire—leur gastrulation—leur état colonial. *Annl. Soc. r. zool. Belg.* **97**: 197–235.

Brien, P. (1967b). Embryogenèse de *Potamolepis stendelli* et de *Spongilla Moorei*. Polyphylétisme des Éponges d'eau douce. *Bull. Acad. r. Belg. Cl. Sci.* **53**: 752–777.

Brøndsted, H. V. (1953). The ability to Differentiate and the Size of Regenerated Cells, after repeated Regeneration in *Spongilla lacustris*. *Q. Jl microsc. Sci.* **94**: 177–184.

Dendy, A. (1914). Observations on the gametogenesis of *Grantia compressa*. *Q. Jl microsc. Sci.* **60**: 313–376.

Duboscq, O. et Tuzet, O. (1937). L'ovogénèse, la fécondation et les premiers stades du développement des éponges calcaires. *Archs Zool. exp. gén.* **79**: 157–316.

Fauré-Fremiet, E. (1932). Involution expérimentale et tension de structure dan les cultures de *Ficulina ficus* L. *Archs Anat. microsc.* **28**: 121–157.

Gatenby, J. B. (1920). The germ-cells, fertilisation and early development of *Grantia* (*Sycon*) *compressa*. *J. Linn. Soc.* (Zool.) **34**: 261–297.

Hadži, J. (1916). Rezultati bioloških istraživanja Jadranskog Mora. Porifera. Calcarea I. *Clathrina blanca* (Miklucho-Maclay). *Prirodsl. Istraž. J.* **9-10**: 1–164.

Huxley, J. S. (1911). Some phenomena of regeneration in *Sycon*. *Phil. Trans. R. Soc.* (*B*) **202**: 165–189.

Huxley, J. S. (1921). Further studies on restitution-bodies and free tissue-culture in *Sycon*. *Q. Jl. microsc. Sci.* **65**: 293–322.

Jörgensen, O. M. (1917). Reproduction in *Grantia compressa*. *Report Dove mar. Lab.* **1916-1917**: 26–32.

Lévi, C. (1956). Étude des *Halisarca* de Roscoff. Embryologie et systématique des Démosponges. *Archs. Zool. exp. gén.* **93**: 1–184.

Maas, O. (1898). Die Keimblätter der Spongien und die Metamorphose von *Oscarella* (*Halisarca*). *Z. wiss. Zool.* **58**: 665–679.

Maas, O. (1906). Ueber die Einwirkung karbonatfreier und kalkfreier Salzlösungen auf erwachsene Kalkschwämme und auf Entwicklungsstadien derselben. *Arch. Entwmech Org.* **22**: 581–599.

Maas, O. (1910). Ueber Involutionsercheinungen bei Schwämmen und ihre Bedeitung für die Auffassung des Spongienkörpers. *Festschrift F. R. Hertwig* **3**: 93–130.

Meewis, H. (1938). Contribution à l'étude de l'embryogénèse des Myxospongidae: *Halisarca lobularis* (Schmidt). *Archs. Biol. Liège* **50**: 1–66.

Meewis, H. (1939). Étude comparative des larves d'Éponges siliceuses. *C. r. Ass. fr. Avanc. Sci.* **63**: 660–669.

Meewis, H. (1941). Contribution à l'étude de l'embryogénèse des Éponges siliceuses. *Annl. Soc. r. zool. Belg.* **72**: 126–149.

Metschinkoff, E. (1886). *Embryologische Studien an Medusen*. Ein Beitrag zur Genealogie der Primitiv-Organe. Wien: Hölder.

Minchin, E. A. (1896). Note on the larva and post larval development of *Leucosolenia variabilis*, H. sp., with remarks on the development of other Asconidae. *Proc. R. Soc. Lond.* **40**: 42–52.

Minchin, E. A. (1900). Porifera, in *Treatise on Zology*. II. 1–188. Lankester R. (Ed.) London: A and C. Black.

Nassonow, N. (1883). Zur Biologie und Anatomie der Clione. *Z. wiss. Zool.* **39**: 295–308.

Okada, Y. (1928). On the development of a Hexactinellid sponge, *Farrea sollasii* *J. Fac. Sci. Tokyo univ.* **2**: 1–27.

Teissier, G. (1931). Étude expérimentale du développement de quelques Hydraires *Annls Sci. Nat.* (*Bot and Zool.*) **14**: 5–60.

Thouveny, Y. (1965). Histological aspects of regeneration in *Polydora flava*: evidence from irradiation and synthetic culture medium. In *Regeneration in*

Animals and related problems: 262–271. Kiortis, V. and Trampusch, H. A. L. (Eds). Amsterdam: North Holland Publ. Comp.

Tuzet, O. et Pavans de Ceccatty, M. (1955). Lam obilisation en amoebocytes des cellules des *Halisarca* (Éponges siliceuses). Les Polyblastes chez les Éponges. *C. r. Séanc. Soc. Biol. Paris* **149**: 799–801.

Vacelet, J. (1964). *Étude monographique de l'Éponge calcaire Pharétronide de Méditerranée, Petrobiona massiliana Vacelet et Lévi. Les Pharétronides actuelles et fossiles*. Thèse Doct. Sci. Nat. Aix-Marseille.

Van de Vyver, G. (1968a). Étude du développement embryonnaire des Hydraires athécates (Gymnoblastiques) à gonophores. II. Formes à actinulas. *Archs Biol. Liège* **79**: 327–364.

Van de Vyver, G. (1968b). Étude du développement embryonnaire des Hydraires athécates (Gymnoblastiques) à gonophores. III. Discussion et conclusions générales. *Archs Biol., Liège* **79**: 365–379.

Watanabe, Y. (1957). Development of *Tethya serica* Lebwohl, a tetraxonian sponge. I. Observations on external changes. *Nat. Sci. Rep. Ochanomizu Univ.* **8**: 97–104.

Watanabe, Y. (1960). Outline of morphological observation on the development of *Tethya serica* Lebwohl. *Bull Mar. biol. Sta. Asamuski* **10**: 145–148.

Weltner, W. (1893). Spongillidenstudien. *Arch. Naturgesch.* **59**: 209–284.

Willmer, E. N. (1960). *Cytology and evolution*. New York–London: Academic Press.

Willmer, E. N. (1967). Morphological problems of cell types, shape, and identification. In *Cells and tissues in culture*. **1**: 143–174. Willmer, E. N. (Ed.). New York–London: Academic Press.

AUTHOR INDEX

Numbers in italics refer to pages in the References at the end of each article.

A

B

C

SYSTEMATIC INDEX

SUBJECT INDEX